NUTRITION AND HEALTH

Adrianne Bendich, PhD, FACN, SERIES EDITOR

For further volumes:
http://www.springer.com/series/7659

Norman J. Temple · Ted Wilson
David R. Jacobs, Jr.
Editors

Nutritional Health

Strategies for Disease Prevention

Third Edition

Foreword by Joan Sabaté

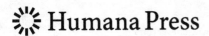 Humana Press

Editors
Norman J. Temple
Centre for Science
Athabasca University
Athabasca, AB, Canada

Ted Wilson
Department of Biology
Winona State University
Winona, MN, USA

David R. Jacobs, Jr.
University of Minnesota
School of Public Health
Division of Epidemiology
    and Community Health
Minneapolis, MN, USA

ISBN 978-1-61779-893-1   ISBN 978-1-61779-894-8 (eBook)
DOI 10.1007/978-1-61779-894-8
Springer New York Heidelberg Dordrecht London

Library of Congress Control Number: 2012943616

Printed on acid-free paper

Humana Press is a brand of Springer
Springer is part of Springer Science+Business Media (www.springer.com)

To Luciana, my wife, with love.

–Norman

To Jack, my youngest son: I am always sorry to remind you that cookies are not a healthy regular part of a 2-year old's diet.

–Ted

There is Maya and Zeke, there is Carlyle and Max. We dedicate our work to the next generation.

–David

Foreword

Given the choice, everyone would choose to be healthy. Given the choice, everyone would also choose to eat foods they like. Although for many these two options are generally available, these choices are often incompatible. The foods we like, the foods we have eaten since childhood, are not necessarily the foods that lead to health. Given the epidemic of obesity and overweight that is so widespread, health promotion does not appear to be a decisive factor in personal diet choices. Yet the sales of medicines and annual expenditures for surgical procedures indicate that health is a vital concern. Understanding this incongruent situation is a challenge.

The motivation for choosing the foods we eat is complex. Among the factors involved, eating food that is familiar, food that was served to us by our mothers and grandmothers is probably the most compelling. These are the foods we like. Some of us are adventuresome eaters, willing to experiment, and try new foods. This may result in the incorporation of foods we didn't eat at home, but for the most part our diet is centered on the foods we grew up eating. With the advent of fast-food restaurants and microwave ovens in the 1960s, family meals veered away from natural and whole foods. Animal protein became standard fare and fresh fruits and vegetables diminished. The foods we like, the foods we're used to, have changed in the last 50 years, without a comprehension of the implications to our health.

Other factors impact our food choices. Time availability is a primary consideration. It takes less time to eat out than to purchase food, cook it, and serve it. Even many foods that are "cooked" at home are microwaved and include ingredients that we can't pronounce. Budgetary concerns also influence what we eat. Foods that are faster to cook are usually less expensive. We buy foods that can stay on the shelf or in the freezer because they are cost-effective and don't require daily trips to the market. Cooking expertise has become a novelty, something to be admired on reality television. And many foods that are recommended for health aren't readily available at the supermarket. If health concerns factor into our thinking about our meals, they have to compete with many other concerns.

For those consumers who do consider health implications regarding their food choices, the situation is especially challenging. A myriad of food containers display health claims, each slightly different. Television commercials entice us with messages of reduced weight and added vigor. The blogosphere offers an ever-changing emphasis from the latest study of the effects of food on our health. Those who have been entrusted with formulating our food policy (i.e., the USDA) have the ambivalent charge of protecting the market for subsidized foods, which may not be the most appropriate for health, while at the same time protecting the health of those purchasing the foods. It is not always clear which of these two roles is taking precedence.

The most reliable source of nutrition information and dietary advice that consumers can access is from their health-care professional. It is imperative therefore that those working in health care and the health science fields be properly informed and educated. For that reason, this book, *Nutritional Health*, is an invaluable resource for nutrition research and health. Beginning with a research overview, the book then addresses the role of diet in relation to several chronic diseases—diabetes, obesity, heart disease, and cancer, among others. Chapters on phytochemicals, food synergy, and food supplements are included. These topics are essential for health-care professionals and for students working towards careers in this area.

However, *Nutritional Health: Strategies for Disease Prevention*, now in its third edition, goes beyond most reference books connecting nutrition with health. Food policies, official nutrition messages, and their relationship to the food industry are also addressed. Food choices have economic, political, social, and environmental consequences that could place improvements to the health of individuals or populations in conflict with other considerations. This treatise offers a comprehensive, integrated review of the scientific knowledge connecting diet and health, and presents some needed food policy strategies to address the current epidemic of chronic degenerative diseases.

Loma Linda, CA, USA   Joan Sabaté

Series Editor Introduction

The great success of the Nutrition and Health Series is the result of the consistent overriding mission of providing health professionals with texts that are essential because each includes (1) a synthesis of the state of the science, (2) timely, in-depth reviews by the leading researchers in their respective fields, (3) extensive, up-to-date fully annotated reference lists, (4) a detailed index, (5) relevant tables and figures, (6) identification of paradigm shifts and the consequences, (7) virtually no overlap of information between chapters, but targeted, inter-chapter referrals, (8) suggestions of areas for future research and (9) balanced, data-driven answers to patient as well as health professionals questions which are based upon the totality of evidence rather than the findings of any single study.

The Series volumes are not the outcome of a symposium. Rather, each editor has the potential to examine a chosen area with a broad perspective, both in subject matter as well as in the choice of chapter authors. The editor(s), whose training(s) is (are) both research and practice oriented, have the opportunity to develop a primary objective for their book, define the scope and focus, and then invite the leading authorities to be part of their initiative. The authors are encouraged to provide an overview of the field, discuss their own research and relate the research findings to potential human health consequences. Because each book is developed de novo, the chapters are coordinated so that the resulting volume imparts greater knowledge than the sum of the information contained in the individual chapters.

Nutritional Health: Strategies for Disease Prevention, Third Edition, edited by Norman J. Temple, Ph.D., Ted Wilson, Ph.D., and David R. Jacobs, Jr., Ph.D. clearly exemplifies the goals of the Nutrition and Health Series. The major objective of this comprehensive volume is to review the growing evidence of the importance of nutrition-related decisions at all stages of life to healthful living.

Nutritional Health, again in this third edition, provides the reader with a comprehensive compilation of the newest data on the ways to conduct nutrition research as well as the many challenges associated with clinical research. It is to the credit of Drs. Temple, Wilson, and Jacobs that they have organized this volume so that it provides an in-depth overview of the role of diet and dietary components in chronic disease prevention; the importance of public health actions, regulatory decisions, and academic research in assuring the safety and efficacy of claims made on foods and dietary supplements.

The volume is organized into 26 comprehensive chapters. The Third Edition contains five completely new chapters, in-depth updates of all chapters from the Second Edition, and several chapters have new authors who provide their perspectives in many areas of nutrition research. The first five chapters include reviews of nutrition research methodologies, beginning with cell culture studies and continuing to clinical research including all aspects of epidemiological research. The recent methods used to develop meta-analyses and systematic reviews are discussed in detail. There is a new chapter on eating disorders that contains important definitions and explanations of anorexia, binge eating, pica, and bulimia and provides information about treatment strategies. There is an updated chapter by Dr. David Barker on recent findings concerning the Barker hypothesis that emphasizes the critical importance of fetal growth and weight gain in the first year of life in predicting diseases in middle age. Dr. Barry Popkin's chapter on the nutrition transition provides an excellent global perspective of how dietary changes affect the health of emerging nations.

Ten chapters are devoted to examining medical nutrition modalities used in the care of patients with diabetes (Types I and II, as well as gestational diabetes), obesity, cardiovascular disease, hypertension, and cancer. Specific sources of bioactive food components such as n-3 polyunsaturated fatty acids, fiber, garlic, carotenoids, phenolics, and foods with low glycemic indices are discussed in detail. Similarly, food components that can adversely affect chronic diseases are also reviewed including *trans* fats, saturated fats, sugar, animal protein, and food contaminants. The importance of physical activity in addition to healthful eating is also discussed in several chapters.

Four chapters are devoted to examining dietary patterns and whole foods and food groups. Alcoholic and nonalcoholic beverages each have their own chapter. The alcohol chapter includes reviews of benefits as well as risks including fetal alcohol syndrome. Included in the nonalcoholic beverage chapter are discussions of water, green tea, coffee, milk, fruit juices, sports drinks, weight loss drinks, and energy drinks. Dr. Ted Wilson, author of this chapter, has coedited an entire well-referenced volume in the "Nutrition and Health" Series entitled "Beverages in Health and Disease."

Seven chapters examine the role of governments and public health issues in the development of national food policies. The major food and dietary supplement companies introduce hundreds of new products into the global marketplace annually and the decision-making process in new product development, label development, and new claims development is explored in several chapters. Important, up-to-date information on the newest sugar, fat, and salt substitutes as well as novel sources of fiber is reviewed.

The final three chapters examine cutting edge topics that are critical to the understanding of the complexity of nutrition research. In one chapter, there is an insightful examination of the importance of biotechnology for the enhancement of food production including genetic modifications for disease and drought resistance. The chapter on nutritional anthropology looks at cultural patterns and societal norms and taboos with relation to food included in diets and compares culturally related eating behaviors to biological requirements for consumption of sources of essential nutrients. The final chapter, written by the volume's editors, outlines nutritional strategies available currently and proposes new strategic global nutrition strategies to reduce overeating and improve overall nutrient density of foods consumed.

The logical sequence of chapters enhances the understanding of the latest information on the current standards of practice for clinicians, related health professionals including the dietician, nurse, pharmacist, physical therapist, behaviorist, psychologist, and others involved in the team effort required for successful treatment of cardiovascular and other relevant diseases as well as conditions that adversely affect normal metabolic processes. This comprehensive volume also has great value for academicians involved in the education of graduate students and postdoctoral fellows, medical students, and allied health professionals who plan to interact with patients whose health is affected by their diet. Of great importance, the editors have provided chapters that balance the most technical information with discussions of its importance for clients and patients as well as graduate and medical students, health professionals, and academicians.

The volume contains 74 detailed tables and figures that assist the reader in comprehending the complexities of the biological significance of nutritional strategies for human health. The overriding goal of this volume is to provide the health professional with balanced documentation and awareness of the newest research and therapeutic approaches including an appreciation of the complexity of this relatively new field of investigation. Hallmarks of the 26 chapters include key words and bulleted key points at the beginning of each chapter, complete definitions of terms with the abbreviations fully defined for the reader, and consistent use of terms between chapters. There are over 1,900 up-to-date references; all chapters include a conclusion to highlight major findings. The volume also contains a highly annotated index.

This unique text provides practical, data-driven resources based upon the totality of the evidence to help the reader understand the basics, treatments, and preventive strategies that are involved in the understanding of the role nutrition interventions may play in healthy individuals as well as those with cardiovascular disease, diabetes, or neurocognitive declines. The overarching goal of the editors is to provide fully referenced information to health professionals, so they may have a balanced perspective on the value of various preventive and treatment options that are available today as well as in the foreseeable future.

In conclusion, *Nutritional Health: Strategies for Disease Prevention*, Third Edition, edited by Norman J. Temple, Ph.D., Ted Wilson, Ph.D., and David R. Jacobs, Jr., Ph.D. provides health professionals in many areas of research and practice with the most up-to-date, well referenced and comprehensive volume on our current understanding of the nutritional sciences. This volume will serve the reader as the most authoritative resource in the field to date and is a very welcome addition to the Nutrition and Health Series.

Adrianne Bendich, PhD., FACN
Series Editor

Preface

The first edition of *Nutritional Health* was published in 2001, and the second edition in 2006. During this first decade of the twenty-first century much water has flown under the bridge of the advancing river that is nutrition research and practice. At the same time huge numbers of new foods are continually appearing in supermarkets, while hundreds of new "breakthrough" diet and health books are published every year. With these large accumulations of developments in the field of nutrition, the need for a new edition of this book is obvious.

During the last century of nutritional advancement, we have frequently been faced with great opportunities that were brilliantly disguised as insoluble problems. Perhaps we are biased but in our eyes problems associated with nutrition are among the most exciting of those in the life sciences. How many other branches of the life sciences offer the promise of slashing the burden of human disease by one third or more?

With a scattering of brilliant exceptions until the 1970s few gave serious consideration to the notion that our diet plays an important role in such chronic diseases as heart disease and cancer. Today, we have a vastly greater understanding of the role of diet in disease; we know, for example, how meat affects colon cancer, how fish fats affect heart disease, and how vitamin D affects osteoporosis. Now, in the early years of the twenty-first century, our vastly improved nutrition knowledge gives us the capability of preventing a sizable fraction of the chronic diseases that afflict the people of our world, but only if we can fully inform its populace about these discoveries.

Ironically, despite overwhelming evidence that nutrition has such enormous potential to improve human well-being—at modest cost and using the knowledge we already have—it still fails to receive the resources it merits. Growth in funding for nutrition research and education remains stunted. By contrast, countless millions of dollars are spent on the glamor areas of biomedical research, such as genetic engineering and gene therapy. But we already know that our genes can only explain a fraction of our disease burden. Even if gene therapy reaches its full potential, it seems most improbable that it will ever achieve a quarter of what nutrition can do for us today.

In the words of Confucius: "The essence of knowledge is that, having acquired it, one must apply it." But a major barrier is that nutrition information often fails to reach the health professionals who most need to apply it, namely the physicians, dietitians, and nurses who represent the front-line workers in health care. How do we bring this information to others who also need this information, such as the nutrition professors who lack the time to read more than a tiny portion of the literature outside of their main area of interest?

Nutritional Health endeavors to address the needs of those who would most benefit from up-to-date information on recent advances in the field of nutrition. Accordingly, our book contains a series of chapters by experts in a diverse range of nutritional areas. Our aim is not so much to cover all the leading edges of nutrition but rather to discuss recent thinking and discoveries that have the greatest capacity to improve human health and nutritional implementation.

Some readers may disagree with some of the opinions presented, but in nutrition differences of opinion are often unavoidable. Owing to the constant changes in our diet, nutrition is by nature in constant dynamic flow, as are our opinions of what constitutes the best nutritional habits. The views expressed in *Nutritional Health* are in many cases particular interpretations by the authors of each chapter on their areas of specialization.

Athabasca, AB, Canada   Norman J. Temple
Winona, MN, USA   Ted Wilson
Minneapolis, MN, USA   David R. Jacobs, Jr.

Contents

Contributors

Kelly C. Allison, PhD Department of Psychiatry, Center for Eating and Weight Disorders, Perelman School of Medicine of the University of Pennsylvania, Philadelphia, PA, USA

David J.P. Barker, MD, PhD, FRS MRC Lifecourse Epidemiology Unit, University of Southampton, Southampton General Hospital, Southampton, UK

Donald C. Beitz, PhD Iowa State University, Ames, IA, USA

Geraldine Cuskelly, PhD Institute for Agri-Food and Land Use, School of Biological Sciences, Queen's University Belfast, Belfast, UK

Cindy D. Davis, PhD Director of Grants and Extramural Activities Office of Dietary Supplements National Institutes of Health, Bethesda, Maryland

Michael R. Flock, BS Department of Nutritional Sciences, The Pennsylvania State University, University Park, PA, USA

Marion J. Franz, MS, RD, CDE Nutrition Concepts by Franz, Inc., Minneapolis, MN, USA

David R. Jacobs, Jr., PhD Division of Epidemiology and Community Health, School of Public Health, University of Minnesota, Minneapolis, MN, USA

Travis J. Knight, PhD Iowa Department of Agriculture and Land Stewardship, Ankeny, IA, USA

Richard B. Kreider, PhD, FACSM, FISSN Exercise and Sport Nutrition Laboratory, Department of Health and Kinesiology, Texas A&M University, TX, USA

Penny M. Kris-Etherton, PhD, RD Department of Nutritional Sciences, The Pennsylvania State University, University Park, PA, USA

Brian Leutholtz, PhD, FACSM Center for Exercise, Nutrition and Preventive Health Research, Department of Health, Human Performance and Recreation, Baylor University, Waco, TX, USA

Rui Hai Liu, MD, PhD Department of Food Science, Cornell University, Ithaca, NY, USA

Claire McEvoy, MPhil Nutrition and Metabolism Research Group, School of Medicine, Dentistry and Biomedical Sciences, Centre for Public Health, Institute of Clinical Science, Queen's University Belfast, Belfast, UK

Michelle C. McKinley, PhD Nutrition and Metabolism Research Group, School of Medicine, Dentistry and Biomedical Sciences, Centre for Public Health, Institute of Clinical Science, Queen's University Belfast, Belfast, UK

John A. Milner, PhD Nutritional Science Research Group, Division of Cancer Prevention, National Cancer Institute, Rockville, MD, USA

Marion Nestle, PhD, MPH Department of Nutrition, Food Studies, and Public Health, New York University, New York, NY, USA

Department of Nutritional Sciences, Cornell University, Ithaca, New York

Gretel H. Pelto, PhD Division of Nutritional Sciences, Cornell University, Ithaca, NY, USA

Barry M. Popkin, PhD Carolina Population Center, School of Public Health, University of North Carolina at Chapel Hill, Chapel Hill, NC, USA

Jill K. Rippe, MS Main Street Ingredients, La Crosse, WI, USA

Neil A. Schwarz, MS Department of Health, Human Performance and Recreation, Baylor University, Waco, TX, USA

Scott P. Segal, PhD Department of Biology, Winona State University, Winona, MN, USA

Nelia P. Steyn, PhD Knowledge Systems, Human Sciences Research Council, Cape Town, South Africa

Norman J. Temple, PhD Centre for Science, Athabasca University, Athabasca, AB, Canada

Claire R. Whittle, PhD Nutrition and Metabolism Research Group, School of Medicine, Dentistry and Biomedical Sciences, Centre for Public Health, Institute of Clinical Science, Queen's University Belfast, Belfast, UK

Ted Wilson, PhD Department of Biology, Winona State University, Winona, MN, USA

Jayne V. Woodside, PhD Nutrition and Metabolism Research Group, School of Medicine, Dentistry and Biomedical Sciences, Centre for Public Health, Institute of Clinical Science, Queen's University Belfast, Belfast, UK

Ian S. Young, MD Nutrition and Metabolism Research Group, School of Medicine, Dentistry and Biomedical Sciences, Centre for Public Health, Institute of Clinical Science, Queen's University Belfast, Belfast, UK

Sera L. Young, PhD Division of Nutritional Sciences, Cornell University, Ithaca, NY, USA

Ali Zirakzadeh, MD, MS Denver Health and Hospital Authority, University of University of Colorado Denver-UCHSC, Denver, CO, USA

Chapter 1
Methods in Nutrition Research

David R. Jacobs, Jr. and Norman J. Temple

Keywords Epidemiological studies • Cohort studies • Case-control studies • Cross-sectional studies • Population studies • Historical studies • Randomized clinical trials • Systematic review • Meta-analysis • Conflict of interest • Ethical approval

Key Points

- Many different research designs are available for investigating nutrition-related questions. Each one has strengths and weaknesses.
- Much nutrition-related research is carried out using epidemiological methods. Major types of epidemiological research include cohort studies, case-control studies, cross-sectional studies, population studies, and historical studies.
- Results from randomized clinical trials (RCTs) are internally valid and usually are more uniformly reliable for causal inference than those from observational epidemiological studies. However, many kinds of nutritional studies do not fit well into the RCT framework. It is therefore important to use observational study logic effectively.
- A recent development is the use of systematic reviews and meta-analyses. First, the literature is reviewed according to strict criteria. Findings from the different studies are then analyzed by making formal pooled risk estimates. The pooled estimate gains power by increasing sample size, but at the expense of loosening

D.R. Jacobs, Jr., PhD(✉)
Division of Epidemiology and Community Health, School of Public Health,
University of Minnesota, 1300 South Second Street, Suite 300, Minneapolis, MN 55454, USA
e-mail: Jacobs@epi.umn.edu

N.J. Temple, PhD
Centre for Science, Athabasca University, Athabasca, AB, T9S 3A3, Canada

N.J. Temple et al. (eds.), *Nutritional Health: Strategies for Disease Prevention*,
Nutrition and Health, DOI 10.1007/978-1-61779-894-8_1,
© Springer Science+Business Media, LLC 2012

definition of variables and concepts, which will vary across studies despite use of strict criteria. This strategy can be particularly helpful in evaluating consistency and in considering the possibility of residual confounding when only observational studies are available.

- Experiments on animals are often used for nutrition-related research and can help as a method to study detailed pathways. However, it is recognized that physiology and pathophysiology in animal models only approximates that in humans.
- Another common strategy is mechanistic research. Here researchers study the details of body physiology and biochemistry, for example by using cell culture. This method can have high throughput and illuminate detailed cellular and molecular relationships, but integrating such findings with the biology of intact humans is problematic. The method does contribute to basic understanding if carefully used.
- There is some evidence of conflict of interest in studies that potentially affect profit of the research funder, most typically in industry funding. The solution to this problem is conceptually simple but practically difficult: create a sufficient pool of research funds, independent of the food industry, to support needed research.
- Research studies involving human subjects require prior approval from an ethics committee.

1.1   Introduction

Excellent research designs are essential for efficiently and accurately answering nutrition-related questions, the answers to which in turn are used in policy making. Such policy would include advice from the perspective of health about what foods people should eat, what foods industry should supply, and whether some compounds should be consumed as dietary supplements. Areas of inquiry range from molecular mechanisms in cells, animals, and humans; to descriptions of what population groups eat; to effects of food and nutrient intake on clinical outcomes and intermediate variables; and to behaviors, policies, and cost structures that influence food preparation and consumption.

Formulating research questions carefully, as an essential part of choosing a research design and interpreting data, can optimize scientific discovery. Coupled with the right methods for a particular issue, these principles can make a research budget go further. However, over the years many findings in the field of nutrition have been reported in scholarly journals, with seemingly strong supporting evidence, only to be disproved by later research. Particularly telling in the category of disproved research are two assumptions which are not likely to be correct. The first questionable assumption is that isolated compounds behave in the same way as they do as part of a food matrix. The second questionable assumption is that all of a macronutrient (i.e., total fat, total carbohydrate, or total protein) behaves in a common way. Disproved research and rescinded health messages lead to widespread confusion in the scientific community and among the general public. Thus, for both positive and negative reasons, an essential tool for a person who has a professional interest in the nutritional sciences is a reasonably good understanding of the methodology used in nutrition research. Such knowledge confers a greater ability to comprehend research studies and to critically evaluate new research findings. That is our primary objective in this chapter.

We present an overview of the major methods used in nutrition research and attempt to explain the strengths and weaknesses of each method. Nutrition is connected to many areas of biomedical science. For that reason the methods described here are equally valid for diverse fields of health. In many cases, the examples given are not directly related to nutrition; rather, they illustrate a method that is also commonly used in nutrition.

We emphasize that this chapter is far from comprehensive. There are excellent texts and articles available that cover this subject in greater detail (e.g., [1–7]).

Much of nutrition research is intended to discover causal pathways, e.g., if a person eats certain foods or nutrients, will there be a health benefit or harm? Because much of this research is observational, the reader is also encouraged to read Bradford Hill's classic 1965 paper [8] for fundamentals of causal inference in the absence of the "mathematical guarantee" for internal validity of causal conclusions afforded in randomized clinical trials (RCTs).

1.2   Fundamental Research Design Principles

The idea of "contrast" is fundamental to all research: to ascertain the effect on some outcome of an exposure of interest contrasted to the outcome under an alternate condition (i.e., when there is less or no such exposure). In nutrition research, the exposure of interest is consumption of food or a food derivative (often an isolated compound found in food and taken as a supplement). There are similarities and differences between nutrition research and pharmacologic research, in which the exposure of interest is a drug. In many ways, pharmacologic research has set a standard for identifying causal pathways; understanding areas in which nutrition and pharmacologic research part ways is therefore important for nutrition scientists. The outcome variable is often a health construct, including death, disability, a clinical disorder, worsening of a clinical disorder, subclinical disease (such as asymptomatic atherosclerosis or neoplastic cells), or an intermediate marker variable (assumed to be on the pathway to disease, such as obesity, blood cholesterol, immune status, or prostate specific antigen). Alternatively, the outcome variable may be maintenance of health, quality of life, or good functioning (as in low absenteeism from work or school). The research target may be cost or other economic aspects. The research may be turned around, in which the outcome is food or nutrient intake and the exposure of interest is individual behavior, industry-related issues (such as advertising, placement of food outlets, and provision of more or less healthy food products), or population policy strategies (such as dietary guidelines intended to influence both food production and consumption).

In all of these cases, the research design adapts the basic principle of contrast: comparing one state of being to another on a level playing field, within a structure of assumed understanding of what the variables mean. The first issue is whether we are measuring what we think we are measuring, and whether that measurement is sufficiently precise to answer the research question? These are questions that deal with concepts such as validity, reliability, accuracy, precision, within person variation, and bias. These ideas hold for both the exposure and the outcome measures. Special to nutrition research is the structure of food intake. Actual food intake varies widely from day to day within an individual, although there is some evidence that this variation is within a more stable pattern, such as vegetarian, Mediterranean/ prudent diet, or Western diet [9]. For acute outcomes, such as the blood glucose response to a meal, the foods would likely be tightly controlled and well defined. For chronic outcomes, however, average food intake over a long period is most salient. Recalled or stated usual intake is highly variable. It may be understated (as is typical of energy intake in obese persons) or overstated (as might be the case for whole grain foods, where the participant knows the desirable answer). Most commonly, the bias of a statement of dietary intake is not great, but consumption is a moving target. The stated amounts may be correct on average over a population of people, but much too low or much too high for any given individual.

The second issue is confounding, defined as correlation of the exposure and the outcome with unmeasured variables. Such correlation may mask true associations or produce false associations. For example, in a general population study, whole grain food intake could relate positively to mortality if whole grain food intake increased sharply with age, because age is almost always strongly positively associated with mortality. Confounding assumes that the correlated variable is not on a causal pathway linking the exposure and the outcome. A similar situation is mediation, in which the correlated variable is in the causal pathway between the intake and the outcome. For example, whole grain food intake may lead to both lower body mass index (BMI) and less incident type 2 diabetes. Here we might conclude that whole grain food intake is associated with incident type 2 diabetes in part because it lowers body weight. Finally, the third issue is effect modification, wherein a given association is not homogeneous across a whole population; for example, when a nutritional factor leads to a good outcome in women but not in men, or leads to a bad outcome only in people with a certain genetic makeup (e.g., phenylalanine in phenylketonuria).

1.3   Errors, Bias, and Within-Person Variation in Dietary Assessment

Methods of self-report of dietary intake are discussed in the following chapter on research challenges in nutritional epidemiology (Chap. 2). Dietary assessment is prone to significant error [10]. Estimates of alcohol and energy intake, for example, are usually substantially underestimated in heavy drinkers and obese persons, respectively. Errors in dietary assessment are not exclusively random and appear to vary with age, BMI, and race/ethnicity [11]. Dietary intake varies greatly from day to day [1]; many days of recall may be needed to arrive at a stable individual estimate of intake [12]. It may be difficult for a participant to properly average intake in a food frequency, whether for issues of memory, ability to synthesize, or rapidly changing intake which does not seem to be amenable to averaging. Nutrients that are derived from few foods (for example, the dairy-specific fatty acids C15:0 and C17:0, or trans fatty acids derived from hydrogenation) may be more precisely measured than those that require precise input from many foods (such as total fat). Foods that are eaten in a regular pattern (breakfast cereal with milk as a typical breakfast) may be more precisely estimated than foods that are eaten irregularly or rarely.

1.4 Epidemiological Studies

Epidemiology is the study of the incidence, distribution, control, and prevention of diseases. Many epidemiological studies share a common objective: to discover the factors that cause or prevent disease, while others aim to describe food or nutrient intake in general or across population subgroups, or to understand factors that lead to intake of a given food. We shall now look at the major types of epidemiological studies. These studies are sometimes referred to as "observational" as they are based on observing relationships rather than on experimental testing under a high level of experimenter control.

1.4.1 Cohort Studies

Cohort studies—also known as prospective, longitudinal, or panel studies—are a highly flexible, utilitarian, and informative research design. Under the right circumstances, they can yield reliable inferences about causal pathways. The first step in a study is the recruitment of large numbers of subjects; typically in the range 30,000–100,000, but sometimes even more, such as half a million. Cohorts of a small number of people are also useful. Participants may be sampled at random from the general population (e.g., the Coronary Artery Risk Development in Young Adults study, CARDIA) [13], may be free of serious disease (e.g., the Multi-Ethnic Study of Atherosclerosis, MESA, in which cardiovascular disease was excluded, although other conditions such as cancer were allowed) [14], or may be specifically selected because they have a disease (e.g., survivors of childhood cancer). Any assemblage of participants can serve as the baseline in a cohort study. They are questioned or examined to determine each person's exposure to the particular factors under study, as well as many other variables that might be of importance as confounders, mediators, or effect modifiers. The subjects are then monitored for a long period of time, typically from 5 to 20 years, by which time a sufficient number have developed the conditions or diseases of interest. A variant would be a life course study, where participants are observed even longer, even from birth to death.

The data can now be analyzed in order to determine the association between dietary and other lifestyle variables and risk of disease. If the participants are known to be free of the disease at study baseline, predictors of incident disease are of particular interest, as it is then known that the exposure and outcome occurred in a temporal ordering consistent with a causal association. The outcome results for low vs. high intake are presented as risk difference (appropriate to computing attributable risk) or relative risk (RR) (appropriate to characterizing disease that develops at a faster rate over time in one group than in another), hazard ratio (HR, used in life table-based methods, or odds ratio (appropriate to methods such as logistic regression that are based on odds (probability of having an event relative to not having it) rather than probability itself). In a format typical for long-term studies of an outcome relative to a diet pattern, a food, or a nutrient, the subjects are divided into five equal sized groups (quintiles) based on their exposure to the variable of interest. It may be more valid and informative to categorize into substantively sensible groupings to capture important contrasts in level of exposure, or to divide into fewer groups if the sample size (especially of outcome events per category) is small or range of intake is limited. The relative risk of the disease in the category with the lowest exposure may be arbitrarily called 1.0. The risk in the other categories is then calculated relative to this reference group. If subjects in the quintile with the highest exposure have a 57% higher risk of the disease, then the RR is stated as 1.57. If subjects in that quintile have a 20% lower risk of the disease, then the RR is stated as 0.80. Typical findings from a cohort study on lifestyle and risk of stroke might be that exercise has a RR of 0.80 while smoking has a RR of 1.9. The RR values are normally given for all quintiles, not just the extreme quintiles. The pattern of absolute risks or of RR values is also examined for linearity and thresholds. Outcome variables may also be continuous in cohort studies. Note that it is good practice to examine both absolute and relative risks to get a good picture of magnitude of association. For example, in a disease that occurs in 1% of the unexposed population, a relative risk of 2 means an excess of 1%, whereas for a condition that occurs in 20% of the unexposed, it would occur in 40% among the exposed. Even a small absolute risk may be important in the face of a substantial relative risk, because the risk may build to considerable proportions over time. The same relative risk of 2 in the context of an incidence of 1% per year followed for 20 years would correspond to approximately 20% in the unexposed group, but 40% in the exposed.

The outstanding strength of cohort studies is that information is collected before the disease develops and this removes a major source of bias, namely that the participant may have started eating the food in question after or because of the disease outcome, in which case the situation would be one of reverse causality, where disease causes dietary exposure. However, there are still significant sources of error. An important one is that people are notoriously inaccurate in describing their diet, as was discussed earlier.

Another significant error is that diet and other aspects of lifestyle, such as smoking and exercise, tend to change over time. It is indeed remarkable that a single assessment in mid-life of smoking, serum cholesterol, and blood pressure in the Framingham and many other studies predicts heart disease incidence far into the future. In contrast, for many variables a single measurement is not accurate or stable enough to predict disease over the long term. Therefore, data recorded at the start of a cohort study may not reflect actual lifestyle a decade later. Some cohort studies endeavor to minimize this source of error by repeating the data collection every few years and accounting in data analysis for these changes. Such dietary changes may be incorporated by changing the dietary assessment every couple of years, for example when the intake is thought to influence disease over the short run only, or by averaging all dietary observations since baseline so that the exposure assessment becomes more and more precise estimate of long-term diet as the study progresses. For example, Stringhini et al. [15] found that repeated measures of diet explained much more of the association of socioeconomic status than did baseline diet as a covariate. Cigarette smoking tends to occur daily over many years, so that a single indication that a person is a smoker anytime during life is indicative of a constant and long-term situation. Perhaps not surprisingly, repeated measures of smoking did not add information in this case.

For nonfatal outcomes, such as body size or insulin resistance, change in the outcome may also be examined according to change in the exposure; this strategy would tend to remove confounders that are constant within an individual, such as race/ethnicity or sex.

The greatest challenge in analyzing data from cohort studies is confounding. Let us suppose one is studying the relationship between dietary factor A and disease X. The findings may indicate that persons with a relatively high intake of dietary factor A have a 50% increased risk of developing disease X. However, further investigation reveals that dietary factor A is associated with smoking which is also associated with disease X. This means that the association between dietary factor A and disease X may be spurious.

Confounding is of great importance in epidemiological studies as lifestyle factors tend to cluster together in the natural setting, that is when the lifestyle associations are not artificially diminished by the action of an experimenter, as in an RCT. Most often this lifestyle clustering takes the form of some people leading a healthy lifestyle with respect to diet, smoking, and exercise, while others lead a generally unhealthy lifestyle. We can illustrate this by looking at findings from a cohort study on the relationship between meat intake and mortality [16]. Comparing men who were in the top quintile for intake of red meat with those in the bottom quintile, the former led a generally less healthy lifestyle: they were more likely to be smokers (14.8% vs. 4.9%), were less likely to engage in regular vigorous physical activity (16.3% vs. 30.7%), had a higher BMI (28.3 vs. 25.9 kg/m^2), and consumed less fiber but more saturated fat (8.8 vs. 13.2 and 12.7 vs. 7.7 g/1,000 kcal, respectively). Thus, even if red meat has no effect on health, we can predict that a relatively high intake of red meat will still be *observed to be associated* with increased risk of death. Another factor associated with a unhealthy lifestyle (and with a high intake of red meat in this study) is having less years of education. It should be noted that this type of clustering of behaviors tends to reflect popular beliefs about what is healthy at the time of data collection. Thus, the postmenopausal women who attempt to follow the healthiest lifestyle may typically also have been more likely to have used hormone replacement therapy decades ago, but not in studies that were initiated more recently. However, lifestyle associations may also reflect causal pathways among behaviors. People may feel less heaviness when eating less meat and therefore want to exercise. A good feeling from exercising may induce certain kinds of diet. Smoking, in contrast, may lead to upper respiratory symptoms that tend to extinguish any desire to do vigorous activity.

Epidemiologists attempt to overcome the error caused by confounding by making statistical corrections. In the above example, men in the top quintile for intake of red meat had a hazard ratio (HR) for death from all causes of 1.44 in comparison with those in the lowest quintile (i.e., they had a 44% increased risk of death during the period of the study) [16]. However, when the data were corrected for the various sources of confounding that were listed above (plus others not mentioned), then the HR was reduced to 1.31. This is known as attenuation.

Two related variations within the cohort design study are the nested case-control design and the case-cohort design. These are based on the collection of biological samples, usually blood, at the time of initial recruitment of the subjects. Years later, when a sufficient number of them have developed a disease of interest, their stored samples are analyzed for selected substances. The difference between the two methods is in how the comparison group is selected. For comparison in the case-control design, samples from randomly selected or matched controls, free of the disease, are also analyzed. For comparison in the case-cohort design, participants randomly selected from the total cohort (which would include some of the cases) are selected. These strategies have two strong advantages over cohort studies as described earlier. First, only a limited number of samples must be analyzed. Second, they allow the researchers to respond to recent research findings. For example, evidence might have been published recently that suggests that phytochemical P is protective against disease D. The original dietary analysis data may not be appropriate for determining the intake of the phytochemical but blood analysis may allow this to be done. There are some advantages and disadvantages between the two designs. The nested case-control design can be highly specific to a disease outcome of interest, with confounding controlled at the outset through matching or post hoc by regression analysis. The case-cohort design is much more flexible: the cohort subset is a miniature of the cohort itself, and can be used for any purpose that the full cohort can be used for, given only the sample size constraint. Only post hoc deconfounding is possible in the case-cohort design.

1.4.2   Case-Control Studies

In general, case-control studies have the same goal as cohort studies, namely to explain why particular individuals within a population suffer from certain diseases.

As noted before for the nested case-control study, the sampling rate is higher in cases (often all available cases) than in controls, which are typically a tiny subset of all existing noncases. Controls should be selected from the source population of the cases. This happens automatically in the nested case-control study (all cases are derived from the cohort participants), but can be difficult to actuate in the nonnested case-control study, in part because it is not always obvious what the source population for cases is. Thus, in all case-control studies a direct comparison is made between healthy subjects and people who already have the disease of interest. Nonnested case control studies lack the advantage of nested case-control studies of known temporality of exposure and outcome, but are often the only alternative in the study of cancer and other rare diseases. Typically, between 50 and 500 people with the disease of interest are recruited. One (sometimes two or more) healthy— but otherwise similar—people are also recruited for comparison with each case. Subjects are asked about their diet at some past time. One may think of the case-control study as being done backwards with respect to research intent. Because selection is within cases, an exposed-to-unexposed ratio can be estimated in cases. The same is true for controls, and a relative exposure ratio can be estimated. Thus, because cases are oversampled compared to noncases, the only excess risk measure available in a case-control design is the odds ratio for being a case in exposed vs. nonexposed participants. Fortunately, for rare diseases, the odds ratio is a close approximation to the relative risk. Absolute risk is not estimable (unless the different sampling rates for cases and for controls are known, in which case a rough reconstruction of the full population could be made).

From a practical point of view, nonnested case-control studies have several advantages over cohort studies. First, far fewer subjects are needed (hundreds rather than tens of thousands). This greatly reduces the cost. Second, the results are generated far more quickly as there is no need to wait years for diseases to develop.

Unfortunately, case-control studies can have serious inherent errors [17]. The first is known as "the healthy volunteer effect." When patients with cancer are asked to participate, typically a high proportion agrees. But recruitment of "matched" controls is usually more problematic as a much lower proportion of healthy subjects agree to participate in the study. Moreover, the healthy subjects tend to include an overrepresentation of health-conscious people. The effect of this is that even if patients with a particular type of cancer have a history of eating an identical diet to the rest of the population, the results of case-control studies will likely indicate that their past diet was relatively unhealthy, such as having a low content of fruits and vegetables. The second problem is recall bias. This refers to the tendency of patients who have recently been diagnosed with cancer to have a distorted recollection of their past diet, usually in the direction of overstating its unhealthy features. Of note, a healthy volunteer bias also can enter the cohort study design, but exclusion of unhealthy, unwilling participants from the cohort baseline would not affect that relationship of exposures to future disease in those who did enter the study.

Studies of the relationship between consumption of fruits and vegetables and risk of cancer demonstrate the systematic bias that can be generated by case-control studies. Well over 200 case-control studies have been published and these consistently indicate that a generous intake of fruits and vegetables are strongly protective against several types of cancer. However, findings from large cohort studies, mostly published after 2000, have reported only minor protective benefits. We can best illustrate this serious inconsistency by looking at the estimate of the decrease in the risk of developing cancer with each extra 100 g/day of fruits and vegetables. The findings from case-control studies suggest a figure of about 10% [18] whereas cohort studies suggest a figure of between zero and 4% [18–20].

It should be remembered that if sampling is well done and exposure information is reliable and valid, the nonnested case-control study is both valid and efficient. Exposures that are easy to recall accurately, such as smoking history or history of exposures at work may meet the needed standard of the case-control study. Furthermore, if the risk following exposure is short term, a cohort study will fail to detect the risk, while a case-control study would likely find such risk. Short-term risk is the rule by definition in identification of triggers of sudden death. A case-control study could be used in this case, asking proxies for sudden deaths and proxies for controls about habitual behavior. However, a better design is appropriate here, related to the case-control study, namely the case crossover study, in which each case serves as its own control. In the sudden death outcome, a proxy would be asked about possibly triggering behavior in the period immediately before death and again the same behavior at a more remote time. This method was used successfully in identifying a bout of physical activity as a trigger for sudden death, much more so in untrained than in trained people [21]. It is interesting that this study successfully decomposed a mixture of situations: trained people had lower overall risk for sudden death, even though their immediate risk increased somewhat for each bout of physical activity.

1.4.3   Cross-Sectional Studies

In the cross-sectional design, the sample is selected to be a snapshot in time, for example selected at random from the general population. As such it is a very general design. Like both cohort and case-control studies, it can investigate conditions that are either present or absent, such as cancer (although only the cohort study can include deaths as outcomes, unless proxy respondents are used; and power to study rare diseases may be low, because they will be encountered in proportion to their existence in the population, i.e., rarely). Cross-sectional studies are often used to look at body parameters that increase by increments, such as BMI and blood pressure. Let us suppose that an investigator wishes to study the relationship between lifestyle and BMI in children. He then recruits several hundred children from the general population and measures both lifestyle factors and BMI. The results might reveal, for example, a positive association between hours of TV watching and BMI.

In principle, the information gathered in a cross-sectional study at a particular time can refer to any time in the past or present. A question like, "When did you start smoking?" is inherently historical. Nevertheless, cross-sectional studies usually attempt to determine current lifestyle, including diet, whereas case-control studies, in line with their particular focus, usually ask about lifestyle at a former time, before diagnosis. Accordingly, cross-sectional studies for the most part reveal nothing about temporal sequence whereas case-control studies attempt to determine this. Because of the paucity of information about temporal sequence typical in the cross-sectional design, it cannot distinguish causal direction, even if other aspects of the data lead to a belief that a causal relation is present. Let us look again at the study of the relationship between lifestyle and BMI in children. If that study reveals that children who spend more hours watching TV tend to have a higher BMI, this could mean that TV watching causes children to become overweight, or, conversely, that overweight children tend to watch more TV. Resolving this problem can only be properly addressed by studying children prospectively (i.e., monitoring a cohort of children over several years). A plausible but inferentially riskier strategy would be to ask people about their past BMI and TV watching, thereby emulating a prospective study, but without the security of measuring BMI and TV watching at their time of occurrence.

As with case-control studies there are significant errors that can easily enter into data collection. Everyone knows whether or not they are obese and this creates the potential error of recall bias. For example, obese people tend to underestimate their energy intake. The healthy volunteer effect may also be a source of error.

It is worth noting that most cohort studies have a cross-sectional design for their baseline.

1.4.4   Population Studies

Population studies—also known as ecological studies—have a long history in epidemiology and have yielded enormous amounts of valuable information. Typically, two dozen or more different countries are compared. Data are collected from each country on exposure to different diets and to possible risk factors and these are compared with disease rates. The findings indicate which variables are most strongly associated with each disease. Population studies take advantage of the enormous variations around the world in many aspects of lifestyle and in disease patterns.

There are several major sources of error in population studies. One is that disease statistics are often unreliable, especially in less developed countries. Food consumption data can also be misleading. In richer countries, for example, much available food may be wasted. Let us take a hypothetical example to illustrate how researchers might easily reach seriously flawed conclusions. They are investigating the relationship between consumption of potatoes and colon cancer. The data reveal that the populations of countries in Group A eat 30% more potatoes than those in Group B and also have a 30% higher incidence of colon cancer. On the surface this suggests that potatoes may cause colon cancer. But closer inspection of the data reveals significant sources of error: first, countries in Group A are rich and waste 30% of their potatoes, while countries in Group B are poor and fail to diagnose 30% of their cases of colon cancer. This means that, in actuality, people in the two groups of countries eat a similar amount of potatoes and have a similar incidence of colon cancer.

As with other types of population studies, confounding presents a serious challenge to the interpretation of findings. In the case of population studies, confounding is a reflection of different populations having many lifestyle factors that cluster together. This is best illustrated by studies that have investigated the relationship between diet and chronic diseases of lifestyle. Since the 1970s many such studies have compared populations in highly developed countries and those still living a more traditional lifestyle (before the arrival of the nutrition transition). What the findings typically reveal is a strong association between a Western diet and many chronic diseases of lifestyle, including colon cancer, breast cancer, coronary heart disease (CHD), and type 2 diabetes. Detailed analysis reveals a strong correlation between all components of the Western diet (such as a high intake of dietary fat, animal protein, and sugar, and a low intake of dietary fiber) and several different diseases.

Population studies have historically been considered to suffer from the "ecologic fallacy." In fact, ecologic variables do not necessarily even have the same substantive meaning as their individual counterparts. For example, gender at the individual level brings to mind a man, a woman, sex hormones, social roles, etc. Ecologically, gender is not very informative; it is generally about 50% male and 50% female. A variable such as the taking of antihypertensive medication means something about high blood pressure and its treatment at the individual level, while the corresponding ecologic variable may have more to do with the structure and focus of the health-care system and prescribing habits [22].

In 1975, Armstrong and Doll [23] published a paper using this methodology in order to explore the relationship between diet and cancer. This paper has become a classic and played a major role in sparking serious interest in the relationship between diet and risk of cancer. According to Google Scholar the paper has been cited 1,763 times (May 2011). The association between diet and breast cancer mortality in 27 countries illustrates the nature of the findings. The correlations (r) were as follows: animal protein 0.83, eggs 0.80, meat 0.74, and milk 0.73. The strongest association was seen for total fat: a correlation of 0.89. Conversely, cereals manifested a correlation of −0.70. Many investigators seized on the impressive association between fat intake and breast cancer and became convinced that there must be at least a small fire amid so much smoke. Alas, the search for solid supporting evidence has been largely futile. With the wisdom of hindsight what these findings really show is that a Western diet (and its associated lifestyle) is strongly *associated* with risk of breast cancer. Making inferences as to *cause-and-effect relationships* for single dietary components should only be done with much caution, particularly when the evidence in favor of the existence of the association is based solely on ecologic data.

Population studies can be done within a country; for example, comparisons of the states of the United States. The degree of variation between states is obviously far less than that seen between highly developed countries and developing countries. This has the advantage of greatly reducing the problem of confounding (i.e., differences in disease rates between states are not likely to be due to such factors as dietary intake of meat or sugar). However, the relatively small state-to-state variation in nutritional factors and disease rates limits the potential use of this technique. This approach has proven useful in situations of strong geographical variation in dietary status. One recent example of this is using regional variation in sunshine exposure as a surrogate marker for vitamin D status. Results from such studies have lent support to the view that vitamin D is protective against cancer [24].

A variant of the population study performs cross-sectional studies in several countries, as was done in Intersalt [25]. This study showed a strong association between 24-h urinary sodium and blood pressure in most of 52 populations and was then able to show the generality of this finding by displaying these slopes across countries. This study greatly reduced the likelihood of confounding by displaying within-country relationships, adjusted for confounding, rather than the simplistic display of country-wide means of urinary sodium arrayed against country-wide means of blood pressure. Another variant of the population study is one in which an integral ecologic characteristic is studied; for example, when an intervention is performed to alter lifestyle behavior with the community as the unit of analysis, or when the question is instances of food-borne bacterial infection as a function of the number of food production sites.

1.4.5   Historical Studies

The relationship between nutritional patterns and changing disease rates often provides valuable clues as to the role played by dietary components in the causation or prevention of different diseases. Such information suffers in the same way as population studies from potential ecologic biases, but can be helpful if used carefully, particularly in conjunction with other evidence.

A strong piece of evidence that supports the role of smoking in causing lung cancer comes from historical studies. The disease was rare at the time of the First World War. At around that time millions of men took up the habit of smoking. Starting around 1930, lung cancer then became progressively more common and is now by far the leading cancer in terms of numbers of deaths. Among women, by contrast, the rapid escalation in the numbers dying from the disease was delayed until around 1960. This much later date accords with the fact that smoking among women became common much later than it did for men.

Another example illustrating this method comes from looking at the relationship between the intake of folate by women and their risk of giving birth to a child with a neural tube defect (NTD). NTDs, such as spina bifida, are a serious congenital disorder.

Various evidence accumulated showing that a low dietary intake of this vitamin increases the risk for NTD. As a result in 1998 fortification of grain products with folic acid became mandatory in Canada and the United States. The effectiveness of this policy was confirmed by the sharp fall in the incidence of NTD in the years following the implementation of this policy [26].

The following example also used this approach to confirm that a particular intervention would deliver the promised benefits. The Pap smear is a test used for the detection of precancerous changes of the uterine cervix. This allows for interventions that help prevent the progression to actual cervical cancer. Different countries or regions have adopted screening programs at different times. Several historical studies from different countries have shown a large fall in deaths from cervical cancer in the years following the implementation of such programs [27]. This evidence provides valuable confirmation that utilization of the Pap smear does indeed substantially reduce the risk of death from the disease.

1.5   Randomized Clinical Trials

Randomized clinical trials (also known as randomized controlled trials; RCTs) are often referred to as the "gold standard." A single pool of participants is selected who would potentially benefit from an intervention; randomization into intervened and control groups breaks up any natural correlations between intervention/control and confounders that would exist if the intervention variable were encountered in an observational setting. Therefore, the RCT has internal validity for comparing intervention to control. Certainly, the findings generated are usually much more reliable than those from epidemiological studies. However, they are still subject to some possible errors and other limitations.

RCTs have the following key features:

- Subjects are recruited who meet predetermined criteria, such as age, gender, willingness to accept and comply with either intervention or control, and equipoise in respect to health status (that they would not clearly benefit or be harmed by either intervention or control treatment).
- Subjects are randomized into at least two groups: a control (i.e., untreated, usual care, placebo, known standard treatment) group and one or more intervention (test, treatment) groups.
- The number of subjects is carefully calculated so that the results will be meaningful. For example, based on the assumption that the intervention will cause a change in the target outcome parameter of 10% over a specified period of observation, the researchers estimate that 50 subjects are needed in both the test and control groups in order to have a 90% chance of the results being statistically significant.
- Studies are carried out "double-blind" when possible. This means that neither the subjects nor the personnel who are in direct contact with them know which group each subject is in. This is important in order to avoid various types of bias.

- Given perfect compliance within the treatment group assignments, inferences about the relation of treatment to outcome are based solely on a mathematical computation; the scientific thinking was done entirely in the setup of the RCT. In the face of noncompliance, data analysis assumes "intention to treat," that is, that the randomized groups are compared to each other, even if there is some non-compliance to the intervention or cross-over between intervention and control during the study. This feature assures that the confounding that was broken up by randomization is not reinstituted by noncompliance. This "rescue" from non-compliance works less and less well, the more the noncompliance (and would be patently absurd if most of the control group received the intervention, while most of the intervention group did not receive it).

Additional characteristics of RCTs in relation to observational cohort studies are provided in Sect. 1.9.

1.6   Systematic Reviews and Meta-analyses

Research methods are constantly evolving and being improved. An excellent example of this is the move towards systematic reviews and meta-analyses. In many fields of health-related research, a dozen or more human studies are carried out on the same topic. In the traditional literature review, experts have then examined each study, one at the time, and made an overall evaluation. A typical conclusion might say: "While the evidence lacks consistency, the majority of studies indicate that a diet with a high content of vitamin X is associated with a reduced risk of disease Y. On balance, this suggests that supplements of vitamin X may reduce risk by 20–30%." This approach has a strong element of value judgment which makes it easy for the reviewer to reach a somewhat biased conclusion.

A trend in recent years has been the publication of systematic reviews and meta-analyses. In a systematic review, the literature is reviewed using search engines according to strict criteria. These criteria might include the date of publication, language, the subject under study, and the type of study. Each paper that meets these criteria is then carefully read and relevant information extracted. The findings from these papers are compiled and then summarized as objectively as possible.

A meta-analysis takes this process one step further. Instead of guesstimating a "batting average" of what the results indicate, the reviewers assume that findings from the different studies are quantitatively poolable. They then make formal pooled risk estimates calculated using the weighted average according to set rules that consider both within-study and between-study variance. By this means larger studies usually receive more weight than smaller ones. While it is clearly helpful in many instances to make a quantitative summary of evidence, it should be recognized that this activity involves a trade-off. The pooled estimate gains power by increasing sample size, but at the expense of loosening definition of variables and concepts, which will vary across studies despite the use of strict criteria.

1.7   Experiments Using Animals

From a statistical design perspective, experiments on animals are comparable to RCTs on humans. In particular, the experimenter sets up different treatment options, paralleling the intervention and control set-up in human RCTs. Compared to RCTs, experiments on animals have obvious advantages and disadvantages. The first advantage is that conditions can be far more rigidly controlled. So if, for example, the investigator wishes to test the effect of a low-fat diet, he can be confident that all the experimental animals have consumed the intended diet, and that there are no other differences between the test and control groups. Such control is rarely possible in studies using human "guinea pigs." The second advantage is that animals can be genetically modified to have or to knock out the capability to produce the biochemical under study, enabling tight control over highly specific pathways. Reporter genes can be introduced to indicate which tissues metabolize the treatment substance. The third advantage is that experimental animals can be subjected to dangerous treatments; later they can be killed and all tissues studied, whereas in humans only easily obtainable tissue can be studied, such as blood, urine, hair, toenails, and occasionally small biopsies.

Experiments on animals do, however, have major disadvantages; this is especially the case in the area of nutrition in relation to health and disease. The dominant problem here is the vast genetic, metabolic, and physiologic gap separating animals from humans. On top of that, humans live a completely different lifestyle to animals in a laboratory. For these reasons, we often find that experimental animals react very differently from humans when their diets are altered. Animals rarely get diseases of interest for humans, and "disease" models that exist are only approximations of human disease. For example, some mice have been bred to become obese and to get diabetes, and in some models diabetes is induced by a drug that greatly harms the pancreas. Whether these forms of "diabetes" are relevant to the diabetes usually encountered in humans is uncertain. Accordingly, while animal models can elucidate biological pathways in living tissues and thereby set some standards of knowledge about how tissue and life work, data generated from studies on animals can only be extrapolated to humans with much caution.

1.8   Mechanistic Research

For many decades, nutritionists have followed the strategy that the road to understanding nutrition is to learn the details of body physiology and biochemistry, such as intermediary metabolism, and thence how the various chemicals in food are involved in health or disease. Certainly, this strategy has revealed a great deal about the role played in the body by vitamins, minerals, amino acids, and other substances, and why deficiencies of them lead to specific symptoms. It is also true that the discovery of several important new drugs has grown out of an understanding of metabolic processes and cell physiology, such as cell receptors.

But when we look at the wider arena of health research, this mechanistic strategy has serious weaknesses, as has been argued previously [28–30]. In particular, in the field of nutrition the strategy is achieving less and less success in recent decades. The reason for this is really quite simple: the human body is so complex that it is extraordinarily difficult to fully comprehend the finer details of the pathways that lead to health and disease. Compounding this, foods have such great complexity that there are vast numbers of possible interactions between food components and body processes. As a result attempts to translate an understanding of intermediary metabolism and cellular function into practical nutritional advice on preventing or treating disease seldom achieve success. This argument perfectly complements the parallel arguments we present in our chapter on food synergy (Chap. 14). Nevertheless, as with animal studies, studies of the composition of food and of the effects of food on performance of cells in vitro can be helpful as part of a total package of knowledge about nutrition, not standing alone, but contributing to inferences about important causal pathways.

All nutrition-oriented research designs contribute knowledge about biology. However, our understanding of human nutrition can be led astray if too much reliance is placed on aspects of intermediary metabolism and cellular function, merely assuming that such metabolism is pertinent to health and disease in intact humans. A better strategy might be to tie such mechanistic studies to metabolic features and nutritional aspects that have already been shown to have an effect in humans, that is, to use mechanistic studies to explain known nutritional associations rather than to find new ones.

1.9   Comparative Strengths and Weaknesses of Clinical Trials and Observational Cohort Studies in Investigations of Drugs vs. Foods

Although the RCT design is highly regarded for its internal validity, and the relevance of internal validity to interpreting a difference between intervention and control as causal, all study designs have strengths and weaknesses. Kish [31] catalogued what he called "the three Rs of research design," namely randomization, representation, and realism. The first design dimension, randomization, referring to aspects of comparison and internal validity, may be rated as excellent in RCTs, but only good in observational mode. The second and third design dimensions, however, favor observational studies. Representation in RCTs is limited, restricted to eligibles in whom treatment is considered safe; in addition, the rigors of compliance to a specific intervention regimen may limit and possibly bias participation. In the observational design, representation tends to be comprehensive. Realism in the RCT is limited by a stylized treatment delivery, designed to maximize internal validity even at the expense of somewhat altering the intervention itself. In the observational design, treatment delivery is naturalistic. Other design criteria also differ between the two research methods. The exposure period in the RCT tends to be relatively short (weeks or years), while in the observational study it is long (e.g., whole prior life, adult life, since onset of condition). Confounding by baseline characteristics is controlled probabilistically in the RCT (that is, randomization to treatment induces a high probability that each measured and unmeasured covariate is equally distributed between treatment groups), but residual confounding is always possible in observational studies. The possibility of an important biasing effect of residual confounding is reduced when many observational studies in many settings concur, and when they also concur with animal or mechanistic evidence. Less well appreciated may be that confounding by changes during the study period are a problem in both designs. Finally, level of investigator control is a strong point of the RCT; clever design and procedures may help in this respect in observational studies. Thus, we conclude that each study design has strong and weak points, at the same time agreeing that the internal validity of RCTs is a short and secure path to causal inference.

Importantly for this chapter on research design in nutrition studies, the solution to the problem of optimal research design differs between the study of drugs and foods. We compare here how typical RCT features play out for drugs compared to foods:

1. Randomization is relatively easy to achieve with a drug, while it is possible, but hard to maintain long term in the case of a food intervention.
2. Double blind is relatively easy with drugs, where look-alike placebos are possible; side effects may reveal the drug regimen. Food, in contrast, is hard to mask. Partial solutions exist, such as incorporation of different fats in margarine or different nuts in muffins or breads, but complete masking is often impossible.
3. Compliance is relatively easy with a drug delivered in a tasteless pill, although side effects of the drug may be limiting. Food, however, is at the center of life. Required drinking or not drinking of coffee, for example, or change in an entire diet pattern can have a major effect on life, several times a day. Factors such as taste, convenience, and effect on laxation are big issues in delivery of food, while they are usually not factors in drug delivery.
4. Long duration of a drug trial may be expensive, but is possible. Food studies are usually weeks long, and long-term food studies are both expensive and likely to be compromised by compliance difficulties.
5. Specificity of intervention is helpful in defining what is being studied. Drugs are often a single compound or a simple combination of compounds. Food is much more complex, being an average over time, due to varietals, growing conditions, various preparations, food groups of similar foods, and additives such salt or emulsifiers added in preparation by the food industry.
6. The outcome variable of most relevance is a clinical event. This outcome is easy to achieve in a drug trial. Food trials have mostly intermediate outcomes due to short duration. The relevance of such outcomes to clinical events may be questionable, particularly where multiple pathways are possible.
7. Control in a drug trial is well defined; it may be a placebo or other comparison drug. There are myriad possibilities for energy-bearing food; for example, saturated fat may be replaced by polyunsaturated fat, any other fat, carbohydrate, or

protein. The possible replacements in terms of foods are also highly varied. The example of a different effect on CHD of saturated fat replaced by phytochemical-poor carbohydrates than by polyunsaturated fats is salient here [32].

For these reasons it is therefore much more difficult to effectively study foods than drugs using an RCT design. Many of the disadvantages of RCTs for food become advantages in an observational setting. For example, food patterns established and relatively consistent over many years are observable, with subsequent long-term follow-up for clinical events. Observed drug taking, however, is beset by indication bias. For example, RCTs have clearly shown the efficacy of blood pressure lowering drugs for reduction of cardiovascular disease rates, but use of blood pressure lowering medication is often observed to be associated with higher risk. The explanation is that drug taking is confounded by the indication for taking the drug, namely that those taking the drug were sicker at the outset. Thus, indication bias can induce reverse causation, even in a prospective observational study design.

1.10   Evaluating the Evidence

Causal inference is easy to make based on results of RCTs, especially when they are repeated and consistent. However, RCTs that investigate food and have a duration of more than a few weeks are rare, and even if done may be difficult to interpret. For this reason, epidemiological studies have tremendous value in revealing diet–disease relationships. But the process of concluding that an observed relationship is causal is much more complex when based on observational data. Such inferences must be made after evaluation of all aspects of the evidence. Convergent evidence across different ways of looking at the problem, including animal and mechanistic studies, multiple observational results in which residual confounding is unlikely to be the same across studies, and short-term RCTs may make causal inference reasonable.

Evidence regularly appears in nutrition science suggesting that a particular food, nutrient, or other substance in food either increases or reduces risk of a particular disease. But often the evidence is weak or inconsistent. How should such evidence be evaluated? We now examine several examples where evidence from different types of research has led to our current level of understanding. Often, some of this evidence has been misleading; this inevitably creates confusion before a clearer picture emerges. These examples are intended to explore what lessons we need to learn from past mistakes. We see, time and time again, that in recent decades the large majority of our information of practical value regarding how diet is related to the prevention or treatment of disease has come from epidemiology and RCTs. In several of these examples, we show that mechanistic studies either produced little of any real value or, in some cases, were actually misleading.

In our first example, we look at Ginkgo biloba. This herb is a popular supplement for aiding the memory. However, the supporting evidence was never strong. Small, early studies showed modest improvements in cognitive function for older adults with dementia. A combination of articles in popular magazines, recommendations from the staff in health food stores, and the placebo effect can easily convince people that the supplement really does boost the memory. However, the results from a RCT published in 2009 provide strong evidence that Ginkgo does nothing to improve memory function [33]. The story of Ginkgo teaches the valuable lesson that people must be skeptical of weak evidence as it is often shot down by findings from well-conducted RCTs. This problem is especially common when those who are supporting a claimed treatment stand to make a financial profit from claiming that it really does work.

The relationship between homocysteine and CHD is an object lesson in the limitations of epidemiology. Numerous studies have revealed that an elevated blood level of homocysteine is a risk factor for CHD [34–36]. It has also been shown that supplements of certain B vitamins (folic acid and vitamins B_6 and B_{12}) lower the blood level of homocysteine. These findings led to the hypothesis that supplements of these B vitamins would act to prevent CHD. But when this was tested in RCTs, there was no reduction of CHD [37–39]. The clear lesson here is that findings coming from epidemiological research should be met with caution and that RCTs are required before statements as to cause-and-effect relationships can be stated with any confidence.

A contrasting example is provided by studies of the relationship between folate status and risk of NTDs. We referred earlier to the efficacy of folic acid supplementation to prevent these birth defects. The original hypothesis was built on epidemiological studies. This evidence was sufficiently strong to inspire researchers to carry out RCTs, first small ones and later larger ones. The findings from these trials provided convincing evidence in support of the hypothesis [40]. Historical studies (the fall in incidence of NTDs in the years following the fortification of grain products with folic acid) gave additional confirmation. Mechanistic research is suggestive of how folate is involved in neural tube development but was never able to reliably predict the dose–response relationship between intake of folate (or folic acid) and risk of NTDs. Therefore, the contribution of mechanistic research to this important discovery was quite minor: it offered a plausible rationale in support of the hypothesis but little more.

We see a similar story in the area of obesity. A combination of epidemiology and RCTs has provided a great deal of valuable information concerning the dietary factors that cause obesity and how the condition can be best prevented. But an altogether different story emerges when we look at mechanistic research: thousands of studies in intermediary metabolism have been of much less practical value.

The relationship between dietary sodium and blood pressure is another clear example of the value of epidemiology and RCTs and of the limited value of mechanistic research. Numerous epidemiological studies, mostly population studies and cross-sectional studies, have reported a strong association between sodium intake and blood pressure [41]. The Intersalt study is one such epidemiological study and was referred to earlier. Numerous RCTs have been carried out and these have generally supported a significant role of excessive dietary sodium in raising blood pressure and causing hypertension [42]. These studies are also of direct relevance to therapy for hypertension. A great deal of research activity has investigated the physiological mechanisms of blood pressure control. These studies have led to the discovery of several new drugs; however, they have produced little of real value as regards the role of diet in either the prevention or therapy of hypertension.

The relationship between dietary vitamin E and CHD provides another informative example. Epidemiological studies had indicated that the vitamin has a protective association with risk of CHD [43]. Mechanistic studies had apparently provided an explanation for this, namely that the vitamin acts as an antioxidant and thereby protects the arterial wall from oxidized lipids [44]. However, contrary to what was predicted based on the epidemiological and mechanistic evidence, RCTs revealed that supplements of the vitamin have little or no preventive action against CHD [45]. The discrepancy may be due to food synergy in that the epidemiologic studies did not look at the foods that provided the vitamin E.

Arguably, the most intriguing story has come from research into the relationship between fruit, vegetables, β-carotene, and cancer. β-Carotene is a vitamin A precursor found in various vegetables. As was discussed earlier, many epidemiological studies, mostly case-control studies, have reported that intake of fruits and vegetables have a strong inverse association with risk of various types of cancer. A similar relationship has been reported for β-carotene and cancer. Based on these findings there was much support for the hypothesis that β-carotene is protective against cancer [46]. β-Carotene is an antioxidant [47, 48] and this provided a ready mechanism to explain the anticarcinogenic action of the vitamin. We see here an obvious parallel with the previous example of vitamin E and CHD. The hypothesis that β-carotene prevents cancer was tested in several RCTs but the results were negative [49].

The story of β-carotene and cancer spans 30 years. Clearly, many mistakes were made. The story can be summarized as follows: two plus two plus two does not equal twelve (or even seven). With the wisdom of hindsight we can see three clear mistakes in how the findings from research studies were interpreted:

- Epidemiological studies indicated that intake of fruits and vegetables have a strong negative association with risk of various types of cancer. Much of this evidence was based on case-control studies. As was discussed earlier, we now know that the strength of the association was exaggerated and that it is really quite weak.
- Epidemiological studies also indicated that β-carotene has a strong negative association with several types of cancer. From this it was reasoned that β-carotene is a factor in fruits and vegetables that deserves much of the credit for the anticarcinogenic action of these foods. However, fruits and vegetables contain hundreds of different substances, many of which may have anticarcinogenic benefits. The negative association between β-carotene and cancer is best explained as a combination of two incorrect interpretations of the literature: (1) that the findings from epidemiological studies are reliable. In reality, the strength of the negative association between β-carotene and cancer was probably much exaggerated for the same reason as seen in the case of case-control studies of

fruit, vegetables, and cancer. (2) Much of the association between β-carotene and cancer was due to confounding. β-Carotene is really little more than a marker for the intake of fruits and vegetables, which may in turn in part be a marker for a generally phytochemical-rich diet.

• Findings from mechanistic studies caused many researchers to have increased confidence in the ability of β-carotene to prevent cancer. While it is an attractive hypothesis that β-carotene prevents cancer by acting as an antioxidant, in reality the supporting evidence was little more than speculation. While antioxidant defense is an important concept, the antioxidant defense system is itself highly complex and localized; free radicals can be both adverse and functional, and antioxidant defense is closely connected to other systems such as inflammatory processes [50]. Basing nutritional inferences solely on antioxidant potential has turned out to be somewhat naïve. Furthermore, cancer is an extremely complex disease process and great caution should be exercised before making bold statements as to how the carcinogenic process can be moderated.

The lesson from these accounts of how various types of study have led to our present knowledge, often after several major errors were made, is that nutrition research is inherently uncertain. Mechanistic research consumes a large share of research resources. While it has led to important new drug discoveries, it has a poor track record in the area of advancing our understanding of how nutrition is related to the prevention or treatment of disease. We are not against mechanistic research but rather against having excessive faith in it as a methodology that will explain the relationship between foods, nutrients, and health. Mechanistic research that puts more emphasis on studying food rather than nutrients may prove to be a valuable form of inquiry. Epidemiology has been of far greater value, but is still prone to throwing up many misleading findings. RCTs are far more reliable. However, as we shall see in the next section, even here serious errors can be made.

1.11   The Problems of Conflict of Interest and of Inadequate Research Funding

In the final analysis, science is based on trust. If a researcher publishes a study, one assumes that he has done what he says he has done and the results are indeed those that were seen. Occasionally, a researcher publishes fraudulent data. This, of course, causes much confusion, wastes the time of many people, and tarnishes the good name of scientists. But such events are fairly rare.

A much more common problem is created by conflict of interest. There is an enormous amount of money tied up in the results of research studies. This problem is most strongly associated with pharmaceutical research [51, 52]. If a drug company can demonstrate that its latest drug is effective, then the potential sales may be worth billions of dollars. The pharmaceutical industry—big pharma—has been highly creative at manipulating the system in order to exaggerate the benefits of its drugs. Methods used include financially supporting researchers who are friendly to the drug industry, designing RCTs in such a way that positive results are more likely, analyzing the data so as to make the conclusions as close as possible to what the drug company wants to hear ("if you torture the data long enough, it will eventually confess"), and only allowing the findings of studies to be published if they report the "right" results. After one or more RCTs have been published, authoritative figures, such as university professors, are then hired to write reviews in journals that paint a very positive impression of the benefits of the drug. The article may even be written by an employee of the drug company (a "ghost writer") and simply passed to the supposedly independent expert (usually along with a generous fee) who then submits it in his own name to a journal. Finally, the committees who write the clinical practice guidelines are loaded up with experts (who coincidentally receive grants from the drug companies) and announce that physicians should prescribe the drug in question.

Fortunately, the situation in nutrition science bears little resemblance to the very sorry state of affairs regularly seen with drug research and development. However, reports have appeared that suggest that conflict of interest does exist and is distorting the findings of some research studies. An analysis was made of studies conducted between 1999 and 2003 on soft drinks, juice, and milk [53]. The findings of each study were classified as being favorable or unfavorable to the industry that sponsored the study. For interventional studies, none of the studies with industry financing reported a finding that was unfavorable whereas 37% of studies with no industry funding did so. For all types of study, including observational studies and reviews, those with industry financing were 7.6 times more likely to report a finding favorable to industry than studies with no industry financing. Much the same has been reported concerning research on olestra, a fat substitute [54].

In 2009, proposals were made for dealing with conflict of interest with regard to studies which have industry funding, particularly in the area of health, nutrition, and food safety [55]. Nevertheless, intentional misreporting, or even shading of findings, is not likely to be the most difficult problem. When profit is a major factor in research about particular products, an even more insidious process may occur. In this process, the literature is reviewed, promising results are translated into products, and a marketing plan is put in place. Further research may be deemed unnecessary once profits start to come in. Alternatively, the researchers associated with a company may do some preliminary research that is deemed "inconclusive and not reportable" and further research in that direction is cut off. Thus omission, namely failure to pursue some lines of research at all, may be a bigger problem than commission, namely shading of results in completed research.

These problems are very large in nutrition research, perhaps more so than in pharmaceutical research, because of funding infrastructure. Stringent government regulations of pharmaceuticals have led to huge expenditures to identify and prove new drugs, followed by a period of cost recovery and profit for the patent-protected "winning drugs." Drugs operate along relatively simple pathways and lead to interesting findings that can be pursued in government-funded basic research studies. Independent scientists serving as peer reviewers tend to like these detailed mechanistic studies, and many are therefore funded. Nearly everything in this scenario is the opposite of food research. Food is considered "safe" and regulation is in areas such as avoiding pest contamination, rather than how food works to promote health or disease. There is no regulatory-induced pot of money directing certain foods into definitive RCTs. Furthermore, studies of food are not as well suited as are drugs to detailed mechanistic studies. Such studies then are de facto less appealing to peer reviewers. For that reason government funding, which is not tainted by a profit motive for the product under study, is hard to come by. The foods researcher is left with one obvious source for funding: the food industry. However, we have arrived back at the beginning of the story: the food industry has no incentive to discover aspects of their product that are counter to profits. Thus, the ultimate solution to the problem of conflict of interest is probably to create a sufficient body of funds to perform foods-based research without such conflict.

1.12   Ethical Approval

In most Western countries, research studies involving human subjects can only be carried out after approval has been obtained from an ethics committee. This safeguard is designed to protect the rights of subjects. The procedure is that the research scientist makes an application to an ethics committee. The application must explain the procedures to be used, the potential risks to the subjects, and what safeguards will be taken. Normally, the degree of scrutiny is in proportion to the potential risk. Therefore, the process is normally much quicker and simpler for a study that poses minimal risk than for the test of, for example, an experimental anticancer drug.

One universal rule is "informed consent." This means that at the time that subjects are recruited, the purpose of the study must be explained as well as any potential hazards. This must be done in a way that the person is able to properly understand. Subjects must be free to refuse to participate without feeling pressured and must be free to withdraw from the study at any time without penalty. Therefore, a professor is not allowed to pressure students into "volunteering" for a study and a physician's patients must not feel concerned that if they decline to participate in a study, this might somehow reduce the quality of the treatment they receive.

A fundamental rule is that the interests of the patient come before the interests of society. Accordingly, even if a researcher sincerely believes that a study is of great importance but will cause harm to the subjects in the study, this does not justify carrying out the study. For example, if the researcher wishes to investigate the effects of consuming a possibly toxic substance, this can only be done by studying people who have accidentally or deliberately consumed the substance; it cannot be done by giving the substance to volunteers. This rule obviously restricts the design of RCTs.

Ethics also covers the protection of patient confidentiality. This means that all personal information collected must be securely stored and can only be viewed by those research personnel who need to see it. Moreover, when the research results are published, no information may be included that identifies any subject (unless the subject gives permission).

Ethical approval must also be obtained in the case of experiments on animals. The rules here are, of course, quite different than with humans. Animals must not be subject to severe pain; there is a blanket ban on such experiments. Moreover, animals must be cared for humanely; they should not be subject to any unnecessary distress except where the experimental procedures being carried out make this unavoidable.

References

1. Willett W. Nutritional epidemiology. New York: Oxford University Press; 1998.
2. Boushey C, Harris J, Bruemmer B, Archer S, Van Horn L. Publishing nutrition research: a review of study design, statistical analysis, and other key elements of manuscript preparation, Part 1. J Am Diet Assoc. 2006;106:89–96.
3. Boushey C, Harris J, Bruemmer B, Archer S. Publishing nutrition research: a review of sampling, sample size, statistical analysis, and other key elements of manuscript preparation, Part 4. J Am Diet Assoc. 2008;108:679–88.
4. Harris J, Boushey C, Bruemmer B, Archer S. Publishing nutrition research: a review of non-parametric methods, Part 3. J Am Diet Assoc. 2008;108:1488–96.
5. Harris J, Gleason P, Sheean P, Boushey C, Beto J, Bruemmer B. An introduction to qualitative research for food and nutrition professionals. J Am Diet Assoc. 2009;109:80–90.
6. Bruemmer B, Harris J, Gleason P, Boushey C, Sheean P, Van Horn L. Publishing nutrition research: a review of epidemiological methods. J Am Diet Assoc. 2009;109:1728–37.
7. Gleason PM, Harris JE, Sheean PM, Boushey CJ, Bruemmer B. Publishing nutrition research: validity, reliability, and diagnostic test assessment in nutrition-related research. J Am Diet Assoc. 2010;110:409–19.
8. Hill AB. The environment and disease: association or causation? Proc R Soc Med. 1965;58: 295–300.
9. Cutler GJ, Flood A, Hannan P, Neumark-Sztainer D. Major patterns of dietary intake in adolescents and their stability over time. J Nutr. 2009;139:323–8.
10. Prentice RL. Dietary assessment and the reliability of nutritional epidemiology research reports. J Natl Cancer Inst. 2010;102:583–5.
11. Neuhouser ML, Tinker L, Shaw PA, et al. Use of recovery biomarkers to calibrate nutrient consumption self-reports in the Women's Health Initiative. Am J Epidemiol. 2008;167: 1247–59.
12. Beaton GH, Milner J, McGuire V, Feather TE, Little JA. Source of variance in 24-hour dietary recall data: implications for nutrition study design and interpretation. Carbohydrate sources, vitamins, and minerals. Am J Clin Nutr. 1983;37:986–95.
13. Friedman GD, Cutter GR, Donahue RP, Hughes GH, Hulley SB, Jacobs Jr DR, et al. CARDIA: study design, recruitment, and some characteristics of the examined subjects. J Clin Epidemiol. 1988;41:1105–16.
14. Bild DE, Bluemke DA, Burke GL, et al. Multi-ethnic study of atherosclerosis: objectives and design. Am J Epidemiol. 2002;156:871–81.
15. Stringhini S, Sabia S, Shipley M, et al. Association of socioeconomic position with health behaviors and mortality. JAMA. 2010;303:1159–66.
16. Sinha R, Cross AJ, Graubard BI, Leitzmann MF, Schatzkin A. Meat intake and mortality: a prospective study of over half a million people. Arch Intern Med. 2009;169:562–71.

17. Willett WC. Fruits, vegetables, and cancer prevention: turmoil in the produce section. J Natl Cancer Inst. 2010;102:510–1.

18. Riboli E, Norat T. Epidemiologic evidence of the protective effect of fruit and vegetables on cancer risk. Am J Clin Nutr. 2003;78(3 Suppl):559S–69.

19. Boffetta P, Couto E, Wichmann J, et al. Fruit and vegetable intake and overall cancer risk in the European Prospective Investigation into Cancer and Nutrition (EPIC). J Natl Cancer Inst. 2010;102:529–37.

20. Hung HC, Joshipura KJ, Jiang R, et al. Fruit and vegetable intake and risk of major chronic disease. J Natl Cancer Inst. 2004;96:1577–84.

21. Siscovick DS, Weiss NS, Fletcher RH, Lasky T. The incidence of primary cardiac arrest during vigorous exercise. N Engl J Med. 1984;311:874–7.

22. Jacobs Jr DR, McGovern PG, Blackburn H. The US decline in stroke mortality: what does ecological analysis tell us? Am J Public Health. 1992;82:1596–9.

23. Armstrong B, Doll R. Environmental factors and cancer incidence and mortality in different countries, with special reference to dietary practices. Int J Cancer. 1975;15:617–31.

24. Giovannucci E, Liu Y, Rimm EB, et al. Prospective study of predictors of vitamin D status and cancer incidence and mortality in men. J Natl Cancer Inst. 2006;98:451–9.

25. Elliott P, Stamler J, Nichols R, et al., for the Intersalt Cooperative Research Group. Intersalt revisited: further analyses of 24 hour sodium excretion and blood pressure within and across populations. BMJ. 1996;312:1249–53.

26. De Wals P, Tairou F, Van Allen MI, et al. Reduction in neural-tube defects after folic acid fortification in Canada. N Engl J Med. 2007;357:135–42.

27. Taylor R, Morrell S, Mamoon H, Wain G, Ross J. Decline in cervical cancer incidence and mortality in New South Wales in relation to control activities (Australia). Cancer Causes Control. 2006;17:299–306.

28. Roberts S, Temple NJ. Medical research: a bettor's guide. Am J Prev Med. 2002;23:231–2.

29. Temple NJ. Medical research: a complex problem. In: Temple NJ, Burkitt DB, editors. Western diseases: their dietary prevention and reversibility. Totowa, NJ: Humana Press; 1994. p. 419–36.

30. Temple NJ. Nutrition and disease: challenges of research design. Nutrition. 2002;18:343–7.

31. Kish L. Statistical design for research. Hoboken, NJ: Wiley; 2004.

32. Kromhout D, Geleijnse JM, Menotti A, Jacobs DR. The confusion about dietary fatty acids recommendations for CHD prevention. Br J Nutr. 2011;106:627–32.

33. Snitz BE, O'Meara ES, Carlson MC, et al. Ginkgo biloba for preventing cognitive decline in older adults: a randomized trial. JAMA. 2009;302:2663–70.

34. Seshadri N, Robinson K. Homocysteine, B vitamins, and coronary artery disease. Med Clin North Am. 2000;84:215–37.

35. Chao CL, Tsai HH, Lee CM, et al. The graded effect of hyperhomocysteinemia on the severity and extent of coronary atherosclerosis. Atherosclerosis. 1999;147:379–86.

36. Stubbs PJ, Al-Obaidi MK, Conroy RM, et al. Effect of plasma homocysteine concentration on early and late events in patients with acute coronary syndromes. Circulation. 2000;102:605–10.

37. Bazzano LA, Reynolds K, Holder KN, He J. Effect of folic acid supplementation on risk of cardiovascular diseases: a meta-analysis of randomized controlled trials. JAMA. 2006;296:2720–6.

38. Albert CM, Cook NR, Gaziano JM, et al. Effect of folic acid and B vitamins on risk of cardiovascular events and total mortality among women at high risk for cardiovascular disease: a randomized trial. JAMA. 2008;299:2027–36.

39. Armitage JM, Bowman L, Clarke RJ, et al. Study of the Effectiveness of Additional Reductions in Cholesterol and Homocysteine (SEARCH) Collaborative Group. Effects of homocysteine-lowering with folic acid plus vitamin B12 vs placebo on mortality and major morbidity in myocardial infarction survivors: a randomized trial. JAMA. 2010;303:2486–94.

40. Czeizel AE, Dudás I. Prevention of the first occurrence of neural-tube defects by periconceptional vitamin supplementation. N Engl J Med. 1992;327:1832–5.

41. He FJ, MacGregor GA. A comprehensive review on salt and health and current experience of worldwide salt reduction programmes. J Hum Hypertens. 2009;23:363–84.

42. Sacks FM, Svetkey LP, Vollmer WM, et al., for the DASH-Sodium Collaborative Research Group. Effects on blood pressure of reduced dietary sodium and the Dietary Approaches to Stop Hypertension (DASH) diet. DASH-Sodium Collaborative Research Group. N Engl J Med. 2001;344:3–10.

43. Knekt P, Ritz J, Pereira MA, et al. Antioxidant vitamins and coronary heart disease risk: a pooled analysis of 9 cohorts. Am J Clin Nutr. 2004;80:1508–20.

44. Princen HM, van Duyvenvoorde W, Buytenhek R, et al. Supplementation with low doses of vitamin E protects LDL from lipid peroxidation in men and women. Arterioscler Thromb Vasc Biol. 1995;15:325–33.

45. Mente A, de Koning L, Shannon HS, Anand SS. A systematic review of the evidence supporting a causal link between dietary factors and coronary heart disease. Arch Intern Med. 2009;169:659–69.

46. Temple NJ, Basu TK. Role of beta-carotene in the prevention of cancer—a review. Nutr Res. 1988;8:685–701.

47. Panayiotidis M, Collins AR. Ex vivo assessment of lymphocyte antioxidant status using the comet assay. Free Radic Res. 1997;27:533–7.

48. Duthie SJ, Ma A, Ross MA, Collins AR. Antioxidant supplementation decreases oxidative DNA damage in human lymphocytes. Cancer Res. 1996;56:1291–5.

49. Druesne-Pecollo N, Latino-Martel P, Norat T, et al. Beta-carotene supplementation and cancer risk: a systematic review and metaanalysis of randomized controlled trials. Int J Cancer. 2010;127:172–84.

50. Hollman PC, Cassidy A, Comte B, et al. The biological relevance of direct antioxidant effects of polyphenols for cardiovascular health in humans is not established. J Nutr. 2011;141:989S–1009.

51. Fraser J. Conflict of interest: a major problem in medical research. In: Temple NJ, Thompson A, editors. Excessive medical spending: facing the challenge. Oxford: Radcliffe Publishing; 2007. p. 20–35.

52. Kassirer JP. On the take: how medicine's complicity with big business can endanger your health. New York: Oxford University Press; 2005.

53. Lesser LI, Ebbeling CB, Goozner M, Wypij D, Ludwig DS. Relationship between funding source and conclusion among nutrition-related scientific articles. PLoS Med. 2007;4:e5.

54. Levine J, Gussow JD, Hastings D, Eccher A. Authors' financial relationships with the food and beverage industry and their published positions on the fat substitute olestra. Am J Public Health. 2003;93:664–9.

55. Rowe S, Alexander N, Clydesdale FM, et al. Funding food science and nutrition research: financial conflicts and scientific integrity. J Nutr. 2009;139:1051–3.

Chapter 2
Challenges in Research in Nutritional Epidemiology

David R. Jacobs, Jr.

Keywords Nutritional epidemiology • Methods of dietary assessment • Dietary recall • Food frequency questionnaire • Biomarkers • Dietary patterns

Key Points

- Diet is a complex aggregate of foods and behaviors. The food is constituted of a wide variety of intended and unintended chemicals which may act singly on human metabolism, but more likely act in groups in a synergistic fashion.
- The study of nutrition and disease in aggregates of human beings—nutritional epidemiology—is hampered by the difficulty in accurately characterizing this complex aggregate, that is, in stating what people are eating. Part of this difficulty is inherent in the large day-to-day variability in what is eaten. Another part of the difficulty relates to finding efficient and accurate ways to collect dietary information, minimizing participant burden, and maximizing utility of the data for investigators.
- Much progress has been made in nutritional epidemiology in recent years owing to the use of food frequency questionnaires, which pose little participant burden and are relatively easy to analyze. However, such data collection instruments are still characterized by high within-person variation and at the same time severely limit collection of important details about diet.
- A critical concept is whether the participant or the researcher synthesizes the dietary information, including issues such as defining the time period over which to average diet, what to do with unusual information, what constitutes a serving,

D.R. Jacobs, Jr., PhD (✉)
Division of Epidemiology and Community Health, School of Public Health,
University of Minnesota, 1300 South Second Street, Suite 300, Minneapolis, MN 55454, USA
e-mail: Jacobs@epi.umn.edu

N.J. Temple et al. (eds.), *Nutritional Health: Strategies for Disease Prevention*,
Nutrition and Health, DOI 10.1007/978-1-61779-894-8_2,
© Springer Science+Business Media, LLC 2012

how foods are grouped (grouping fruit juice and fruit drink together, or not, for example), and what emphasis to put on brand names.

- The author speculates on protocol changes and computer technology advances that might allow more complete and accurate diet data collection.
- It is important to study foods, food groups, and food patterns as well as nutrients and other chemicals contained in food. Food is what people eat. Where many chemical constituents of a food act synergistically, an association will be found with the food but none will be found with individual constituents. The associations of food patterns with risk provide feedback to policy makers on the likely success of nutritional pronouncements.

Much has been written about the practice and challenges of research in nutritional epidemiology. For general details concerning this topic, the reader is referred to existing and extensive source materials, including *Design Concepts in Epidemiology*, edited by Margetts and Nelson [1] and *Nutritional Epidemiology* by Willett [2]. These books provide myriad technical details on the goals of nutritional epidemiology and the conduct and interpretation of studies, with discussion of potential pitfalls. This chapter focuses on two issues that are particularly challenging in nutritional epidemiology: (1) how to find out what people eat, and (2) how to think about the effect of diet on health.

2.1   How to Find Out What People Eat

2.1.1   The Nature of Dietary Information

A full characterization of a person's diet would consist of a large number of discrete pieces of information. There are thousands of foods, prepared in myriad ways, and eaten in various amounts and combinations. Even a single "food" such as a carrot [2] or an onion [3] presents a challenge, as there are many varieties and genetic variation; growing conditions are influential in food composition. The timing and context of eating, as well as the number of meals eaten, may all contribute to metabolism of food. Willett [2] spends an entire chapter showing that actual consumption varies widely from day to day. It may take months for individual diets to settle down to a steady state average.

Each food supplies myriad chemicals. Among these chemicals, Willett [2] lists essential nutrients (vitamins, minerals, lipids, amino acids), major energy sources (fat, protein, carbohydrate, alcohol), additives (preservatives, flavorings), agricultural contaminants (pesticides, growth hormones), microbial toxins (aflatoxins), inorganic contaminants (cadmium, lead), chemicals formed in the cooking or processing of food (nitrosamines), natural toxins (natural pesticides), and other natural compounds (including DNA, enzymes, and enzyme inhibitors, many of which he says are thought of as "incidental to the human diet"). Energy content and nutrients, along with a few natural compounds, are readily available in a variety of food tables, while assessment of the remaining categories requires specialized databases. All of these chemicals pertain to each food eaten and can be summarized over the entire diet. The complete characterization of diet, foods, and the chemicals eaten, is clearly formidable. At some point in the research process, this large volume of information must be synthesized to be used in data analysis, that is statistical variables such as food groups and nutrients must be defined based on the available information.

2.1.2   Methods of Dietary Assessment

Two primary classes of methods have been used historically to assemble individual dietary information and synthesize it into something usable in data analysis, described in detail by Willett [2]. The first method includes dietary recalls and records. Dietary recalls are obtained by an interviewer assisting the participant to remember precisely what was eaten, usually over the past 24 h. Dietary records, on the other hand, are obtained by having the participant write down what was eaten, shortly after it was consumed; in practice, participants often wait until the end of the day to do their recording, so that the record easily transmutes to a self-administered recall. Variations in these methods include weighing foods before eating; collecting a duplicate portion of the food for subsequent chemical analysis; and recording onto partially precoded forms. Dietary recalls may differ in how intensively they inquire about different aspects of diet; for example, an interviewer may inquire deeply and pointedly, to a greater or lesser extent, for hard-to-obtain full information on such topics as alcoholic beverages drunk, salt-containing condiments used, or brand names of products eaten. Timing of eating may be obtained so that the integrity of individual meals can be maintained in the database. In both recalls and records, the data consist of a description of the food eaten and its portion size, perhaps with notes on brand names and preparation methods. The fact that a hamburger and a bun were eaten will generally be maintained in the database, but it is fairly common not to maintain whether the two were eaten as a sandwich.

The second method is a food frequency questionnaire (FFQ), characterized by asking the participant general questions about diet. A typical question would be: "Do you eat hamburgers, and if so, how often and in what portion size?" Other kinds of general questions are also common. For example, one might ask: "When you eat a hamburger, is it usually a low-fat variety?" The scope of questions may include related aspects, for example: "Do you prefer white bread or whole wheat bread?" An important aspect is that foods are often grouped: "How often do you eat apples or pears?" FFQs come in several varieties, e.g., from 12 to 250 questions, with and without information about portion size. Those that ask about portion size are called semi-quantitative FFQs. In a popular variant, the Willett-style questionnaire, a portion size is given for each food and frequency of portions is queried. In the other popular variant, the Block-style questionnaire, frequency of eating occasions is queried for each food, with a separate question about portion size. Additional variants exist, for example in which pictures or food models are provided to facilitate food recognition and portion size estimation.

The dietary history method is closely related to the FFQ. Here, time is spent in general discussion of the diet prior to recording answers to the formal questions; this discussion is thought to improve the context of the interview and help the participant to put together the information needed. In the diet history, the close-ended questions may be general, e.g., "Do you eat red meat?," with an open-ended elicitation of foods eaten for those who answer affirmatively. The CARDIA Diet History [4–6] is of this form: 1,609 food codes or recipes were endorsed by at least one of over 5,000 participants in one of two administrations of this questionnaire through 1993. Due to expansion of the specific products supplied by industry, the number of food codes endorsed expanded dramatically in the 2005–2006 administration of this questionnaire.

It is probably a coincidence of history that the primary approach to dietary assessment used in cardiovascular disease epidemiology in most major studies through the early 1980s was 24-h recalls; used, for example, in the Lipid Research Clinics [7, 8] and Multiple Risk Factor Intervention Studies [9, 10]. Special attention was paid to translating the myriad pieces of information into energy and nutrient intake. The synthesis of the data proved quite difficult and relatively little work was done to study the associations of individual foods or food groups on long-term health outcomes. Where food grouping was done, it was done inflexibly, so only certain combinations of foods could be examined. Examination of nutrients within food groups (e.g., monounsaturated fat from plant vs. animal foods) has received little attention. In principle, the data are available for such analyses, but it is unlikely that anyone will ever have the time, money, and study connections for such purposes. In contrast, cancer epidemiologists have long used FFQs [11]. This choice may be related to the traditional use of the case–control design for rare cancers. The desired information was the diet before diagnosis, and this would not be obtainable by recording or recalling current diet. In the cancer epidemiology field, much more has been written about foods and food groups than in the cardiovascular disease epidemiology field. In contrast to analyses of dietary recall data, nutrient analyses within food groups are fairly common. On the other hand, the FFQ obtains much less information than does the recall/record method. For example, information about "yellow and green leafy vegetables" may be all that is collected; therefore, no information is obtained regarding which vegetables were eaten.

An example of a local effort that addresses this issue is the foods and nutrient database maintained in the Department of Nutrition at the University of Oslo, which has long had a food grouping code for each food. Therefore, foods analysis has been available independent of the nature of the method of dietary data collection. Such analysis has been performed fruitfully, also allowing diet pattern analysis to take place [12]. Further, partially addressing this issue, the Nutrition Coordinating Center in the early 2000s added a food grouping system with 166 food subgroups in its Nutrition Data System for Research (NDS-R) diet analysis system (University of Minnesota, Minneapolis, MN, http://www.ncc.umn.edu/index.html, accessed May 6, 2011). The CARDIA database added these for its diet history data in 1985–1986,

1992–1993, and 2005–2006, which has enabled substantial food group analyses. Nevertheless, a great number of details in the CARDIA diet database remain inaccessible, primarily for reasons of cost in pulling those data (other than the preformulated food group) from those massive databases.

2.1.3   Ability to Represent Usual Diet

Two major conceptual differences exist between the recall/record and FFQ methods. The first relates to representativeness of usual diet. The strength of the recall/record method is that it can collect accurate and detailed information about actual consumption of particular meals. However, the particular day or meal is rarely of interest in nutritional epidemiology. It is well agreed that a single day's recall or record is inadequate as a representation of typical intake [2]. The general experience has been that the recall/record method has not worked well in studies of diet and chronic disease outcomes. Nevertheless, multiple days of recalls or records can represent the typical diet quite accurately, as in the Framingham Children's Study [13, 14]. However, it is rare for large studies to undertake more than one or possibly 2 days of recalls.

The FFQ class of methods, in contrast, asks about the typical dietary pattern during a longer time frame, typically the past year. Many studies have found associations of nutrients and/or food groups with chronic disease outcomes using this method [15]. An even more powerful method uses repeated FFQ assessments during follow-up in a cohort study [16]. When the typical diet is not changing greatly over several years, averaging results from repeated FFQ assessments can be quite powerful.

2.1.4   Who Synthesizes Dietary Information?

The second major conceptual difference between the recall/record and FFQ methods relates to how the myriad dietary details get synthesized into data analytic variables. This refers to the acts of summarizing, as an average, or otherwise characterizing, such as eating or not, discounting or upweighting unusual days or periods, dealing with unusual items, setting defaults for portion size and other aspects that are not specifically known, such as in restaurant eating, making fine distinctions, such as between fruit, fruit juice, and fruit drink, focusing on brand names, or not, and how to deal with waste. In the recall/record method, a huge database is created with near infinite flexibility. The researcher is responsible for putting this information together in a manner that is usable in data analysis. In practice, this synthesis is often limited to energy and nutrient intake analysis; however, it is quite possible that the inherent flexibility of this method may be better utilized in coming years as computer technology continues to improve; for example, as indicated above this has occurred in the interactions between the Nutrition Coordinating Center and the CARDIA study.

In the FFQ class of methods, the participant synthesizes the information. Much potential detail, and therefore flexibility, is lost, but the vastly reduced amount of information collected tends to make it a small job to create arbitrary combinations of food and nutrient variables. It seems likely that the investigators' formal synthesis of multiple recalls or records would be more accurate than the participant's informal synthesis. However, especially if the investigators' synthesis never gets done, the participant's synthesis is not without merit, despite variability in synthetic capability across participants and difficulty in defining typical patterns. For example, if a person actually drank 20 glasses of milk in a month, including one stretch of 5 days in which 10 of the glasses were drunk, one might say that the typical pattern is two-thirds glass per day. A recall could easily be done on a day when no milk or two glasses were drunk, thus getting the wrong answer, but it is easy for a person to summarize their pattern into something like a glass every other day.

Some cleverness may be needed in the FFQ mode to get at nutritional concepts with which the public is less familiar, such as whole grain bread. A prime example is the use by Willett of the term "dark bread" to elicit breads that were most likely to have at least moderate whole grain content. Although "dark bread" is a somewhat oblique reference, asking directly about whole grain bread might not have been well understood by participants, and most breads containing a substantial amount of whole grain are darker than American white bread. Dark bread is oblique due to exceptions popular in the US, including pumpernickel cooked with molasses and rye bread made with refined rye. Despite these potential problems, the reference to "dark bread" succeeded in eliciting breads that were inversely associated with coronary heart disease mortality in the Iowa Women's Health Study [15]. Another interesting Willett innovation in an attempt to get at an important detail, and also used in the Iowa Women's Health Study, was the additional query of the brand name of the usual breakfast cereal eaten [15]. Despite the fact that many people eat more than one breakfast cereal, this detail provided the ability to categorize brands, a great boon in the study of whole grains and health. Similarly, the CARDIA Diet History was innovative in that it intended to blend recall and synthesis. It asked for the last 30 days of typical intake, recent enough for some level of recall to assist the participant in synthesizing. It also allowed tremendous detail in the participant's self-assessment of typical intake by prompting the participant with 100 general food categories (e.g., eggs), then asking the participant to name all foods consumed within each category. The question, "How often do you eat at fast food restaurants?," while not specifically asking about foods consumed, falls within the FFQ type of query. It has been used fruitfully in finding, for example, that fast food intake appears to promote obesity and insulin resistance [17–19], while eating at "slow food restaurants" does not have the same effect [17, 18].

2.1.5   Can Accurate Dietary Information Be Obtained?

A great deal of progress has been made in understanding the relationship of diet with chronic disease, based mostly on FFQs. Nevertheless, validation studies of FFQs against 1–4 weeks of food diaries are somewhat discouraging. It is difficult for most people to summarize their diet accurately. There are several reasons for this including: that such summarization requires considerable quantitative ability; that most people simply eat, without making habitual summaries of what they are eating; that diet varies considerably and what is typical for the past month might be different from what is typical for the past year; and that the researchers' questions might not be the optimal formulation for eliciting particular dietary facts. Criterion measures have revealed correlations in the range of 0.3–0.6 between the two methods [5, 20–22]. The resulting within-person error leads to serious problems in interpretation of dietary data [1, 2, 23, 24].

Certain data analytic and interpretive approaches can be helpful. Cautious statements and consistency checks are called for. For example, an assertion that a nutrient is related to incident disease will be stronger if all the foods that contain the nutrient are individually also related to that disease, given that different foods contain different mixes of nutrients [2]. Conversely, if an apparent relationship of disease with a nutrient exists only for a single food that was eaten often and is high in the nutrient, that would be more consistent with the concept that the food, not the nutrient, is causally related to incident disease. Then the causal pathway might rely on a synergy of the components of the food or on a different single nutrient. An example of this type of finding was that phosphorous from dairy, but not from other sources, was related to future hypertension [25]. While this type of finding could reflect synergy of some type, other possible explanations include selective misclassification of the nutrient across the food groups (e.g., phosphate may be preferentially underestimated for processed foods) or introduction of new confounding. Meta-analysis showing consistency of findings across studies can also be helpful [26, 27]. Nevertheless, the FFQ method appears to have intrinsic limitations in how precisely it can define individual intake. Among possibilities for improvement of the FFQ method are increasing precision and innovation of questions; repeated administrations of the questionnaire with averaging to reduce the influence of within-person variation in intake; and enhancing dietary awareness of participants, for example by encouraging or requiring the participant to keep informal dietary records for a few days prior to filling out the questionnaire or by giving advance instruction in portion size determination.

A single recall or record does not accurately represent typical dietary information because of intrinsic day-to-day variation [2]. In contrast, in the Framingham Children's Study the clarity of findings in only 95 children with repeated diet assessments is impressive [13, 14], but they obtained many more diet records than is typical of studies in nutritional epidemiology. The detail obtained from many dietary records is seductive from the research perspective. This approach, in its flexibility for the researcher, far outstrips the already successful studies, for example at Harvard and the University of Minnesota, that have relied on FFQs. The multiple diet record method is a powerful cohort study design indeed that obtains unlimited accurate dietary characterization and follow-up for many different chronic disease outcomes. However, even with added power from such a large number of diet records, it is probable that thousands of participants would be needed in studies of remote and rare chronic disease outcomes. In most practical epidemiological situations, the possibilities are limited for obtaining four to twelve 24-h diet records per year in the assembly line fashion that would be needed for a cohort study of a chronic disease. Given present methodologies it is unlikely that many studies will achieve this standard. Nevertheless, we can dream.

The success of the internet and the surge in computer power means that one might optimistically hope for better methods in the future. In particular, one could imagine widespread collection of self-administered dietary information on the internet, with full software including help and dialog boxes that would simulate the support currently given by an interviewer. Thus the dietary collection instrument could even be a mixture of recall and synthesis. The open-ended methods of the CARDIA Diet History might be helpful, combined with some aspects of artificial intelligence. Branching logic for finding food codes could be employed, similar to that currently used by the NDS-R, a "Windows-based software package incorporating a time-tested, highly accurate database with an up-to-date interface," released in 1998 by the Nutrition Coordinating Center of the University of Minnesota [28]. One could even envision questionnaires filled out over the telephone, with automated voice prompts to assist in accuracy. As questionnaires accrued, the foods database could automatically expand in line with what was reported by participants. Thus a participant could repeatedly and at their convenience do a 24-h recall or report typical intake over the past week with verbal or online prompts that help find correct food codes and pointed questions to help improve the quality of the information obtained.

A requisite for exploiting this type of ambitious scheme would be correspondingly simple-to-use programs to extract nutrients, foods, food groups, and food group-specific nutrients. The researcher would require package programs to assemble the data, to formulate and reformulate food groups, and to compute nutrient values. As new information comes along, it could be added to the food table, to simplify study of novel compounds.

These schemes are perhaps dreamlike, but maybe not completely out of the question. Who would have imagined only a few years ago the internet, or, to cite one important application, millions of journal abstracts and articles themselves available at the touch of a few computer keystrokes? Or, for that matter, "telephones" that are really personal computers with highly specialized "apps" that enable highly individualized and detailed participant contact. In the near term, however, it is most likely that nutritional studies of chronic disease outcomes will continue to be based on the FFQ class of methodologies, bolstered by findings from short-term human and animal studies and the native ingenuity of the scientists doing the research.

Willett [2] comments on another method that has promise, but also pitfalls: correlation of food intake with biomarkers. A biomarker is a chemical measured in some biological sample, commonly blood or urine, but others as well, for example feces, hair, toenails, cheek cells, adipocytes, and skin scrapings. Minerals reside in toenails, which grow over several months; therefore this measure represents an average intake over several months. This technique has been used in studies on the relationship between selenium status and risk of cancer [29]. Urinary nitrogen is a marker of nitrogen and therefore protein intake. Sodium and potassium intake are mirrored quite rapidly (over ~2 days) in urinary sodium and potassium. Serum carotenoids and ascorbic acid are highly responsive to both dietary and supplemental intake of the same substances. Freedman and coworkers have suggested methods for combining biomarker and diet information to improve accuracy [30, 31]. Nevertheless, biomarkers have limitations as indicators of dietary intake. Each tissue and substance has its own half life and metabolism. Some tissues store substances, and some utilize them rapidly. The amount of a substance in blood may not be representative of its occurrence throughout the body. Substances may be maintained homeostatically, or may be partially under dietary and partially under homeostatic control. There may be changes in nutrients consumed prior to storage, for example, elongation of fatty acids. For all these reasons, biomarkers are rarely perfect representations of intake. An example of this is the imperfect relationship between serum carotenoids and total antioxidant intake [32]. Furthermore, biomarkers tell us nothing about dietary behaviors. Still, biomarkers have a future in dietary assessment. Research should continue to identify and better understand biomarkers in relation to dietary intake.

2.2 What Element of Diet Should Be Studied?

In Sect. 2.1.1, following Willett [2], the kind of chemicals that are dietary components was cited. The number and kind of such components present a very complex picture. Diet can also be described in terms of food, food groups, or dietary patterns. The early history of nutrition research focused primarily on chemicals, with some justification according to Willett. The existence of deficiency diseases such as scurvy (ascorbic acid), rickets (vitamin D), beriberi, pellagra, and neural tube defects (B vitamins) points to one class of nutritional problems. Willett cites a model of Mertz [33] that begins with death and deficiency disease at sufficiently low level of a nutrient, complemented by similarly severely reduced function at levels that are sufficiently high. Also in the model is reduced function at modestly reduced or elevated levels of the nutrient. Willett calls this "subclinical dysfunction," a view much in line with the slow, mostly subclinical development of diseases such as cancer and cardiovascular disease. There is also a broad plateau at highest function across a wide range of intake of the nutrient.

Willett [2] further thinks that the focus on major energy sources is justified because they are quantitatively important in the diet and manifestly vary markedly across human populations. These focuses on nutrients have led to the development of extensive tables of energy and of these dietary chemicals. Furthermore, there is a strong tendency among basic scientists toward reductionism: the belief that worthwhile knowledge consists of simple pathways linking single nutrients to bodily function and pathogenesis [34, 35], what Willett calls "linkage to our fundamental knowledge of biology." An excellent example is the protective association of folate with neural tube defects [36], as is improvement in insulin function and metabolic control in diabetics with supplemental magnesium [37]. Much remains to be studied regarding the composition of foods. The tabulated nutrient composition of a food does not fully describe the physiological effect of that food, whether because of differential bioavailability or unknown constituents. There are thousands of untabulated or unidentified compounds in foods, including many phytochemicals. Additionally, a relatively undeveloped aspect of diet characterization is that of food function. For example, Blomhoff and colleagues [32, 38] analyzed thousands of food samples for their total antioxidant content, measured as the molar content of donatable electrons using the ferric reducing ability of plasma, FRAP; those data are available as a dietary exposure measure. A similar functional assessment in the idea stage is the ability of a given food to prevent cell proliferation in in vitro incubation with cancer cells, A la work by Eberhardt et al. [39].

Foods themselves should also be studied even if that does not immediately lead to additional knowledge of specific biological pathways. Foods are what people eat; findings regarding foods are directly applicable to people's diets. Most importantly, it is quite likely that there are synergies among food constituents and between foods [34, 35]; studies of individual chemical constituents may never find the relevant pathways because they are more complex than the researchers imagined. In a nondeficiency state, despite findings that foods containing antioxidants are associated with better long-term health, consumption of isolated nutrients or chemicals does not fare so well. The most striking example is that of supplementary β-carotene, which has been administered in several large, long-term clinical trials, with the effect of increasing disease [40]. Higher antioxidant nutrient intake was associated with more diabetic retinopathy in one study [41]. Other provocative examples from the author's observational work include that supplemental vitamin C in diabetics was associated with increased coronary heart disease [42], and that supplemental iron in association with breakfast cereal intake (which is often fortified with supplemental iron) was associated with an increased rate of distal colon cancer [43].

These findings are supportive of the concept that food synergies are important: the compounds in question are part of foods that appear to be healthy, but do not work outside their food matrix. The food matrix arises from a living organism consisting of thousands of compounds with checks and balances among those compounds to maintain homeostasis and life by preventing the action of any one compound from getting out of control. It is likely that some of this multiplicity of function is retained during human metabolism of the food. For example, whole grain breakfast cereals are associated with reduced risk of chronic disease [14, 44, 45], as are fruits and vegetables [46], which are high in β-carotene and vitamin C, among a wide variety of phytochemicals. The concept of food synergy is discussed at greater length in Chap. 14.

In a very simple example of food synergy, vitamin E functions as an antioxidant by accepting electrons, after which it exists in an oxidized state, that is, as a pro-oxidant. To reduce the risk that it will cause damage, it must be reduced, which is done by vitamin C. The vitamin C is then oxidized and must be reduced, and so on until the cycle reaches an end. One important in vitro study was suggestive of the influence of balancing substances in food by showing that cell proliferation in a cancer cell line was much lower when incubated with apple or apple skin than it was when incubated with an amount of isolated vitamin C that had an equivalent total antioxidant capacity [39].

A final aspect of diet that has been successfully studied is food patterns. Dietary patterns have been discovered using factor analysis. For example, Hu et al. [47, 48] identified a "prudent" pattern associated with reduced incidence of cardiovascular disease and a "Western" pattern associated with increased incidence. Many other authors have followed a similar strategy, generally finding support for the general prudent pattern [49]. The association of a food pattern with incident disease is suggestive of a synergy between foods. There has been much advice about a diet that has potential to prevent chronic disease; the lower risk associated with the "prudent" pattern suggests that many people have apparently taken that advice and that the advised diets do have merit in risk reduction.

2.3   Summary

Two particularly challenging issues in nutritional epidemiology were discussed in editorial fashion. Concerning how to find out what people eat, nutritional epidemiologists use variants of two basic methods. In the first, the participant records or recalls extensive detail about recent intake. The investigator then synthesizes this information into analytically usable variables. This method does not represent typical diet well unless multiple recalls/records are obtained. In the second method, the participant synthesizes his/her dietary information by responding to general questions about diet, such as how often a particular class of foods is eaten. This method does determine the typical diet, but fails to obtain details that are necessary for many types of analysis. It is hoped that advances in technology will enable simpler and more extensive collection and processing of dietary intake data.

Concerning how to think about the effect of diet on health, I suggest that simple nutrient pathways are inadequate for a full understanding of diet. It is proposed that considerable attention be paid to the foods and food patterns that people eat, as well as to the relationships of these foods and food patterns with disease outcomes.

References

1. Margetts BM, Nelson M, editors. Design concepts in nutritional epidemiology. Oxford: Oxford University Press; 1997.
2. Willett W. Nutritional epidemiology. 2nd ed. New York: Oxford University Press; 1998.

3. Yang J, Meyers KJ, van der Heide J, Liu RH. Varietal differences in phenolic content and antioxidant and antiproliferative activities of onions. J Agric Food Chem. 2004;52:6787–93.
4. McDonald A, Van Horn L, Slattery M, et al. The CARDIA dietary history: development, implementation, and evaluation. J Am Diet Assoc. 1991;91:1104–12.
5. Liu K, Slattery M, Jacobs Jr DR, et al. A study of the reliability and comparative validity of the CARDIA dietary history. Ethn Dis. 1994;4:15–27.
6. Liu K, Slattery M, Jacobs DR Jr. Is the dietary recall the method of choice in black populations? Ethn Dis. 1994;4:12–4 (letter to the editor).
7. Prewitt TE, Haynes SG, Graves K, Haines PS, Tyroler HA. Nutrient intake, lipids, and lipoprotein cholesterols in black and white children: the Lipid Research Clinics Prevalence Study. Prev Med. 1988;17:247–62.
8. Dennis BH, Zhukovsky GS, Shestov DB, Davis CE, Deev AD, Kim H, et al. The association of education with coronary heart disease mortality in the USSR Lipid Research Clinics Study. Int J Epidemiol. 1993;22:420–7.
9. Dolecek TA, Johnson RL, Grandits GA, Farrand-Zukel M, Caggiula AW. Nutritional adequacy of diets reported at baseline and during trial years 1–6 by the special intervention and usual care groups in the Multiple Risk Factor Intervention Trial. Am J Clin Nutr. 1997;65 (1 Suppl):305S–13.
10. Dolecek TA, Stamler J, Caggiula AW, Tillotson JL, Buzzard IM. Methods of dietary and nutritional assessment and intervention and other methods in the Multiple Risk Factor Intervention Trial. Am J Clin Nutr. 1997;65(1 Suppl):196S–210.
11. Graham S, Mettlin C, Marshall J, Priore R, Rzepka T, Shedd D. Dietary factors in the epidemiology of cancer of the larynx. Am J Epidemiol. 1981;113:675–80.
12. Lockheart MS, Steffen LM, Rebnord HM, et al. Dietary patterns, food groups and myocardial infarction: a case-control study. Br J Nutr. 2007;98:380–7.
13. Singer MR, Moore LL, Garrahie EJ, Ellison RC. The tracking of nutrient intake in young children: the Framingham Children's Study. Am J Public Health. 1995;85:1673–7.
14. Moore LL, Singer MR, Bradlee ML, et al. Intake of fruits, vegetables, and dairy products in early childhood and subsequent blood pressure change. Epidemiology. 2005;16:4–11.
15. Jacobs DR, Meyer KA, Kushi LH, Folsom AR. Whole grain intake may reduce risk of coronary heart disease death in postmenopausal women: The Iowa Women's Health Study. Am J Clin Nutr. 1998;68:248–57.
16. Willett WC, Stampfer MJ, Manson JE, et al. Intake of trans fatty acids and risk of coronary heart disease among women. Lancet. 1993;341:581–5.
17. Pereira MA, Kartashov AI, Ebbeling CB, et al. Fast-food habits, weight gain, and insulin resistance (the CARDIA study): 15-year prospective analysis. Lancet. 2005;365:36–42. Erratum in Lancet. 2005;365:1030.
18. Duffey KJ, Gordon-Larsen P, Jacobs Jr DR, Williams OD, Popkin BM. Differential associations of fast food and restaurant food consumption with 3-y change in body mass index: the Coronary Artery Risk Development in Young Adults Study. Am J Clin Nutr. 2007;85:201–8.
19. Duffey KJ, Gordon-Larsen P, Steffen LM, Jacobs Jr DR, Popkin BM. Regular consumption from fast food establishments relative to other restaurants is differentially associated with metabolic outcomes in young adults. J Nutr. 2009;139:2113–8.
20. Munger RG, Folsom AR, Kushi LH, Kaye SA, Sellers TA. Dietary assessment of older Iowa women with a food frequency questionnaire: nutrient intake, reproducibility, and comparison with 24-hour dietary recall interviews. Am J Epidemiol. 1992;136:192–200.
21. Willett WC, Sampson L, Browne ML, et al. The use of self-administered questionnaire to assess diet four years in the past. Am J Epidemiol. 1988;127:188–99.
22. Feskanich D, Rimm EB, Giovannucci EL, et al. Reproducibility and validity of food intake measurements from a semiquantitative food frequency questionnaire. J Am Diet Assoc. 1993;93:790–6.
23. Fraser GE. Diet, life expectancy, and chronic disease. New York: Oxford University Press; 2003. p. 265–76.

24. Schatzkin A, Kipnis V. Could exposure assessment problems give us wrong answers to nutrition and cancer questions? J Natl Cancer Inst. 2004;96:1564–5.
25. Alonso A, Nettleton JA, Ix JH, et al. Dietary phosphorus, blood pressure, and incidence of hypertension in the atherosclerosis risk in communities study and the Multi-Ethnic Study of Atherosclerosis. Hypertension. 2010;55:776–84.
26. Pereira MA, O'Reilly E, Augustsson K, et al. Dietary fiber and risk of coronary heart disease: a pooled analysis of cohort studies. Arch Intern Med. 2004;164:370–6.
27. Knekt P, Ritz J, Pereira MA, et al. Antioxidant vitamins and coronary heart disease risk: a pooled analysis of 9 cohorts. Am J Clin Nutr. 2004;80:1508–20.
28. http://www.ncc.umn.edu. Accessed 11 May 2011.
29. Zhuo H, Smith AH, Steinmaus C. Selenium and lung cancer: a quantitative analysis of heterogeneity in the current epidemiological literature. Cancer Epidemiol Biomarkers Prev. 2004;13:771–8.
30. Freedman LS, Tasevska N, Kipnis V, et al. Gains in statistical power from using a dietary biomarker in combination with self-reported intake to strengthen the analysis of a diet-disease association: an example from CAREDS. Am J Epidemiol. 2010;172:836–42.
31. Freedman LS, Kipnis V, Schatzkin A, Tasevska N, Potischman N. Can we use biomarkers in combination with self-reports to strengthen the analysis of nutritional epidemiologic studies? Epidemiol Perspect Innov. 2010;7:2.
32. Svilaas A, Sakhi AK, Andersen LF, et al. Intakes of antioxidants in coffee, wine, and vegetables are correlated with plasma carotenoids in humans. J Nutr. 2004;134:562–7.
33. Mertz W. The essential trace elements. Science. 1981;213:1332–8.
34. Messina M, Lampe JW, Birt DF, et al. Reductionism and the narrowing nutrition perspective: time for reevaluation and emphasis on food synergy. J Am Diet Assoc. 2001;101:1416–9.
35. Jacobs DR, Steffen LM. Nutrients, foods, and dietary patterns as exposures in research: a framework for food synergy. Am J Clin Nutr. 2003;78 Suppl 3:508S–13.
36. Stover PJ. Physiology of folate and vitamin B12 in health and disease. Nutr Rev. 2004;62 (6 Pt 2):S3–12; discussion S13.
37. Rodriguez-Moran M, Guerrero-Romero F. Oral magnesium supplementation improves insulin sensitivity and metabolic control in Type 2 diabetic subjects: a randomized double blind controlled trial. Diabetes Care. 2003;26:1147–52.
38. Halvorsen BL, Holte K, Myhrstad MC, et al. A systematic screening of total antioxidants in dietary plants. J Nutr. 2002;132:461–71.
39. Eberhardt MV, Lee CY, Liu RH. Antioxidant activity of fresh apples. Nature. 2000;405:903–4.
40. Clarke R, Armitage J. Antioxidant vitamins and risk of cardiovascular disease. Review of large-scale randomised trials. Cardiovasc Drugs Ther. 2002;16:411–5.
41. Mayer-Davis EJ, Bell RA, Reboussin BA, Rushing J, Marshall JA, Hamman RF. Antioxidant nutrient intake and diabetic retinopathy: the San Luis Valley Diabetes Study. Ophthalmology. 1998;105:2264–70.
42. Lee DH, Aaron R, Folsom AR, Harnack L, Halliwell B, Jacobs DR. Does supplemental vitamin C increase cardiovascular disease risk in women with diabetes? Am J Clin Nutr. 2004;80:1194–200.
43. Lee DH, Jacobs DR, Folsom AR. A hypothesis: interaction between supplemental iron intake and fermentation affecting the risk of colon cancer. The Iowa Women's Health Study. Nutr Cancer. 2004;48:1–5.
44. Jacobs DR, Gallaher DD. Whole grain intake and cardiovascular disease: a review. Curr Atheroscler Rep. 2004;6:415–23.
45. Liu S, Sesso HD, Manson JE, Willett WC, Buring JE. Is intake of breakfast cereals related to total and cause-specific mortality in men? Am J Clin Nutr. 2003;77:594–9.
46. Key TJ, Schatzkin A, Willett WC, Allen NE, Spencer EA, Travis RC. Diet, nutrition and the prevention of cancer. Public Health Nutr. 2004;7(1A):187–200.

47. Hu FB, Rimm E, Smith-Warner SA, et al. Reproducibility and validity of dietary patterns assessed with a food-frequency questionnaire. Am J Clin Nutr. 1999;69:243–9.
48. Hu FB, Rimm EB, Stampfer MJ, Ascherio A, Spiegelman D, Willett WC. Prospective study of major dietary patterns and risk of coronary heart disease in men. Am J Clin Nutr. 2000;72:912–21.
49. Newby PK, Tucker KL. Empirically derived eating patterns using factor or cluster analysis: a review. Nutr Rev. 2004;62:177–203.

Chapter 3
Eating Disorders

Kelly C. Allison

Keywords Anorexia nervosa • Bulimia nervosa • Binge-eating disorder • Purging disorder • Night eating syndrome • Pica • Rumination • Avoidant/restrictive food intake disorder

Key Points

- Eating disorder diagnoses consist of pica, rumination, avoidant/restrictive food intake disorder, anorexia nervosa (AN; restricting type and binge-eating/purging type), bulimia nervosa (BN), binge-eating disorder (BED), and feeding and eating conditions not elsewhere classified (including atypical AN, subthreshold BN, purging disorder, and night eating syndrome [NES]).
- Physical complications of AN affect most major systems in the body and are caused by starvation and the effects of purging. Most physical complications of BN are due to purging.
- Overweight and obesity are linked with BED and NES. Patients typically request that weight loss be addressed with treatment.
- AN is difficult to treat and may need initial inpatient treatment for refeeding. Subsequently, family therapy is recommended for patients still living with their families.
- Cognitive behavioral therapy (CBT) is the first line of therapy recommended for BN, purging disorder, BED, and NES. Interpersonal therapy has also been shown effective for BN and BED with similar efficacy as CBT at 12-months posttreatment.

K.C. Allison, PhD (✉)
Department of Psychiatry, Center for Eating and Weight Disorders, Perelman School of Medicine of the University of Pennsylvania, 3535 Market Street, Suite 3027, Philadelphia, PA 19104-3309, USA
e-mail: kca@mail.med.upenn.edu

N.J. Temple et al. (eds.), *Nutritional Health: Strategies for Disease Prevention*, Nutrition and Health, DOI 10.1007/978-1-61779-894-8_3,
© Springer Science+Business Media, LLC 2012

- Selective serotonin reuptake inhibitors have been shown effective for treating BN, BED, and NES, as has topiramate. Medication trials have not identified a drug that effectively reduces the severity of AN or that maintain gains established in inpatient treatment.
- Prevention studies for eating disorders are in their infancy, but dissonance-based programs have shown promise.

3.1   Introduction

Eating disorders can affect persons of all shapes and sizes. From extremes in underweight, across the spectrum of normal weight and overweight, and up to the extremes of obesity, underlying nutritional, psychological, and medical effects of eating disorders can be found. Disordered eating encompasses restriction of food as well as binge eating, inappropriate compensation with purging or excessive exercise, and patterns of eating that disrupt normal, daily functioning. Nutritional counseling is often an integral part of treatment for these complex, psychiatric disorders, typically in the context of a team approach.

Current diagnostic criteria for eating disorders that are outlined in the Diagnostic and Statistical Manual IV-TR (DSM-IV-TR) from the American Psychiatric Association [1] are under revision for DSM 5 (www.dsm5.org), to be released in 2013. The criteria and proposed changes for eating disorders will be discussed in this chapter. These include pica, rumination disorder, avoidant/restrictive food intake disorder (ARFID), anorexia nervosa (AN), bulimia nervosa (BN), binge-eating disorder (BED), and feeding and eating conditions not elsewhere classified (FEC-NEC; atypical anorexia, subthreshold BN, purging disorder, and night eating syndrome [NES]). Additionally, this chapter will provide a brief overview of prevalence, assessment issues, treatment, and prevention efforts, where available.

3.2   Pica, Rumination Disorder, and Avoidant/Restrictive Food Intake Disorder

These eating disorders were classified under a section entitled Feeding and Eating Disorders of Infancy or Early Childhood in the DSM-IV-TR. They will all be included in the eating disorders section in DSM-5, as pica and rumination can often occur in adulthood. These disorders are not as widely studied as the other disorders reviewed here, so assessment and treatments are still somewhat limited.

3.2.1   Pica

Pica involves: (1) persistent eating of nonnutritive, nonfood substances over a period of at least 1 month; (2) this eating is inappropriate to the developmental level of the individual; (3) this eating is not part of a culturally sanctioned practice; and (4) if the eating behavior occurs in the context of another mental disorder, it is sufficiently severe to warrant independent clinical attention [2]. Pica often occurs comorbidly with mental retardation, pervasive developmental disorder, or schizophrenia. It is within these contexts that pica is typically first noted. In adulthood, pica more typically emerges during pregnancy, and noted more recently, after gastric bypass surgery. Kushner et al. note that the occurrence of pica during these times may be related to iron-deficiency anemia [3].

3.2.2 Rumination Disorder

Rumination disorder is described as repeated regurgitation of food over a period of at least 1 month. The regurgitated food may be re-chewed, re-swallowed, or spit out. Additionally, there is no evidence that a gastrointestinal or other medical condition is the sole source of the behavior. Rumination must not occur exclusively during the course of another eating disorder, and if they occur in the context of another mental disorder, such as mental retardation, they are sufficiently severe to require independent clinical attention [2]. Re-chewing food most often occurs in children. In adults, reflux and nausea can be related to this behavior. Among other mental disorders, rumination most often co-occurs with AN and BN, but it has also been noted among those with depression, obsessive-compulsive disorder, and other forms of anxiety [2].

3.2.3 Avoidant/Restrictive Food Intake Disorder

The disorder formerly known as feeding and eating disorders of infancy or early adulthood is now proposed under the name avoidant/restrictive food intake disorder (ARFID). ARFID consists of an eating or feeding disturbance in which there is a persistent failure to meet expected nutritional and/or energy needs associated with at least one of these features: (1) significant weight loss or disturbance in growth curves in growing children; (2) significant nutritional deficiency; (3) dependence on enteral feeding; or (4) marked interference with psychosocial functioning. There appear to be three main presentations, including lack of interest in eating, avoidance related to the sensory characteristics of food, or concern about aversive consequences of eating [4]. If the main presentation involves problems with the texture of food or other sensory properties, assessment may be warranted for autism spectrum disorder or pervasive developmental disorders. The food avoidance must not be related to food insecurity or cultural practices, and there should be no evidence of disturbance in the perception of body weight and shape, as there is so centrally in AN and BN. As this is a newly proposed group of disorders, research specific to the ARFID diagnosis is rare.

3.3   Anorexia Nervosa

AN was first noted in the scientific community in the late seventeenth century and first appeared in the DSM-III in 1980 as a diagnostic entity. In its current conceptualization, the first core feature is a refusal to maintain a minimally normal body weight for age and height. There is variability across individuals by body type, ethnicity, and gender for what is a "minimally acceptable weight," and the current guidelines suggest a cutoff of weighing less than 85% of expected weight for height or at or below a body mass index (BMI) of 17.5 kg/m². This criterion will be loosened in DSM 5, with more clinical judgment allowed regarding what constitutes a "significantly low weight." For adolescents and children, lack of weight gain, rather than active weight loss, would also be an appropriate measure of this criterion. The Centers for Disease Control growth charts [5] or the World Health Organization's growth charts [6] should be reviewed to assess if a child or adolescent has fallen significantly below his or her original weight trajectory.

The second criterion describes an intense fear of gaining weight or persistent behavior that prevents weight gain. The third requires a distortion in the way that body weight and shape are viewed or a "persistent lack of recognition of the seriousness of the low body weight" [4]. Persons with AN evaluate their self-worth almost entirely by their perceptions of their body weight and shape, and these distorted beliefs help maintain the severe caloric deficits necessary to sustain their low weight [7–10]. Amenorrhea lasting >3 months has been a diagnostic criterion in the past. However, women on hormonal birth control, girls who have not reached menarche, and men are not able to apply this criterion; therefore, it will not be included in DSM 5.

There are two subtypes of AN. The restricting type is classified by the strict use of caloric restriction and excessive exercise as a means of controlling their weight. The binge eating/purging subtype describes those who engage in binge eating or inappropriate compensatory measures, such as vomiting or misuse of laxatives, diuretics, or enemas. Those with the AN, binge eating/purging subtype differ from persons with BN because of their extremely low body weight. Thus, a diagnosis of AN supersedes a diagnosis of BN.

Almost every physical system is negatively impacted by AN; this is due to starvation and, when present, the effects of purging. Resulting abnormalities include bradycardia, arrhythmia, hypothyroidism, low bone density, constipation, infertility, and perinatal complications [11]. Gray matter volume in the brain is decreased. Atrophied neural networks may maintain sufferers' psychological delusions regarding their fears of fat and beliefs that they are not thin enough, as well as obsessions and compulsive rituals with food. Despite the gravity of their symptoms, those with AN do not typically complain of their ailments and deny the seriousness of their physical and psychological states [12]. With this denial, many sufferers refuse medical treatment until they have been seriously medically compromised.

Paradoxically, excessive exercise and movement are observed in AN, perhaps due to lowered leptin levels that increase the drive for movement. This drive for food was likely once associated with food-finding behaviors built for evolutionary purposes to avoid starvation [13]. Later in the illness, physical restlessness could be related to a biological drive to increase core body temperature. When excessive movement subsides and fatigue sets in, this may indicate severe depression, electrolyte imbalance, or severe dehydration. Cardiac functioning may also be poor at that point, and suicidal ideation and intent should be assessed [14]. For both of these reasons, AN is considered the deadliest psychiatric disorder.

3.4 Bulimia Nervosa

The core features of BN are binge eating and the subsequent use of inappropriate compensatory behaviors. These behaviors are used in an attempt to attain a low body weight or prevent weight gain. As with AN, there is undue influence of weight and shape on self-evaluation and self-concept. DSM 5 criteria require that the binge-eating episodes and inappropriate compensatory behaviors occur at least once/week for at least 3 months.

Inappropriate compensatory behaviors consist of various forms of purging, restricting, and excessive exercise. Purging behaviors most often consist of vomiting, used in 80–90% of cases [1, 15]; laxative abuse is the second most common type of purging. Many persons with BN become skilled at inducing vomiting so that they can vomit at will and no longer need to use their fingers or another instrument. Four common signs associated with vomiting include "Russell's sign" (scarring on the back of the knuckles due to self-induced vomiting), swollen cheeks associated with parotid gland enlargement, dental enamel erosion, and receding gums. Laxative abuse is commonly associated with peripheral edema and bloating. Constipation results when laxatives abuse is discontinued, but it generally resolves in less than a month with exercise and gradual increases in fluids and fiber. Both vomiting and laxative use are associated with electrolyte imbalance, fatigue, heart arrhythmias, and gastrointestinal problems, such as gastroesophageal reflux disease (GERD) [16].

Most persons with BN have a BMI in the healthy weight range, with some in the overweight and obese ranges. Individuals with BN feel free of their binge food after purging and consequently experience psychological relief (if only temporarily), but, in reality, many of the calories from their binge episodes are absorbed and metabolized. Studies have shown that purging by laxatives eliminates the absorption of about 10% of calories from a binge [17, 18], while about half, up to an apparent cap of about 1,200 kcal, are eliminated from vomiting [19]. With laxative use, laxatives affect the colon after much of the food has been absorbed through the intestine. The lack of effect with vomiting is likely related to the length of the binge and the volume rejected.

Persons with BN may also restrict between binge episodes and exercise, but not to the extent that is observed with AN, binge eating/purging subtype. Malnutrition may still occur in BN, but most of the medical complications in this disorder are caused by the purging behaviors. While these medical complications are not as severe as those observed in AN, persons with BN generally are less tolerant of their physical symptoms. Those with BN typically have more insight into their disorder than those with AN, often feeling guilt and shame related to their binge-eating and purging behaviors. However, those with BN also have an increased risk for suicide, so this should be closely monitored [14].

3.5   Binge-Eating Disorder

The hallmark of BED is eating large amounts of food, accompanied by a loss of control. Additionally, at least three of the following five signs must be present during binge-eating episodes: (1) eating more rapidly than normal; (2) eating until uncomfortably full; (3) eating when not physically hungry; (4) eating alone due to embarrassment; and (5) feeling disgusted, depressed, or markedly guilty after an episode. Diagnosis requires that distress regarding the binge eating must be present, and the episodes must occur, on average, at least once/week for 3 months [4].

Most individuals with BED are overweight or obese, and many present primarily for weight loss. Persons with BN typically restrict more consistently between binges than do persons with binge-eating disorder, but in laboratory studies those with BN consume more energy during binges than those with BED. Persons with BED typically engage in binge episodes in addition to eating normal- to large-sized meals throughout the day. This general pattern of overeating coupled with the lack of compensatory behaviors contributes to weight gain (see ref. [20] for review).

3.6   Feeding and Eating Disorders: Not Elsewhere Classified

There are many forms of disordered eating that are serious and cause psychological and physical distress but that do not fit the diagnostic criteria for BN or AN. These are captured in the Feeding and Eating Disorders: Not Elsewhere Classified (FED-NEC) category and include atypical AN, subthreshold BN, purging disorder, and NES. Each of these is described below.

3.6.1   Atypical Anorexia Nervosa

This diagnosis would be used when all of the criteria for AN are met except for the significantly low weight criterion. Examples of this include persons who have lost a significant amount of weight through behavioral weight loss, inappropriate compensatory behaviors, or bariatric surgery. In these cases, weight may still fall in the obese or overweight range but preoccupation with body shape and size, along with the other core features of AN, is present. They could also be medically compromised and show bradycardia and other hallmark features of AN, thus warranting clinical attention.

3.6.2 Subthreshold Bulimia Nervosa

Here, all criteria for BN are met except that the frequency of binge purge episodes occur, on average, less than once/week and/or the duration is <3 months.

3.6.3 Purging Disorder

Purging disorder is defined as the regular occurrence of purging behaviors (e.g., vomiting, laxative use, or diuretic misuse) in order to influence weight or shape [4]. The purging continues in the absence of regular binge-eating episodes and without a significantly low body weight [21]. As in BN, there is also undue influence of weight and shape on self-evaluation. Thus, persons with purging disorder generally feel distressed after eating anywhere from a typical meal to a small snack and have an overwhelming urge to purge afterwards.

The effects of purging, as in the case of BN, can be dangerous and debilitating. A feeding study has shown that women with the disorder reported more postprandial fullness and gastrointestinal discomfort after a standardized meal than those with BN, and greater release of cholecystokinin (CCK) [22], suggesting that physiological cues may contribute to the purging behavior. Persons with purging disorder report higher levels of body dissatisfaction and dietary restraint, but lower impulse control problems than persons with BED [23].

3.6.4 Night Eating Syndrome

The NES was first described in 1955 as a disorder of morning anorexia, evening hyperphagia, and insomnia, usually accompanied by a depressed mood and stressful life circumstances [24]. It did not receive much research or clinical attention until the 1990s. This renewed attention was likely influenced by the rise of the prevalence of obesity and the search for correlates and contributors of excessive weight gain. In 1999, awakenings with ingestions (*nocturnal ingestions*) were added to the provisional set of criteria [25]. However, as research advanced our understanding of the condition, different criteria sets were increasingly used, making comparisons across studies difficult.

The following diagnostic criteria were reached by consensus at the First International Night Eating Symposium in 2008 [26]. First, the daily pattern of eating must show greatly increased intake in the evening and/or night time, as manifested by one or both of the following: (a) at least 25% of food intake is consumed after the evening meal; and (b) at least two eating episodes occur each week upon awakening during the night. Second, the clinical picture is characterized by at least three of five of the following features: (a) a lack of desire to eat in the morning and/or breakfast is omitted on four or more mornings per week; (b) the presence of a strong urge to eat between dinner and bedtime and/or during the night; (c) sleep onset and/or sleep maintenance insomnia are present four or more nights/ week; (d) presence of a belief that one must eat in order to get to sleep; and (e) mood is frequently depressed and/or it worsens in the evening.

Persons who meet these criteria must also have awareness and recall of the evening and nocturnal eating episodes to distinguish the behavior from sleep-related eating disorder, which is a parasomnia marked by impaired consciousness and the consumption of unusual food or nonedible objects. Diagnosis requires that the night eating behaviors must be present for at least 3 months, and there must be distress or impairment of functioning present in relation to the night eating.

One epidemiological and two clinical studies have shown a link between NES and obesity. However, other studies have failed to verify this. Average caloric intake consumed during nocturnal ingestions is similar to regular snacks (approximately 300–400 kcal) [25, 27]. An early report suggested that carbohydrates dominate nocturnal food choices [25], but a subsequent report showed no difference in the proportion of macronutrient content of foods consumed during the night vs. the day [28]. However, the repeated and persistent nature of the disorder likely contributes to weight gain among its sufferers.

3.7 Prevalence

Hudson et al. provided recent and rigorous lifetime prevalence estimates for the most commonly studied of the eating disorders. Estimates for women and men, respectively, are as follows: AN had occurred in 0.9 and 0.3%; BN in 1.5 and 0.5%; and BED in 3.5 and 2.0% [29]. Subthreshold BED was also assessed (which did not include the five descriptors, such as eating more rapidly than usual, or the distress criteria); this yielded prevalence estimates of 0.6% of women and 1.9% of men.

Large-scale survey studies of women that assessed the prevalence of purging disorder revealed rates of 5.3% in an Australian twin cohort, 1.1% in an Italian cohort, and 0.85% in an adolescent Portuguese cohort [21]. The relative frequency of these rates, as compared to the other eating disorders, has varied, with some studies finding purging disorder more common and others less common than BN and AN. Estimates of NES in the general population of the United States, Germany, Sweden, and Australia range from 1.5 up to 6% [30–33].

There are no prevalence estimates for the new, more inclusive criteria for pica, rumination, and ARFID. Relatively little research had been completed on these disorders when they were relegated to disorders of childhood, and with the inclusion of adults, prevalence rates have yet to be established.

3.8   Risk Factors for Eating Disorders

As indicated by the prevalence estimates, women are at higher risk for most of the eating disorders, although the rates are much closer between genders for BED and NES. Traditionally, non-Hispanic white women are at higher risk for AN and BN, although minority rates have been rising [34]. Certain medical conditions may also increase risk for disordered eating, such as type 1 diabetes, where insulin manipulation becomes a means of inappropriate compensation for eating [35].

Certain psychological factors have been identified as increasing risk for developing an eating disorder. Elevated dietary restraint has long been recognized as a risk factor for most of the eating disorders [36]. Increased attention to societal pressure to be thin and body image preoccupation have been linked to increased risk of BN and subthreshold BN [37]. Further, high levels of life stress have been associated with the onset of BED [38] and NES [24].

Using a variety of known risk factors in their model, Stice et al. recently reported an 8-year prospective study of nearly 500 adolescent girls identifying pathways to the onset of eating disorders [39]. They identified three prominent paths. First, the most powerful predictor was body dissatisfaction; the upper 24% were at a fourfold risk for developing any eating disorder. Second, those with body dissatisfaction plus elevated depressive symptoms were at a further 2.9 increased incidence rate. Third, a group with low body dissatisfaction but in the highest 12% for dieting showed a 3.6 increased incidence for eating disorders.

Periods of transition are often identified as vulnerable periods for the onset of eating disorders. Clinician's awareness of family transitions may help provide extra support and decrease risk of onset. Such events may be onset of new school, death of a family member, relationship changes, home and job transitions, illness or hospitalization (start losing weight and get reinforcement for it), and abuse/sexual assault/incest [40].

3.9   Treatment

Much progress has been made in treating BN, BED, and NES, with early studies of purging disorder suggesting it is responsive to treatments for BN. Treatments for AN that have long-term effectiveness lag behind, but family therapy seems to be promising. Table 3.1 provides an overview of effective treatment modalities. The first step in assigning treatment is to assess how medically compromised a patient may be [41]. With AN, inpatient hospitalization may warrant refeeding. The next step down is residential treatment, followed by partial hospitalization or day treatment programs. These treatments typically involve a multidisciplinary team of professionals, including physicians, dietitians, psychologists, and, in some cases, art therapists and occupational therapists. Interventions include both group and individual treatments.

Table 3.1 Effective treatments for eating disorders

Disorder	Cognitive behavioral therapy	Interpersonal therapy	SSRIs	Other
Anorexia nervosa	Mixed	Mixed	No	Inpatient/residential multidisciplinary treatment; family therapy; no medications proven effective
Bulimia nervosa	Yes	Yes	Yes	Topiramate, not buproprion
Binge eating disorder	Yes	Yes	Yes	Topiramate, behavioral weight loss
Night eating syndrome	Yes	Not tested	Yes	Topiramate (case reports only)

Note: Other disorders are not included because specific treatment studies have not been reported

Therapeutic meals are included where patients are challenged to eat nutritionally balanced meals and snacks at regular intervals each day, typically every 3–4 h. Patients are encouraged to gain approximately 1–2 lb/week, at an initial intake of about 1,500 kcal/day (30–40 kcal/kg/day), increasing by 70–80 kcal/kg/day [42]. Liquid meal supplements are often used to help patients reach this goal. One risk of this process is "refeeding syndrome," a potentially fatal shift of fluid and electrolytes that may occur among severely undernourished patients [43, 44]. Those who have been fasting for more than 10 days, have significant alcohol intake, abuse laxatives, diuretics, or insulin, or who have had rapid weight loss may be at particular risk. Careful medical monitoring of patients at risk is therefore warranted during refeeding.

Patients must be carefully monitored after meals during inpatient or other intensive treatment, particularly in the bathroom and their rooms, to prevent purging. BN can typically be treated on an outpatient basis, but persistent or very severe cases require residential or partial hospitalization treatment. Pica, rumination, BED, purging disorder, and NES are also most commonly treated on an outpatient basis. In patients presenting with pica, iron-deficiency anemia should be assessed, as treatment of the anemia may diminish pica behaviors [3]. However, on the whole, most of the treatment literature on pica, rumination, and ARFID consist of case studies, so their treatments will not be reviewed here. More rigorous treatment trials for these disorders are needed.

3.9.1 Psychotherapy

The most effective outpatient psychotherapy approach for eating disorders is cognitive behavioral therapy (CBT). A 20-session course of treatment is effective for BN and BED [45]. Sessions occur twice weekly for the first 2 weeks, followed by weekly sessions. Maintenance sessions are encouraged after the initial 20-week course. CBT produces abstinence from binge-eating and purging behaviors in varying proportions of study participants with BN, ranging from 24 to 71% [46]. Similarly, CBT in binge-eating disorder produces abstinence in binge eating ranging from 37 to 79% of study participants [20]. However, weight is not significantly reduced among persons with BED, despite large reductions in binge episodes. Less impressive results have been reported using CBT for active AN, although it may be helpful in maintaining treatment gains. Only an uncontrolled pilot study among patients with NES has been tested to date with significant reductions in nocturnal ingestions and evening eating.

Interpersonal psychotherapy has been tested by several groups of researchers and applied successfully to BN and BED [47]. However, AN has not responded as robustly. As persons with eating disorders typically experience interpersonal or social dysfunction, interpersonal therapy for eating disorders focuses on how these social deficits contribute to binge-eating and purging behaviors. Interpersonal therapy focuses on one of four areas of interpersonal functioning, including unresolved grief, role transition (e.g., graduating high school or college), role dispute (e.g., problems in communicating with a boyfriend or parent), and interpersonal deficit. Interpersonal psychotherapy is not generally recommended as a first-line approach because it relieves symptoms at a slower pace than CBT. However, the treatment outcomes are similar at 1-year follow-up [20, 46].

Family therapy has been shown to be the only effective psychotherapeutic approach for AN [48], and it is also effective among those with BN [49]. It works particularly well for younger patients living with their families. The Maudsley Approach is the most well-validated family therapy approach for AN and is intended to reduce the need for inpatient treatment and to help parents successfully refeed their child, which is the first goal. The second goal is for the adolescent to start to take control again of eating and weight gain, at a level appropriate to maturational status. Finally, an overview of normal adolescent development is covered with the family, and the therapist helps to identify any other outstanding social–emotional issues for which the family may still need help.

Finally, behavioral weight loss therapy reduces binge eating and produces weight loss in persons with BED [50]. However, abstinence rates from binge eating are not as high as those produced through CBT. Thus, if weight loss is strongly desired by a patient and their other psychiatry comorbidities, such as major depression, substance abuse, or an anxiety disorder, are not causing noticeable impairments in functioning, then a behavioral weight loss program may suit those patients best.

3.9.2 Psychotropic Medications

Antidepressants are widely prescribed for the treatment of eating disorders. This is for two reasons: they are effective in reducing binge-eating and purging behaviors, and they improve comorbid mood and anxiety symptoms. Unfortunately, they have not been shown to reduce the core symptom of AN, i.e., refusal to maintain a healthy body weight. Thus, there are currently no efficacious medications used for the treatment or maintenance of AN [51, 52]. However, antidepressants may still relieve comorbid depression or anxiety, when present.

Tricyclic antidepressants, monoamine oxidase inhibitors (MAOIs), and selective serotonin reuptake inhibitors (SSRIs) have all shown efficacy over placebos in reducing binge eating and purging [20, 46]. SSRIs are now most commonly used, with typical reductions of 45–65% in binge eating. Sertraline, a SSRI, has also been shown to significantly reduce evening hyperphagia and nocturnal ingestions among those with NES [53]. Topiramate decreases binge eating and purging as compared to placebo treatment and is associated with weight loss; however, cognitive side effects may be intolerable for some users [54]. Case reports of topiramate in the treatment of NES have also shown significant reductions in evening hyperphagia, nocturnal ingestions, and weight [55]. Buproprion is not indicated for those who engage in purging behaviors as it has been associated with increased risk of seizures.

3.10   Prevention

Prevention programs aimed at reducing the incidence of eating disorders have been designed for children, adolescents, and college students. Dissonance-based interventions have been tested most rigorously and have been shown to have the greatest effect on reduction of eating disorder risk factors, symptoms, risk of onset, and future risk of development of obesity [56]. Cognitive dissonance programs involve having participants speak or behave in a manner that is opposite to their beliefs. As applied to eating disorders, women would be challenged to voice active criticism of the thin ideal; this is because internalization of the thin ideal is a risk factor for developing AN and BN. Among college students, peer-led dissonance-based interventions have been shown effective among women considered at high and low risk for developing AN or BN. These programs are likely most applicable in a school setting as they may be difficult to administer in a community setting.

Other approaches have focused on media literacy and advocacy, but more evidence is needed, particularly in light of the superior effectiveness of the dissonance-based programs. In peer-led programs, the media advocacy intervention is effective in reducing risk for disordered eating among high-risk women, but not those at low risk [57]. One trial of a prevention program based on CBT delivered via the internet showed reductions in the onset of eating disorders in two subgroups, namely those who were overweight and a subset of those who reported preexisting purging behaviors [58]. Finally, programs that focus on body shape and weight acceptance for children and adolescents have also been used in school programming, but little or no formal testing of the effects of these programs has been reported. As with other prevention approaches, more controlled studies are needed to test these interventions and to compare their efficacy with the dissonance-based and media advocacy approaches.

Childhood and adult obesity are major medical concerns and warrant increased attention. As a result various obesity prevention programs have been established. These include programs at schools that change foods that are offered in cafeterias or send information home to parents about their children's BMI status; laws that mandate posting calorie information at restaurants; and national efforts such as Michelle Obama's *Let's Move* campaign. However, some professionals believe that these programs have a detrimental effect on those with or prone to eating disorders, but studies that would measure their direct impact are difficult to design and complete. As a safeguard some professional groups, such as the Academy for Eating Disorders, have called for messages to be focused on health, good nutrition, and well-being instead of specifically focusing on weight [59].

3.11 Conclusion

Eating disorders range from severe caloric restriction to severe overeating. Extreme dissatisfaction with weight and shape is present across the different diagnoses. In most diagnoses, there is also an uncontrollable urge to binge eat. When medical complications are severe, inpatient treatment is warranted, particularly in AN. For most cases of BN, BED, purging disorder, and NES, outpatient psychotherapy is the first line of treatment. Psychotropic medications, most recently SSRIs, have also been proven effective in treating BN, BED, and NES. Family therapy is the most efficacious treatment for AN. More research is needed on pica, rumination, and ARFID to establish effective treatment approaches. Prevention programs using dissonance-based interventions are promising for decreasing the incidence of eating disorders among college-age students, but other programs that target children and adolescents need to be formally evaluated.

References

1. American Psychiatric Association. Diagnostic and statistical manual of mental disorders. 4th ed. Washington, DC: American Psychiatric Association; 2000.
2. Bryant-Waugh R, Markham L, Kreipe RE, Walsh BT. Feeding and eating disorders in childhood. Int J Eat Disord. 2010;43:98–111.
3. Kushner RF, Gleason B, Shanta-Retelny V. Reemergence of pica following gastric bypass surgery for obesity: a new presentation of an old problem. J Am Diet Assoc. 2004;104: 1393–7.
4. American Psychiatric Association. DSM-5: the future of psychiatric diagnosis. www.dsm5. org. Accessed 26 May 2011.
5. 2000 Centers for Disease Control Growth Charts for the United States. www.cdc.gov/growth charts/. Accessed 26 May 2011.
6. WHO. Child growth standards: methods and development growth velocity based on weight, length and head circumference. Geneva: World Health Organization; 2009. http://www.who. int/childgrowth/en/. Accessed 26 May 2011.

7. Killen JD, Taylor CB, Hayward C, et al. Weight concerns influence the development of eating disorders: a 4-year prospective study. J Consult Clin Psychol. 1996;64:936–40.
8. Field AE, Camargo CA, Taylor CB, Berkey CS, Roberts SB, Colditz GA. Peer, parent, and media influences on the development of weight concerns and frequent dieting among preadolescent and adolescent girls and boys. Pediatrics. 2001;107:54–60.
9. Neumark-Sztainer DR, Wall MM, Haines JI, Story MT, Sherwood NE, van den Berg PA. Shared risk and protective factors for overweight and disordered eating in adolescents. Am J Prev Med. 2007;33:359–69.
10. Fairburn CG, Shafron R, Cooper Z. A cognitive behaviour theory of anorexia nervosa. Behav Res Ther. 1999;37:1–13.
11. Katzman DK. Medical complications in adolescents with anorexia nervosa: a review of the literature. Int J Eat Disord. 2005;37:S52–9.
12. Touyz SW, Beumont PJ, Collins JK, McCabe M, Jupp J. Body shape perception and its disturbance in anorexia nervosa. Br J Psychiatry. 1984;144:167–71.
13. Nogueira JP, Maraninchi M, Lorec AM, et al. Specific adipocytokines profiles in patients with hyperactive and/or binge/purge form of anorexia nervosa. Eur J Clin Nutr. 2010;64:840–4.
14. Crow SJ, Peterson CB, Swanson SA, et al. Increased mortality in bulimia nervosa and other eating disorders. Am J Psychiatry. 2009;166:1342–6.
15. Binford RB, le Grange DL. Adolescents with bulimia nervosa and eating disorder not otherwise specified-purging only. Int J Eat Disord. 2005;38:157–61.
16. Mitchell JE, Specker SM, de Zwaan M. Comorbidity and medical complications of bulimia nervosa. J Clin Psychiatry. 1991;52(Suppl):13–20.
17. Bo-Linn GW, Santa Ana CA, Morawski SG, Fordtran JS. Purging and calorie absorption in bulimic patients and normal women. Ann Intern Med. 1983;99:14–7.
18. Lacey JH, Gibson E. Does laxative abuse control body weight? A comparative study of purging and vomiting bulimics. Hum Nutr Appl Nutr. 1985;39:36–42.
19. Kaye WH, Wltzin TE, Hsu LKG, McConaha CW, Bolton B. Amount of calories retained after binge eating and vomiting. Am J Psychiatry. 1993;150:969–71.
20. Mitchell JE, Devlin MJ, de Zwaan M, Crow SJ, Peterson CB. Eating behavior, psychobiology, medical risks, and pharmacotherapy of binge-eating disorder. In: binge-eating disorder: clinical foundations and treatment. New York: Guilford Press; 2008.
21. Keel PK. Purging disorder: subthreshold variant or full-threshold eating disorder? Int J Eat Disord. 2007;40:589–94.
22. Keel PK, Wolfe BE, Liddle RA, DeYoung KP, Jimerson DC. Clinical features and physiological response to a test meal in purging disorder and bulimia nervosa. Arch Gen Psychiatry. 2007;64:1058–66.
23. Keel PK, Holm-Denoma JM, Crosby RD. Clinical significance and distinctiveness of purging disorder and binge eating disorder. Int J Eat Disord. 2011;44:311–6.
24. Stunkard AJ, Grace WJ, Wolff HG. The night-eating syndrome: a pattern of food intake among certain obese patients. Am J Med. 1955;19:78–86.
25. Birketvedt G, Florholmen J, Sundsfjord J, et al. Behavioral and neuroendocrine characteristics of the night-eating syndrome. JAMA. 1999;282:657–63.
26. Allison KC, Lundgren JD, O'Reardon JP, et al. Proposed diagnostic criteria for night eating syndrome. Int J Eat Disord. 2010;43:241–7.
27. Allison KC, Stunkard AJ, Thier SL. Overcoming night eating syndrome: a step-by-step guide to breaking the cycle. Oakland, CA: New Harbinger; 2004.
28. Allison KC, Ahima RS, O'Reardon JP, et al. Neuroendocrine profiles associated with energy intake, sleep, and stress in the night eating syndrome. J Clin Endocrinol Metab. 2005;90:6214–7.
29. Hudson JI, Hiripi E, Pope HG, Kessler RC. The prevalence and correlates of eating disorders in the National Comorbidity Survey replication. Biol Psychiatry. 2007;61:348–58.
30. Rand CSW, MacGregor AM, Stunkard AJ. The night eating syndrome in the general population and among post-operative obesity surgery patients. Int J Eat Disord. 1997;22:65–9.

31. Colles SL, Dixon JB, O'Brien PE. Night eating syndrome and nocturnal snacking: association with obesity, binge eating and psychological distress. Int J Obes. 2007;31:1722–30.

32. Tholin S, Lindroos AK, Tynelius P, et al. Prevalence of night eating in obese and nonobese twins. Obesity. 2009;17:1050–5.

33. Lamerz A, Kuepper-Nybelen J, Bruning N, et al. Prevalence of obesity, binge eating and night eating in a cross sectional field survey of 6-year-old children and their parents in a German urban population. J Child Psychol Psychiatry. 2005;46:385–93.

34. Franko D. Race, ethnicity, and eating disorders: considerations for DSM-V. Int J Eat Disord. 2007;40:S31–4.

35. Ackard DM, Vikc N, Neumark-Sztainer D, Schmitz KH, Hannan P, Jacobs Jr DR. Disordered eating and body dissatisfaction in adolescents with type 1 diabetes and a population-based comparison sample: comparative prevalence and clinical implications. Pediatr Diabetes. 2008;9(pt 1):312–9.

36. Fairburn CG, Cooper Z, Doll HA, Davies BA. Identifying dieters who will develop an eating disorder: a prospective, population-based study. Am J Psychiatry. 2005;162:2249–55.

37. McKnight I. Risk factors for the onset of eating disorders in adolescent girls: results of the McKnight longitudinal risk factor study. Am J Psychiatry. 2003;160:248–54.

38. Striegel-Moore RH, Dohm F, Kraemer H, Schreiber GB, Taylor CB, Daniels S. Risk factors for binge eating disorders: an exploratory study. Int J Eat Disord. 2007;40 Suppl 6:481–7.

39. Stice E, Marti CN, Durant S. Risk factors for onset of eating disorders: evidence of multiple risk pathways from an 8-year prospective study. Behav Res Ther. 2011;49(10):622–7. doi:10.1016/j.brat.2011.06.009.

40. Berge JM, Loth K, Hanson C, Croll-Lampert J, Neumark-Sztainer D. Family life cycle transitions and the onset of eating disorders: a retrospective grounded theory approach. J Clin Nurs. 2012;21:355–63.

41. Berkman ND, Bulik CM, Brownley KA, et al. Evidence report/technology assessment, number 135, 2006. Agency for Healthcare Research and Quality U.S. Department of Health and Human Services. www.ahrq.gov. Accessed 26 May 2011.

42. Halmi KA. Management of anorexia nervosa in inpatient and partial hospitalization settings. In: Yager J, Powers PS, editors. Clinical manual of eating disorders. Washington, DC: American Psychiatric Publishing; 2007. p. 113–25.

43. Mehanna HM, Moledina J, Travis J. Refeeding syndrome: what it is, and how to prevent and treat it. BMJ. 2008;336:1495–8.

44. Academy for Eating Disorders Medical Care Standards Taskforce. Eating disorders: critical points for early recognition and medical risk management in the care of individuals with eating disorders. 2nd ed. AED Report 2011. www.aedweb.org/Medical_Care_Standards. Accessed 26 May 2011.

45. Fairburn CG, Marcus MD, Wilson GT. Cognitive-behavioral therapy for binge eating and bulimia nervosa: a comprehensive treatment manual. In: Fairburn CG, Wilson GT, editors. Binge eating: nature, assessment, and treatment. New York: Guilford Press; 1993. p. 361–404.

46. Mitchell JE, Steffen KJ, Roerig JL. Management of bulimia nervosa. In: Yager J, Powers PS, editors. Clinical manual of eating disorders. Washington, DC: American Psychiatric Publishing; 2007. p. 171–93.

47. Tanofsky-Kraff M, Wilfley DE. Interpersonal psychotherapy for bulimia nervosa and binge-eating disorder. In: Grilo CM, Mitchell JE, editors. The treatment of eating disorders: a clinical handbook. New York: Guilford Press; 2010. p. 271–93.

48. Dare C, Eisler I. Family therapy for anorexia nervosa. In: Garner DM, Garfinkel P, editors. Handbook of treatment for eating disorders. New York: Guilford Press; 1997. p. 307–24.

49. le Grange D, Lock J, Dymek M. Family-based therapy for adolescents with bulimia nervosa. Am J Psychother. 2003;57:237–51.

50. Stunkard AJ, Allison KC. Binge eating disorder: disorder or marker? Int J Eat Disord. 2003;34:S107–16.

51. Crow SJ, Mitchell JE, Roerig JD, Steffen K. What potential role is there for medication treatment in anorexia nervosa? Int J Eat Disord. 2009;42:1–8.

52. Walsh BT, Kaplan AS, Attia E. Fluoxetine after weight restoration in anorexia nervosa: a randomized controlled trial. JAMA. 2006;14(295):2605–12.
53. O'Reardon JP, Allison KC, Martino NS, Lundgren JD, Heo M, Stunkard AJ. A randomized, placebo-controlled trial of sertraline in the treatment of night eating syndrome. Am J Psychiatry. 2006;163:893–8.
54. McElroy SL, Guerdjikova AI, Martens B, Keck Jr PE, Pope HG, Hudson JI. Role of antiepileptic drugs in the management of eating disorders. CNS Drugs. 2009;23:139–56.
55. Howell MJ, Schenck CH, Crow SJ. A review of nighttime eating disorders. Sleep Med Rev. 2009;13:23–34.
56. Stice E, Shaw H, Becker CB, Rohde P. Dissonance-based interventions for the prevention of eating disorders: using persuasion principals to promote health. Prev Sci. 2008;9:114–28.
57. Becker CB, Bull S, Schaumberg K, Cauble A, Franco A. Effectiveness of peer-led eating disorders prevention: a replication trial. J Consult Clin Psychol. 2008;76:347–54.
58. Taylor CB, Bryson S, Luce KH, et al. Prevention of eating disorders in at-risk college-age women. Arch Gen Psychiatry. 2006;63:881–8.
59. Daníelsdóttir S, Burgard D, Oliver-Pyatt W. AED guidelines for childhood obesity prevention programs. www.aedweb.org/AM/Template.cfm?Section=Advocacy&Template=/CM/Content Display.cfm&ContentID=1659. Accessed 26 May 2011.

Chapter 4
The Developmental Origins of Chronic Disease in Later Life

David J.P. Barker

Keywords Birthweight • Fetal programming • Fetal growth • Placenta • Chronic disease • Cardiovascular disease

Key Points

- Low birthweight, as a result of slow fetal growth, is associated with increased rates of coronary heart disease and the related disorders stroke, hypertension, and type 2 diabetes: these associations extend across the normal range of birthweight.
- They are thought to be consequences of developmental plasticity, the phenomenon by which one genotype can give rise to a range of different physiological or morphological states in response to environmental conditions during development. This is often referred to as programming.
- There is increasing evidence that a wide range of chronic diseases are programed.
- People who were small at birth may be vulnerable to later disease, because they have reduced functional capacity in key organs, such as the kidney, or altered settings of hormones and metabolism, or altered responses to adverse influences in the postnatal environment.
- Slow growth in infancy and rapid weight gain after the age of 1 year further increase the risk of later disease.
- Slow fetal growth is the product of the mother's body composition and diet before and during pregnancy, together with her metabolism.
- The placenta is complicit in the growth of the fetus and hence in the programing of disease.

D.J.P. Barker, MD, PhD, FRS (✉)
MRC Lifecourse Epidemiology Unit, University of Southampton,
Southampton General Hospital, Mailpoint 95, Southampton SO16 6YD, UK
e-mail: djpb@mrc.soton.ac.uk

N.J. Temple et al. (eds.), *Nutritional Health: Strategies for Disease Prevention*,
Nutrition and Health, DOI 10.1007/978-1-61779-894-8_4,
© Springer Science+Business Media, LLC 2012

4.1 Developmental Origins

The recent discovery that people who develop coronary heart disease (CHD) grew differently compared to other people during fetal life and childhood has led to a new "developmental" model for chronic disease [1, 2]. The model proposes that nutrition during fetal life, infancy, and early childhood establishes functional capacity, metabolic competence, and responses to the later environment by changing gene expression [3]. There is now clear evidence that the pace and pathway of early growth is a major risk factor for the development of chronic disease in adult life.

To explore the developmental origins of chronic disease required studies of a kind that had not hitherto been carried out. It was necessary to identify groups of men and women, now in middle or late life, whose size at birth had been recorded. Their birthweight could thereby be related to the later occurrence of chronic disease. In the county of Hertfordshire, UK, from 1911 onwards, when women had their babies they were attended by a midwife who recorded the birthweight. A health visitor went to the baby's home at intervals throughout infancy, and the weight at 1 year was recorded. Table 4.1 shows the findings in 10,636 men born during 1911–1930 [1, 4]. Standardized mortality ratios for CHD fell with increasing birthweight. There were stronger trends with weight at 1 year. A subsequent study confirmed a similar trend with birthweight among women [4]. Table 4.2 shows the findings for a sample of 370 men aged 64 years in 1989 [5]. The percentage with impaired glucose tolerance or type 2 diabetes fell steeply with increasing birthweight and with weight at 1 year. There were similar trends with birthweight among women.

Table 4.1 Hazard ratios (95% confidence intervals [CI]) for death from coronary heart disease (CHD) according to weight at birth and at age 1 year in 10,636 men in Hertfordshire

Weight (lb)	Death from CHD	
	Before 65 years	All ages
At birth		
≤5.5	1.50 (0.98–2.31)	1.37 (1.00–1.86)
−6.5	1.27 (0.89–1.83)	1.29 (1.01–1.66)
−7.5	1.17 (0.84–1.63)	1.14 (0.91–1.44)
−8.5	1.07 (0.77–1.49)	1.12 (0.89–1.40)
−9.5	0.96 (0.66–1.39)	0.97 (0.75–1.25)
≥10	1.00	1.00
p for trend	0.001	0.005
Age 1 year		
≤18	2.22 (1.33–3.73)	1.89 (1.34–2.66)
−20	1.80 (1.11–2.93)	1.58 (1.15–2.16)
−22	1.96 (1.23–3.12)	1.66 (1.23–2.25)
−24	1.52 (0.95–2.45)	1.36 (1.00–1.85)
−26	1.36 (0.82–2.26)	1.29 (0.93–1.78)
≥27	1.00	1.00
p for trend	<0.001	<0.001

Table 4.2 Percentage of men aged 64 with impaired glucose tolerance or diabetes according to weight at birth and at age 1 year in 370 men in Hertfordshire

Weight (lb)	% of men with 2-h glucose of ≥7.8 mmol/L	Odds ratio (95% CI)[a]
At birth		
≤5.5	40	6.6 (1.5–28)
–6.5	34	4.8 (1.3–17)
–7.5	31	4.6 (1.4–16)
–8.5	22	2.6 (0.8–8.9)
–9.5	13	1.4 (0.3–5.6)
>9.5	14	1.0
p for trend	<0.001	
Age 1 year		
≤18	43	8.2 (1.8–38)
–20	32	4.8 (1.2–19)
–22	30	4.2 (1.1–16)
–24	18	2.1 (0.5–7.9)
–26	19	2.1 (0.5–9.0)
≥27	13	1.0
p for trend	<0.001	

[a]Adjusted for BMI

The association between low birthweight and CHD has now been replicated among men and women in Europe, North America, and India [6–12]. The association between low weight gain in infancy and CHD in men has been confirmed in Helsinki [13]. Low birthweight has been shown to predict altered glucose tolerance in studies around the world [14–18].

4.2 Confounding Variables

These findings suggest that influences linked to early growth have an important effect on the risk of CHD and type 2 diabetes. It has been argued, however, that people whose growth was impaired in utero and during infancy may continue to be exposed to an adverse environment in childhood and adult life, and it is this later environment that produces the effects attributed to intrauterine influences. There is now strong evidence that this argument cannot be sustained.

In a number of studies, data on lifestyle, including smoking habits, employment, alcohol consumption, and exercise, were collected. In the Nurses' Health Study in the USA, allowance for these influences had little effect on the association between birthweight and CHD [8]. Similar results came from Sweden and the UK [5, 10]. In studies of type 2 diabetes and blood pressure, the associations with size at birth are again independent of social class, cigarette smoking, and alcohol consumption. Adult lifestyle, however, adds to the effects of early life: for example, the prevalence of impaired glucose tolerance is highest in people who had low birthweight but became obese as adults [5, 14–18]. As described later in this chapter, slow fetal growth may also alter the body's response to socioeconomic influences in later life. Associations between low birthweight and altered glucose tolerance and raised blood pressure have been found in numerous studies of children, which is a further argument against these associations being the product of confounding variables in adult life.

4.3 Biological Basis

Like other living creatures in their early life, human beings are "plastic" and able to adapt to their environment. The development of the sweat glands provides a simple example of this. All humans have similar numbers of sweat glands at birth, but none of them function. In the first 3 years after birth, a proportion of the glands become functional, depending on the temperature to which the child is exposed. The hotter the conditions, the greater the number of sweat glands that are programmed to function. After 3 years the process is complete and the number of sweat glands is fixed. Thereafter, the child who has experienced hot conditions will be better equipped to adapt to similar conditions in later life because people with more functioning sweat glands cool down faster.

This brief description encapsulates the essence of developmental plasticity: a critical period when a system is plastic and sensitive to the environment, followed by loss of plasticity and a fixed functional capacity. For most organs and systems, the critical period occurs in utero. There are good reasons why it may be advantageous, in evolutionary terms, for the body to remain plastic during development. It enables the production of phenotypes that are better matched to their environment than would be possible if the same phenotype was produced in all environments. Developmental plasticity is defined as the phenomenon by which one genotype can give rise to a range of different physiological or morphological states in response to different environmental conditions during development [19]. Plasticity during intrauterine life enables animals, and humans, to receive a "weather forecast" from their mothers that prepares them for the type of world in which they will have to live [20]. If the mother is poorly nourished, she signals to her unborn baby that the environment it is about to enter is likely to be harsh. The baby responds to these signals by adaptations, such as reduced body size and altered metabolism, which help it to survive a shortage of food after birth. In this way, plasticity gives a species the ability to make short-term adaptations, within one generation, in addition to the long-term genetic adaptations that come from natural selection. Since, as Mellanby noted many years ago, the ability of a human mother to nourish her baby is partly determined when she herself is in utero, and by her childhood growth, the human fetus is receiving a "weather forecast" based not only on conditions at the time of the pregnancy but on conditions a number of decades before [3, 21]. This may be advantageous in populations which experience periodic food shortages.

Until recently, we have overlooked a growing body of evidence that systems of the body which are closely related to adult disease, such as the regulation of blood pressure, are also plastic during early development. In animals it is surprisingly easy to produce lifelong changes in the blood pressure and metabolism of a fetus by minor modifications to the diet of the mother before and during pregnancy [22, 23].

The different size of newborn human babies exemplifies plasticity. The growth of babies has to be constrained by the size of the mother, otherwise normal birth could not occur. Small women have small babies: in pregnancies after ovum donation they have small babies even if the woman donating the egg is large [24]. Babies may be small because their growth is constrained in this way or because they lack the nutrients for growth. As McCance wrote long ago, "The size attained in utero depends on the services which the mother is able to supply. These are mainly food and accommodation" [25]. Since the mother's height or pelvic dimensions are generally not found to be important predictors of the baby's long-term health, research into the developmental origins of disease has focused on the nutrient supply to the baby, while recognizing that other influences, such as hypoxia and stress, also influence fetal growth. This focus on fetal nutrition was endorsed in a recent review [26].

4.4 Fetal Origins Hypothesis

The fetal origins hypothesis proposed that CHD, type 2 diabetes, stroke, and hypertension originate in developmental plasticity, in response to undernutrition during fetal life and infancy [2, 27]. Why should fetal responses to undernutrition lead to disease in later life? The general answer is clear: "life history theory," which embraces all living things, states that, during development, increased allocation of energy to one trait, such as brain development, necessarily reduces allocation to one or more other traits, such as tissue repair processes. Smaller babies, who have had a lesser allocation of energy, must incur higher costs and these, it seems, include disease in later life. A more specific answer to the question is that people who were small at birth are vulnerable to later disease through three kinds of process. First, they have fewer cells in key organs, such as the kidney. One theory holds that hypertension is initiated by the reduced number of glomeruli found in people who were small at birth [28]. A reduced number necessarily leads to increased blood flow through each glomerulus. Over time, this hyperfiltration is thought to lead to the development of glomerulosclerosis which, combined with the loss of glomeruli that accompanies normal aging, leads to accelerated age-related loss of glomeruli, and a self-perpetuating cycle of rising blood pressure and glomerular loss.

Another process by which slow fetal growth may be linked to later disease is in the setting of hormones and metabolism. An undernourished baby may establish a "thrifty" way of handling food. Insulin resistance, which is associated with low birthweight, may be viewed as persistence of a fetal response by which blood glucose concentrations were maintained for the benefit of the brain, but at the expense of glucose transport into the muscles and muscle growth [29].

Table 4.3 Hazard ratios (95% CI) for CHD among 3,676 men in Helsinki according to ponderal index at birth (birthweight/length^3) and taxable income in adult life

Household income in lb sterling/year	Hazard ratios	
	Ponderal index ≤26.0 kg/m^3 ($n=1,475$)	Ponderal index >26.0 kg/m^3 ($n=2,154$)
>15,700	1.00	1.19 (0.65–2.19)
15,700	1.54 (0.83–2.87)	1.42 (0.78–2.57)
12,400	1.07 (0.51–2.22)	1.66 (0.90–3.07)
10,700	2.07 (1.13–3.79)	1.44 (0.79–2.62)
≤8,400	2.58 (1.45–4.60)	1.37 (0.75–2.51)
p for trend	<0.001	0.75

A third link between low birthweight and later disease is that people who were small at birth are more vulnerable to adverse environmental influences in later life. Observations on animals show that the environment during development permanently changes not only the body's structure and function but also its responses to environmental influences encountered in later life [20]. Table 4.3 shows the effect of low income in adult life on CHD among men in Helsinki [30]. As expected, men who had a low taxable income had higher rates of the disease. There is no agreed explanation for this, but the association between poverty and CHD is a major component of the social inequalities in health in many Western countries. Among the men in Helsinki the association was confined to men who had slow fetal growth and were thin at birth, defined by a ponderal index (birthweight/length^3) of less than 26 kg/m^3 (see Table 4.3). Men who were not thin at birth showed no association between CHD and income, suggesting that they were resilient to the biological effects of low income.

One explanation for these findings emphasizes the psychosocial consequences of a low position in the social hierarchy, as indicated by low income and social class, and suggests that perceptions of low social status and lack of success lead to changes in neuroendocrine pathways and hence to disease [31]. The findings in Helsinki seem consistent with this. People who were small at birth are known to have persisting alterations in responses to stress, including raised serum cortisol concentrations [32]. It is suggested that persisting small elevations of cortisol concentrations over many years may have effects similar to those seen when tumors lead to more sudden, large increases in glucocorticoid concentrations. People with Cushing's syndrome, the result of over-activity of the adrenal cortex, are insulin resistant and have raised blood pressure, both of which predispose to CHD.

4.5   Childhood Growth and Coronary Heart Disease

Figure 4.1 shows the growth of 357 men who were either admitted to hospital with CHD or died from it [33]. They belong to a cohort of 4,630 men who were born in Helsinki, Finland. Their mean height, weight, and body mass index (BMI,

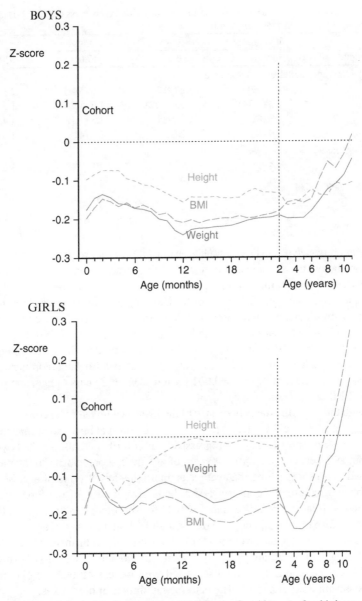

Fig. 4.1 Height, weight, and body mass index (BMI) in the first 11 years after birth among boys and girls who had coronary heart disease as adults. The mean values for all boys and all girls are set at 0, with deviations from the mean expressed as standard deviations (z scores)

weight/height^2) at each month from birth to 2 years of age, and at each year from 2 to 11 years of age, are expressed as standard deviations (z scores). The mean z score for the cohort is set at 0 and a boy maintaining a steady position as tall or short, or fat or thin, in relation to other boys would follow a horizontal path on the figure.

Table 4.4 Hazard ratios for CHD according to birthweight and body mass index (BMI) at age 11 years among 13,517 men and women in Helsinki

| Birthweight (kg) | BMI at age 11 | | | |
	<15.7	−16.6	−17.6	>17.6
Hospital admissions and deaths (1,235 cases)				
<3.0	1.4 (991)	1.6 (719)	1.8 (581)	2.1 (560)
−3.5	1.3 (1,394)	1.5 (1,422)	1.5 (1,264)	1.6 (1,246)
−4.0	1.3 (827)	1.4 (984)	1.3 (1,122)	1.4 (1,110)
>4.0	1.0 (167)	1.2 (254)	1.1 (413)	1.0 (463)
Deaths (480 cases)				
<3.0	1.4	1.8	2.1	3.0
−3.5	1.4	1.9	2.2	2.7
−4.0	1.9	1.8	1.7	1.6
>4.0	1.0	1.4	1.6	1.3

The mean body size of the boys who later had CHD was approximately 0.2 standard deviations below the average and they were thin. Between birth and 2 years of age, mean z scores for each measurement fell, so that at 2 years the boys were thin and short. After 2 years of age their z scores for BMI began to increase and continued to do so. In a simultaneous regression, both low BMI at 2 years of age and high BMI at 11 years of age were associated with later coronary events ($p < 0.001$ and $p = 0.05$, respectively). When BMI at birth was added to the model, the measurements of body size at each of the three ages were associated with later coronary events ($p = 0.04$ for low BMI at birth, $p = 0.001$ for low BMI at 2 years of age, and $p = 0.03$ for high BMI at 11 years of age).

As with the boys, the mean body size of the 87 girls who later had coronary events was below the average (Fig. 4.1). They tended to be short at birth rather than thin, but their mean z scores for BMI fell progressively after birth so that, like the boys, they were thin at 2 years of age. After 4 years of age the z scores began to increase and continued to do so, reaching the average at approximately 8 years of age. Similarly to the boys, in a simultaneous regression body size at each of the three ages was associated with later coronary events ($p = 0.02$ for short length at birth, $p = 0.002$ for low BMI at 2 years of age, and $p = 0.02$ for high BMI at 11 years of age).

Table 4.4 shows the hazard ratios for CHD according to birthweight and fourths of BMI at age 11 years among 13,517 men and women in Helsinki born during 1924–1944 [27]. The risk of the disease fell with increasing birthweight and rose with increasing BMI at age 11. The pattern was similar in both sexes.

4.6   Type 2 Diabetes and Hypertension

People who were small at birth remain biologically different from people who were larger, and these differences include an increased susceptibility to type 2 diabetes and hypertension. Table 4.5 is based on the same cohort of men and women shown

Table 4.5 Odds ratios (95% confidence intervals) for type 2 diabetes and hypertension according to birthweight and BMI at age 11 years among 13,517 men and women in Helsinki

Birthweight (kg)	BMI at age 11 year			
	<15.7	–16.6	–17.6	>17.6
Type 2 diabetes (698 cases)				
<3.0	1.3 (0.6–2.8)[a]	1.3 (0.6–2.8)	1.5 (0.7–3.4)	2.5 (1.2–5.5)
–3.5	1.0 (0.5–2.1)	1.0 (0.5–2.1)	1.5 (0.7–3.2)	1.7 (0.8–3.5)
–4.0	1.0 (0.5–2.2)	0.9 (0.4–1.9)	0.9 (0.4–2.0)	1.7 (0.8–3.6)
>4.0	1.0	1.1 (0.4–2.7)	0.7 (0.3–1.7)	1.2 (0.5–2.7)
Hypertension (2,997 cases)				
<3.0	2.0 (1.3–3.2)	1.9 (1.2–3.1)	1.9 (1.2–3.0)	2.3 (1.5–3.8)
–3.5	1.7 (1.1–2.6)	1.9 (1.2–2.9)	1.9 (1.2–3.0)	2.2 (1.4–3.4)
–4.0	1.7 (1.0–2.6)	1.7 (1.1–2.6)	1.5 (1.0–2.4)	1.9 (1.2–2.9)
>4.0	1.0	1.9 (1.1–3.1)	1.0 (0.6–1.7)	1.7 (1.1–2.8)

[a]Odds ratios adjusted for sex and year of birth

in Table 4.4, and shows odds ratios for type 2 diabetes and hypertension according to birthweight and fourths of BMI at age 11 years. The two disorders are associated with the same general pattern of growth as CHD [27]. Risk of disease falls with increasing birthweight and rises with increasing BMI.

Associations between low birthweight and type 2 diabetes, shown in Table 4.2, have been found in other studies [5, 14–18]. The association with hypertension has also been found elsewhere [34]. There is a substantial literature showing that birthweight is associated with differences in blood pressure and insulin sensitivity within the normal range [5, 14, 18, 35]. These differences are found in children and adults but they tend to be small. A 1-kg difference in birthweight is associated with around 3 mmHg difference in systolic pressure. The contrast between this small effect and the large effect on hypertension (see Table 4.5) suggests that lesions that accompany poor fetal growth and that tend to elevate blood pressure, and which may include a reduced number of glomeruli, have a small influence on blood pressure within the normal range because counter-regulatory mechanisms maintain normal blood pressure levels. As the lesions progress through, for example, hyper-filtration of the reduced number of glomeruli and consequent glomerulosclerosis, these mechanisms are no longer able to maintain homeostasis and, as a result, blood pressure rises. This may initiate a cycle of rise in blood pressure resulting in further progression of the lesions and further rise in blood pressure [28, 36]. Direct evidence in support of this has come from a study of the kidneys of people killed in road accidents. Those being treated for hypertension had fewer, but larger, glomeruli [37]. Evidence to support the development of self-perpetuating cycles comes from a study of elderly people in Helsinki among whom the effect of birthweight on blood pressure was confined to those being treated for hypertension [38]. Despite their treatment, the blood pressures of those who had low birthweight were markedly higher, whereas among the normotensive subjects birthweight was unrelated to blood pressure. Whether measured in the clinic or by ambulatory methods, there

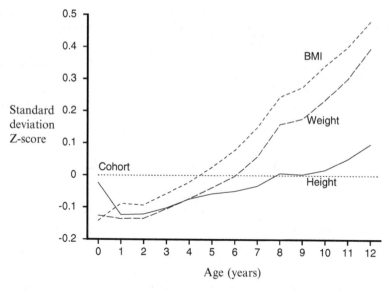

Fig. 4.2 Mean standard deviation scores (z scores) for height, weight, and BMI during childhood in 290 boys and girls who later developed type 2 diabetes within a cohort of 8,760 children. At any age, the mean z-score for the cohort is set at 0 while the standard deviation is set at 1

was >20 mmHg difference in systolic pressure between those who weighed <2,500 g (5.5 lb) at birth and those who weighed >4,000 g (8.8 lb). An inference is that by the time they reach old age most of the people with lesions acquired in utero have developed clinical hypertension. Studies in South Carolina bear on this issue. They show that among 3,236 hypertensive patients the blood pressures of those with low birth-weight tended to be more difficult to control with medication [39].

Figure 4.2 shows the growth of boys and girls who later developed type 2 diabetes. They had below average body size at birth and at 1 year, after which their weight and BMI rose progressively to exceed the average [40]. Table 4.6 shows the relation between age at "adiposity rebound" and later type 2 diabetes. After the age of 2 years the degree of obesity of young children, as measured by BMI, decreases to a minimum around 6 years of age before increasing again—the so-called adiposity rebound. The age at adiposity rebound ranges from around 3 to 8 years or more. Table 4.6 shows that early adiposity rebound is strongly related to a high BMI in later childhood, as has previously been shown [41]. It also predicts an increased incidence of type 2 diabetes in later life. This new observation has been replicated in a longitudinal study in Delhi, India [42]. In both studies an early adiposity rebound was also associated with thinness at birth and at 1 year [40, 42]. It is not therefore the young child who is overweight that is at the greatest risk of type 2 diabetes but the one who is thin but subsequently gains weight rapidly. There is, however, another path of growth that leads to type 2 diabetes: babies born to mothers with gestational diabetes are obese at birth and are overweight as children.

Table 4.6 BMI at age 11 years and cumulative incidence of type 2 diabetes according to age at adiposity rebound in 8,760 men and women in Helsinki

Age at adiposity rebound (year)	Mean BMI at age 11 All	Cumulative incidence of diabetes % (n)		
		Men	Women	All
≤4	19.7	8.1 (86)	8.9 (112)	8.6 (198)
5	17.6	6.2 (904)	2.5 (864)	4.4 (1,768)
6	17.0	3.7 (1,861)	2.5 (1,456)	3.2 (3,317)
7	16.8	2.4 (249)	2.1 (243)	2.2 (492)
≥8	16.7	3.0 (135)	0.7 (150)	1.8 (285)
p for trend	<0.001	<0.001	0.002	<0.001

4.7 Compensatory Growth

When undernutrition during early development is followed by improved nutrition, many animals and plants stage accelerated or "compensatory" growth [27]. This has costs, however, which in animals include reduced lifespan [43]. A recent paper from Helsinki describes how compensatory growth among boys is associated with increased mortality from all causes and therefore a reduced lifespan [44]. There are a number of processes by which, in humans, undernutrition and small size at birth followed by rapid childhood growth could lead to cardiovascular disease and type 2 diabetes in later life [13, 16]. Rapid growth may be associated with persisting hormonal and metabolic changes. Larger body size may increase the demand on functional capacity that has been reduced by slow early growth—fewer glomeruli, for example. Rapid weight gain may lead to an unfavorable body composition. Babies that are small and thin at birth lack muscle, a deficiency that will persist as the critical period for muscle growth occurs in utero and there is little cell replication after birth [45]. If they develop a high body mass during later childhood, they may have a disproportionately high fat mass in relation to lean body mass, which will lead to insulin resistance [46].

4.8 Pathways to Disease

New studies, especially the Helsinki studies with their detailed information on child growth and socioeconomic circumstances, increasingly suggest that the pathogenesis of CHD and the disorders related to it depend on a series of interactions occurring at different stages of development. To begin with, the effects of the genes acquired at conception may be conditioned by the early environment. Table 4.7 is based on a study of 476 elderly people in Helsinki [47]. It shows mean fasting plasma insulin concentrations according to which of two polymorphisms of the PPAR-γ (peroxisome proliferator-activated receptor) gene was present. The Pro12Ala polymorphism is known to be associated with insulin resistance,

Table 4.7 Mean (number of subjects) fasting insulin concentrations (pmol/L) in 476 elderly people in Helsinki according to PPAR-γ gene polymorphism and birthweight

	Birthweight (g)			
	<3,000	3,000–3,500	>3,500	p for difference
Pro12Pro	84 (56)	71 (161)	65 (107)	0.003
Pro12Ala/Ala12Ala	60 (37)	60 (67)	65 (48)	0.31
p for difference	0.008	0.02	0.99	

Table 4.8 Cumulative incidence % of hypertension according to birthweight and father's social class in 8,760 men and women in Helsinki

	Father's social class			
Birthweight (g)	Laborer	Lower middle class	Upper middle class	p for trend
<3,000	22.2	20.2	10.5	0.002
–3,500	18.8	15.2	10.6	<0.001
–4,000	14.5	12.5	10.3	0.04
>4,000	11.1	15.6	15.7	0.11
p for trend	<0.001	0.05	0.79	

indicated by elevated fasting plasma insulin concentrations. Table 4.7 shows, however, that this effect occurs only among men and women who had low birthweight. Conversely, low birthweight has been consistently linked to later insulin resistance [18], but Table 4.7 shows that this effect occurs only among people with the Pro12Ala polymorphism. As birthweight serves as a marker of fetal nutrition [26], this gene–birthweight interaction may reflect a gene–nutrient interaction during development.

The effects of the intrauterine environment on later disease are conditioned not only by events at conception but by events after birth. Tables 4.4 and 4.5 show how the effects are conditioned by childhood weight gain. Table 4.3 shows that the effects of low ponderal index at birth are conditioned by living conditions in adult life. Table 4.8 shows how the effects of low birthweight on later hypertension are conditioned by living conditions in childhood, indicated by the occupational status of the father [48]. Among all the men and women low birthweight was associated with an increased incidence of hypertension, as has been shown before [34]. This association, however, was present only among those who were born into families where the father was a laborer or of lower middle class.

It seems that the pathogenesis of cardiovascular disease and type 2 diabetes cannot be understood within a model in which risks associated with adverse influences at different stages of life add to each other [49]. Rather, disease is the product of branching paths of development. The environment triggers the branchings and these determine the vulnerability of each individual to what lies ahead. The pathway to CHD can originate either in slow fetal growth as a consequence of undernutrition or in poor infant growth as a consequence of poor living conditions.

The effects of slow fetal growth and low birthweight, and the effects of postnatal development, depend on the environmental influences and paths of development that precede and follow them. Low birthweight, or any other single influence, does not have "an" effect that is best estimated by a pooled estimate from all published studies. A recent pooled estimate led to the conclusion that because the effects of birthweight on blood pressure within the normal range are small, the effects on disease are also small [50]. Such a conclusion is biologically fallacious for the reasons already described under "developmental origins hypothesis." It is also statistically fallacious because it discounts interactions of the kind described. As Dubos [51] wrote: "The effects of the physical and social environments cannot be understood without knowledge of individual history." Unraveling disease causation, and hence the way to prevent it, will therefore require an understanding of heterogeneity.

4.9   Strength of Effects

Low birthweight, though a convenient marker in epidemiological studies, is an inadequate description of the phenotypic characteristics of a baby that determine its long-term health. The wartime famine in the Netherlands produced lifelong insulin resistance in babies who were in utero at the time, but there was little alteration in birthweight [52]. In babies, as in children, slowing of growth is a response to a poor environment, especially undernutrition, but body weight at birth does not adequately describe the long-term morphological and physiological consequences of undernutrition. The same birthweight can be attained by many different paths of fetal growth and each is likely to be accompanied by different gene–environment interactions [26]. Nevertheless, birthweight provides a basis for estimating the magnitude of the effects of the fetal phase of development on later disease, though it is likely to underestimate them.

Because the risk of cardiovascular disease is influenced both by small body size at birth and during infancy and by rapid weight gain in childhood, estimation of the risk of disease attributable to early development requires data on fetal, infant, and childhood growth. Currently, the Helsinki studies are the main source of information [27]. Table 4.4 shows that men and women who had birthweights >4 kg (8.8 lb) and whose prepubertal BMI was in the lowest fourth had around half the risk of CHD when compared with people who had birthweights <3 kg (6.6 lb) but whose BMI at age 11 years was in the highest fourth. The hazard ratios for admissions and deaths were 0.80 (95% CI 0.72–0.90) for each kilogram increase in birthweight and 1.06 (1.03–1.10) for each kg/m^2 increase in BMI at age 11 years. The hazard ratios for deaths alone were 0.83 (0.69–0.99) and 1.10 (1.04–1.16), respectively.

In Table 4.9, subjects were divided according to thirds of body size at birth and whether their standard deviation score for BMI decreased or increased between ages 3 and 11 years. In both men and women, the highest incidence of CHD occurred in those who were in the lowest third of birthweight and whose standard deviation score for BMI increased between 3 and 11 years. Among men ponderal index at birth was more strongly related to CHD than birthweight, while among women length at birth was stronger. From these data it can be calculated that if each man

Table 4.9 Cumulative incidence % of CHD according to body size at birth and change in standard deviation score for BMI between ages 3 and 11 years among 6,345 men and women in Helsinki

Birth size	Change in standard deviation score for BMI between ages 3 and 11	
	Decrease	Increase
Men		
Birthweight (kg)		
<3.2	8.8 (512)	9.0 (476)[a]
−3.6	6.9 (662)	11.3 (512)
>3.6	5.9 (740)	8.6 (521)
Ponderal index (kg/m^3)		
<25	8.0 (411)	11.7 (394)
−27	7.6 (649)	10.8 (556)
>27	6.2 (838)	7.2 (539)
Women		
Birthweight (kg)		
<3.2	1.6 (563)	3.8 (604)
−3.6	1.5 (612)	2.5 (438)
>3.6	0.7 (450)	3.6 (334)
Birth length (cm)		
<49	1.5 (543)	4.2 (520)
−50	1.5 (452)	3.3 (338)
>50	0.8 (609)	2.6 (496)

[a] Cumulative incidence, % (n) of coronary heart disease (hospital admissions and deaths). There were 279 cases in men and 66 in women

in the cohort had been in the highest third of ponderal index at birth, and each woman in the highest third of birth length, and if each man or woman had decreased their BMI score between ages 3 and 11 years, the incidence of CHD would have been reduced by 25% in men and 63% in women [27].

Table 4.5 showed that men and women who had birthweights >4 kg and whose prepubertal BMI was in the lowest fourth had around half the risk of type 2 diabetes and hypertension when compared with people who had birthweights <3 kg but whose BMI was in the highest fourth. The odds ratio for type 2 diabetes was 0.67 (95% CI 0.58–0.79) for each kilogram increase in birthweight and 1.18 (95% CI 1.13–1.23) for each kg/m^2 increase in BMI at age 11 years. The corresponding figures for hypertension were 0.77 (95% CI 0.71–0.84) and 1.07 (95% CI 1.04–1.09), respectively.

In Table 4.10, subjects are again divided into six groups according to thirds of birthweight and whether their standard deviation score for BMI decreased or increased between ages 3 and 11 years. For both type 2 diabetes and hypertension, there were independent effects of birthweight and change in BMI score. The patterns of odds ratios and incidence shown in Tables 4.5 and 4.10 were similar in the two sexes. If each individual in the cohort had been in the highest third of birthweight and had decreased their standard deviation score for BMI between ages 3 and 11 years, the incidence of type 2 diabetes would have been reduced by 52% and the incidence of hypertension by 25% [27].

Table 4.10 Cumulative incidence % (number of subjects) of type 2 diabetes and hypertension according to birthweight and change in standard deviation score for BMI between ages 3 and 11 years among 6,424 men and women in Helsinki

Birthweight (kg)	Change in standard deviation score for BMI between ages 3 and 11	
	Decrease	Increase
Type 2 diabetes (227 cases)		
<3.2	3.1 (1,075)	5.5 (1,080)
–3.6	2.4 (1,274)	4.3 (950)
>3.6	1.5 (1,190)	5.4 (855)
Hypertension (1,036 cases)		
<3.2	15.9 (1,075)	21.3 (1,080)
–3.6	14.8 (1,274)	19.4 (950)
>3.6	12.0 (1,190)	13.9 (855)

4.10 Maternal Influences on Fetal Nutrition

Size at birth is the product of the fetus's trajectory of growth, which is set at an early stage in development, and the materno-placental capacity to supply sufficient nutrients to maintain that trajectory. In Western communities, randomized controlled trials of maternal macronutrient supplementation have indicated relatively small effects on birthweight [53]. This has led to the view that regulatory mechanisms in the maternal and placental systems act to ensure that human fetal growth and development is little influenced by normal variations in maternal nutrient intake, and that there is a simple relationship between a woman's body composition and the growth of her fetus. Recent experimental studies in animals and observational data in humans challenge these concepts [54]. They suggest that a mother's own fetal growth and her dietary intakes and body composition can exert major effects on the balance between the fetal demand for nutrients and the materno-placental capacity to meet that demand. Specific issues that have not yet been adequately addressed include: (1) maternal effects on the trajectory of fetal growth; (2) intergenerational effects; (3) paradoxical effects on placental growth; and (4) the importance of the mother's body composition and the balance of macronutrients in her diet.

4.10.1 The Fetal Growth Trajectory

A rapid trajectory of growth increases the demand of the fetus for nutrients. This demand is greatest late in pregnancy but the trajectory is thought to be primarily determined by genetic and environmental effects in early gestation. Experimental studies of pregnant ewes have shown that, although a fast growth trajectory is generally associated with larger fetal size and improved neonatal survival, it renders the fetus more vulnerable to a reduced materno-placental supply of nutrients in late gestation. Thus, maternal undernutrition during the last trimester adversely affects the development of rapidly growing fetuses with high requirements, while having little effect on those growing more slowly [55]. Rapidly growing fetuses were found to make a series of adaptations in order to survive, including fetal wasting and placental oxidation of fetal amino acids to maintain lactate output to the fetus [54]. Experiments in animals have shown that alterations in the maternal diet around the time of conception can change the fetal growth trajectory. In a recent study, rats were fed a 9% casein low-protein diet in the periconceptional period. This led to structural changes at the blastocyst stage of embryonic development, reduced fetal growth rates, small size at birth, and raised blood pressure in the offspring during adult life [56]. The sensitivity of the human embryo to its environment is being increasingly recognized with the development of assisted reproductive technology [57]. The trajectory of fetal growth is thought to increase with improvements in periconceptional nutrition and is faster in male fetuses. The consequent greater vulnerability of male fetuses to undernutrition may contribute to the shorter lives of men [58].

4.10.2   Intergenerational Effects

Experimental studies in animals have shown that undernutrition can have effects on reproductive performance which may persist for several generations. Among rats fed a protein-deficient diet over 12 generations, there was a progressive fall in fetal growth rates. When restored to a normal diet, it took three generations before growth and development were normalized [59].

Strong evidence for major intergenerational effects in humans has come from studies showing that a woman's birthweight influences the birthweight of her offspring [60]. A study in the UK showed that whereas low-birthweight mothers tended to have thin infants with a low ponderal index, the father's birthweight was unrelated to ponderal index at birth [61]. The effect of maternal birthweight on thinness at birth is consistent with the hypothesis that in low-birthweight mothers the fetal supply line is compromised and unable to meet fetal nutrient demand. Potential mechanisms underlying this effect include alterations in the uterine or systemic vasculature, changes in maternal metabolism, and impaired placentation.

4.10.3   Maternal Diet and Body Composition

Direct evidence supporting a long-term effect on the fetus of levels of maternal nutrient intake during pregnancy has come from a follow-up study in Argentina of children whose mothers took part in a randomized controlled trial of calcium supplementation in pregnancy [62]. Supplementation was associated with lowering of the offspring's blood pressure in childhood, even though it was not associated with any change in birthweight. Follow-up studies after the Dutch famine of 1944–1945 found that severe maternal caloric restriction at different stages of pregnancy was variously associated with obesity, dyslipidemia, and insulin resistance in the offspring, and there is preliminary evidence of an increased risk of CHD [52, 63, 64]. Again, these effects were largely independent of size at birth.

In the Dutch studies, famine exposure per se was not associated with raised blood pressure in the offspring, but there was an effect of macronutrient balance. Maternal rations with a low protein density were associated with raised blood pressure in the adult offspring [65]. This adds to the findings of studies in Aberdeen, UK, which show that maternal diets with either a low or a high ratio of animal protein to carbohydrate were associated with raised blood pressure in the offspring during adult life [66].

In the Aberdeen study, maternal diets with a high protein density were not only associated with raised blood pressure in the offspring but also with insulin deficiency and impaired glucose tolerance [67]. While it may seem counter-intuitive that a high-protein diet should have adverse effects, these findings are consistent with the results of controlled trials of protein supplementation in pregnancy which show that high protein intakes are associated with reduced birthweight [68]. The Aberdeen findings have been replicated in a follow-up study of men and women in Motherwell, UK, whose mothers were advised to eat a diet high in meat protein and low in carbohydrate during pregnancy [69]. Those whose mothers had high intakes of meat and fish in late pregnancy but low intakes of carbohydrate had raised blood pressure, particularly if the mother also had a low intake of green vegetables. Although raised blood pressure was also related to low birthweight, taking account of birthweight had little effect on the relation between the maternal diet and the offspring's blood pressure. One possibility is that the effect on blood pressure may be a consequence of the metabolic stress imposed on the mother by an unbalanced diet in which high intakes of essential amino acids are not accompanied by the micronutrients required to utilize them. These high intakes create an excess which is potentially toxic unless they are degraded and oxidized. The degradation of essential amino acids consumes nonessential amino acids, which are synthesized by the body. Their synthesis requires cofactors, especially folate and other B vitamins [3]. Direct evidence of metabolic stress in the offspring comes from analysis of their fasting plasma cortisol concentrations [70]. Men and women whose mothers had high intakes of meat and fish and low intakes of green vegetables had raised cortisol concentrations.

The fetus does not live on the mother's diet alone: that would be too dangerous a strategy. It also lives off stored nutrients and the turnover of protein and fat in the mother's tissues [71]. Maternal size and body composition account for up to 20% of the variability in birthweight [72]. Gestational diabetes is known to be associated with adverse long-term outcomes in the offspring [73]. More recently, studies in Europe and India have shown that high maternal weight and adiposity are associated with adult development of insulin deficiency, type 2 diabetes, and CHD in the offspring [9, 16, 74]. Of great importance is an increasing body of consistent evidence showing strong links between low maternal weight or BMI and insulin resistance in the adult offspring [52, 69, 75]. Table 4.11 shows plasma glucose and insulin concentrations in Chinese men and women aged around 45 years following

Table 4.11 Mean 2-h plasma glucose and insulin concentrations according to maternal BMI in late pregnancy in 584 Chinese men and women

	Maternal BMI at 38 weeks of pregnancy				
	≤23	−24.5	−26	>26	p for trend
2-h glucose (mmol/L)	7.6	6.6	6.7	5.7	0.003
2-h insulin (pmol/L)	304	277	282	177	0.007

a standard oral glucose challenge. Low maternal BMI at 38 weeks of pregnancy was associated with raised plasma glucose and insulin concentrations [75]. Results for maternal BMI in early pregnancy, around 15 weeks, were stronger. In contrast to these associations between maternal BMI and insulin resistance, thin maternal skin-fold thicknesses and low pregnancy weight gain have been consistently associated with raised blood pressure in the offspring [76–79]. One of the metabolic links between maternal body composition and birth size is protein synthesis. Women with a greater lean body mass have higher rates of protein synthesis in pregnancy [80]. Variation in rates of maternal protein synthesis explains around a quarter of the variability in birth length.

4.10.4 Placental Transfer

A baby's birthweight also depends on the placenta's ability to transport nutrients to it from its mother. The placenta seems to act as a nutrient sensor regulating the transfer of nutrients to the fetus according to the mother's ability to deliver them and the demands of the fetus for them [81]. The size, weight, and shape of the placenta are all subject to wide variations [82]. Its size reflects its ability to transfer nutrients [83]. Small babies generally have small placentas but, in some circumstances, an undernourished baby can expand its placental surface to extract more nutrients from the mother [84].

In the last century, the surface of the placenta was described as being either "oval" or "round" [85]. In order to describe the extent to which the surface was more oval than round, two so-called diameters of the surface were routinely recorded in some hospitals, a maximal diameter (the length of the surface), and a lesser one bisecting it at right angles (the breadth) [86].

Preeclampsia is associated with reduced placental size. The abnormal placentation is a result of impaired invasion of the maternal spiral arteries by the trophoblast at implantation [87]. In the Helsinki Birth Cohort, placentas from pregnancies complicated by preeclampsia had a more oval surface than those from normotensive pregnancies because of a disproportionate reduction in the breadth [88]. The relation with the breadth was graded: the shorter the breadth, the greater the risk for, and severity of, preeclampsia. This led to the conclusion that placental growth is polarized from the time of implantation, so that growth along the major axis, the length, is qualitatively different to growth along the breadth. One possibility is that

Table 4.12 Mean systolic blood pressure mmHg among men and women aged 50, born at term

Birth weight (lb)	Placental weight (lb)			
	≤1.0	−1.25	−1.5	>1.5
≤6.5	149	152	151	167
−7.5	139	148	146	159
>7.5	131	143	148	153

growth along the major axis is aligned with the rostro-caudal axis of the embryo, while tissue along the minor axis may be the nutrient sensor.

Low placental weight is associated with an increased risk of hypertension in later life [89]. However, a study of men and women born in a maternity hospital in Preston, UK, showed that high placental weight in relation to birthweight is also associated with later hypertension [90]. Table 4.12 shows that, as expected, at any placental weight lower birthweight was associated with higher systolic pressure; but at any birthweight higher placental weight was associated with higher systolic pressure. The highest systolic pressure was in people who had the lowest birthweights but the highest placental weights. This observation has been replicated, and high placental weight in relation to birthweight has also been shown to predict CHD [91]. Observations in sheep show that in response to undernutrition in mid-gestation the fetus is able to extend the area of the placenta by expanding the individual cotyledons [84]. This increases the area available for nutrient and oxygen exchange, and results in a larger lamb than there would otherwise have been. This is profitable for the farmer, and manipulation of placental size by changing the pasture of pregnant ewes is standard practice in sheep farming. There is evidence for a similar phenomenon in humans [86].

It is becoming apparent that the shape and size of the placental surface at birth is a new marker for chronic disease in later life. The predictions of later disease depend on the combination of placental size/shape and mother's body size. Particular combinations have been shown to predict hypertension [86], chronic heart failure [92], CHD [93], and certain forms of cancer [94]. Presumably, the mother's body size determines whether a placental phenotype programs disease because it reflects the availability of nutrients at the placental surface and because it affects the development and function of the placenta.

The list of chronic diseases whose origins lie in early development now extends beyond cardiovascular disease and type 2 diabetes. There is, for example, strong evidence that osteoporosis is another of the body's "memories" of undernutrition at a critical early stage of development [95]. This is perhaps unsurprising given that rickets has served as a long-standing example of the persisting structural changes induced by early undernutrition. The intrauterine origins of cancer are being explored [94, 96, 97].

4.11 Research Challenges

Further research has to address two overarching questions.

4.11.1 Environmental Influences

What are the environmental influences which, acting through the mother, or directly on the infant and young child, alter gene expression and thereby permanently change the body's structure and function? Research on maternal influences will need to address: (1) effects on the fetal growth trajectory; (2) effects on placental growth; (3) intergenerational effects; and (4) maternal dietary pattern during periconception period.

4.11.2 Pathogenesis

How do gene–environment interactions during development translate into chronic disease? Through a combination of clinical and experimental studies, progress is being made in understanding the developmental origins of altered glucose–insulin and lipid metabolism, stress responses, blood pressure, and renal function.

4.12 Disease Prevention

The evidence presented in this review indicates that prevention of a substantial proportion of chronic diseases, including cardiovascular disease, type 2 diabetes, and osteoporosis, may depend on interventions at a number of stages of development. Strategies which target infants and young children may give the most immediate benefit but improving the intrauterine environment is an important long-term goal. Despite current levels of nutrition in Western countries, the nutrition of many fetuses and infants remains suboptimal because their mothers' diets are unbalanced or because their delivery is constrained by maternal metabolism or inadequate placental development. There is sufficient knowledge to implement preventive programs now. More research is needed, however, to increase the effectiveness of these programs.

Mother:

- An optimal diet begins before pregnancy. In developing countries micronutrients in the diet may be limiting factors in fetal growth, whereas in Western countries macronutrient balance, especially between protein and carbohydrate, seems likely to be important.
- Women require an optimal body composition before pregnancy, with avoidance of excessive thinness or overweight.
- It is not known whether the greatest benefits for the next generation will come from improving the nutrition of adult women, adolescent girls, or girl children. Any rational policy needs to address all three.

Infant:

- The growth in weight and length during the first year after birth needs to be protected by good infant feeding practices, including breastfeeding and avoidance of recurrent infections.

Child:

- Young children who were small or thin at birth should not increase their centile score for BMI after the age of 2 years.

Adult:

- People who were small at birth are more vulnerable to adverse influences acting in adult life. At present these are known to include obesity and aspects of psychosocial stress.

4.13   Conclusions

Low birthweight is now known to be associated with increased rates of CHD and the related disorders, namely stroke, hypertension, and type 2 diabetes. These associations have been extensively replicated in studies in different countries and are not the result of confounding variables. They extend across the normal range of birthweight and depend on lower birthweights in relation to the duration of gestation rather than the effects of premature birth. The associations are thought to be consequences of developmental plasticity, the phenomenon by which one genotype can give rise to a range of different physiological or morphological states in response to different environmental conditions during development. Recent observations have shown that impaired growth in infancy and rapid childhood weight gain exacerbate the effects of impaired prenatal growth. CHD and the disorders related to it arise through a series of interactions between environmental influences and the pathways of development that preceded them. These diseases are the product of branching pathways of development in which the branchings are triggered by the environment before and after birth.

Variations in the normal processes of early human development have implications for health throughout life. This may be because these variations affect key processes that determine vulnerability to disease, including the quality of stem cells, antioxidant defenses, tissue repair, immune competence, and chronic inflammation. This new research is prompting a reevaluation of the regulation of early development. Impetus has been added to this reevaluation by recent findings showing that a woman's dietary balance and body composition in pregnancy are related to levels of cardiovascular risk factors and the risk of CHD in her offspring in adult life without necessarily affecting size at birth. These observations challenge the view that the fetus is little affected by variations in maternal nutrition, except in the extreme circumstance of famine. There is an increasing body of evidence that a woman's own fetal growth, and her diet and body composition through childhood up to the time of her pregnancy, play a major role in determining the future health of her children. A new vision of optimal early human development is emerging which takes account of both short- and long-term outcomes.

References

1. Barker DJP, Osmond C, Winter PD, Margetts B, Simmonds SJ. Weight in infancy and death from ischaemic heart disease. Lancet. 1989;2:577–80.
2. Barker DJP. Fetal origins of coronary heart disease. BMJ. 1995;311:171–4.
3. Jackson AA. All that glitters. Br Nutr Found Nutr Bullet. 2000;25:11–24.
4. Osmond C, Barker DJP, Winter PD, Fall CHD, Simmonds SJ. Early growth and death from cardiovascular disease in women. BMJ. 1993;307:1519–24.
5. Hales CN, Barker DJP, Clark PMS, et al. Fetal and infant growth and impaired glucose tolerance at age 64. BMJ. 1991;303:1019–22.
6. Frankel S, Elwood P, Sweetnam P, Yarnell J, Davey Smith G. Birthweight, body mass index in middle age, and incident coronary heart disease. Lancet. 1996;348:1478–80.
7. Stein CE, Fall CHD, Kumaran K, Osmond C, Cox V, Barker DJP. Fetal growth and coronary heart disease in South India. Lancet. 1996;348:1269–73.
8. Rich-Edwards JW, Stampfer MJ, Manson JE, et al. Birth weight and risk of cardiovascular disease in a cohort of women followed up since 1976. BMJ. 1997;315:396–400.
9. Forsén T, Eriksson JG, Tuomilehto J, Teramo K, Osmond C, Barker DJP. Mother's weight in pregnancy and coronary heart disease in a cohort of Finnish men: follow up study. BMJ. 1997;315:837–40.
10. Leon DA, Lithell HO, Vagero D, et al. Reduced fetal growth rate and increased risk of death from ischaemic heart disease: cohort study of 15 000 Swedish men and women born 1915–29. BMJ. 1998;317:241–5.
11. Forsen T, Eriksson JG, Tuomilehto J, Osmond C, Barker DJP. Growth in utero and during childhood among women who develop coronary heart disease: longitudinal study. BMJ. 1999;319:1403–7.
12. Forsen T, Osmond C, Eriksson JG, Barker DJP. Growth of girls who later develop coronary heart disease. Heart. 2004;90:20–4.
13. Eriksson JG, Forsen T, Tuomilehto J, Osmond C, Barker DJP. Early growth and coronary heart disease in later life: longitudinal study. BMJ. 2001;322:949–53.
14. Lithell HO, McKeigue PM, Berglund L, Mohsen R, Lithell UB, Leon DA. Relation of size at birth to non-insulin dependent diabetes and insulin concentrations in men aged 50–60 years. BMJ. 1996;312:406–10.
15. McCance DR, Pettitt DJ, Hanson RL, Jacobsson LTH, Knowler WC, Bennett PH. Birth weight and non-insulin dependent diabetes: thrifty genotype, thrifty phenotype, or surviving small baby genotype? BMJ. 1994;308:942–5.
16. Forsén T, Eriksson J, Tuomilehto J, Reunanen A, Osmond C, Barker D. The fetal and childhood growth of persons who develop type 2 diabetes. Ann Intern Med. 2000;133:176–82.
17. Rich-Edwards JW, Colditz GA, Stampfer MJ, et al. Birthweight and the risk for type 2 diabetes mellitus in adult women. Ann Intern Med. 1999;130:278–84.
18. Newsome CA, Shiell AW, Fall CHD, Phillips DIW, Shier R, Law CM. Is birthweight related to later glucose and insulin metabolism—a systematic review. Diabet Med. 2003;20: 339–48.
19. West-Eberhard MJ. Phenotypic plasticity and the origins of diversity. Ann Rev Ecol Syst. 1989;20:249–78.
20. Bateson P, Martin P. Design for a life: how behaviour develops. London: Jonathan Cape; 1999.
21. Mellanby E. Nutrition and child-bearing. Lancet. 1933;2:1131–7.

22. Widdowson EM, McCance RA. The effect of finite periods of undernutrition at different ages on the composition and subsequent development of the rat. Proc R Soc Lond B Biol Sci. 1963;158:329–42.

23. Gluckman P, Hanson M, editors. Developmental origins of health and disease. Cambridge: Cambridge University Press; 2006.

24. Brooks AA, Johnson MR, Steer PJ, Pawson ME, Abdalla HI. Birth weight: nature or nurture? Early Hum Dev. 1995;42:29–35.

25. McCance RA. Food, growth and time. Lancet. 1962;2:621–6.

26. Harding JE. The nutritional basis of the fetal origins of adult disease. Int J Epidemiol. 2001;30:15–23.

27. Barker DJP, Eriksson JG, Forsén T, Osmond C. Fetal origins of adult disease: strength of effects and biological basis. Int J Epidemiol. 2002;31:1235–9.

28. Brenner BM, Chertow GM. Congenital oligonephropathy: an inborn cause of adult hypertension and progressive renal injury? Curr Opin Nephrol Hypertens. 1993;2:691–5.

29. Phillips DIW. Insulin resistance as a programmed response to fetal undernutrition. Diabetologia. 1996;39:1119–22.

30. Barker DJP, Forsén T, Uutela A, Osmond C, Eriksson JG. Size at birth and resilience to the effects of poor living conditions in adult life: longitudinal study. BMJ. 2001;323:1273–6.

31. Marmot M, Wilkinson RG. Psychosocial and material pathways in the relation between income and health: a response to Lynch et al. BMJ. 2001;322:1233–6.

32. Phillips DIW, Walker BR, Reynolds RM, et al. Low birth weight predicts elevated plasma cortisol concentrations in adults from 3 populations. Hypertension. 2000;35:1301–6.

33. Barker DJP, Osmond C, Forsén TJ, Kajantie E, Eriksson JG. Trajectories of growth among children who have coronary events as adults. N Engl J Med. 2005;353:1802–9.

34. Curhan GC, Chertow GM, Willett WC, et al. Birth weight and adult hypertension and obesity in women. Circulation. 1996;94:1310–5.

35. Huxley RR, Shiell AW, Law CM. The role of size at birth and postnatal catch-up growth in determining systolic blood pressure: a systematic review of the literature. J Hypertens. 2000;18:815–31.

36. Ingelfinger JR. Is microanatomy destiny? N Eng J Med. 2003;348:99–100.

37. Keller G, Zimmer G, Mall G, Ritz E, Amann K. Nephron number in patients with primary hypertension. N Eng J Med. 2003;348:101–8.

38. Ylihärsilä H, Eriksson JG, Forsén T, Kajantie E, Osmond C, Barker DJP. Self-perpetuating effects of birth size on blood pressure levels in elderly people. Hypertension. 2003;41:446–50.

39. Lackland DT, Egan BM, Syddall HE, Barker DJP. Associations between birthweight and anti-hypertensive medication in black and white Americans. Hypertension. 2002;39:179–83.

40. Eriksson JG, Forsen T, Tuomilehto J, Osmond C, Barker DJP. Early adiposity rebound in childhood and risk of type 2 diabetes in adult life. Diabetologia. 2003;46:190–4.

41. Rolland-Cachera MF, Deheeger M, Guilloud-Bataille M, Avons P, Patois E, Sempe M. Tracking the development of adiposity from one month of age to adulthood. Ann Hum Biol. 1987;14:219–29.

42. Bhargava SK, Sachdev HS, Fall CHD, et al. Relation of serial changes in childhood body mass index to impaired glucose tolerance in young adulthood. N Eng J Med. 2004;350:865–75.

43. Metcalfe NB, Monaghan P. Compensation for a bad start: grow now, pay later? Trends Ecol Evol. 2001;16:254–60.

44. Barker DJP, Kajantie E, Osmond C, Thornburg K, Eriksson JG. How boys grow determines how long they live. Am J Hum Biol. 2011;23:412–6.

45. Widdowson EM, Crabb DE, Milner RDG. Cellular development of some human organs before birth. Arch Dis Child. 1972;47:652–5.

46. Eriksson JG, Forsen T, Jaddoe VWV, Osmond C, Barker DJP. Effects of size at birth and childhood growth on the insulin resistance syndrome in elderly individuals. Diabetologia. 2002; 45:342–8.

47. Eriksson JG, Lindi V, Uusitupa M, et al. The effects of the Pro12Ala polymorphism of the peroxisome proliferator-activated receptor-γ2 gene on insulin sensitivity and insulin metabolism interact with size at birth. Diabetes. 2002;51:2321–4.

48. Barker DJP, Forsén T, Eriksson JG, Osmond C. Growth and living conditions in childhood and hypertension in adult life: longitudinal study. J Hypertens. 2002;20:1951–6.
49. Kuh D, Ben-Shlomo Y. A life-course approach to chronic disease epidemiology. Oxford: Oxford University Press; 1997.
50. Huxley R, Neil A, Collins R. Unravelling the fetal origins hypothesis. Lancet. 2002; 360:2074–5.
51. Dubos R. Mirage of health. London: Allen and Unwin; 1960.
52. Ravelli ACJ, van der Meulen JHP, Michels RPJ, et al. Glucose tolerance in adults after prenatal exposure to famine. Lancet. 1998;351:173–7.
53. Kramer MS. Effects of energy and protein intakes on pregnancy outcome: an overview of the research evidence from controlled clinical trials. Am J Clin Nutr. 1993;58:627–35.
54. Barker DJP. Mothers, babies and health in later life. Edinburgh: Churchill Livingstone; 1998.
55. Harding JE, Liu L, Evans P, Oliver M, Gluckman P. Intrauterine feeding of the growth-retarded fetus: can we help? Early Hum Dev. 1992;29:193–7.
56. Kwong WY, Wild A, Roberts P, Willis AC, Fleming TP. Maternal undernutrition during the pre-implantation period of rat development causes blastocyst abnormalities and programming of postnatal hypertension. Development. 2000;127:4195–202.
57. Walker SK, Hartwick KM, Robinson JS. Long-term effects on offspring of exposure of oocytes and embryos to chemical and physical agents. Hum Reprod Update. 2000;6:564–7.
58. Eriksson JG, Kajantie E, Osmond C, Thornburg K, Barker DJ. Boys live dangerously in the womb. Am J Hum Biol. 2010;22:330–5.
59. Stewart RJC, Sheppard H, Preece R, Waterlow JC. The effect of rehabilitation at different stages of development of rats marginally malnourished for ten to twelve generations. Br J Nutr. 1980;43:403–12.
60. Emanuel I, Filakti H, Alberman E, Evans SJW. Intergenerational studies of human birthweight from the 1958 birth cohort. I. Evidence for a multigenerational effect. Br J Obstet Gynaecol. 1992;99:67–74.
61. Godfrey KM, Barker DJP, Robinson S, Osmond C. Maternal birthweight and diet in pregnancy in relation to the infant's thinness at birth. Br J Obstet Gynaecol. 1997;104:663–7.
62. Belizan JM, Villar J, Bergel E, et al. Long term effect of calcium supplementation during pregnancy on the blood pressure of offspring: follow up of a randomised controlled trial. BMJ. 1997;315:281–5.
63. Roseboom TJ, van der Meulen JH, Osmond C, Barker DJ, Ravelli AC, Bleker OP. Plasma lipid profiles in adults after prenatal exposure to the Dutch famine. Am J Clin Nutr. 2000;72:1101–6.
64. Roseboom TJ, van der Meulen JH, Osmond C, et al. Coronary heart disease after prenatal exposure to the Dutch famine, 1944–45. Heart. 2000;84:595–8.
65. Roseboom TJ, van der Meulen JH, van Montfrans GA, et al. Maternal nutrition during gestation and blood pressure in later life. J Hypertens. 2001;19:29–34.
66. Campbell DM, Hall MH, Barker DJP, Cross J, Shiell AW, Godfrey KM. Diet in pregnancy and the offspring's blood pressure 40 years later. Br J Obstet Gynaecol. 1996;103:273–80.
67. Shiell AW, Campbell DM, Hall MH, Barker DJ. Diet in late pregnancy and glucose-insulin metabolism of the offspring 40 years later. Br J Obstet Gynaecol. 2000;107:890–5.
68. Rush D. Effects of changes in maternal energy and protein intake during pregnancy, with special reference to fetal growth. In: Sharp F, Fraser RB, Milner RDG, editors. Fetal growth. London: Royal College of Obstetricians and Gynaecologists; 1989. p. 203–33.
69. Shiell AW, Campbell-Brown M, Haselden S, Robinson S, Godfrey KM, Barker DJP. A high meat, low carbohydrate diet in pregnancy: relation to adult blood pressure in the offspring. Hypertension. 2001;38:1282–8.
70. Herrick K, Phillips DIW, Haselden S, Shiell AW, Campbell-Brown M, Godfrey KM. Maternal consumption of a high-meat, low-carbohydrate diet in late pregnancy: relation to adult cortisol concentrations in the offspring. J Clin Endocrinol Metab. 2003;88:3554–60.
71. James WPT. Long-term fetal programming of body composition and longevity. Nutr Rev. 1997;55:S41–3.
72. Catalano PM, Husten LP, Thomas AJ, Fung CM. Effect of maternal metabolism on fetal growth and body composition. Diabetes Care. 1998;21:B85–90.

73. Silverman BL, Purdy LP, Metzger BE. The intrauterine environment: implications for the offspring of diabetic mothers. Diabetes Rev. 1996;4:21–35.
74. Fall CHD, Stein CE, Kumaran K, et al. Size at birth, maternal weight, and type 2 diabetes in South India. Diabet Med. 1998;15:220–7.
75. Mi J, Law C, Zhang K-L, Osmond C, Stein C, Barker DJP. Effects of infant birthweight and maternal body mass index in pregnancy on components of the insulin resistance syndrome in China. Ann Intern Med. 2000;132:253–60.
76. Margetts BM, Rowland MGM, Foord FA, Cruddas AM, Cole TJ, Barker DJP. The relation of maternal weight to the blood pressures of Gambian children. Int J Epidemiol. 1991; 20:938–43.
77. Godfrey KM, Forrester T, Barker DJP, et al. Maternal nutritional status in pregnancy and blood pressure in childhood. Br J Obstet Gynaecol. 1994;101:398–403.
78. Clark PM, Atton C, Law CM, Shiell A, Godfrey K, Barker DJP. Weight gain in pregnancy, triceps skinfold thickness and blood pressure in the offspring. Obstet Gynecol. 1998;91:103–7.
79. Adair LS, Kuzawa CW, Borja J. Maternal energy stores and diet composition during pregnancy program adolescent blood pressure. Circulation. 2001;104:1034–9.
80. Duggleby SL, Jackson AA. Relationship of maternal protein turnover and lean body mass during pregnancy and birth length. Clin Sci. 2001;101:65–72.
81. Jansson T, Powell TL. Role of the placenta in fetal programming: underlying mechanisms and potential interventional approaches. Clin Sci. 2007;113:1–13.
82. Hamilton WJ, Boyd JD, Mossman HW. Human embryology. Cambridge: W. Heffer & Sons; 1945.
83. Sibley CP. The pregnant woman. In: Case RM, Waterhouse JM, editors. Human physiology: age, stress, and the environment. Oxford: Oxford University Press; 1994. p. 3–27.
84. McCrabb GJ, Egan AR, Hosking BJ. Maternal undernutrition during mid-pregnancy in sheep; variable effects on placental growth. J Agric Sci. 1992;118:127–32.
85. Burton GJ, Barker DJP, Moffett A, Thornburg K, editors. The placenta and human developmental programming. Cambridge: Cambridge University Press; 2010. p. 5.
86. Barker DJP, Thornburg KL, Osmond C, Kajantie E, Eriksson JG. The surface area of the placenta and hypertension in the offspring in later life. Int J Dev Biol. 2010;54:525–30.
87. Roberts JM, Cooper DW. Pathogenesis and genetics of preeclampsia. Lancet. 2001;357: 53–6.
88. Kajantie E, Barker DJP, Osmond C, Kajantie E, Eriksson JG. In pre-eclampsia the placenta grows slowly along its minor axis. Int J Dev Biol. 2010;54:469–73.
89. Eriksson J, Forsen T, Toumilheto J, Osmond C, Barker D. Fetal and childhood growth and hypertension in adult life. Hypertension. 2000;36:790–4.
90. Barker DJP, Bull AR, Osmond C, Simmonds S. Fetal and placental size and risk of hypertension in adult life. Br Med J. 1990;301:259–62.
91. Martyn CN, Barker DJP, Osmond C. Mothers' pelvic size, fetal growth, and death from stroke and coronary heart disease in men in the UK. Lancet. 1996;348:1264–8.
92. Barker DJ, Gelow J, Thornburg K, Osmond C, Kajantie E, Eriksson JG. The early origins of chronic heart failure: impaired placental growth and initiation of insulin resistance in childhood. Eur J Heart Fail. 2010;12:819–25.
93. Eriksson JG, Kajantie E, Thornburg KL, Osmond C, Barker DJP. Mother's body size and placental size predict coronary heart disease in men. Eur Heart J. 2011;32:2297–303.
94. Barker DJ, Thornburg KL, Osmond C, Kajantie E, Eriksson JG. The prenatal origins of lung cancer. II. The placenta. Am J Hum Biol. 2010;22:512–6.
95. Javaid MK, Eriksson JG, Kajantie E, et al. Growth in childhood predicts hip fracture risk in later life. Osteoporos Int. 2011;22:69–73.
96. Barker DJ, Osmond C, Thornburg KL, Kajantie E, Forsén TJ, Eriksson JG. A possible link between the pubertal growth of girls and breast cancer in daughters. Am J Hum Biol. 2008;20:127–31.
97. Barker DJ, Osmond C, Thornburg KL, Kajantie E, Erikkson JG. A possible link between the pubertal growth of girls and ovarian cancer in their daughters. Am J Hum Biol. 2008;20: 659–62.

Chapter 5
The Nutrition Transition Is Speeding Up: A Global Perspective

Barry M. Popkin

Keywords Nutrition transition • Nutrition-related noncommunicable disease • Physical activity • Vegetable oils • Sugar • Obesity • Prevention of obesity

Key Points

- The speed of increase in the prevalence of overweight and obesity in the developing world is greater than that seen in higher-income developed countries. This very much relates to a mismatch between our biology and modern technology.
- The major dietary changes are: increased intake of edible oil (an increase that is affordable by the world's poor in the majority of even the lower income countries); increased intake of caloric sweetener (particularly in sweetened beverages in some countries but also in many other processed food sources); and also a rapid increase in the total intake of animal source foods.
- A marked upward shift in the technologies available to the developing world for work, transportation, home production, and leisure are combining to rapidly increase sedentarianism.
- There is emerging research that indicates that there might be important biological differences between the populations found in Asia, Africa, and Latin America that might predispose many of them to higher risk of many nutrition-related noncommunicable diseases at lower BMI levels than heretofore found in the United States and Europe.
- The United Kingdom has created an exemplary program to reduce the prevalence of obesity while in the low income and developing world only Mexico today has an ongoing systematic effort underway.

B.M. Popkin, PhD (✉)
Carolina Population Center, School of Public Health, University of North Carolina at Chapel Hill, CB # 8120 University Square, Chapel Hill, NC 27516-3997, USA
e-mail: popkin@unc.edu

N.J. Temple et al. (eds.), *Nutritional Health: Strategies for Disease Prevention*,
Nutrition and Health, DOI 10.1007/978-1-61779-894-8_5,
© Springer Science+Business Media, LLC 2012

5.1   Introduction: What Is the Nutrition Transition?

The world is witnessing rapid shifts in diet and body composition, with resultant important changes in health profiles. In many ways, these shifts are a continuation of large-scale changes that have occurred repeatedly over time; the changes facing low- and moderate-income countries today, however, appear to be occurring very rapidly. Broad shifts in, first, population size and its age composition and second, in disease patterns are occurring around the world. These two sets of dynamic shifts are termed the demographic and epidemiological transitions. Third, large shifts have occurred in dietary and physical activity and inactivity patterns. These changes are reflected in nutritional outcomes, such as changes in average stature and body composition. These dietary and physical activity changes, reflected in nutritional outcomes such as changes in average stature and body composition, are referred to as the nutrition transition.

Human diet and activity patterns and nutritional status have undergone a sequence of major shifts, defined as broad patterns of food use and corresponding to nutrition-related disease. Over the last three centuries, the pace of dietary and activity change appears to have accelerated, to varying degrees in different regions of the world. In particular, changes have accelerated more in the past decade. These dietary, activity, and body composition distribution changes are paralleled by major changes in health status, as well as by major demographic and socioeconomic changes. Obesity emerges early in the shift as does the level and age composition of morbidity and mortality. We can think of five broad nutrition patterns. They are not restricted to particular periods of human history. For convenience, the patterns are outlined as historical developments; however, "earlier" patterns are not restricted to the periods in which they first arose, but continue to characterize certain geographic and socio-economic subpopulations.

5.1.1   Pattern 1: Collecting Food

This diet, which characterizes hunter-gatherer populations, is high in carbohydrates and fiber and low in fat, especially saturated fat [1, 2]. Meat from wild animals has a significantly higher proportion of polyunsaturated fat than does meat from modern domesticated animals [3]. Activity patterns are very high and little obesity is found among hunter-gatherer societies. It is important to note that much of the research on hunter-gatherers is based on modern hunter-gatherers as there is much less evidence on prehistoric people. As we shift from hunter-gather populations to more sedentary communities, this is as much about the development of farming and the development of agriculture and how it affected our diet and overall life-style. Famine accompanied the shift toward less-varied sources of food based on agriculture.

5.1.2   Pattern 2: Famine

The diet becomes much less varied and subject to periods of acute scarcity of food. These changes related to a shift toward settlements and cultivation first of crops and later also livestock and poultry. These dietary changes are hypothesized to be associated with nutritional stress and a reduction in stature (estimated by some at about 4 in.) [4, 5]. During the later phases of this pattern, social stratification intensifies, and dietary variation according to gender and social status increases [6]. The pattern of famine (as with each of the patterns) has varied over time and space. Some civilizations are more successful than others in alleviating famine and chronic hunger, at least for their more privileged citizens [7]. The types of physical activities changed but there was little change in activity levels during this period.

5.1.3   Pattern 3: Receding Famine

The consumption of fruits, vegetables, and animal protein increases, and starchy staples become less important. Many earlier civilizations made great progress in reducing chronic hunger and famines, but only in the last third of the last millennium have these changes become widespread, leading to marked shifts in diet. However, famines continued well into the eighteenth century in portions of Europe and remain common in some regions of the world. Activity patterns start to shift and inactivity and leisure becomes a part of the lives of more people.

5.1.4   Pattern 4: Nutrition-Related Noncommunicable Disease

A diet high in total fat, cholesterol, sugar, and other refined carbohydrates and low in polyunsaturated fatty acids and fiber, often accompanying an increasingly sedentary life, is characteristic of most high-income societies (and increasingly of portions of the population in low-income societies). This results in increased prevalence of obesity and contributing to the degenerative diseases that characterize Omran's final epidemiologic stage [8]. Omran's epidemiologic transition moves from a pattern of high prevalence of infectious diseases and malnutrition to a pattern where chronic and degenerative diseases predominate.

5.1.5   Pattern 5: Behavioral Change

A new dietary pattern appears to be emerging as a result of changes in diet evidently associated with the desire to prevent or delay degenerative diseases and prolong health. Whether these changes, instituted in some countries by consumers and in

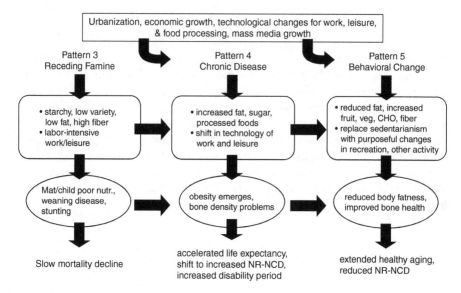

Fig. 5.1 Stages of the nutrition transition

others as a result of prodding by government policy, will constitute a large-scale transition in dietary structure and body composition remains to be seen [9–11]. If such a new dietary pattern takes hold, it may be very important in enhancing "successful aging," that is, postponing infirmity and increasing the disability-free life expectancy [12, 13].

Our focus is increasingly on patterns 3–5, in particular on the rapid shift in much of the world's low- and moderate-income countries from the stage of receding famine to nutrition-related noncommunicable disease (NR-NCD). Figure 5.1 presents this focus. The concern on this period is so great that the term "Nutrition Transition" is synonymous for many with this shift from Pattern 3 to 4.

5.2 What Are Some Critical Dimensions?

5.2.1 Biological Mismatch

Many scholars, including this author, have discussed one or more ways that our evolved biology is mismatched with modern food and drink and the technologies of marketing and distribution [14, 15]. This perspective in many ways argues that obesity is a selective adaptation. However, it is the components that I note below that are critical to energy imbalance. There are four major domains for this mismatch. They very much related to the ways modern technology clash with our evolved biology to profoundly affect overall energy imbalance.

Our evolved biology	Modern technology
• Sweet preferences • Thirst vs. hunger/satiety mechanisms are not linked • Fatty food preference • Desire to reduce exertion	• Cheap caloric sweeteners, use of sugars in food processing • Caloric beverage revolution • Edible oil revolution, cheap edible oils • Technology in all phases of movement and exertion

The area most discussed recently is the issue of beverage compensation. Mattes and others have led a revolution in our understanding of the way the body deals differentially with food and beverages of any composition [16–20]. The history of beverages is such that only recently have we shifted from water to these caloric beverages [21]. We do not fully understand why we do not compensate for sugar-sweetened and any other caloric beverage by reducing food intake; however, this has been shown. Coupled with modern production and marketing of sugar-sweetened beverages, juices, and other equivalent high-calorie fatty and/or sweet drinks, a large shift in consumption of these beverages is occurring globally. The subject of beverages is examined in Chap. 16 by Wilson.

The other areas are documented by scholars in the field and are not addressed in this chapter [3, 4, 14, 15, 22–27].

5.2.2   The Speed of Change Is Greater Today

Is there anything about the great rapidity of change in diet, activity, and body composition that matters? What does the high prevalence of the undernutrition and overweight combination in the same household mean in this context [28]? While there is no study that clearly explores these points, available data would lead us to believe that the rates of change in diet, activity, and obesity today in the developing world are far beyond that experienced earlier in Europe and the United States [29]. In a very short time many low- and middle-income countries have attained rates of overweight and obesity equal to or greater than that of the US and Western Europe [30, 31].

The pace of the rapid nutrition transition shifts in diet and activity patterns from the period, termed the receding famine pattern, to one dominated by NR-NCDs seems to be accelerating in the lower- and middle-income transitional countries. I use nutrition rather than diet so that the term NR-NCD incorporates the effects of diet, physical activity, and body composition rather than solely focusing on dietary patterns and their effects. This is based partially on information that seems to indicate that the prevalence of obesity and a number of NR-NCDs are increasing far faster in the lower- and middle-income world than it has in the West [30, 31]. Another element is that the rapid changes in urban populations are much greater than those experienced a century or less ago in the West; yet another element is the shift in occupation structure and the rapid introduction of the modern mass media. Underlying such changes is a general concern for rapid globalization as the root cause.

Clearly, there are quantitative and qualitative dimensions to these changes. From the quantitative perspective, changes toward a high-density diet, reduced-complex carbohydrates, and other important elements, and inactivity may be proceeding faster than in the past. Similarly, the shift from labor-intensive occupations and leisure activities toward more capital-intensive, less strenuous work and leisure is also occurring faster. On the other hand, the qualitative dimensions are the complex ways our overall eating pattern (the sources and types of foods) and activity pattern interrelate. Social and economic stresses people face and feel as these changes occur might also be included.

At the start of the new millennium, scholars often feel as if the pace and complexity of life, reflected in all aspects of work and play, are increasing exponentially; there are also unanticipated developments, new technologies, and the impact of a very modern, high-powered communications system. It is this sense of rapid change that makes it so important to understand what is happening and anticipate the way changes in patterns of diet, activity, and body composition are occurring. While the penetration and influence of modern communications, technology, and economic systems—related to what is termed globalization—have been a dominant theme of the last few decades, there seem to be some unique issues that have led to a rapid increase of "globalization and its impact."

Placing the blame on globalization is, on the one hand, focusing on broad and vaguely measured sets of forces; this ignores the need to be focused and specific which would allow us to develop potentially viable policy options. It is difficult to measure each element of this globalization equation and its impact. These processes certainly have been expanded, as indicated by enhanced free trade, a push toward reduction of trade barriers in the developing world, and the increasing penetration of international corporations into the commerce of each country (measured by share of GNP or manufacturing). Similarly, other economic issues related to enhanced value given to market forces and international capital markets are important. Equally, the increasing access to Western media, the removal of communication barriers enhanced by the World Wide Web, cable TV, mobile telephone systems, etc. is important. The accelerated introduction of Western technology into the manufacturing, basic sectors of agriculture, mining, and services is also a key element.

Another way to consider the types of changes the developing world is facing is to consider an urban squatter's life and a rural villager's life in China 25 year ago and today. During the 1970s, food supply concerns still existed, there was no television, limited bus and mass transportation, little food trade, minimal availability of processed food, and most rural and urban occupations were very labor-intensive. Today, work and life activities have changed: small gas-powered tractors are available, modern industrial techniques are multiplying, offices are quite automated, soft drinks and many processed foods are found everywhere, TVs are found in about 89% of households (at least a fifth of whom are linked to Hong Kong Star and Western advertising and programming), and mass transit has become heavily used (to the point where children younger than 12 are not allowed even to ride bicycles). Multiply such changes by similar ones occurring in much of Asia, North Africa, the Middle East, Latin America, and many areas (particularly cities) in sub-Saharan

Africa and it is evident that the shift from a subsistence economy to a modern, industrialized one occurred in a mere 10–20 year, whereas in Europe and other industrialized high-income societies this occurred over many decades or centuries.

To truly measure and examine these issues, we would need to compare changes in the 1980–2000 period, for countries that are low and middle income, with changes that occurred a half century earlier for the developing world. However, data on diet and activity patterns are not available and there are only minimal data on NR-NCDs and obesity.

The elements of the nutrition transition that we know to be linked with NR-NCDs are:

• Obesity
• Adverse dietary changes (e.g., shifts in the structure of diet toward a greater content of higher fat and added sugar foods, and away from whole grains, nuts, legumes, and fruit and vegetable intake, and with an overall reduction in fiber intake, greater saturated fat intake, and greater energy density)
• Reduced physical activity in work and leisure

We focus on these first, and then a few select underlying factors. The causes of these elements of the nutrition transition are not as well understood as are the trends in each of them. In fact, there are few studies attempting to study the causes of such changes and there are only a few data sets that are equipped to allow such crucial policy analyses to be undertaken.

5.2.3   Edible Oil

The edible vegetable oils story is particularly important as its effects have been quite profound. Until the decade following World War II, the majority of fats available for human consumption were animal fats obtained from milk, butter, and meat. Subsequently, a revolution in the production and processing of oilseed-based fats occurred. Principal vegetable oils include soybean, sunflower, rapeseed (as consumed in Asia vs. the genetically modified version consumed in North America called canola oil), palm, and peanut oil. Technological breakthroughs in the development of high-yield oilseeds and in the refining of high-quality vegetable oils greatly reduced the cost of baking and frying fats, margarine, butter-like spreads, salad oils, and cooking oils in relation to animal-based products [32]. Worldwide demand for vegetable fats was fueled by health concerns regarding the consumption of animal fats and cholesterol. Furthermore, a number of major economic and political initiatives led to the development of oil crops, not only in Europe and the United States, but also in South East Asia (palm oils), in Brazil, and in Argentina (soybean oils). The net effect was that from 1945 to 1965, there was almost a fourfold increase in the US production of vegetable oils, while animal fat production increased by only 11% [33].

In developing nations, one of the earliest shifts toward a higher-fat diet began with major increases in the domestic production and import of oilseeds and vegetable oils, rather than increased import of meat and milk. At this stage, vegetable oils contributed far more energy to the human food supply than meat or animal fats [34]. Global availability of most vegetable oils (i.e., soybean, sunflower, rapeseed, and palm, though not peanut oil) has approximately tripled from 1961 to 1990. Soybeans now account for the bulk of vegetable oil consumption worldwide. It is also important to note that many of these processed oils are not well regulated and some of the new edible oils are highly pathogenic [35].

5.2.4   Caloric Sweetener Revolution

Sugar is the world's predominant sweetener. It is not clear exactly when sugar became the world's principal sweetener—most likely in the seventeenth or eighteenth century, as the New World began producing large quantities of sugar at reduced prices [36, 37]. Sugar use has since been linked with industrialization, and with the proliferation of processed foods and beverages that have sugar added to them (e.g., tea, coffee, cocoa). We use the term caloric sweetener instead of added sugar as there are such a range of nonsugar products used today. High-fructose corn syrup is a prime example as it is the sweetener used in all US soft drinks. There are two major sugar crops: sugar beets and sugar cane. Sugar and syrups are also produced from the sap of certain species of maple trees, from sweet sorghum when cultivated explicitly for making syrup, and from sugar palm. Under the name sweeteners, the Food and Agriculture Organization (FAO) includes products used for sweetening that are either derived from sugar crops, cereals, fruits, milk, or produced by insects. This category includes a wide variety of monosaccharides (glucose and fructose) and disaccharides (sucrose and saccharose), which exist either in a crystallized state as sugar or in thick liquid form as syrups. Included in sweeteners are maple sugar and syrups, caramel, golden syrup, artificial and natural honey, maltose, glucose, isoglucose (also known as high-fructose corn syrup), other types of fructose, sugar confectionery, and lactose. In the last several decades, increasingly larger quantities of cereals (primarily maize) have been used to produce sweeteners derived from starch.

We provide detailed information elsewhere on the caloric sweetener revolution [38]. In 2000, there were 74 more kcal per capita per day of caloric sweetener consumed than in 1962. This represents an increase of 32% in the percentage of energy from caloric sweeteners and a 21% increase in the proportion of carbohydrates from refined caloric sweeteners.

Drewnowski and Popkin [39] also show the profound changes that occurred between the 1960s and the past decade, and also the large urbanization effect. This regression model predicts that rapid urbanization, usually associated with greater incomes and economic growth, can have independent effects on diet structure. In 1962 there was very little sweetener consumed by populations in lower-income countries. At lower-income levels, according to the regression model, urbanization can more than double the amount of sweeteners in the diet. The model confirms previous observations that people living in urban areas consume diets distinct from those of their rural counterparts [40, 41]. The potential impact of urbanization in flattening the income-sweetener relationship deserves further analysis; however, it is clear that the increased urbanization of lower-income nations is accelerating the shift to increased consumption of sweeteners and fats.

5.2.5   Animal Source Foods

The animal source foods (ASF) revolution refers to the increase in demand and production of meat, fish, and milk in low-income developing countries. Dr. Christopher Delgado from the International Food Policy Research Institute has studied this issue extensively in a number of seminal reports and papers [42]. Most of the world's growth in production and consumption of these foods is coming from the developing countries. Although cereal production is increasing, this is mainly used to feed animals that produce ASFs (in other words a derived demand for additional ASF).

The ASF revolution is driven by demand as ASF are substituted into starch-based diets. As income rises, there is a large demand for ASF in China, South Korea, and Morocco [43, 44]. Over time, we expect to see a larger share of household food expenditures on animal products. Red meat has what economists describe as possessing a very high-income elasticity; this means that price has a strong effect on sales. Price elasticity for cereals, on the other hand, tends to be low. Price changes impacts low-income groups faster than higher-income groups. Since 1970 relative prices of food have dropped considerably, most dramatically for beef. Due to market saturation and technological changes that increase productivity, the *increase in demand for* ASF is projected to peak by 2020. As relative commodity prices decrease and income increases, people tend to both increase the diversity of their diet and to shift into higher priced foods that are more highly processed. While average income growth explains overall growth in total food expenditures, the greater increase in ASF demand in developing countries is explained by urbanization and population growth.

As Delgado has shown, 31% of food fish is from aquaculture, but 60% of world aquaculture is in China, as well as 40% of the world's pigs [42]. China is trying to promote beef consumption because beef requires less grain than pork. Projections to 2020 include huge increases in China's food fish consumption, as well as 31% of world beef production. India is the largest producer of milk, although 42% of the population still claim to not eat any ASF, and only the Muslim segment of the population consumes beef. India already has food safety systems in place, especially in urban areas. Large amounts of sweets containing mostly milk and sugar are processed and distributed in India.

Whether it is a curse or a blessing depends on policy and technology. The ASF revolution is driven by demand, substituting into starch-based diets. In 2020 developing countries will produce 63% of meat and 50% of milk. It is a global food activity, transforming the grain markets for animal feed. It also leads to resource degradation, rapid increase in feed grain imports, rapid concentration of production and consumption, and promotion of social change.

5.3   Increased Sedentarianism Is Occurring Concurrently

There are several important changes in physical activity taking place. These include reductions in activity at work, a shift from active to inactive transport (e.g., from biking or walking to taking a bus), and increased sedentariness in leisure. This topic is examined briefly as it is not central to this chapter.

The first dominant change has been a shift toward the service sector and away from high-energy expenditure activities, such as farming, mining, and forestry. We have described this large effect elsewhere [45].

A second change is in the level of energy expenditure in the same occupation. From the 1980s onwards in urban areas there have been quite pervasive shifts in the technology for performing work. This appears to be the most important factor affecting physical activity [46]. The proportion of urban adults (male and female) working in occupations where they participate in vigorous activity patterns has decreased, and, conversely, increased where the activity pattern is light. In rural areas, however, there has been a shift toward increased physical activity linked to holding multiple jobs and more intensive effort. For rural women there is a shift toward a larger proportion engaged in more energy-intensive work, but there are also some for whom the shift is toward a lighter work effort. In contrast, for rural men there is a small decrease in the proportion engaged in light work effort. We have also shown how shifts in activity at an occupation represent a significant determinant of the increase in obesity in China [47].

Other major changes relate to mode of transportation and activity patterns during leisure hours. Fourteen percent of households in China acquired a motorized vehicle between 1989 and 1997. In one study in that country we showed that the odds of being obese were 80% higher ($p < 0.05$) for men and women in households who owned a motorized vehicle compared to those who did not [48]. Compared to those whose vehicle ownership did not change, men who acquired a vehicle experienced a 1.8 kg greater weight gain ($p < 0.05$) and were twice as likely to become obese.

The final pattern is a type of leisure activity. Television ownership has skyrocketed in China, leading to greater inactivity during leisure time [49]. The consequences of this shift, not only in TV ownership but also in usage patterns for TV watching, remain to be fully studied.

5.4   Is the Biology Different?

There are a number of different ways in which this question could be answered in the affirmative. One is if the body composition and other unmeasured race-ethnic factors affect susceptibility to NR-NCDs. Another might be if previous disease patterns (e.g., the presence of malaria or other tropical diseases) led to disease patterns that predisposed the population to certain problems. There is limited but strong evidence that there are potentially important biological differences between those from the developing world and those from Europe or European genetic backgrounds.

Do we need different body mass index (BMI) cut points for subpopulation groups, and is this based on biological differences or just adiposity measurement that is missed with the use of BMI? Both are probably true. There is evidence from a range of body composition and BMI-disease studies that would lead us to believe that Asians, Africans, and Latin Americans are more likely than whites in the United States and Europe to have larger amounts of body fat and central fat for the same BMI; in addition, the former groups have a higher likelihood of experiencing cardiovascular disease (CVD) at lower BMI levels.

Another pathway of possible importance is related to the role of previous health problems. Examples of this are malnutrition causing virus mutation, parasitic infections affecting long-term absorption patterns, or a parasite linked with an unknown genotype—comparable to sickle cell anemia and its evolutionary linkage with malaria. We have no basis for speculation about this potential pathway and no real documentation of its impact.

However, the final pathway—the effect of fetal and infant insults on subsequent metabolic function—is one that appears to be a critical area. If the rapid shifts toward positive energy imbalance are occurring concurrently with previously high levels of low birth-weight in a population, then this becomes a very critical issue to understand. In the developing world, where intrauterine malnutrition rates are high and a high prevalence of nutrition insults during infancy exist, the work of Barker and many others portends important potential effects on the prevalence of NR-NCDs in coming decades. There an emerging consensus that fetal insults, in particular with regard to thin, low birth-weight infants who subsequently face a shift in the stage of the transition and become overweight, are linked with increased risk of the NR-NCDs. In addition, infancy may be a period of high vulnerability in that poor growth or stunting during that time period may have comparable effects to fetal insults. Termed fetal origins or infant programming, this literature suggests that a rapid shift toward energy imbalance, preceded by high levels of thin babies and infant stunting, will have important long-term effects in increasing the probability that the subsequent energy imbalance leads to CVD and various associated conditions. This is discussed in more detail in Chap. 4 by Barker.

5.5 Can We Turn Back the Clock or Modify the Adverse Dynamics? Program and Policy Issues

The most interesting development over the past decade has been the emergence of a systematic and seemingly quite successful effort in the United Kingdom to not only prevent more obesity but also to reduce the prevalence of obesity across all age groups while enhancing dietary and activity patterns. The UK began with the Foresight Obesity Project, the goal of which was to produce a sustainable response to obesity in the UK over a 40-year period [50–52]. This systematic government effort began with quantitative modeling of the growth of obesity, the economic effect of this, and the effect on the national health system [53]. It then created a fairly complex systems-map of the causes of energy imbalance which laid out societal as well as individual causes of food consumption and activity. Out of this came a broad examination of all potential leverage points with weighting of the causal linkages.

The project organizers established a goal of being the first major nation to reverse the rising tide of obesity. They also focused attention and action on the environmental causes. Actions taken provide some sense of the process and, significantly, funding that went with all activities and actions:

- Vending machines banned in schools, and drinking water promoted; schools record annual assessments of all activities and changes
- Advertisers banned from advertising to children and banned from advertising unhealthy foods in the media
- Children aged 6–8 required to receive weekly cooking classes to learn about food and its preparation and handling
- Government accountability created across all agencies
- Set of initiatives and coalitions created, focused on Change4Life, Coalition for Better Health, and the Healthy Food Code (e.g., traffic light system)
- Power to restrict fast food restaurants near schools and parks
- Project to provide fresh fruit and vegetables to small stores in deprived areas
- Putting in systematic surveillance and monitoring

There were many other actions related to children. These began with midwife and nursery training, school workforce and nurse training, parent support advisers, training of school caterers, an array of people linked to supporting youth activity in and out of the schools, and many private sector activities. The organizers continue to study causes and solutions and remain most active in addressing obesity across the life cycle. They did not see this as one action but a long series of actions, evaluations, studies, and new actions.

In the developing world, initial steps were taken by Brazil a decade ago but these have essentially not been followed with systematic meaningful change. Matsudo et al. [54] presented some of the key elements used to launch a mass promotion of physical activity in São Paulo, Brazil. Termed the Agita São Paulo, it began as a multilevel, community-wide intervention designed to increase the knowledge about benefits and the level of physical activity in a mega-population of 34 million of São Paulo State. It is slowly being expanded into a global effort; however, the documentation of changes related to this is limited. This Agita movement began after the reduction in obesity noted in the paper by Monteiro et al. [55]. In Brazil, a more coordinated and systematic effort created a number of important legislative and regulatory policies, revised one very large school feeding program, and has done much to focus on the national policy environment. At the same time efforts are underway at mass communication directly via the mass media and through schools and food stores.

More recently, Mexico has become one of the few large countries to have systematically begun to address obesity. Research on beverage patterns and trends in Mexico were used to guide the Mexican Beverage Guidance Panel—established by the Minister of Health [56]. Based on this and related research, the Minister of Health and the Mexican Beverage Guidance Panel shifted from whole milk to skim milk and are working now to reduce overall consumption of sugar added to juice, soda, flavored water, and water plus juice. As part of their recommendations, they are actively considering taxation as one option to reduce intake of whole milk and sugar-sweetened beverages [56]. The government has already shifted milk provided under various welfare programs serving about 20 million people from whole to 1.5%, is working to do the same for the schools, and is considering other activities along with taxation to reduce energy intake from beverages in Mexico. A second component is front of the package labeling. There are many other components of their obesity prevention agenda that are slowly being initiated.

Earlier work in South Korea also showed how that country used a variety of educational and other nutrition programs to promote the traditional high-vegetable and low-fat diet [57, 58].

5.6   Summary

This chapter explores shifts in nutrition transition from the period termed the receding famine pattern to one dominated by NR-NCDs. It examines the speed of these changes, summarizes dietary and physical activity changes, and provides some sense of the health effects and economic costs. The focus is on the lower- and middle-income countries of Asia, Africa, the Middle East, and Latin America. The article shows that changes are occurring at great speed and at earlier stages of countries' economic and social development. The burden of disease from NR-NCDs is shifting towards the poor and the costs are also becoming greater than those for undernutrition. Policy options are identified.

References

1. Truswell AS. Diet and nutrition of hunter-gatherers. Health and diseases in tribal societies. Ciba Foundation Symposium 149. Elsevier: Amsterdam; 1977.
2. Harris D. The prehistory of human subsistence: a speculative outline. In: Walcher DN, Kretchmer N, editors. Food, nutrition and evolution: food as an environmental factor in the genesis of human variability. New York: Masson; 1981.
3. Eaton SB, Shostak M, Konner M. The Paleolithic prescription: a program of diet and exercise and a design for living. New York: Harper & Row; 1988.
4. Eaton SB, Konner M. Paleolithic nutrition: a consideration on its nature and current implications. N Engl J Med. 1985;312:283–9.
5. Vargas LA. Old and new transitions and nutrition in Mexico. In: Swedlund AC, Armelagos GJ, editors. Disease in populations in transition: anthropological and epidemiological perspectives. New York: Bergin and Garvey; 1990.
6. Gordon KD. Evolutionary perspectives on human diet. In: Johnston FE, editor. Nutritional anthropology. New York: Liss; 1987.

7. Newman LF, Crossgrove W, Kates R, Matthews R, Millman S. Hunger in history: food shortage, poverty, and deprivation. Cambridge, MA: Blackwell; 1990.
8. Omran AR. The epidemiologic transition. A theory of the epidemiology of population change. Milbank Mem Fund Q. 1971;49:509–38.
9. Milio N. Nutrition policy for food-rich countries: a strategic analysis. Baltimore: The Johns Hopkins University Press; 1990.
10. Popkin BM, Haines PS, Reidy KC. Food consumption trends of US women: patterns and determinants between 1977 and 1985. Am J Clin Nutr. 1989;49:1307–19.
11. Popkin BM, Haines PS, Patterson RE. Dietary changes in older Americans, 1977–1987. Am J Clin Nutr. 1992;55:823–30.
12. Manton KG, Soldo BJ. Dynamics of health changes in the oldest old: new perspectives and evidence. Milbank Mem Fund Q Health Soc. 1985;63:206–85.
13. Crimmins EM, Saito Y, Ingegneri D. Changes in life expectancy and disability-free life expectancy in the United States. Pop Devel Rev. 1989;15:235–67.
14. Gluckman PD, Hanson MA. Living with the past: evolution, development, and patterns of disease. Science. 2004;305:1733–6.
15. Trevantham W, Smith EO, McKenna JJ, editors. Evolutionary medicine. New York: Oxford University Press; 1999.
16. Mourao DM, Bressan J, Campbell WW, Mattes RD. Effects of food form on appetite and energy intake in lean and obese young adults. Int J Obes (Lond). 2007;31:1688–95.
17. Mattes R. Fluid calories and energy balance: the good, the bad, and the uncertain. Physiol Behav. 2006;89:66–70.
18. DiMeglio DP, Mattes RD. Liquid versus solid carbohydrate: effects on food intake and body weight. Int J Obes Relat Metab Disord. 2000;24:794–800.
19. Flood J, Roe L, Rolls B. The effect of increased beverage portion size on energy intake at a meal. J Am Diet Assoc. 2006;106:1984–90.
20. DellaValle DM, Roe LS, Rolls BJ. Does the consumption of caloric and non-caloric beverages with a meal affect energy intake? Appetite. 2005;44:187–93.
21. Wolf A, Bray GA, Popkin BM. A short history of beverages and how our body treats them. Obes Rev. 2008;9:151–64.
22. Power ML, Schulkin J. The evolution of obesity. Baltimore: The Johns Hopkins University Press; 2009.
23. Eaton SB, Eaton III SB, Konner MJ. Paleolithic nutrition revisited: a twelve-year retrospective on its nature and implications. Eur J Clin Nutr. 1997;51:207–16.
24. Jonsson T, Ahren B, Pacini G, Sundler F, Wierup N, Steen S. A Paleolithic diet confers higher insulin sensitivity, lower C-reactive protein and lower blood pressure than a cereal-based diet in domestic pigs. Nutr Metab (Lond). 2006;3:39.
25. Cordain L, Eaton SB, Miller JB, Mann N, Hill K. The paradoxical nature of hunter-gatherer diets: meat-based, yet non-atherogenic. Eur J Clin Nutr. 2002;56:S42–52.
26. Cordain L, Eaton SB, Sebastian A, et al. Origins and evolution of the Western diet: health implications for the 21st century. Am J Clin Nutr. 2005;81:341–54.
27. Cordain L, Miller JB, Eaton SB, Mann N, Holt SH, Speth JD. Plant-animal subsistence ratios and macronutrient energy estimations in worldwide hunter-gatherer diets. Am J Clin Nutr. 2000;71:682–92.
28. Doak CM, Adair LS, Bentley M, Monteiro C, Popkin BM. The dual burden household and the nutrition transition paradox. Int J Obes. 2005;29:129–36.
29. Popkin BM. The shift in stages of the nutrition transition in the developing world differs from past experiences! Public Health Nutr. 2002;5:205–14.
30. Popkin BM, Adair LS, Ng SW. Global nutrition transition and the pandemic of obesity in developing countries. Nutr Rev. 2012;70:3–21.
31. Jones-Smith JC, Gordon-Larsen P, Siddiqi A, Popkin BM. Cross-national comparisons of time trends in overweight inequality by socioeconomic status among women using repeated cross-sectional surveys from 37 developing countries, 1989–2007. Am J Epidemiol. 2011;173:667–75.

32. Williams G. Development and future direction of the world soybean market. Q J Int Agr. 1984;23:319–37.
33. U.S. Department of Agriculture. 1996 U.S. fats and oils statistics. U.S. Department of Agriculture Statistical Bulletin No. 376. Washington, DC: ERS; 1966. p. 1909–1965.
34. Morgan N. World vegetable oil consumption expands and diversifies. Food Rev. 1993;16: 26–30.
35. Wallingford JC, Yuhas R, Du S, Zhai F, Popkin BM. Fatty acids in Chinese edible oils: evidence for unexpected impact in changing diet. Food Nutr Bull. 2004;25:330–6.
36. Galloway J. Sugar. In: Kiple KF, Ornelas KC, editors. The Cambridge world history of food. New York: Cambridge University Press; 2000. p. 437–49.
37. Mintz SW. Time, sugar and sweetness. Marx Perspect. 1979;2:56–73.
38. Popkin BM, Nielsen SJ. The sweetening of the world's diet. Obes Res. 2003;11:1325–32.
39. Drewnowski A, Popkin BM. The nutrition transition: new trends in the global diet. Nutr Rev. 1997;55:31–43.
40. Popkin B, Bisgrove EZ. Urbanization and nutrition in low-income countries. Food Nutr Bull. 1988;10:3–23.
41. Solomons NW, Gross R. Urban nutrition in developing countries. Nutr Rev. 1995;53:90–5.
42. Delgado CL. Rising consumption of meat and milk in developing countries has created a new food revolution. J Nutr. 2003;133:3907S–10.
43. Popkin BM, Du S. Dynamics of the nutrition transition toward the animal foods sector in China and its implications: a worried perspective. J Nutr. 2003;133:3898S–906.
44. Guo X, Mroz TA, Popkin BM, Zhai F. Structural changes in the impact of income on food consumption in China, 1989–93. Econ Dev Cult Change. 2000;48:737–60.
45. Popkin BM. Urbanization, lifestyle changes and the nutrition transition. World Dev. 1999;27:1905–16.
46. Popkin BM. The world is fat–the fads, trends, policies, and products that are fattening the human race. New York: Avery-Penguin Group; 2008.
47. Bell AC, Ge K, Popkin BM. Weight gain and its predictors in Chinese adults. Int J Obes Relat Metab Disord. 2001;25:1079–86.
48. Bell AC, Ge K, Popkin BM. The road to obesity or the path to prevention: motorized transportation and obesity in China. Obes Res. 2002;10:277–83.
49. Du S, Lu B, Zhai F, Popkin BM. The nutrition transition in China: a new stage of the Chinese diet. In: Caballero B, Popkin B, editors. The nutrition transition: diet and disease in the developing world. London: Academic; 2002. p. 205–22.
50. Kopelman P. Symposium 1: overnutrition: consequences and solutions Foresight Report: the obesity challenge ahead. Proc Nutr Soc. 2010;69:1–6.
51. McPherson K, Marsh T, Brown M. Foresight report on obesity. Lancet. 2007;370:1755; author reply 1755.
52. King D. Foresight report on obesity. Lancet. 2007;370:1754; author reply 1755.
53. Foresight. Tackling obesities: future choices-project report. 2nd ed. Government Office for Science, London. 2007. http://www.foresight.gov.uk/OurWork/ActiveProjects/Obesity/Obesity. asp. Accessed 3 June 2010.
54. Matsudo V, Matsudo S, Andrade D, et al. Promotion of physical activity in a developing country: the Agita Sao Paulo experience. Public Health Nutr. 2002;5:253–61.
55. Monteiro CA, D'A Benicio MH, Conde WL, Popkin BM. Shifting obesity trends in Brazil. Eur J Clin Nutr. 2000;54:342–6.
56. Rivera JA, Muñoz-Hernández O, Rosas-Peralta M, Aguilar-Salinas CA, Popkin BM, Willett WC. Consumo de bebidas para una vida saludable: Recomendaciones para la población (beverage consumption for a healthy life: Recommendations for the mexican population). Salud Publica Mex. 2008;50:173–95.
57. Kim S, Moon S, Popkin BM. The nutrition transition in South Korea. Am J Clin Nutr. 2000; 71:44–53.
58. Lee MJ, Popkin BM, Kim S. The unique aspects of the nutrition transition in South Korea: the retention of healthful elements in their traditional diet. Public Health Nutr. 2002;5:197–203.

Chapter 6
Medical Nutrition Therapy for Diabetes: Prioritizing Recommendations Based on Evidence

Marion J. Franz

Keywords Diabetes nutrition therapy • Evidence-based diabetes nutrition recommendations • Glycemic index • Carbohydrate counting • Diabetes medications • Diabetes monitoring

Key Points

- Long-term clinical trials have documented the importance of metabolic control of glucose, lipids, and blood pressure in persons with diabetes. Although new oral medications and insulin preparations are now available, nutrition therapy continues to be essential if desired medical goals are to be achieved. Successful nutrition therapy is an ongoing process.
- Clinical trials and outcomes studies have documented the effectiveness of nutrition therapy. In general, nutrition therapy provided by registered dietitians can lower hemoglobin A1c (A1C) by approximately 1–2%, depending on the type and duration of diabetes, LDL cholesterol by 19–25 mg/dL (0.46–0.65 mmol/L), and blood pressure by 5 mmHg.
- For persons with type 1 diabetes, the first priority is to identify a food/meal plan that can be used to integrate an insulin regimen into the person's lifestyle. Physiological insulin regimens consisting of basal and bolus insulins or insulin pumps provide flexibility in timing and frequency of meals, amount of carbohydrate eaten, and timing of physical activity. Insulin-to-carbohydrate ratios are used to adjust the bolus insulin dose needed to cover the planned carbohydrate intake.
- Type 2 diabetes is a progressive disease beginning with insulin resistance. However, glucose levels remain normal if adequate insulin is available; it is not until insulin deficiency (β-cell failure) develops that hyperglycemia occurs.

M.J. Franz, MS, RD, CDE(✉)
Nutrition Concepts by Franz, Inc., 6635 Limerick Drive, Minneapolis, MN 55439, USA
e-mail: MarionFranz@aol.com

N.J. Temple et al. (eds.), *Nutritional Health: Strategies for Disease Prevention*,
Nutrition and Health, DOI 10.1007/978-1-61779-894-8_6,
© Springer Science+Business Media, LLC 2012

Therefore, for persons with type 2 diabetes, nutrition therapy progresses from improving insulin resistance and preventing diabetes by modest weight loss and regular physical activity to contributing to improved metabolic control by carbohydrate counting and reduced energy intake.

• Monitoring of glucose, lipids, and blood pressure is essential to assess the outcomes of nutrition therapy and/or to determine if additional changes in nutrition therapy or medications are necessary.

6.1 Introduction

Diabetes mellitus is group of diseases characterized by elevated glucose concentrations resulting from insulin deficiency. Without sufficient insulin, hyperglycemia occurs causing both the acute and long-term complications of diabetes. A primary function of insulin, a hormone produced by the β-cells of the pancreas, is the use or storage of body fuels. In diabetes abnormalities in the metabolism of carbohydrate, protein, and fat are present.

Worldwide, the number of persons with diabetes and those who are at risk for diabetes is increasing at an alarming rate. According to the International Diabetes Federation (IDF), 285 million people worldwide have diabetes and the IDF predicts the total number of people with diabetes will exceed 435 million if the current rate of growth continues. India has 50.8 million people living with diabetes and close behind is China with 43.2 million. The IDF reported that diabetes costs the world economy at least $376 billion in 2010—11.6% of the total health-care expenditure and projects this number to exceed $490 billion by 2030 [1].

In 2011, approximately 26 million people in the United States—about 8.3% of the population—are reported to have diabetes, and an additional 79 million adults—about 35% of US adults—are estimated to have prediabetes, a risk factor for type 2 diabetes, heart disease, and stroke [2]. The burden of diabetes is extensive, including serious microvascular and macrovascular complications, quality-of-life impairments, and escalating costs of care.

6.2 Types of Diabetes

6.2.1 Type 1 Diabetes

Immune-mediated type 1 diabetes accounts for 5–10% of all diagnosed cases of diabetes. The primary defect is a cellular-mediated autoimmune destruction of pancreatic β-cells, usually leading to absolute insulin deficiency and resulting in hyperglycemia, polyuria, polydipsia, weight loss, dehydration, electrolyte disturbance, and ketoacidosis. The capacity of a healthy pancreas to secrete insulin is far in excess of what is needed normally; therefore, the clinical onset of diabetes may be preceded by an extensive asymptomatic period of months to years, during which β-cells are undergoing gradual destruction. Persons with type 1 diabetes are dependent on exogenous insulin to prevent ketoacidosis and death. Although type 1 diabetes can occur at any age, even in the eighth and ninth decades of life, most cases are diagnosed in people younger than 30 years, with peak incidence at around ages 10–12 years in girls and 12–14 years in boys.

The etiology involves a genetic disposition and an autoimmune destruction of the islet β-cells that produce insulin. At diagnosis, 85–90% of persons with type 1 diabetes have one or more circulating autoantibodies. Antibodies identified as contributing to the destruction of β-cells are as follows:

1. Islet cell autoantibodies (ICAs).
2. Insulin autoantibodies (IAAs), which may occur in persons who have never received insulin therapy.
3. Autoantibodies to glutamic acid decarboxylase autoantibodies (GAD$_{65}$), a protein on the surface of β-cells which appears to provoke an attack by the T cells (killer T lymphocytes) and to destroy the β-cells.
4. Autoantibodies to tyrosine phosphatases IA-2 and IA-2β.

The disease also has strong human leukocyte antigen (*HLA*) associations, with linkage to the *DQA* and *DQB* genes, and influenced by the *DRB* genes. These *HLA-DR/DQ* alleles can be either predisposing or protective [3].

Frequently, after diagnosis and the correction of hyperglycemia, metabolic acidosis, and ketoacidosis, there is a recovery of endogenous insulin secretion. During this "honeymoon phase," exogenous insulin requirements decrease dramatically. However, the need for exogenous insulin is inevitable, and within 8–10 years after clinical onset, β-cell loss is complete and insulin deficiency is absolute. These patients are also prone to other autoimmune disorders such as Graves' disease, Hashimoto's thyroiditis, Addison's disease, celiac sprue, and pernicious anemia.

Latent autoimmune diabetes of adults (LADA) describes a minority of adult-onset persons with diabetes. Typically, patients are positive for GAD antibodies, >35 years, nonobese, and present without ketosis and weight loss. Although many maintain good glycemic control for several years with sulfonylureas, these persons become "insulin-dependent" more rapidly than antibody-negative persons with type 2 diabetes [4].

6.2.2   Type 2 Diabetes

Type 2 diabetes accounts for 90–95% of all diagnosed cases of diabetes and is a progressive disease that is often present long before it is diagnosed. Persons may or may not experience the classic symptoms of uncontrolled diabetes and they are not prone to develop ketoacidosis. The progressive loss of β-cell secretory function means that persons with type 2 diabetes will require additional medication(s) over time to maintain the same level of glycemic control and eventually exogenous insulin will be required [5]. Insulin is also required sooner during periods of stress-induced hyperglycemia, such as during illness or surgery.

Risk factors include a strong genetic predisposition and the risk increases with age, obesity, and physical inactivity. Obesity alone causes some degree of insulin resistance. Even persons who develop diabetes and are not obese by traditional weight criteria usually have an increased percentage of intraabdominal body fat [3].

Type 2 diabetes results from a combination of insulin resistance and insulin deficiency (β-cell failure). Endogenous insulin levels may be normal, depressed, or elevated; but they are inadequate to overcome the concomitant insulin resistance and as a result hyperglycemia occurs. Insulin resistance is first demonstrated in target tissues, mainly muscle, liver, and adipose cells. Initially there is a compensatory increase in insulin secretion, which maintains glucose concentrations in the normal or prediabetic range, but in many persons their pancreas is unable to continue to produce adequate insulin, hyperglycemia occurs, and the diagnosis of diabetes is made. Therefore, hyperglycemia does not develop until there is inadequate insulin available (insulin deficiency) to control elevated blood glucose levels.

Hyperglycemia is first exhibited as an elevation of postprandial glucose due to insulin resistance at the cellular level. As insulin secretion decreases, hepatic glucose production increases causing elevations in fasting glucose levels. The insulin response is also inadequate in suppressing α-cell glucagon secretion, resulting in glucagon hypersecretion and increased hepatic glucose production. Compounding the problem is the deleterious effect of hyperglycemia itself—glucotoxicity—on both insulin sensitivity and insulin secretion, hence the importance of achieving near-euglycemia in persons with type 2 diabetes.

Insulin resistance is also demonstrated in adipocytes, leading to lipolysis and an elevation in circulating free fatty acids. Increased free fatty acids further cause a decrease in insulin sensitivity at the cellular level, impair insulin secretion by the pancreas, and augment hepatic glucose production (lipotoxicity). To slow the progressive nature of type 2 diabetes requires that both conditions (hyperglycemia and lipotoxicity) be corrected [6].

6.2.3 Gestational Diabetes Mellitus

For many years gestational diabetes mellitus (GDM) was defined as any degree of glucose intolerance with onset or first recognition during pregnancy. However, because of the increase in type 2 diabetes in women of childbearing age, the number of women with undiagnosed type 2 diabetes has increased. Therefore, based on recommendations from the International Association of Diabetes in Pregnancy Study Group, the American Diabetes Association recommended that high-risk women be screened for diabetes at their initial prenatal visit using the criteria in Table 6.1 and if they meet diagnostic criteria receive a diagnosis of overt, not gestational, diabetes [3, 7]. Pregnant women not known to have diabetes should be

Table 6.1 Criteria for the diagnosis of diabetes

Hemoglobin A1c (A1C) ≥6.5%. The test should be performed in a laboratory using a method that is certified and standardized to the Diabetes Control and Complications Trial (DCCT) assay[a]

or

Fasting plasma glucose (FPG) ≥126 mg/dL (7.0 mmol/L). Fasting is defined as no caloric intake for at least 8 h[a]

or

2-h Plasma glucose (PG) ≥200 mg/dL (11.1 mmol/L) during an oral glucose tolerance test (OGTT). This test uses the equivalent of 75 g anhydrous glucose dissolved in water[a]

or

Classic symptoms of diabetes and casual PG ≥200 mg/dL (11.1 mmol/L)

Adapted from ref. [7]

[a]In the absence of unequivocal hyperglycemia, results should be confirmed with repeat testing

Table 6.2 Screening for and diagnosis of gestational diabetes mellitus (GDM)

In women not previously diagnosed with overt diabetes, screen at 24–28 weeks of gestation using a 75-g OGTT with plasma glucose testing fasting and at 1 and 2 h

- OGTT performed in the morning after an overnight fast of at least 8 h

Diagnosis of GDM is made when any of the following plasma glucose values are exceeded

- Fasting ≥92 mg/dL (5.1 mmol/L)
- 1 h ≥180 mg/dL (10.0 mmol/L)
- 2 h ≥153 mg/dL (8.5 mmol/L)

From ref. [7]

screened for GDM between 24 and 28 weeks of gestation, using a 75-g 2-h oral glucose tolerance test (OGTT) and the diagnosis criteria in Table 6.2. It is during the second and third trimester that insulin-antagonist hormones increase and insulin resistance occurs. However, the American College of Obstetricians and Gynecologists (ACOG) continue to support the previous two-step approach for screening and diagnosing diabetes [8].

Using the new diagnostic criteria for gestational diabetes will increase the proportion of women with a diagnosis of GDM. An international multicenter study found that 18% of pregnancies were affected by gestational diabetes [2]. Women with GDM should be screened for persistent diabetes 6–12 weeks postpartum and should have lifelong screening for diabetes or prediabetes at least every 3 years [7]. They have a 35–60% chance of developing diabetes in the next 10–20 years [2].

6.2.4 Other Types of Diabetes

Diabetes associated with or secondary to other conditions may occur in about 2% of all disorders comprising the syndrome of diabetes. These conditions include genetic defects of the β-cell, genetic defects in insulin action, other genetic syndromes, pancreatic disease, hormonal disease, and drug- or chemical-induced hyperglycemia [7].

Table 6.3 Categories of increased risk for diabetes (prediabetes)[a]

A1C 5.7–6.4%
or
FPG 100–125 mg/dL (5.6–6.9 mmol/L): impaired fasting glucose (IFG)
or
2-h PG in the 75-g OGTT 140–199 mg/dL (7.8–11.0 mmol/L): impaired glucose tolerance (IGT)

From ref. [7]

[a]For all three tests, risk is continuous, extending below the lower limit of the range and becoming disproportionately greater at higher ends of the range

6.2.5 Prediabetes

Individuals with impaired glucose tolerance (IGT), impaired fasting glucose (IFG), and/or hemoglobin A1c (A1C) levels between 5.7 and 6.4% have prediabetes and have an increased risk for diabetes. They are also at high risk for future cardiovascular disease (CVD). Prediabetes is associated with obesity, especially intraabdominal obesity, dyslipidemia with high triglycerides and/or low high-density lipoprotein (HDL) cholesterol, and hypertension (often referred to as metabolic syndrome). Lifestyle interventions for the prevention of type 2 diabetes are the subject of the following chapter.

6.3 Diagnosis and Testing for Prediabetes and Diabetes

Four tests to diagnosis diabetes are available and in the absence of unequivocal hyperglycemia with acute metabolic decompensation, each must be confirmed on a subsequent day. Criteria for the diagnosis of diabetes are listed in Table 6.1 [7]. Categories of increased risk for diabetes (prediabetes) are listed in Table 6.3 [7]. Both categories have added A1C as an accepted diagnostic test. When hemoglobin and other proteins are exposed to glucose, the glucose becomes attached in a slow, nonenzymatic, and concentration-dependent manner. In persons without diabetes, A1C values are 4–6%. These values correspond to mean glucose levels of ~70–126 mg/dL (3.9–7.0 mmol/L).

Testing for prediabetes and diabetes should be considered in all adults who are overweight (BMI ≥25 kg/m², although at-risk BMI may be lower in some ethnic groups) and who have one or more of the following risk factors for diabetes:

- Physical inactivity
- First-degree relative with diabetes
- Members of a high-risk race/ethnic population (e.g., African American, Latino, Native American, Asian American, Pacific Islander)
- Women who delivered a baby weighing >9 lb (4.1 kg) or were diagnosed with GDM
- Hypertension (blood pressure ≥140/90 mmHg or on therapy for hypertension)

- HDL cholesterol level ≤0.90 mmol/L (≤35 mg/dL) and/or a triglyceride level ≥2.82 mmol/L (≥250 mg/dL)
- Women with polycystic ovarian syndrome (PCOS)
- A1C ≥5.7%, IGT, or IFG on previous testing
- Other clinical considerations associated with insulin resistance (e.g., severe obesity and acanthosis nigricans—gray-brown skin pigmentations)
- History of CVD

Testing of those without these risk factors should begin at age 45. If tests are normal, repeat testing should be carried out at least at 3-year intervals [7].

Children, and youth at increased risk for type 2 diabetes should be tested if they are both:

- Overweight (BMI >85th percentile for age and sex, weight for height >85th percentile, or >120% of ideal weight for height)
- Have two of the following risk factors present—family history of type 2 diabetes in first- or second-degree relative, member of high-risk ethnic populations, signs of insulin resistance or conditions associated with insulin resistance (acanthosis nigricans, hypertension, dyslipidemia, PCOS, or small-for-gestational-age at birth)

Testing should be initiated at 10 years of age or at onset of puberty, if puberty occurs at a younger age, and, if normal, repeated every 3 years [7].

6.4   Metabolic Goals

Landmark studies have documented the importance of glycemic control in persons with diabetes. The Diabetes Control and Complications Trial (DCCT) demonstrated beyond a doubt the clear link between glycemic control and microvascular complications in persons with type 1 diabetes. Patients who achieve control similar to the intensively treated patients in the study can expect a 50–75% reduction in the risk of progression to retinopathy, nephropathy, and neuropathy [9]. The long-term follow-up of DCCT participants demonstrated that intensive diabetes treatment also significantly reduced the risk of long-term cardiovascular complications [10].

The United Kingdom Prospective Diabetes Study (UKPDS) demonstrated conclusively that elevated blood glucose levels cause long-term complications in type 2 diabetes just as in type 1 diabetes [11]. In the intensive therapy group, the microvascular complications rate decreased significantly by 25% and the risk of macrovascular disease decreased by 16%. Aggressive treatment of even mild-to-moderate hypertension was also beneficial in both groups [12].

The UKPDS also illustrates the progressive nature of type 2 diabetes. Before randomization into intensive or conventional treatment, subjects received individualized intensive nutrition therapy for 3 months. During this run-in period, the mean A1C decreased by 1.9% (from ~9 to ~7%) and patients lost an average of 3.5 kg (8 lb) [13]. UKPDS researchers concluded that a reduction of energy intake was at least as important, if not more important than, the actual weight lost in determining glucose levels. An important lesson learned from the UKPDS is that therapy needs to be intensified overtime and that as the disease progresses, nutrition therapy alone is not enough to keep the majority of the patients' A1C levels at 7%. Medication(s), and for many individuals eventually insulin, needs to be added to the treatment regimen. It is not the "diet" failing, but instead is the pancreas failing to secrete enough insulin to maintain adequate glucose control.

The ACCORD (Action to Control Cardiovascular Risk in Diabetes) trial was designed to determine whether targeting an A1C below 6% would reduce risk of serious cardiovascular events in middle-aged and elderly people with type 2 diabetes compared to targeting an A1C of 7–7.9% [14]. The patients in the intensive treatment group were treated with multiple medications to reach the 6% goal. This part of the trial was discontinued after a mean of 3.7 years because of increased mortality among persons with advanced type 2 diabetes and a high risk of CVD. Use of intensive therapy for 3.7 years reduced 5-year nonfatal myocardial infarctions of this group but increased 5-year mortality [15]. Therefore an A1C goal of <6% should not be used in high-risk patients with advanced type 2 diabetes. The American Diabetes Association recommends that most people with diabetes should strive for an A1C of <7%, but as always stresses individualization of treatment goals [7].

6.5 Medical Nutrition Therapy: Evidence-Based Nutrition Recommendations

The Academy of Nutrition and Dietetics (formerly the American Dietetic Association) published evidence-based nutrition practice guidelines for type 1 and type 2 adults in adults [16] and American Diabetes Association nutrition recommendations are summarized in their annual standards of care [7] and in a position statement [17].

6.5.1 Goals and Outcomes of Medical Nutrition Therapy for Diabetes

Improving health through food choices and physical activity is the basis of all nutrition recommendations for diabetes. The goals of nutrition therapy emphasize improving glucose control, lipid and lipoprotein profiles, and blood pressure. However, because lifestyle modifications impact almost immediately on glycemia, glucose goals are often the first focus.

Research supports nutrition therapy as an effective therapy. Randomized controlled trials and observational studies of diabetes nutrition therapy provided by registered dietitians (RDs) have demonstrated decreases in A1C of approximately 1–2% (range=−0.5 to −2.6%), depending on the type and duration of diabetes [18]. These outcomes are similar to those from glucose-lowering medications.

An example of the effectiveness of nutrition therapy for type 1 diabetes is the Dose Adjusted for Normal Eating (DAFNE) trial [19]. Conventional insulin therapy in the United Kingdom at the time of the study was to determine the insulin regimen first. This required individuals with type 1 diabetes to eat according to the times of action of the insulin they had been prescribed. This conventional therapy was compared to use of insulin-to-carbohydrate ratios. Dietitians taught patients how to bolus insulin doses based on planned meal carbohydrate intake (DAFNE training). In the DAFNE training group, A1C levels were lowered by approximately 1% with no significant increase in hypoglycemia, along with positive effects on quality of life, despite an increase in insulin injections (but not in the total insulin amount) and in blood glucose monitoring. In a similar type of nutrition intervention, improvements in A1C and quality of life were sustained for 1 year [20].

Although nutrition therapy has been shown to be effective at any time in the type 2 disease process, it appears to have its greatest impact earlier in the course of the disease. In individuals with newly diagnosed type 2 diabetes, decreases in A1C of ~2% are reported [13, 21], whereas in individuals with an average duration of diabetes of 4 years, average decreases in A1C of ~1% are reported [21, 22] Of interest is a randomized controlled trials of subjects with average duration of type 2 of 9 years who had A1C levels >7% despite optimized drug therapy. The intervention group received intensive individualized nutrition therapy. The difference in A1C between the intervention and control groups at 6 months (−0.4%) was highly significant as were changes in anthropometric measurements. This study demonstrates the effectiveness of nutrition therapy even in diabetes of long duration [23].

Outcomes of nutrition therapy interventions are evident by 6 weeks to 3 months. Central to these interventions are multiple encounters to provide education and counseling initially and on a continued basis. A variety of nutrition interventions have been shown to improve glycemic control. Interventions used included reduced energy/fat intake, carbohydrate counting, simplified meal plan, healthy food choices, individualized meal-planning strategies, exchange lists, insulin-to-carbohydrate ratios, physical activity, and behavioral strategies [16].

Nutrition therapy is reported to lower LDL cholesterol by 15–25 mg/dL or by 9–12% compared to baseline values or to a Western diet [24]. Nutrition therapy for hypertension is reported to decrease both systolic and diastolic blood pressure ~5 mmHg [25].

6.5.2 *Evidence-Based Nutrition Recommendations for Diabetes*

Although numerous studies have attempted to identify ideal percentages of macronutrients that would assist in achieving metabolic goals, it is unlikely that one such combination of macronutrients exits [16, 17]. If guidance is needed, the dietary reference intakes (DRIs) for healthy eating (45–65% of total energy from carbohydrate, 20–35% from fat, and 10–35% from protein) are encouraged [26].

6.5.2.1   Carbohydrate

Because carbohydrate is the nutrient that most affects postprandial glycemia and is the major determinant of bolus insulin doses, it is addressed first. Foods containing carbohydrate—grains, fruits, vegetables, low-fat milk—are important components of a healthful diet and should be included in the food/meal plan of persons with diabetes. This recommendation reflects the concern that low-carbohydrate diets eliminate many foods that are important for all persons to eat as part of a healthy lifestyle.

Monitoring total grams of carbohydrate, whether by carbohydrate counting, exchanges, or experienced based estimation is a key strategy for achieving glycemic control [17]. Numerous studies have reported that when subjects are allowed to choose from a variety of starch and sugars, the glycemic response is similar, as long as the total amounts of carbohydrate is kept constant [16]. Day-to-day consistency in the amount of carbohydrate eaten at meals and snacks has been shown to improve glycemic control in persons on nutrition therapy alone, glucose-lowering medications, or on fixed insulin regimens. In persons with type 1 or type 2 diabetes who adjust their bolus insulin doses, insulin doses should be adjusted to match carbohydrate intake (insulin-to-carbohydrate ratios) [16].

The glycemic index (GI) of food was developed to compare the physiologic effects of carbohydrate on glucose. The GI measures the relative area under the postprandial glucose curve of 50 g of digestible carbohydrates compared with 50 g of a standard food, either glucose or food. The GI does not measure how rapidly blood glucose levels increase. The peak glucose response for individual foods [27] and meals [28] occurs at approximately the same time. The estimated glycemic load (GL) of foods, meals, and dietary patterns is calculated by multiplying the GI by the amount of carbohydrates in each food and then totaling the values for all foods in a meal or dietary pattern. Thus, whereas GI indicates the glycemic response to a quantity of a particular food containing 50 g of carbohydrate, the GL attempts to measure the predicted impact on the body's glycemic response to the actual meal or diet.

A major problem with the GI is the variability of response to a specific carbohydrate food. For example, the mean glycemic response and standard deviation of 50 g of carbohydrate from white bread tested in 23 subjects was 78±73 with an individual coefficient of variation (CV) of 94%. Although the average GI of bread from three tests was 71%, the range of GI values was broad, ranging from 44 to 132 and a CV 34% [29].

Although different carbohydrates do have different glycemic responses, there is limited evidence to show long-term glycemic benefit when low-GI diets vs. high-GI diets are implemented [17]. Two trials, each 1 year in duration, reported no significant differences in A1C levels from low-GI vs. high-GI diets [30] or American Diabetes Association diets [31]. Therefore, for a primary nutrition therapy intervention, an approach documented to have the greatest impact on metabolic outcomes should be selected and information on the glycemic responses of foods can perhaps best be used for fine-tuning glycemic control [17].

As for the general population, people with diabetes are encouraged to consume a variety of fiber-containing foods, but there is no reason to recommend that people with diabetes eat a greater amount of fiber than other Americans. Diets containing 44–50 g/day fiber have been shown to improve glycemia; however, more usual fiber intake (up to 24 g/day) has not shown beneficial effects on glycemia [16, 17]. However, consuming foods containing 25–30 g/day fiber, with special emphasis on soluble fiber sources (7–13 g) is recommended as part of cardioprotective nutrition therapy [16].

Even though sucrose restriction cannot be justified on the basis of its glycemic effect, people with diabetes, as is the general public, are advised to avoid foods containing large amounts of sucrose. Reduced calorie sweeteners approved by the US Food and Drug Administration (FDA) include sugar alcohols (erythritol, sorbitol, mannitol, xylitol, isomalt, lactitol, and hydrogenated starch hydrolysates) and tagatose. They produce a lower glycemic response and, on average, 2 kcal/g. Although their use appears safe, gastric discomfort and diarrhea, after ingestion of large quantities is often a problem, especially in children.

Saccharin, aspartame, neotame, acesulfame potassium, and sucralose are nonnutritive sweeteners approved for use by the FDA. All such products undergo rigorous testing by the manufacturer and scrutiny by the FDA. An acceptable daily intake (ADI) is defined as the amount of a food additive that can be safely consumed on a daily basis over a person's lifetime. The ADI generally includes a 100-fold safety factor that greatly exceeds average intake levels. The risk to health of these substances is extremely small, especially when compared to that from added sugars. In 2008, the FDA stated that the stevia-derived sweetener, Rebaudioside A, is generally recognized as safe (GRAS) and it is currently being marketed.

6.5.2.2   Protein

There is no evidence to suggest that usual intake of protein (15–20% of energy intake) be changed in people who do not have renal disease. In patients with diabetic nephropathy, a protein intake of <1.0 g/kg/day is recommended. Energy and protein intake must be monitored to ensure adequate intake [16].

Usual protein intake has minimal acute effects on glucose and no long-term effect on insulin requirements. Although nonessential amino acids undergo gluconeogenesis, in well-controlled diabetes, the glucose produced does not enter into the general circulation; however, ingested protein is just as potent a stimulant of acute insulin release as dietary carbohydrate. Patients are often told that eating protein can slow the absorption of carbohydrate. However, research has clearly shown that ingested protein does not slow the absorption of carbohydrates. Furthermore, patients are often told that adding protein to the treatment of hypoglycemia will prevent subsequent hypoglycemia but research has also shown that this is not the case [32]. There is also no evidence that adding protein to bedtime snacks is helpful. The long-term effects of a diet higher in protein and lower in carbohydrate in persons with diabetes on regulation of energy intake, satiety, and weight loss have not been adequately studied.

6.5.2.3 Dietary Fat

Limiting intake of saturated fats, *trans* fatty acids, and dietary cholesterol is recommended, especially in individuals with LDL cholesterol ≥100 mg/dL (2.6 mmol/L). Research in persons with diabetes supporting these guidelines is limited. However, persons with diabetes are considered to be at a CVD risk similar to persons with a past history of CVD. Therefore, after focusing on glycemic control, cardioprotective nutrition interventions should be implemented in the initial education series [16].

There is evidence from the general population that foods containing *n*-3 polyunsaturated fatty acids are beneficial and two to three servings of fish per week are recommended [17]. Studies in persons with diabetes using *n*-3 supplements have shown beneficial effects on lowering triglycerides and platelet reactivity, whether they also reduce blood pressure, leukocyte reactivity, and arrhythmias in these patients similarly to what occurs in other patient groups remains to be established. The majority of the evidence suggests that any side effects from consuming *n*-3 supplements are minor and overall they are safe to ingest [33].

6.5.2.4 Micronutrients

There is no evidence of benefit from vitamin or mineral supplementation in persons with diabetes who do not have underlying deficiencies [17]. In select groups such as the elderly, pregnant or lactating women, strict vegetarians, or those on calorie-restricted diets, a multivitamin supplement may be needed.

Routine supplementation of the diet with antioxidants such as vitamin C and E and carotene has not proven beneficial [34, 35]. Routine supplementation is not advised because of lack of effectiveness and concern related to long-term safety [17].

Several small studies suggested a role for chromium supplementation. However, a systematic review found no significant effects of chromium on lipid or glucose metabolism in people without diabetes and inconsistent effects in people with diabetes. Therefore, chromium supplementation is not recommended [17].

Recent attention on the prevalence of vitamin D deficiency has raised questions about intake recommendations and supplemental needs. In addition questions have been raised about its possible role in diabetes [36, 37]. The latest research indicates that the association with diabetes is unclear [36].

6.5.2.5 Alcohol

Recommendations for alcohol intake are similar to those for the general public. Alcoholic drinks should be limited to less than two per day for men and less than one per day for women. One drink is defined as a 12 oz beer, 5 oz wine, or 1.5 oz of distilled spirits, each of which contains ~15 g alcohol. The type of alcoholic beverage does not make a difference. Moderate amounts of alcohol when ingested with food have minimal, if any, effect on blood glucose and insulin concentrations.

Table 6.4 Carbohydrate servings[a]

Starch	Milk
1 Slice of bread (1 oz)	1 Cup skim/reduced-fat milk
⅓ Cup cooked rice or pasta	⅔ Cup fat-free fruited yogurt sweetened with
¾ Cup dry cereal	nonnutritive sweetener or fructose (6 oz)
4–6 Crackers	
½ Large baked potato with skin (3 oz)	
¾ oz Pretzels, potato, or tortilla chips	
Fruit	Sweets and desserts
1 Small fresh fruit (4 oz)	2 Small cookies
½ Cup fruit juice	1 Tablespoon jam, honey, syrup
¼ Cup dried fruit	½ Cup ice cream, frozen yogurt, or sherbet

[a]One serving contains 15 g of carbohydrate

For individuals using insulin or insulin secretagogues, if alcohol is consumed, it should be consumed with food to prevent hypoglycemia. Excessive amounts of alcohol (three or more drinks per day on a consistent basis) can contribute to hyperglycemia but this improves as soon as alcohol use is discontinued [38].

In persons with type 2 diabetes, light to moderate amounts of alcohol are associated with a decreased risk of atherosclerosis and coronary heart disease, perhaps due to increases in HDL cholesterol and improved insulin sensitivity [39]. In observational and short-term studies, moderate amounts of alcohol do not increase triglyceride levels and have beneficial effects on blood pressure levels. However, because the available evidence on alcohol use is primarily observational, it does not support recommending alcohol consumption to persons who do not currently drink. Abstention from alcohol should be advised for people with a history of alcohol abuse or dependency, women during pregnancy, and people with medical conditions such as liver disease, pancreatitis, advanced neuropathy, or severe hypertriglyceridemia [17]. The relationship between alcohol and various aspects of health is discussed in Chap. 15.

6.5.3   Carbohydrate Counting

Carbohydrate counting is based on the evidence that it is the balance between ingested carbohydrate and available insulin that determines postprandial glucose levels. All persons with diabetes can benefit from basic information about carbohydrates—what foods contain carbohydrates (starches, fruit, starchy vegetables, milk, sweets), average 15 g portion sizes, and how many servings to select for meals (and snacks if desired). Table 6.4 lists some examples of a carbohydrate serving. One of the first decisions for food and meal planning is the total number of carbohydrate servings the person with diabetes chooses to eat at meals or for snacks.

Women often do well with 3 or 4 carbohydrate servings per meal and 1–2 for a snack. Men may need 4–5 carbohydrate servings per meal and 1–2 for a snack.

Food records along with blood glucose monitoring data can then be used to evaluate if treatment goals are being met, or if there is a need for additional lifestyle and/or medication changes. Learning how to use Nutrient Facts on food labels is also useful. Individuals first should take note of the serving size and the total amount (grams) of carbohydrate. The total grams of carbohydrate are then divided by 15 to determine the number of carbohydrate servings in the serving size.

Carbohydrate counting does not mean that meat and fat portions can be ignored. Individuals with diabetes must also know the approximate number of meat and fat servings they should select for meals and snacks. Weight control is important as is the maintenance of a healthy balance of food choices.

6.5.4   Weight Management and Diabetes

A major challenge in diabetes care is the high prevalence of overweight and obesity in persons with type 2 diabetes. Weight loss is an important goal for overweight and obese persons as it assists in the prevention of diabetes and reduces other risk factors for CVD. Clinicians need to be alert to any weight increases and/or weight issues and intervene early to assist their patients in weight management. Unfortunately, in clinical practice, weight loss interventions are often recommended too late in the natural progression of diabetes. A moderate weight loss of 5–10% can reduce risk of developing type 2 diabetes and may improve glycemia, particularly in individuals who are primarily insulin resistant, including those with prediabetes and early-onset type 2 diabetes. However, as the disease progresses and individuals become more insulin deficient, weight loss may fail to significantly improve glycemia [40, 41].

The Academy of Nutrition and Dietetics Association reviewed 11 studies which included a minimum of one weight-loss diet group and reported weight loss and A1C values at 12 months. Weight loss diet groups in five studies reported improvement in A1C; however, weight loss diet groups in six studies reported no improvement in A1C despite fairly similar weight losses [16]. Of interest is the Look AHEAD (Action for Health in Diabetes) trial being conducted in 16 centers around the United States in an attempt to determine the effectiveness of intentional weight loss in reducing CVD. Over 1 year participants randomized to the intensive lifestyle intervention consisting of meal replacements or structured food plans, 175 min/week of physical activity, and three to four education/counseling sessions per month, experienced an average weight loss of 8.6% and a decrease in A1C of 0.6% [42]. At the end of 4 years, an average weight loss of 6.1% and a decrease in A1C of 0.4% was reported [43]. Participants will be followed for up to 11.5 years to determine if these outcomes result in a reduction in CVD events.

Energy restriction, with or without weight loss, can decrease insulin resistance. Weight loss strategies that use a behavioral approach, combining a reduction in energy intake with an increase in physical activity, are most effective. Individuals lose, on average, 5–10% of their starting weight by 6 months at which time weight loss reaches a plateau. Adaptive mechanisms occur with a reduced energy intake (e.g., hormonal regulation, adaptive thermogenesis leading to a decrease in voluntary energy expenditure, and decline in basal energy metabolism) and, at this point, maintenance of the weight lost should become the focus of the ongoing care and intervention. In studies extending to 48 months, 3–6% of weight loss was maintained [44].

Evidence supports the fact that weight loss is primarily associated with energy deficit, diet adherence, and enthusiasm of the counselor, but not with nutrient composition [45, 46]. Two 1-year trials in subjects with type 2 diabetes following a low-carbohydrate diet compared to subjects following a high-monounsaturated fat or low-fat diet reported no beneficial effects on weight, A1C, LDL cholesterol, or blood pressure from the low-carbohydrate diet [47, 48]. Advice for reducing energy intake should be individualized to account for food preferences and the individual's preferred approach for reducing energy intake. A variety of behavioral strategies can be employed that have been shown to be successful.

6.6 Physical Activity/Exercise

The lifestyle factor most consistently reported to improve insulin resistance and glucose tolerance is physical training [49]. Several long-term studies reported improved insulin action in exercise-trained obese individuals, independent of weight loss [50, 51]. Exercise also decreases blood pressure and triglyceride levels. Although reported to increase HDL cholesterol in the general population, it is not clear if this is also the case in people with insulin resistance. Physical activity in the absence of calorie restriction has only a modest, if any, effect on weight loss, but is useful as an adjunct to other weight loss strategies and is important in long-term maintenance of weight loss. In persons with diabetes, regular exercise has been shown to improve blood glucose control, reduce CVD risk factors, contribute to weight management, and improve well-being [7]. It is especially noteworthy that among men with diabetes those who are overweight but fit are reported to have a lower risk of mortality from chronic disease than those who are lean but unfit [52].

People with diabetes should be advised to perform at least 150 min/week of moderate-intensity aerobic physical activity (50–70% of maximum heart rate) or at least 75 min/week of vigorous aerobic exercise (>70% of maximum heart rate). The physical activity should be distributed over at least 3 days/week and with no more than 2 consecutive days without physical activity. In the absence of contraindications, people with type 2 diabetes should also be encouraged to perform resistance exercise three times a week. There is an additive benefit of combined aerobic and resistance training in adults with type 2 diabetes [7].

As the benefits of screening asymptomatic persons with diabetes for coronary heart disease are unclear, it is recommended that providers should assess patients for conditions that might contraindicate certain types of exercise or predispose to injury. High-risk patients should be encouraged to start with short periods of low-intensity activity and increase the intensity and duration slowly [7].

6.7   Medications

6.7.1   Insulin

Persons with type 1 diabetes depend on insulin to survive. In persons with type 2 diabetes, insulin is frequently needed to restore glycemia to normal. Circumstances that require the use of insulin in type 2 include failure to achieve adequate control with use of glucose-lowering medications; periods of acute injury, infection, or surgery; or pregnancy.

Insulin regimens are designed to mimic normal insulin action. Approximately 50% of the total daily insulin dose is required to provide for basal or background insulin needed in the post-absorptive state to restrain endogenous glucose output primarily from the liver and to limit lipolysis and excess flux of free fatty acids to the liver. Glargine or detemir are frequently used 24-h peakless insulins and are given at bedtime (or at any consistent time during the day). Detemir can last up to 24 h, though its duration of action is thought to be less than that of glargine. NPH insulin is another basal insulin. It must be administered twice a day, usually before breakfast and the evening meal, but because of its peaking action often results in erratic glucose levels and overnight hypoglycemia.

Rapid-acting insulin is given before meals and is referred to a bolus or mealtime insulin. It is divided among the meals either proportionate to the amount of carbohydrate planned to be eaten or by giving 1.0–1.5 units/10–15 g of carbohydrate. The larger amount is usually needed to cover breakfast carbohydrate. An insulin-to-carbohydrate ratio can be established for an individual that will determine the amount of bolus insulin to inject.

Insulin pump therapy provides basal rapid-acting insulin pumped continuously by a mechanical device in micro amounts through a subcutaneous catheter that is monitored 24 h a day. Boluses of the insulin are then given before meals.

The administration of basal insulin once or twice a day may suffice for persons with type 2 diabetes who still have significant endogenous insulin. Once daily glargine, determir, or NPH at bedtime or a premixed insulin before the evening meals is a commonly used regimen. The rationale is that supplementing with overnight insulin will control fasting hyperglycemia. A concern with evening NPH is nocturnal hypoglycemia. Oral agents may be continued during the day to prevent worsening of daytime glycemia. However, many individuals with type 2 diabetes also will eventually require an insulin regimen that better mimics the physiological actions of insulin.

6.7.2   Glucose-Lowering Medications for Type 2 Diabetes

A consensus statement on management of hyperglycemia has been published by the American Diabetes Association and the European Association for the Study of

Diabetes [53]. Interventions at the time of diagnosis include nutrition therapy and metformin. If A1C is ≥7%, the next well-validated therapies are to add either a sulfonylurea or basal insulin. The alternative path is to add therapies of pioglitazone or an incretin glucagon-like peptide-1 (GLP-1) agonist, but these are not as well validated. A recent meta-analysis reported that the addition of an oral glucose-lowering agent in addition to current therapy caused an additional decrease in A1C levels of ~1 to 1.25% with most of the treatment effect evident by 3–6 months of initiating therapy. The effect was fairly consistent between classes with sulfonylureas and thiazolidinediones (TZDs) having the greatest reduction in A1C [54].

Biguanides, such as metformin, are categorized as insulin sensitizers because they reduce insulin-stimulated liver glucose production. This is manifested as improvement in fasting glucose values.

Sulfonylureas, such as glimepiride, glipizide, and glyburide, are known as insulin secretagogues and cause insulin release from the β-cells of the pancreas. They depend on the concentration of intracellular calcium. Sulfonylureas bind to receptors on pancreatic β-cell surfaces to enhance the entry of calcium into β-cells with a resultant increase in insulin secretion.

The TZDs reduce insulin resistance by binding to a nuclear peroxisome-proliferation activated receptor γ (PPAR-γ) in muscle and adipose cells. This changes transcription of genes mediating carbohydrate and lipid metabolism resulting in an increase in insulin stimulated uptake of glucose by skeletal muscle cells. TZDs also have a favorable effect on lipids and do not independently cause hypoglycemia. Adverse effects include increased incidence of heart failure, weight gain, edema, and bone fractures in postmenopausal women. Pioglitazone is the most commonly prescribed TZD.

The incretins are hormones secreted by the small intestine that mediate the effects of oral glucose and are primarily driven by GLP-1 and glucose-dependent insulinotropic peptide (GIP). GLP-1 stimulates insulin secretion, inhibits hyperglucagonemia in hyperglycemia, slows gastric emptying, reduces appetite, and improves satiety. It has a very short-life in plasma of 1–2 min because of degradation by the enzyme dipeptidyl peptidase IV (DPP-4). Exenatide (Byetta), the first incretin mimetic, mimics the action of GLP-1 but is resistant to DPP-4 degradation. Liraglutide (Victoza) activates the GLP-1 receptor leading to insulin release in the presence of elevated glucose concentrations. Both drugs are associated with modest weight loss and must be injected subcutaneously. The most common adverse effect is nausea which tends to wane over time.

α-Glucosidase inhibitors, such as acarbose and miglitol, unlike other classes of medications do not target the mechanisms underlying insulin resistance or deficient insulin secretion. Instead, they competitively inhibit the intestinal epithelial enzyme that catalyzes the conversion of polysaccharides and disaccharides to monosaccharides in the small intestine, thereby delaying carbohydrate absorption and reducing postprandial glucose levels.

The meglitinides (repaglinide and nateglinide) differ from the sulfonylureas in that they have short metabolic half-lives, which result in brief episodic stimulation of insulin. They are given before meals decreasing postprandial glucose levels. Nateglinide only works in the presence of glucose.

Numerous DPP-4 inhibitors are being developed. Sitagliptin and saxagliptin are the first available and produce an approximately twofold increase in GLP-1 and GIP. They are associated with an A1C reduction of ~0.7%, whether used as monotherapy or in combination with metformin, sulfonylureas, or TZDs, are weight neutral, and are not associated with nausea.

Another injectable drug is an analog of human amylin, a hormone made in the β-cells of the pancreas. It is called Symlin (Pramlinitide) and is used as adjunctive therapy to insulin to help control postprandial blood glucose in patients with type 1 and type 2 diabetes who have failed to achieve desired glucose control despite optimal insulin therapy. Small reductions in weight compared to persons on insulin alone are also reported.

6.8 Monitoring

It is important that persons with diabetes know their target blood glucose, lipid, and blood pressure goals. See Table 6.5 for a listing of target goals. Self-monitoring of blood glucose (SMBG) is used on a day-to-day basis to adjust treatment regimens; the frequency of monitoring depending on the type of diabetes and overall therapy. However, A1C measurement provides the best available index of overall diabetes control. The healthcare team, including the individual with diabetes, should work together to implement blood glucose monitoring and establish individual target blood glucose goals.

Table 6.5 Glucose, lipids, and blood pressure recommendations for adults with diabetes

Glycemic control	
A1C	<7.0%[a]
Preprandial plasma glucose	70–130 mg/dL (3.9–7.2 mmol/L)
Postprandial plasma glucose	<180 mg/dL (<10.0 mmol/L)
Blood pressure	<130/80 mmHg
Lipids	
LDL cholesterol	<100 mg/dL (<2.6 mmol/L)
Triglycerides	<150 mg/dL[b] (<1.7 mmol/L)
HDL cholesterol	>40 mg/dL[c] (>1.1 mmol/L)

From ref. [7]

[a]Referenced to a nondiabetic range of 4.0–6.0% using a DCCT-based assay

[b]In individuals with overt cardiovascular disease (CVD), a lower LDL cholesterol goal of <70 mg/dL (1.8 mmol/L), using a high dose of a statin, is an option

[c]For women, it has been suggested that the HDL cholesterol goal be increased by 10 mg/dL

6.8.1 A1C

Long-term glycemic control is assessed by use of A1C results. An A1C test should be performed at least two times a year in patients who are meeting treatment goals and have stable glycemic control. It should be performed quarterly in patients who therapy has changed or who are not meeting glycemic goals [7].

6.8.2 Self-Monitoring of Blood Glucose

SMBG can be performed up to eight times per day—before breakfast, lunch, and dinner; at bedtime; 1–2 h after the start of meals; during the night; or when needed to determine causes of hypoglycemia or hyperglycemia. For patients using multiple insulin injections or insulin pump therapy, SMBG is recommended three or more times daily, generally before each meal and at bedtime [7]. Recommendations for SMBG for patients using less frequent insulin injections, noninsulin therapies, or nutrition therapy alone are controversial. The American Diabetes Association states that for these individuals SMBG may be useful as a guide to the success of therapy [7]. The American Dietetic Association reported that using SMBG was associated with greater improvement in A1C when it was a part of structured education programs and individuals used the information to make changes in their management. Evidence on optimum frequency and duration of SMBG was inconclusive [16]. It is often recommended that these individuals perform SMBG one to four times a day, before breakfast and before and 2 h after the largest meal, but only 3 or 4 days/week. When adding to or modifying therapy, patients with type 1 and type 2 diabetes should test more often than usual.

In evaluating blood glucose monitoring records, it should be remembered that factors other than food affect blood glucose concentrations. An increase in blood glucose can be the result of insufficient insulin or insulin secretagogues, too much food, or increases in glucagon and other counterregulatory hormones as a result of stress, illness, or infection. Factors contributing to hypoglycemia include too much insulin or insulin secretagogues, not enough food, unusual amounts of exercise, and skipped or delayed meals.

The next generation of glucose monitors is the continuous glucose monitoring (CGM) devices, which measure interstitial blood glucose and provide readings every 5–10 min. They also have alarms for glucose highs and lows and the ability to download data and track trends over time. Persons experiencing unexplained elevations in A1C or unexplained hypoglycemia and/or hyperglycemia may benefit from the use of CGM or more frequent SMBG. Use of CGM is likely to increase in the future and is also the next step in the development of closed loop insulin therapy.

6.8.3   Lipids and Blood Pressure

Other parameters besides glucose that must be regularly monitored are lipids and blood pressure (see Table 6.5 for lipid and blood pressure goals). In most adults with diabetes, lipid tests should be done at least annually. In adults with low-risk values, lipid assessments may be repeated every 2 years [7]. Lifestyle interventions, including reduction in saturated fat, *trans* fat, and cholesterol intake; increase of *n*-3 fatty acids, viscous fiber, and plant stanols/sterols, weight loss (if indicted); and increased physical activity, are recommended. Statin therapy, however, should be added to nutrition therapy, regardless of baseline lipid values for almost all patients [6]. In individuals without overt CVD, the primary goals is an LDL cholesterol to <2.6 mmol/L (<100 mg/dL) and for individuals with overt CVD, a lower LDL cholesterol (<1.8 mmol/L; 70 mg/dL) is recommended. Glucose control can also beneficially modify plasma lipid levels, while smoking cessation to decrease the risk of CVD is also important.

A goal blood pressure of <130/<80 mmHg is appropriate for most patients with diabetes [7]. The role of nutrition therapy in the control of blood pressure is examined in Chap. 11. If lifestyle modifications—weight loss, if overweight; Dietary Approaches to Stop Hypertension (DASH)-style eating pattern including reducing sodium; moderation of alcohol consumption; and increased physical activity levels—do not achieve these goals in 3 months, medication should be added until blood pressure goals are met. Multiple drug therapy (two or more agents at proper doses) is generally required to achieve blood pressure targets [7].

6.8.4   Nephropathy Screening

The American Diabetes Association recommends that an annual screening to assess urine albumin excretion be performed in patients with type 1 diabetes ≥5 year duration and in all patients with type 2 diabetes starting at diagnosis [7]. Creatinine should also be measured at least annually in all adults with diabetes regardless of the degree of urine albumin excretion. The serum creatinine is used to estimate glomerular filtration rate and stage the level of chronic kidney disease, if present.

6.8.5   Ketone Tests

Urine or blood testing can be used to detect ketones. Testing for ketonuria or ketonemia should be performed regularly during periods of illness and when blood glucose levels consistently exceed 240 mg/dL (13.3 mmol/L). The presence of persistent, moderate, or large amounts of ketones, along with elevated blood glucose levels, requires insulin adjustments. Persons with type 2 diabetes seldom have ketosis. However, ketone testing should be done in the presence of a serious illness.

6.9   Prioritizing Nutrition-Related Interventions for Type 1 and Type 2 Diabetes

Evidence from metabolic goals, nutrition therapy recommendations, physical activity/exercise, medications, and monitoring can be used to prioritize nutrition-related interventions for type 1 and type 2 diabetes.

6.9.1   Type 1 Diabetes

The first priority for persons requiring insulin therapy is to integrate an insulin regimen into the patient's lifestyle. The food/meal plan is developed first and is based on the individual's appetite, preferred foods, and usual schedule of meals and physical activity. After the dietitian, working with the individual with diabetes, develops a food plan, this information is shared with the professional determining the insulin regimen. Insulin therapy can then be integrated into food and physical activity schedules.

6.9.1.1   Insulin-to-Carbohydrate Ratios

Persons are encouraged to learn carbohydrate counting and to calculate their insulin-to-carbohydrate (I:C) ratio which is used to adjust bolus insulin doses based on the planned carbohydrate content of the meal. To determine the I:C ratio, the individual can either (1) eat a fixed amount of carbohydrate at a meal, adjust the bolus insulin dose to meet postmeal glucose goals, and then determine the ratio, or (2) start with an estimated ratio and adjust it based on the resulting patterns of postprandial glucose levels. Different I:C ratios may be needed for each meal [55].

6.9.1.2   Correction Factors

A correction factor (CF) is used to correct preprandial glucose levels that are outside of target goal ranges. The insulin CF is defined as the estimated number of mg/dL (mmol/L) blood glucose levels will drop over a 2–4 h period following administration of 1 unit of rapid-acting insulin. The CF is determined using the "1,700 rule"; some centers use a 1,800 rule. The CF is 1,700 (or 1,800) divided by the total daily insulin dose. For example, if an individual's total daily insulin dose is 60 units of insulin (basal and bolus), the CF = 1,700/60 = 28. For this individual, I unit of insulin should lower their blood glucose by 28 mg/dL (1.6 mmol/L). The CF is added to the bolus insulin dose. Because of overlapping dosing effect, the CF insulin dose should only be given at 4 h intervals. The CF can also be used in sick-day management for correcting hyperglycemia [55].

6.9.2 Type 2 Diabetes

As individuals move from being insulin resistant to insulin deficient, priorities of nutrition therapy shift from weight loss to glucose, lipid, and blood pressure control. The goal of therapy is to achieve euglycemia in order to slow β-cell exhaustion. Although moderate weight loss may be beneficial for some individuals, primarily those who are still insulin resistant, for most it is too late for weight loss to improve hyperglycemia. At later stages of the disease when medications—including insulin—need to be combined with nutrition therapy, weight gain often occurs and preventing this weight gain becomes a factor. However, glycemic control must still take precedence over concern about weight.

Teaching individuals how to make appropriate food choices (usually by means of carbohydrate counting) and using data from blood glucose monitoring to evaluate short-term effectiveness are important components of successful nutrition therapy for type 2 diabetes. Regular physical activity is encouraged, primarily for the benefits associated with enhanced cardiorespiratory fitness that are independent of weight. The impact of nutrition interventions on glycemic control is evident by 6 weeks to 3 months. At this point, it can be determined if medications need to be added (or adjusted). When insulin is required, consistency in timing of meals and of their carbohydrate content becomes important.

Many individuals with type 2 diabetes also have dyslipidemia and hypertension, so decreasing intakes of saturated fat, cholesterol, and sodium should also be a priority.

6.10 Summary

Nutrition therapy begins by developing a rapport with the individual with diabetes. Whether provided individually or in groups, nutrition therapy involves a common process: (1) assessment (evaluation at follow-up visits) for determining nutrition-related problems, (2) nutrition interventions including self-management education and counseling, and (3) monitoring and evaluation of nutrition-related outcomes. Planned follow-up and ongoing education and support are also important.

It is important that all health-care providers understand nutrition issues and guide the individual's efforts by referring patients with prediabetes or diabetes for nutrition therapy as soon as the diagnosis is made, by promoting and reinforcing the importance of lifestyle modifications, and by providing support for the nutrition therapy intervention process.

References

1. International Diabetes Federation. IDF diabetes atlas. 4th ed. 2009. http://www.diabetesatlas. org. Accessed 6 Feb 2011.
2. Centers for Disease Control and Prevention. National diabetes fact sheet. 2011. http://www. cdc.gov/diabetes/pubs/factsheet11.htm. Accessed 6 Mar 2011.

3. American Diabetes Association. Diagnosis and classifications of diabetes mellitus. Diabetes Care. 2011;34 Suppl 1:S62–9.
4. Rosario PWS, Reis JS, Amim R, Fagundes TA, Calsolari MR, Silva SC, Purisch S. Comparison of clinical and laboratory characteristics between adult-onset type 1 diabetes and latent auto-immune diabetes in adults. Diabetes Care. 2005;28:1803–4.
5. DeWitt DE, Hirsch IB. Outpatient insulin therapy in type 1 and type 2 diabetes mellitus. JAMA. 2003;289:2254–64.
6. Defronzo RA. Banting lecture. From the triumvirate to the ominous octet: a new paradigm for the treatment of type 2 diabetes mellitus. Diabetes Care. 2009;58:773–95.
7. American Diabetes Association. Standards of medical care in diabetes—2011. Diabetes Care. 2011;34 Suppl 1:S11–61.
8. The American College of Obstetricians and Gynecologists Committee on Obstetric Practice. Screening and diagnosis of gestational diabetes mellitus. Obstet Gynecol. 2011;118: 751–3.
9. Diabetes Control and Complications Trial Research Group. The effect of intensive treatment of diabetes on the development and progression of long-term complications in insulin-dependent diabetes mellitus. N Engl J Med. 1993;339:977–86.
10. Diabetes Control and Complications Trial/Epidemiology of Diabetes Interventions and Complications (DCCT/EDIC) Study Research Group. Intensive diabetes treatment and cardiovascular disease in patients with type 1 diabetes. N Engl J Med. 2005;353:2643–53.
11. U.K. Prospective Diabetes Study Group. Intensive blood-glucose control with sulphonylureas or insulin compared with conventional treatment and risk of complications in patients with type 2 diabetes (UKPDS 34). Lancet. 1998;352:854–65.
12. U.K. Prospective Diabetes Study Group. Tight blood pressure control and risk of macrovascular complications in type 2 diabetes (UKPDS 38). BMJ. 1998;317:703–13.
13. UKPDS Group. UK Prospective Study 7: response of fasting glucose to diet therapy in newly presenting type II diabetic patients. Metabolism. 1990;39:905–12.
14. The Action to Control Cardiovascular Risk in Diabetes Study Group. Effects of intensive glucose lowering in type 2 diabetes. N Engl J Med. 2008;358:2545–59.
15. The ACCORD Study Group. Long-term effects of intensive glucose lowering on cardiovascular outcomes. N Engl J Med. 2011;364:818–28.
16. Franz MJ, Powers MA, Leontos C, et al. The evidence for medical nutrition therapy for type 1 and type 2 diabetes in adults. J Am Diet Assoc. 2010;110:1852–89.
17. American Diabetes Association. Nutrition recommendations and interventions for diabetes (position statement). Diabetes Care. 2008;31 Suppl 1:S61–78.
18. Franz MJ, Boucher JL, Green-Pastors J, Powers MA. Evidence-based nutrition practice guidelines for diabetes and scope and standards of practice. J Am Diet Assoc. 2008;108:S52–8.
19. DAFNE Study Group. Training in flexible, intensive insulin management to enable dietary freedom in people with type 1 diabetes: dose adjustment for normal eating (DAFNE) randomized trial. BMJ. 2002;325:746–52.
20. Lowe J, Linjawi S, Mensch M, James K, Attia J. Flexible eating and flexible insulin dosing in patients with diabetes: results of an intensive self-management management course. Diabetes Res Clin Pract. 2008;80:439–43.
21. Franz MJ, Monk A, Barry B, et al. Effectiveness of medical nutrition therapy provided by dietitians in the management of non-insulin-dependent diabetes mellitus: a randomized controlled trial. J Am Diet Assoc. 1995;95:1009–17.
22. Lemon CC, Lacey K, Lohse B, Hubachr D, Klawitter B, Palta M. Outcomes monitoring of health, behavior, and quality of life after nutrition intervention in adults with type 2 diabetes. J Am Diet Assoc. 2004;104:1805–15.
23. Coppel KJ, Kataoka M, Williams SH, Chisholm AW, Vorgers SM, Mann JI. Nutritional intervention in patients with type 2 diabetes who are hyperglycaemic despite optimized drug treatment—Lifestyle Over and Above Drugs in Diabetes (LOADD) study: randomized controlled trial. BMJ. 2010;341:c337.
24. Van Horn L, McCoin M, Kris-Etherton PM, et al. The evidence for dietary prevention and treatment of cardiovascular disease. J Am Diet Assoc. 2008;108:287–331.

25. American Dietetic Association Evidence Library. Effectiveness of MNT for hypertension. http://www.adaevidencelibrary.com/conclusion.cfm?conclusion_statement_id=251204. Accessed 6 Feb 2011.

26. Institute of Medicine. Dietary reference intakes; energy, carbohydrate, fiber, fat, fatty acids, cholesterol, protein, and amino acids. Washington, DC: National Academies Press; 2002.

27. Brand-Miller JC, Stockmann K, Atkinson F, Petocz P, Denyer G. Glycemic index, postprandial glycemia, and the shape of the curve in healthy subjects: analysis of a database of more than 1000 foods. Am J Clin Nutr. 2009;89:97–105.

28. Rizkalla SW, Boillot J, Taghrid L, et al. Improved plasma glucose control, whole-body glucose utilization, and lipid profile on a low-glycemic index diet in type 2 diabetic men: a randomized controlled trial. Diabetes Care. 2004;27:1866–72.

29. Vega-Lopez S, Ausman LM, Griffith JL, Lichtenstein AH. Interindividual variability and intra-individual reproducibility of glycemic index values for commercial white bread. Diabetes Care. 2007;30:1412–7.

30. Wolever TMS, Gibbs AL, Mehling C, et al. The Canadian Trial of Carbohydrates in Diabetes (CCD), a 1-y controlled trial of low glycemic index dietary carbohydrate in type 2 diabetes. No effect on glycated hemoglobin but reduction in C-reactive protein. Am J Clin Nutr. 2008;87:114–25.

31. Ma Y, Olendzki BC, Merriam PA, et al. A randomized clinical trial comparing low-glycemic index versus ADA dietary education among individuals with type 2 diabetes. Nutrition. 2008;24:45–56.

32. Gray RO, Butler PC, Beers TR, Kryshak EJ, Rizza RA. Comparison of the ability of bread versus bread plus meat to treat and prevent subsequent hypoglycemia in patients with insulin-dependent diabetes mellitus. J Clin Endocrinol Metab. 1996;81:1508–11.

33. Caterina R, Madonna R, Bertolotto A, Schmiddt EB. n-3 Fatty acids in the treatment of diabetic patients. Biological rationale and clinical data. Diabetes Care. 2007;30:1012–26.

34. Lonn E, Yusuf S, Hoogwerf B, et al.; on behalf of the Heart Outcomes Prevention Evaluation (HOPE) investigators. Effects of vitamin E on cardiovascular and microvascular outcomes in high-risk patients with diabetes. Diabetes Care. 2002;25:1919–27.

35. Vivekananthan DP, Penn MS, Sapp SK, Hsu A, Topol EJ. Use of antioxidant vitamins for the prevention of cardiovascular disease: meta-analysis randomized trials. Lancet. 2003;361:2017–23.

36. Pittas AG, Lau J, Hu FB, Dawson-Hughes B. The role of vitamin D and calcium in type 2 diabetes. A systematic review and meta-analysis. J Clin Endocrinol Metab. 2007;92:2017–29.

37. Pittas AG, Chung M, Trikalinos T, et al. Systematic review: vitamin D and cardiometabolic outcomes. Ann Intern Med. 2010;152:307–14.

38. Howard AA, Arnsten JH, Gourevitch MN. Effect of alcohol consumption on diabetes mellitus. A systematic review. Ann Intern Med. 2004;140:211–9.

39. O'Keefe JH, Bybee KA, Lavie CJ. Alcohol and cardiovascular health: the razor-sharp double-edged sword. J Am Coll Cardiol. 2007;50:1009–14.

40. Watts NB, Spanheimer RG, DiGirolamo M, et al. Prediction of glucose response to weight loss in patients with non-insulin-dependent diabetes mellitus. Arch Intern Med. 1990;150:803–6.

41. Wolf AM, Conaway MR, Crowther JQ, et al. Translating lifestyle interventions to practice in obese patients with type 2 diabetes. Diabetes Care. 2004;27:1570–6.

42. Look AHEAD Research Group; Pi-Sunyer X, Blackburn G, Brancati FL, et al. Reduction in weight and cardiovascular disease risk factors in individuals with type 2 diabetes: one-year results of the Look AHEAD Trial. Diabetes Care. 2007;30:1374–83.

43. The Look AHEAD Research Group. Long-term effects of a lifestyle intervention on weight and cardiovascular risk factors in individuals with type 2 diabetes mellitus. Four-year results of the Look AHEAD Trial. Arch Intern Med. 2010;170:1566–75.

44. Franz MJ, VanWormer JJ, Crain LA, et al. Weight loss outcomes: a systematic review and meta-analysis of weight loss clinical trials with a minimum 1-year follow-up. J Am Diet Assoc. 2007;107:1755–67.

45. Dansinger ML, Gleason JA, Griffith JL, et al. Comparison of the Atkins, Ornish, Weight Watchers, and Zone diets for weight loss and heart disease risk reduction: a randomized trial. JAMA. 2005;293:43–53.
46. Sacks FM, Gray GA, Carey VJ, et al. Comparison of weight-loss diets with different compositions of fat, protein, and carbohydrate. N Engl J Med. 2009;360:859–73.
47. Brehm BJ, Lattin BL, Summer SS, et al. One-year comparison of a high-monounsaturated fat diet with a high-carbohydrate diet in type 2 diabetes. Diabetes Care. 2009;32:215–20.
48. Davis NJ, Tomuta N, Schechter C, et al. Comparative study of the effects of a 1-year dietary intervention of a low-carbohydrate diet versus a low-fat diet on weight and glycemic control in type 2 diabetes. Diabetes Care. 2009;32:1147–52.
49. Colberg SR, Sigal RJ, Fernhall B, et al. Exercise and type 2 diabetes: the American College of Sports Medicine and the American Diabetes Association: joint position statement. Diabetes Care. 2010;33:2692–6.
50. Duncan GE, Perri MG, Theriaque DW, Hutson AD, Eckel RH, Stacpoole PW. Exercise training without weight loss, increases insulin sensitivity and postheparin plasma lipase activity in previously sedentary adults. Diabetes Care. 2003;26:557–62.
51. Jeon CY, Lokken RP, Hu FB, van Dam RM. Physical activity of moderate intensity and risk of type 2 diabetes. A systematic review. Diabetes Care. 2007;30:744–52.
52. Church TS, Cheng YJ, Earnest CP, et al. Exercise capacity and body composition as predictors of mortality among men with diabetes. Diabetes Care. 2004;27:83–8.
53. Nathan DM, Buse JB, Davidson MB, Ferrannini E, Holman RR, Sherwin R, et al.; American Diabetes Association, European Association for Study of Diabetes. Medical management of hyperglycemia in type 2 diabetes: a consensus algorithm for the initiation and adjustment of therapy: a consensus statement from the American Diabetes Association and the European Association for the Study of Diabetes. Diabetes Care. 2009;32:193–203.
54. Sherifali D, Nerenberg K, Pullenayegum E, Cheng JE, Gerstein HC. The effect of oral antidiabetic agents on A1C levels. Diabetes Care. 2010;33:1850–64.
55. Kaufman FR. Medical management of type 1 diabetes. 5th ed. Alexandria: American Diabetes Association; 2008.

Chapter 7
Diet and the Prevention of Type 2 Diabetes

Norman J. Temple and Nelia P. Steyn

Keywords Type 2 diabetes • Dietary fiber • Cereal fiber • Glycemic index
• Prevention of diabetes

Key Points

- The prevalence of type 2 diabetes is increasing rapidly in all parts of the world.
- Obesity and physical inactivity are major modifiable risk factors.
- Whereas cohort studies indicate that cereal fiber, a major source of insoluble fiber, is most potent for preventing diabetes, intervention studies indicate that soluble fiber is most effective for improving glycemic control.
- Evidence from cohort studies shows an association between the glycemic index and the glycemic load of the diet and the risk of diabetes.
- Saturated fat appears to increase the risk of diabetes whereas polyunsaturated fat appears to be protective.
- Diets poor in magnesium may increase the risk of diabetes.

By far the most common type of diabetes is type 2. The focus of this chapter is the role of diet in the prevention of that condition. The previous chapter by Marion Franz focused on the therapy of both type 1 and type 2 diabetes. For the past several decades the prevalence of type 2 diabetes has been rising rapidly around the world. We can confidently ascribe this to changes in lifestyle, most notably on diet, a reduction in physical activity, and the resulting obesity epidemic.

N.J. Temple, PhD
Centre for Science, Athabasca University, Athabasca, AB, T9S 3A3, Canada

N.P. Steyn, PhD (✉)
Knowledge Systems, Human Sciences Research Council,
Private Bag X9182, Cape Town 8000, South Africa
e-mail: npsteyn@hsrc.ac.za

N.J. Temple et al. (eds.), *Nutritional Health: Strategies for Disease Prevention*,
Nutrition and Health, DOI 10.1007/978-1-61779-894-8_7,
© Springer Science+Business Media, LLC 2012

7.1   Classification and Diagnosis

Type 2 diabetes was formerly known as non-insulin-dependent diabetes (NIDDM). As many as half of all people with the disease have not been diagnosed. For this reason unless systematic testing of a population is carried out, estimates of the prevalence of the disease are likely to be serious underestimates. Furthermore, because of the differences in criteria, comparisons of rates from recent and earlier studies must be made with caution.

The fasting plasma glucose concentration (FPG) is now given greater importance as a criterion for diagnosis. The FPG considered diagnostic of diabetes is ≥7.0 mmol/L (≥126 mg/dL) [1, 2]. This value was introduced in the late 1990s and is a decrease from the former value of 7.8 mmol/L (140 mg/dL). Diagnosis can also be made based on an oral glucose tolerance test using 75 g of glucose. A 2-h postload glucose level of ≥11.1 mmol/L (200 mg/dL) indicates diabetes. If the oral glucose tolerance test reveals a level above normal (7.8 mmol/L or 140 mg/dL) but below the cut-off for diabetes, this is indicative of impaired glucose tolerance (IGT). An additional category, impaired fasting glycemia (IFG), has been introduced. This refers to a FPG level that is above normal but below the cut-off for diabetes, i.e., 6.1 mmol/L (110 mg/dL) to <7.0 mmol/L (125 mg/dL). Only a minority of individuals with IGT have IFG, and conversely, only a minority of those with IFG have IGT. The above values are WHO criteria. The criteria used by the American Diabetes Association (ADA) are almost the same (see Table 6.1 in the previous Chap. 6). The one notable difference is that the ADA uses a slightly lower cut-off for IFG: a FPG of ≥5.6 mmol/L (≥100 mg/dL).

7.2   Epidemiology

7.2.1   Prevalence and Incidence of Type 2 Diabetes

Based on the National Health Interview Survey, the estimated lifetime risk of developing diabetes for an individual born in the USA in 2000 is 32.8% for men and 38.5% for women [3]. However, the authors of this estimate note that the true risk may well be higher. The following chapter gives prevalence data for the USA and other countries.

The lowest rates are found in rural communities where people retain traditional lifestyles. But dramatic increases in the prevalence and incidence of type 2 diabetes have been observed in communities where the diet has shifted from a traditional indigenous diet to a typical "Western" diet. There is growing evidence that the disease has reached epidemic proportions in many developing and newly industrialized countries [4].

The prevalence of type 2 diabetes differs considerably between different ethnic groups living in the same country in apparently similar environments. For example, in the United Kingdom and other countries, Asian Indians have high prevalence rates of diabetes compared with the indigenous populations [5, 6]. Similarly, the prevalence in Hispanic Americans is slightly higher than in African-Americans, while the white population has a much lower prevalence [3]. Native American populations have the highest prevalence rates [7, 8]. The Pima Indians of Arizona have the unfortunate distinction of having the highest prevalence rates in the world; roughly 50% of all adults have the condition [9]. The one other population with a comparable prevalence (about 40%) is the Micronesians of the Pacific island of Nauru [10].

7.2.2   Mortality

Age-adjusted mortality rates among persons with type 2 diabetes are 1.5–2.5 times higher than in the general population; this translates to a reduction of life expectancy of approximately 10 year [11]. Other researchers estimated that for an individual diagnosed at age 40 with the condition, the number of years of life lost will be 11.6 in men and 14.3 in women; for quality-adjusted life years the loss will be 18.6 and 22.0 year, respectively [3]. This is strongly related to the considerably higher risk of cardiovascular disease, especially coronary heart disease [12], as well as of renal disease. This risk is greatest where other risk factors are also present, notably hyperlipidemia, hypertension, and smoking, and if the diabetes has been present for a long period of time.

7.3   Metabolic Changes During Development of Type 2 Diabetes: Insulin Resistance and Impaired Glucose Homeostasis

The underlying metabolic defect in type 2 diabetes is resistance to the action of insulin [13, 14]. The beta cells of the pancreas respond to this by increasing output of insulin in order to maintain normal blood glucose levels. Therefore, insulin resistance is indicated by elevated insulin concentrations, either in the fasting state or after a glucose load.

In the early stages of the disease process glucose tolerance may be normal. However, hypersecretion of insulin is usually insufficient to maintain normal glucose levels indefinitely and, as a result there is progression to IGT and IFG. At this stage insulin secretion declines at a variable rate and the disease process advances from IGT and IFG to overt diabetes. The early abnormalities of glucose metabolism are steps on the road to type 2 diabetes [13, 14]. This occurs either as a result of an inherent defect of the beta cell or because of glucotoxicity whereby the beta cell is damaged as blood glucose levels rise. As will be discussed in Sect. 6, we have growing evidence that this sequence of events can be slowed or even reversed by appropriate intervention, especially lifestyle change.

IGT, IFG, and type 2 diabetes are often associated with a cluster of clinical and metabolic abnormalities including central obesity, hypertension, hyperuricemia, raised levels of plasminogen activator inhibitor-1, and an abnormal blood lipid pattern, namely raised levels of triglyceride and reduced levels of high-density lipoprotein cholesterol. This constellation of features constitutes the "syndrome" now known as the metabolic syndrome or insulin resistance syndrome [15]. These features all contribute to the increased risk of cardiovascular disease associated with IGT and type 2 diabetes and may coexist with hyperinsulinemia before abnormalities of blood glucose are detectable.

A wide range of lifestyle-related factors have been implicated, including physical inactivity, several aspects of the diet, and the subsequent development of overweight and obesity. These factors may be associated both with the development of insulin resistance as well as with its progression to impaired glucose metabolism (IFG, IGT) and eventually to type 2 diabetes. The nutritional status of a person before birth as well as in infancy and childhood appears to be a significant factor as discussed in Chap. 4 by Barker.

7.4 Irreversible Risk Factors

7.4.1 Genetic Factors

As discussed earlier the prevalence of type 2 diabetes varies considerably among different ethnic groups living in apparently similar environments. While differences in lifestyle undoubtedly account for some of this, inherent genetic differences in susceptibility to the disease are also probably involved.

The importance of genetic factors is further strongly suggested by studies of family history as a risk factor. The risk of having type 2 diabetes is increased two to sixfold if a parent or sibling has the disease [16]. At present it is impossible to quantify the relative contributions of genetic and environmental factors. The disease is believed to be polygenic; some relevant genes have been identified, but by no means all.

7.4.2 Age

Among Caucasians in the developed world the prevalence of type 2 diabetes increases with age, at least into the seventies. But in developing countries many cases occur in younger adults. A newly emerging feature of type 2 diabetes is its occurrence in adolescence [17]. This has been observed in many ethnic groups in recent years. In Japan, for example, increasing numbers of cases are seen in school children so that the disease is now more common than type 1 diabetes [18]. Caucasians, however, have largely escaped this trend. As in adults, type 2 diabetes in adolescents is frequently asymptomatic and is detected mainly by screening.

7.5 Modifiable Risk Factors

7.5.1 Obesity

Many studies have documented that obesity is a major risk factor. For example, data from the Nurses' Health Study revealed that the risk of diabetes is 20-fold greater in obese women (BMI 30–35) as compared with slim women (BMI<23) [19]. Even being at the high end of normal may carry significant risk: the risk ratio was 2.7 when comparing those with a BMI of 23–25 with those with a BMI of <23 [19]. Obesity has increased rapidly in many populations in recent years and this is associated with a parallel rise in the prevalence of type 2 diabetes.

The distribution of body fat may be a more reliable predictor than BMI of the risk of developing diabetes. Several studies indicate that a high waist circumference or a high waist-to-hip ratio (an apple shape) is an important risk factor [20]. For example, in a study of Japanese-American men, the anthropometric measure that best predicted the disease was the amount of intra-abdominal fat [21]. As there are relatively little data available concerning the waist circumference or waist-to-hip ratio in different populations, BMI is still the most common tool used to measure risk associated with excess adiposity and to provide guidelines. However, future studies may permit guidelines to be established that utilize waist circumference or waist-to-hip ratio. Waist circumference is generally considered a more useful parameter than waist-to-hip ratio as it is easier to measure and has good predictive ability [22].

7.5.2 Physical Inactivity

Physical activity has repeatedly been shown to have a protective association with the risk of diabetes [23–25]. The relationship is most pronounced in subjects who also have other risk factors, such as obesity, hypertension, or parental diabetes.

Guidelines used today typically recommend moderate exercise on at least 5 day/week but do not specify heart rate targets. However, more recent evidence suggests that vigorous exercise may be required to improve insulin sensitivity. This was shown in a study conducted by McAuley et al. [26] on normoglycemic, insulin-resistant adults. Insulin sensitivity improved in those who engaged in vigorous exercise but not in those who participated in moderate exercise. The vigorous exercise program required participants to train five times/week for at least 20 min/session at an intensity of 80–90% of predicted maximum heart rate.

7.5.3   Carbohydrates and Dietary Fat

A debate that has been ongoing for decades is the question of whether high intakes of either carbohydrate or fat predispose to diabetes [27]. One important piece of indirect evidence is that a relatively high intake of dietary fat increases the risk of obesity, a subject that is reviewed in Chap. 8 by Whittle, McKinley, and Woodside. By extension this also implicates dietary fat in the causation of diabetes. The same applies to sugar, as discussed below.

In experimental animals diets rich in any type of fat (with the exception of n-3 fatty acids) have been shown to result in insulin resistance relative to high-carbohydrate diets [28–30]. The data from epidemiological studies are less consistent. While some studies have suggested that a high-fat or low-carbohydrate intake may hasten the progress of diabetes [31–35], most studies, especially large cohort studies, have shown little association between the quantity of fat or carbohydrate in the diet and the risk of the disease [36–39]. Thus a wide range of carbohydrate intakes may be acceptable in terms of achieving a low risk of type 2 diabetes. We now examine a more important question, namely the impact of the type of carbohydrate-containing foods, especially the fiber content, and also the type of fat.

7.5.4   Dietary Fiber

During the 1970s Trowell proposed that dietary fiber had an important role in the prevention and management of a range of diseases, including type 2 diabetes [40, 41]. Subsequent research studies have clearly shown that dietary fiber is one of the factors that influence postprandial glucose and insulin response.

Cross-sectional studies have revealed an inverse relationship between fiber intake and blood insulin levels [33, 35, 42, 43], implying that fiber improves insulin sensitivity. Three large cohort studies, the Health Professionals Follow-up Study carried out on men aged 40–75 [36], the (original) Nurses' Health Study carried out on women aged 40–65 [44], the Nurses' Health Study II carried out on women aged 24–44 [45], and the Iowa Women's Health Study carried out on women aged 55–69 [38], have studied the effects of fiber on the risk of developing diabetes. All four studies clearly showed that a relatively low intake of dietary fiber poses a significantly increased risk for the disease. The association was found to be strongest for cereal fiber; comparison of the extreme quintiles revealed a risk ratio of 0.64–0.72, after correcting for confounding variables such as age, BMI, smoking, and physical activity. Later studies have lent confirmation to these findings [46]. These findings suggest that insoluble fiber is strongly protective against diabetes. By contrast, a much weaker protective association has been seen for sources of soluble fiber.

As the main dietary source of cereal fiber is whole grain products, these findings strongly imply that whole grains are protective against the development of diabetes. There are many substances contained in whole grains that may deserve the credit for this. We must be cautious, therefore, before bestowing the credit on dietary fiber. Other possibilities include vitamin E, antioxidants, phytochemicals (such as isoflavins), and lignans. Since many of these factors occur together in cereals, it is extremely difficult to determine the precise benefits of each.

Dietary intervention studies have been conducted in which supplements of dietary fiber have been given for several weeks. This has been shown to lower both postprandial glycemia and insulin levels [47] and to lead to an overall improvement in glycemic control as measured by HbA_{1C}. This is seen in both normal subjects and in those with type 2 diabetes. The effects are most pronounced when soluble fiber is given, regardless of whether it is taken as a supplement or in food. Lesser effects have been reported using insoluble fiber. We therefore see a clear contrast between what has been observed in the above cohort studies and in intervention studies: the former indicate that insoluble fiber from cereals is most potent for preventing diabetes whereas the latter indicate that soluble fiber is most effective for improving glycemic control.

The studies discussed here leave little doubt that the great majority of people consume insufficient dietary fiber, especially cereal fiber; this is a significant factor in increasing the risk of diabetes while also worsening glycemic control in those with the disease. The required intake is likely to be at least as great as the median amount consumed by people in the highest quintile in the cohort studies discussed earlier. That amount was 24.1 g [44] and 26.5 g [38] in women and 29.7 g [36] in men. This is similar to the Dietary Reference Intakes for fiber published in the USA which recommended an intake of 21–25 g for women and 30–38 g for men.

As there is still conjecture as to both the ideal intake and most beneficial type of fiber, it makes most sense to emphasize appropriate carbohydrate sources rather than to specify quantities of fiber. In other words, people should be encouraged to eat generous amounts of whole grain cereals, legumes, vegetables, and fruit.

7.5.5 Sugar

There had been much speculation over the past several decades that a high intake of sugar is a factor in the causation of diabetes, but the supporting evidence was never convincing. However, evidence that has emerged in recent years has changed the picture. Findings from cohort studies have revealed that intake of sugar-sweetened beverages (SSBs) significantly increase the risk of developing type 2 diabetes. Comparing extreme quantiles of intake of SSBs (typically 1–2 servings/day vs. <1/month) persons with the highest intake had a 26% greater risk of developing type 2 diabetes than those in the lowest quantile [48]. A major reason that SSBs cause diabetes is because they induce weight gain [49].

7.5.6 Glycemic Index and Glycemic Load

Different forms of carbohydrate-rich foods cause very different postprandial glycemic responses. In recognition of this, the term glycemic index (GI) was coined in 1981 [50]. It is defined as the glycemic response elicited by a 50 g carbohydrate portion of a food expressed as a percentage of that elicited by a 50 g portion of a reference food (glucose or white bread). Low-GI foods have lower 2-h areas under the glucose curve than the reference food, while high GI foods have higher areas.

GI is a valuable means for comparing different types of food. However, an important limitation of the term is that it ignores the quantity of carbohydrate in the food. For instance, carrots have a high GI but this is misleading as their content of carbohydrate is low. For that reason a newer concept is glycemic load (GL). This is the mean GI of the diet multiplied by its carbohydrate content. Thus a diet rich in foods that have both a high content of carbohydrate and a high GI will have a high GL.

Foods with a low GL include pasta, bran cereals, beans, nuts, apples, apple juice, milk, and yogurt. Intermediate foods include shredded wheat, muesli, banana, pineapple, orange juice, and ice cream. High-GL foods include white bread, rye bread, instant rice, cornflakes, and potatoes. Soft drinks generally have a low GI as the carbohydrate they contain is sucrose which is 50% fructose. In brief, a low-GI diet is plentiful in fruit, vegetables, legumes, bran-rich cereals, and stone-ground bread, while limited in potatoes and white flour. A low-GI diet is really quite similar to a high-fiber diet.

Numerous cohort studies have examined whether GI or GL is associated with the risk of diabetes [51, 52]. Overall, persons who consume a diet with a high GI or high GL value have a 30–40% elevated risk of diabetes.

While the exact mechanisms by which a diet with a high GL may accelerate the development of type 2 diabetes are not fully understood, there is evidence that the following is a reasonably accurate picture of the etiology of the disease [53]. A diet with a high GL repeatedly causes hyperglycemia and thence hyperinsuline-mia. The next step in the chain is insulin resistance (though other factors may also cause this). The cycle of hyperinsulinemia and insulin resistance places the beta cells of the pancreas under long-term increased demand. Eventually, the beta cells start to lose their ability to function properly and the body is now on the road from IGT to type 2 diabetes.

The evidence is quite strong that a diet with a high GI and GL significantly increases the risk of diabetes. Much the same applies to cardiovascular disease [52]. However, this does not constitute a strong reason to incorporate this infor-mation into dietary recommendations. One reason for this is that there is a lack of strong evidence for an association with other health problems, such as obesity [54]. Another reason is that methodological problems means that GI values for foods, especially mixed dishes, are poorly defined. Additionally, it is confusing for people to follow recommendations for a low-GI diet [54]. It makes most sense for people to be encouraged to simply follow the well-established dietary recommendations; it is unsure whether anything more is achieved by adding specific recommendations for a low-GI diet. In the following chapter Franz makes the same argument with respect to therapy for persons who already have diabetes.

7.5.7   Type of Dietary Fat

There is evidence that the type of dietary fat may modify glucose tolerance and insulin sensitivity [55, 56]. In epidemiological studies intake of saturated fat has been associated with poor glucose tolerance [31] and higher fasting levels of both glucose [57, 58], and insulin [33, 59]. Conversely, the intake of polyunsaturated fat has been associated with lower fasting and 2-h glucose concentrations [58, 60] as well as lower risk of type 2 diabetes [37, 61, 62]. There is some evidence that n-3 fatty acids may be protective, but these findings are not consistent [63–66]. In a short-term intervention study substitution of saturated fat by unsaturated fat improved glucose tolerance in young healthy women [67]. Longerterm studies (3 month) [68] showed that replacing saturated fat with monounsaturated fat significantly improves insulin sensitivity in healthy subjects. These findings parallel the situation with heart disease: saturated fat appears to accelerate the metabolic process leading to diabetes whereas polyunsaturated fat is protective.

7.5.8   Magnesium

Several large cohort studies have reported that intake of magnesium shows a strong inverse association with the risk of type 2 diabetes [46]. A meta-analysis reported that the risk ratio is 0.77 when comparing the extreme quintiles, after correcting for confounding variables such as age, BMI, smoking, and physical activity. Some of the studies also adjusted for cereal fiber, an important source of magnesium. When this was done, the strength of the association was reduced but still remained strong and statistically significant. However, in the absence of any randomized trials it would be premature to make specific recommendations concerning magnesium intake.

7.5.9   Meat, Vegetables, and Nuts

Epidemiological evidence has reported that several different foods affect the risk of development of type 2 diabetes. In a recent systematic review, consumption of 50 g/day of processed meat was associated with a 19% higher risk of diabetes, while consumption of unprocessed red meat was associated with a nonsignificant trend toward a higher risk of diabetes [69]. By contrast, the following two foods appear to have a protective association with risk of diabetes. People with a higher intake of green leafy vegetables were reported to have a 14% reduction in risk [70]. The Nurses' Health Study observed a lower risk in regular consumers of nuts [71].

7.5.10   Alcohol

Consumption of moderate amounts of alcohol has a protective association with the risk of diabetes. This is discussed further in the Chap. 15 by Temple.

7.6   Lifestyle Modifications and Risk Reduction

One of the most dramatic demonstrations of the power of lifestyle change to prevent diabetes took place in Cuba [72]. That country experienced a severe economic crisis from 1989 to 2000 following the collapse of the Soviet Union. Because of shortages of both food and fuel there was an estimated 1,040 kcal/day decrease in energy intake while at the same people had to engage in more physical activity. These changes were followed by a decrease in BMI of 1.5 units. During this period deaths from diabetes fell by half.

Several randomized intervention trials (RCTs) have been conducted that have examined the efficacy of lifestyle intervention to prevent or at least slow the progression of IGT to diabetes. Two RCTs on overweight subjects with IGT demonstrated that weight loss achieved by an increase in physical activity and dietary change, including a reduction in total and saturated fat and increased dietary fiber, can reduce the incidence of type 2 diabetes. The Finnish Diabetes Prevention Study (FDPS) assessed the efficacy of an intensive diet and exercise program in 522 adults [73]. The intervention group received individual counseling with respect to diet, weight loss, and physical activity. Weight loss was about 3 kg greater in the intervention group than in the control group. The cumulative incidence of diabetes after 4 year was 23% in the control group and 11% in the intervention group. A similar intervention program, the Diabetes Prevention Program, was conducted in the United States on a larger sample of 3,234 adults [74]. During the 3 year of follow-up about 29% of the control group developed diabetes but only 14% of the diet and exercise group (a second intervention group was treated with a drug). In both studies the estimated risk reduction was about 58%.

The Da Qing Study, conducted in China, was undertaken over a longer intervention period (6 year) than the above studies [75]. The 577 participants with IGT were randomized by clinic rather than as individuals, into a control group or into one of three lifestyle interventions: diet only (including weight loss if the BMI was >25), exercise only, or both diet and exercise. The cumulative incidence of diabetes after 6 year was 68% in the control group, 44% in the diet group, 41% in the exercise group, and 46% in the diet plus exercise group. After adjusting for differences in baseline BMI and fasting glucose, reduction in risk was 31% with diet alone, 46% with exercise alone, and 42% with both diet and exercise.

Follow-up continued for a further 14 year after the 6 year of active intervention had ended [76]. A large decrease in diabetes risk was seen over the 20-year period: compared with control participants, those in the combined lifestyle intervention groups had a 43% lower incidence. The average annual incidence of diabetes was 7% for intervention participants vs. 11% in control participants.

In addition to the above three studies several smaller RCTs have been carried out in recent years. Gillies et al. carried out a systematic review and meta-analysis [77]. The pooled results indicate that diet reduces the risk of IGT progressing to diabetes by 33%, exercise by 51%, and the two interventions together by 51%. This suggests that exercise has a stronger impact than diet.

These studies provide strong evidence that for adults who are at high risk of developing type 2 diabetes, changes in lifestyle can be highly protective. This was achieved even though the average amount of weight lost was relatively small. This emphasizes the importance of even a small degree of weight loss in conjunction with an increase in physical activity in the prevention of diabetes. While these lifestyle intervention studies show that quite modest changes can substantially reduce the progression of IGT to diabetes, it is not clear whether it will be possible to achieve this success in the general population or to maintain these lifestyle changes for longer periods.

While these trials are of enormous importance, it is nevertheless equally important to appreciate that even in the intensive intervention groups, many of the participants go on to develop diabetes, and this steadily worsens over the years. It is not clear whether this is due to an inability to sustain the necessary intensive lifestyle interventions or whether there is an inevitable deterioration in beta cell function. It is well established that the ability to secrete insulin has already started to decrease even during the phase of IGT and then declines progressively as the disease process continues, regardless of treatment. Thus the best hope of truly "preventing" type 2 diabetes probably lies not in focusing on those with IGT but rather in implementing lifestyle intervention programs in obese people in the general population and others at elevated risk, especially in populations where the disease is common.

7.7 Conclusions

Type 2 diabetes was previously a disease of the middle-aged and elderly, but in recent years it has escalated in younger age groups and the condition is now seen in adolescence, especially in high-risk populations. The disease has a devastating health impact.

The fast growing epidemic of the disease has occurred in parallel with a dramatic rise in the frequency of overweight and obesity. There is convincing evidence that obesity, particularly when centrally distributed, increases insulin resistance and thence leads to IGT, and eventually to diabetes. The risk of diabetes is seen to increase with weight, even within the normal range of BMI. An optimum BMI is therefore at the lower end of the normal range (i.e., around 21). Weight loss in the overweight and obese has been convincingly shown to reduce insulin resistance and diabetes risk.

Lack of physical activity is also a major factor in the causation of diabetes. Ideally, people should engage in moderate or vigorous physical activity for at least 1 h every day. Vigorous activity appears to be more effective than moderate activity. Physical activity that results in weight loss is likely to prove especially effective.

Much research indicates that while both carbohydrates and fats are associated with diabetes, it is the type of carbohydrate-containing food and the type of fat that is probably critical, rather than the quantity. A generous intake of dietary fiber, especially the soluble type, appears to improve insulin sensitivity and glycemic control in type 2 diabetes. However, cohort studies have revealed that cereal fiber (i.e., insoluble fiber) is most closely associated with a reduced risk of the disease, consonant with a general benefit of whole grain foods. Other evidence indicates that replacement of foods with a high GI by foods with a lower GI, thereby lowering the GL of the diet, is another beneficial step.

Saturated fat has been shown to be associated with higher fasting glucose and insulin levels, an increased risk of IGT, and increased rates of progression from IGT to diabetes. Replacing an appreciable proportion of dietary saturated fat with unsaturated fat, within the usual range of total fat intake, is associated with improved glucose tolerance and improved insulin sensitivity.

A dietary pattern that minimizes the risk of IGT and diabetes is therefore one that emphasizes foods with a low GL and that are rich in fiber but low in saturated fat. However, the efficacy of this dietary strategy has not yet been conclusively established using intervention studies. Nevertheless, it should be noted that low rates of type 2 diabetes are seen in groups and populations consuming diets rich in whole grain cereals, legumes, fruit, and vegetables, and with low intakes of foods rich in saturated fat.

In view of the enormous and growing economic, social, and personal cost of the disease, it seems prudent that primary prevention should be a major priority. Accordingly, there is an urgent need to tackle the epidemic of overweight and obesity, and to encourage greater participation in physical activity. These are the two central components of an antidiabetes strategy. Simultaneously, a healthy diet should be a key feature of such a lifestyle approach.

The dietary factors and lifestyle changes that help prevent diabetes are also effective in therapy. In that regard diabetes resembles the conditions that are closely associated with it, namely obesity, hypertension, hyperlipidemia, and cardiovascular disease. Indeed, the similarities with these conditions go much further: the lifestyle approach outlined above for the prevention of diabetes will, in general, help prevent all of them.

References

1. Gavin JR, Alberti KGMM, Davidson MB, et al. Report of the expert committee on the diagnosis and classification of diabetes mellitus. Diabetes Care. 1997;20:1183–97.
2. WHO Consultation Group. Definition, diagnosis and classification of diabetes mellitus and its complications. Part 1: diagnosis and classification of diabetes mellitus. Geneva: World Health Organisation; 1999.
3. Narayan KM, Boyle JP, Thompson TJ, Sorensen SW, Williamson DF. Lifetime risk for diabetes mellitus in the United States. JAMA. 2003;290:1884–90.
4. Shaw JE, Sicree RA, Zimmet PZ. Global estimates of the prevalence of diabetes for 2010 and 2030. Diabetes Res Clin Pract. 2010;87:4–14.

 5. Mather HM, Chaturvedi N, Fuller JH. Mortality and morbidity from diabetes in South Asians and Europeans: 11- year follow-up of the Southall Diabetes survey, London, UK. Diabet Med. 1998;15:53–9.
 6. Omar MA, Seedat MA, Dyer RB, et al. South African Indians show a high prevalence of NIDDM and bimodality in plasma glucose distribution patterns. Diabetes Care. 1994;17: 70–3.
 7. Gohdes D. Diabetes in North American Indians and Alaska Natives. In: Harris MI, Cowie CC, Stern MP, et al., editors. Diabetes in America. Bethesda, MD: National Institutes of Health; 1995.
 8. Knowler WC, Pettitt DJ, Saad MF, et al. Diabetes mellitus in the Pima Indians: incidence, risk factors and pathogenesis. Diabetes Metab Rev. 1990;6:1–27.
 9. Bennett PH. Type 2 diabetes among the Pima Indians of Arizona: an epidemic attributable to environmental change? Nutr Rev. 1999;57(5 Pt 2):S51–4.
10. King H, Rewers M. Global estimates for prevalence of diabetes mellitus and impaired glucose tolerance in adults. WHO Ad Hoc Diabetes Reporting Group. Diabetes Care. 1993;16: 157–77.
11. Gu K, Cowie CC, Harris MI. Mortality in adults with and without diabetes in a national cohort of the U.S. population, 1971–1993. Diabetes Care. 1998;21:1138–45.
12. Stamler J, Vaccaro O, Neaton JD, et al. Diabetes, other risk factors, and 12-yr cardiovascular mortality for men screened in the Multiple Risk Factor Intervention Trial. Diabetes Care. 1993;16:434–44.
13. Edelstein SL, Knowler WC, Bain RP, et al. Predictors of progression from impaired glucose tolerance to NIDDM: an analysis of six prospective studies. Diabetes. 1997;46:701–10.
14. Weyer C, Bogardus C, Mott DM, et al. The natural history of insulin secretory dysfunction and insulin resistance in the pathogenesis of type 2 diabetes mellitus. J Clin Invest. 1999;104: 787–94.
15. Klein BE, Klein R, Lee KE. Components of the metabolic syndrome and risk of cardiovascular disease and diabetes in Beaver Dam. Diabetes Care. 2002;25:1790–4.
16. Everhart JE, Knowler WC, Bennett PH. Incidence and risk factors for noninsulin-dependent diabetes. In: Harris MI, Hamman RF, editors. Diabetes in America, Diabetes Data Compiled 1984. Washington, DC: US Dept of Health and Human Services. NIH Publication No. 85–1468;1985.
17. Dabelea D, Hanson RL, Bennett PH, et al. Increasing prevalence of type II diabetes in American Indian children. Diabetologia. 1998;41:904–10.
18. Kitagawa T, Owada M, Urakami T, et al. Increased incidence of non-insulin dependent diabetes mellitus among Japanese schoolchildren correlates with an increased intake of animal protein and fat. Clin Pediatr (Phila). 1998;37:111–5.
19. Hu FB, Manson JE, Stampfer MJ, et al. Diet, lifestyle, and the risk of type 2 diabetes mellitus in women. N Engl J Med. 2001;345:790–7.
20. Chan JM, Rimm EB, Colditz GA, et al. Obesity, fat distribution, and weight gain as risk factors for clinical diabetes in men. Diabetes Care. 1994;17:961–9.
21. Boyko EJ, Fujimoto WY, Leonetti DL, et al. Visceral adiposity and risk of type 2 diabetes: a prospective study among Japanese Americans. Diabetes Care. 2000;23:465–71.
22. Lean ME, Han TS, Seidell JC. Impairment of health and quality of life in people with large waist circumference. Lancet. 1998;351:853–6.
23. Manson JE, Rimm EB, Stampfer MJ, et al. Physical activity and incidence of non-insulin-dependent diabetes mellitus in women. Lancet. 1991;338:774–8.
24. Manson JE, Nathan DM, Krolewski AS, et al. A prospective study of exercise and incidence of diabetes among US male physicians. JAMA. 1992;268:63–7.
25. Helmrich SP, Ragland DR, Leung RW, et al. Physical activity and reduced occurrence of non-insulin-dependent diabetes mellitus. N Engl J Med. 1991;325:147–52.
26. McAuley KA, Williams SM, Mann JI, et al. Intensive lifestyle changes are necessary to improve insulin sensitivity: a randomized controlled trial. Diabetes Care. 2002;25:445–52.
27. Grundy SM. The optimal ratio of fat-to-carbohydrate in the diet. Annu Rev Nutr. 1999;19: 325–41.

28. Storlien LH, Baur LA, Kriketos AD, et al. Dietary fats and insulin action. Diabetologia. 1996;39:621–31.
29. Hedeskov CJ, Capito K, Islin H, Hansen SE, Thams P. Long-term fat-feeding-induced insulin resistance in normal NMRI mice: postreceptor changes of liver, muscle and adipose tissue metabolism resembling those of type 2 diabetes. Acta Diabetol. 1992;29:14–9.
30. Storlien LH, Jenkins AB, Chisholm DJ, Pascoe WS, Khouri S, Kraegen EW. Influence of dietary fat composition on development of insulin resistance in rats. Relationship to muscle triglyceride and omega-3 fatty acids in muscle phospholipid. Diabetes. 1991;40:280–9.
31. Feskens EJM, Virtanen SM, Räsänen L, et al. Dietary factors determining diabetes and impaired glucose tolerance. A 20-year follow-up of the Finnish and Dutch cohorts of the Seven Countries Study. Diabetes Care. 1995;18:1104–12.
32. Marshall JA, Hoag S, Shetterly S, Hamman RF. Dietary fat predicts conversion from impaired glucose tolerance to NIDDM. The San Luis Valley Diabetes Study. Diabetes Care. 1994;17:50–6.
33. Marshall JA, Bessesen DH, Hamman RF. High saturated fat and low starch and fiber are associated with hyperinsulinemia in a non-diabetic population: the San Luis Valley Diabetes Study. Diabetologia. 1997;40:430–8.
34. Mayer EJ, Newman B, Quesenberry CP, Selby JV. Usual dietary fat intake and insulin concentrations in healthy women twins. Diabetes Care. 1993;16:1459–69.
35. Lovejoy J, DiGirolamo M. Habitual dietary intake and insulin sensitivity in lean and obese adults. Am J Clin Nutr. 1992;55:1174–9.
36. Salmeron J, Ascherio A, Rimm EB, et al. Dietary fiber, glycemic load and risk of NIDDM in men. Diabetes Care. 1997;20:545–50.
37. Salmeron J, Hu FB, Manson JE, et al. Dietary fat intake and risk of type 2 diabetes in women. Am J Clin Nutr. 2001;73:1019–26.
38. Meyer KA, Kushi LH, Jacobs DR, Slavin J, Sellers TA, Folsom AR. Carbohydrates, dietary fiber, and incident type 2 diabetes in older women. Am J Clin Nutr. 2000;71:921–30.
39. Bessesen DH. The role of carbohydrates in insulin resistance. J Nutr. 2001;131:2782S–6S.
40. Trowell HC. Dietary fiber, ischaemic heart disease and diabetes mellitus. Proc Nutr Soc. 1973;32:151–7.
41. Trowell HC. Dietary fiber hypothesis of the etiology of diabetes mellitus. Diabetes. 1975;24:762–5.
42. Feskens EJ, Loeber JG, Kromhout D. Diet and physical activity as determinants of hyperinsulinaemia: the Zutphen Elderly Study. Am J Epidemiol. 1994;140:350–60.
43. Ludwig DS, Pereira MA, Kroenke CH, et al. Dietary fiber, weight gain, and cardiovascular disease risk factors in young adults. JAMA. 1999;282:1539–46.
44. Salmeron J, Manson JE, Stampfer MJ, Colditz GA, Wing AL, Willett WC. Dietary fiber, glycemic load, and risk of non-insulin-dependent diabetes mellitus in women. JAMA. 1997;277:472–7.
45. Schulze MB, Liu S, Rimm EB, Manson JE, Willett WC, Hu FB. Glycemic index, glycemic load, and dietary fiber intake and incidence of type 2 diabetes in younger and middle-aged women. Am J Clin Nutr. 2004;80:348–56.
46. Schulze MB, Schulz M, Heidemann C, Schienkiewitz A, Hoffmann K, Boeing H. Fiber and magnesium intake and incidence of type 2 diabetes: a prospective study and meta-analysis. Arch Intern Med. 2007;167:956–65.
47. McIntosh M, Miller C. A diet containing food rich in soluble and insoluble fiber improves glycemic control and reduces hyperlipidemia among patients with type 2 diabetes mellitus. Nutr Rev. 2001;59:52–5.
48. Malik VS, Popkin BM, Bray GA, Després JP, Hu FB. Sugar-sweetened beverages, obesity, type 2 diabetes mellitus, and cardiovascular disease risk. Circulation. 2010;121:1356–64.
49. Malik VS, Popkin BM, Bray GA, Després JP, Willett WC, Hu FB. Sugar-sweetened beverages and risk of metabolic syndrome and type 2 diabetes: a meta-analysis. Diabetes Care. 2010;33:2477–83.
50. Jenkins DJ, Wolever TM, Taylor RH, et al. Glycemic index of foods: a physiological basis for carbohydrate exchange. Am J Clin Nutr. 1981;34:362–6.

51. Barclay AW, Petocz P, McMillan-Price J, et al. Glycemic index, glycemic load, and chronic disease risk—a meta-analysis of observational studies. Am J Clin Nutr. 2008;87:627–37.
52. Chiu CJ, Liu S, Willett WC, Wolever TM, Brand-Miller JC, Barclay AW, et al. Informing food choices and health outcomes by use of the dietary glycemic index. Nutr Rev. 2011;69: 231–42.
53. Ludwig DS. The glycemic index: physiological mechanisms relating to obesity, diabetes, and cardiovascular disease. JAMA. 2002;287:2414–23.
54. Hare-Bruun H, Nielsen BM, Grau K, Oxlund AL, Heitmann BL. Should glycemic index and glycemic load be considered in dietary recommendations? Nutr Rev. 2008;66:569–90.
55. Lichtenstein AH, Schwab US. Relationship of dietary fat to glucose metabolism. Atherosclerosis. 2000;150:227–43.
56. Hu FB, van Dam RM, Liu S. Diet and risk of type II diabetes: the role of types of fat and carbohydrate. Diabetologia. 2001;44:805–17.
57. Feskens EJ, Kromhout D. Habitual dietary intake and glucose tolerance in euglycaemic men: the Zutphen Study. Int J Epidemiol. 1990;19:953–9.
58. Trevisan M, Krogh V, Freudenheim J, et al. Consumption of olive oil, butter and vegetable oils and coronary heart disease risk factors. The Research Group ATS-RF2 of the Italian National Research Council. JAMA. 1990;263:688–92.
59. Parker DR, Weiss ST, Troisi R, Cassano PA, Vokonas PS, Landsberg L. Relationship of dietary saturated fatty acids and body habitus to serum insulin concentrations: the Normative Aging Study. Am J Clin Nutr. 1993;58:129–36.
60. Mooy JM, Grootenhuis PA, de Vries H, et al. Prevalence and determinants of glucose intolerance in a Dutch caucasian population. The Hoorn Study. Diabetes Care. 1995;18:1270–3.
61. Colditz GA, Manson JE, Stampfer MJ, Rosner B, Willett WC, Speizer FE. Diet and risk of clinical diabetes in women. Am J Clin Nutr. 1992;55:1018–23.
62. Meyer KA, Kushi LH, Jacobs DR, Folsom AR. Dietary fat and incidence of type 2 diabetes in older Iowa women. Diabetes Care. 2001;2(4):1528–35.
63. Villegas R, Xiang YB, Elasy T, et al. Fish, shellfish, and long-chain n-3 fatty acid consumption and risk of incident type 2 diabetes in middle-aged Chinese men and women. Am J Clin Nutr. 2011;94:543–51.
64. Nanri A, Mizoue T, Noda M, et al.; The Japan Public Health Center–based Prospective Study Group. Fish intake and type 2 diabetes in Japanese men and women: the Japan Public Health Center-based Prospective Study. Am J Clin Nutr. 2011;94:884–91.
65. Brostow DP, Odegaard AO, Koh WP, Duval S, Gross MD, Yuan JM, et al. Omega-3 fatty acids and incident type 2 diabetes: the Singapore Chinese Health Study. Am J Clin Nutr. 2011;94:520–6.
66. Djoussé L, Gaziano JM, Buring JE, Lee IM. Dietary omega-3 fatty acids and fish consumption and risk of type 2 diabetes. Am J Clin Nutr. 2011;93:143–50.
67. Uusitupa M, Schwab U, Mäkimattila S, et al. Effects of two high-fat diets with different fatty acid compositions on glucose and lipid metabolism in healthy young women. Am J Clin Nutr. 1994;59:1310–6.
68. Vessby B, Uusitupa M, Hermansen K, et al. Substituting dietary saturated for monounsaturated fat impairs insulin sensitivity in healthy men and women: the KANWU study. Diabetologia. 2001;44:312–9.
69. Micha R, Wallace SK, Mozaffarian D. Red and processed meat consumption and risk of incident coronary heart disease, stroke, and diabetes mellitus: a systematic review and meta-analysis. Circulation. 2010;121:2271–83.
70. Carter P, Gray LJ, Troughton J, Khunti K, Davies MJ. Fruit and vegetable intake and incidence of type 2 diabetes mellitus: systematic review and meta-analysis. BMJ. 2010;341:c4229.
71. Jiang R, Manson JE, Stampfer MJ, Liu S, Willett WC, Hu FB. Nut and peanut butter consumption and risk of type 2 diabetes in women. JAMA. 2002;288:2554–60.
72. Franco M, Orduñez P, Caballero B, et al. Impact of energy intake, physical activity, and population-wide weight loss on cardiovascular disease and diabetes mortality in Cuba, 1980–2005. Am J Epidemiol. 2007;166:1374–80.

73. Tuomilehto J, Lindstrom J, Eriksson JG, et al. Prevention of type 2 diabetes mellitus by changes in lifestyle among subjects with impaired glucose tolerance. N Engl J Med. 2001;344:1343–50.

74. Knowler WC, Barrett-Connor E, Fowler SE, et al. Reduction in the incidence of type 2 diabetes with lifestyle intervention or metformin. N Engl J Med. 2002;346:393–403.

75. Pan XR, Li GW, Hu YH, Wang JX, Yang WY, An ZX, et al. Effects of diet and exercise in preventing NIDDM in people with impaired glucose tolerance. Diabetes Care. 1997;20:537–44.

76. Li G, Zhang P, Wang J, et al. The long-term effect of lifestyle interventions to prevent diabetes in the China Da Qing Diabetes Prevention Study: a 20-year follow-up study. Lancet. 2008;371:1783–9.

77. Gillies CL, Abrams KR, Lambert PC, et al. Pharmacological and lifestyle interventions to prevent or delay type 2 diabetes in people with impaired glucose tolerance: systematic review and meta-analysis. BMJ. 2007;334:299–307.

Chapter 8
Diet in the Prevention and Treatment of Obesity

Claire R. Whittle, Michelle C. McKinley, and Jayne V. Woodside

Keywords Obesity • Energy density • Portion size • Low-carbohydrate diet • Low-fat diet • Moderate-protein diet

Key Points

- Prevalence rates of obesity have escalated worldwide over the past 3 decades, both in developed and developing countries.
- Dietary composition is a key factor in total energy intake, with energy-dense foods providing less satiety than foods with a low energy density; this can encourage passive overconsumption.
- Fat is the most energy-dense macronutrient providing 9 kcal/g compared to 4 kcal/g for carbohydrate and protein.
- Portion sizes and the availability of cheap energy-dense foods have increased over time, resulting in easy access to excessive energy intakes.
- There is controversy over which is the most effective and safe dietary weight-loss strategy in the long term. The strongest evidence supports a low-fat diet with a low energy density combined with control over portion sizes.

C.R. Whittle, PhD (✉) • M.C. McKinley, PhD • J.V. Woodside, PhD
Nutrition and Metabolism Research Group, School of Medicine, Dentistry and Biomedical Sciences, Centre for Public Health, Institute of Clinical Science, Queen's University Belfast, Block B, Grosvenor Road, Belfast BT12 6BJ, UK
e-mail: c.whittle@qub.ac.uk

N.J. Temple et al. (eds.), *Nutritional Health: Strategies for Disease Prevention*, Nutrition and Health, DOI 10.1007/978-1-61779-894-8_8, © Springer Science+Business Media, LLC 2012

8.1 Introduction

Prevalence rates of obesity have increased dramatically worldwide over the past 3 decades in adults [1, 2]. The condition is caused by an accumulation of excess body fat and has now become the major nutrition-related disease, due to its association with a host of debilitating and life-threatening disorders. Overweight and obesity are associated with the development of hypertension, type 2 diabetes, hyperinsulinemia, dyslipidemia, atherosclerosis, and certain types of cancer (Table 8.1) among other conditions. This is already placing a substantial burden on health-care systems [3], and this is projected to escalate. However, obesity is generally preventable by making lifestyle changes, in particular dietary changes.

8.1.1 Definition of Obesity

As the direct measure of body fat is difficult, body mass index (BMI) is a simple ratio which describes relative weight for height and is significantly correlated with total body fat content. BMI is calculated as weight in kilograms divided by height in meters squared. The World Health Organization (WHO) classifies underweight, normal weight, overweight, and obesity according to categories of BMI (Table 8.2). Unusually large muscle mass, as in trained athletes, such as body builders, cyclists, and rugby players, can increase BMI to 30 kg/m^2, but rarely to >32. This height-independent measure of weight allows comparisons to be made more readily within and between populations of the same ethnic origins.

Table 8.1 Health consequences of obesity

Greatly increased risk (relative risk >3)
    Type 2 diabetes
    Hypertension
    Dyslipidemia
    Breathlessness
    Sleep apnoea
    Gallbladder disease

Moderately increased risk (relative risk about 2–3)
    Coronary heart disease or heart failure
    Osteoarthritis (knees)
    Hyperuricemia and gout
    Complications of pregnancy (e.g., pre-eclampsia)

Increased risk (relative risk about 1–2)
    Cancer (many cancers in men and women)
    Impaired fertility/polycystic ovary syndrome
    Low back pain
    Increased risk during anesthesia
    Fetal defects arising from maternal obesity

Based on ref. [137]

Table 8.2 Interpretation of adult body mass index (BMI) in white individuals

	BMI (kg/m^2)
Classification	Principal cut-off points
Underweight	<18.5
Normal range	18.5–24.9
Overweight	25.0–29.9
Obese	30.0–39.9
Morbidly obese	≥40.0

Table 8.3 Waist circumference and level of associated health risk in white men and women

Level	Men	Women	Health risk[a]
Below action level 1	<94	<80	Low
Action level 1–2	≥94–101.9	≥80–87.9	Increased
Above action level 2	≥102	≥88	High

[a]Risk for type 2 diabetes, coronary heart disease, or hypertension

In recent years there has been much debate whether there is a need to develop different BMI cut-off points for different ethnic groups, such as Asian populations, as a result of accumulating evidence that the associations between BMI, percentage of body fat, and body fat distribution differ across populations. Therefore, for certain ethnic groups, the health risks may actually increase below the overweight cut-off point of 25. A WHO Expert Consultation concluded that the proportion of Asian people with a high risk of type 2 diabetes and cardiovascular disease is substantial at BMIs lower than the existing WHO cut-off point for overweight. However, data available to date do not indicate a clear BMI cut-point for increased risk. The Consultation, therefore, recommended that the current WHO BMI cut-off points should be retained as the international classification [4].

Current guidelines with respect to obesity also recommend the measurement of waist circumference in individuals with a BMI between 25 and 35 [5], as the disproportionate accumulation of abdominal fat relative to total body fat is an independent predictor of disease risk and morbidity. Waist circumference is positively correlated with abdominal fat. There are sex-specific cut-offs which can be used to identify an increased relative risk for the development of obesity-associated diseases in adults, such as type 2 diabetes, hypertension, and coronary heart disease (Table 8.3). Thus, in men and women with a BMI between 25 and 35, high risk is considered as having a waist circumference >102 cm (>40 in.) and 88 cm (>35 in.), respectively [5].

8.1.2 Overweight and Obesity: Prevalence and Trends Over Time

Obesity rates have increased worldwide in recent years, with the condition now being referred to as an epidemic or even a pandemic, because almost every country and age group has been affected. It is estimated by the International Obesity Task

Force that approximately one billion adults are currently overweight and a further 475 million are obese. The US consistently has the greatest prevalence rates and most significant increases of overweight and obesity over time, with rates of obesity having doubled since 1970 [1]. In 2007–2008, the prevalence of adults who were overweight was 68%, of whom 32% were obese [1]. Europe is following the same pathway, although in most cases the prevalence rates are somewhat lower. Approximately 53% of European adults are overweight, of which 17.2% are obese [6]. In total it is estimated that over 143 million Europeans are overweight, with 68.5 million obese [6]. There are variations within Europe, with the higher prevalence rates in Central, Eastern, and Southern Europe [7]. In all countries, variations have been documented for racial and ethnic groups. Obesity is now becoming a problem even in some of the poorest countries of the world [8]; in the past, obesity would have generally been a condition affecting more affluent and developed countries; however, nowadays prevalence rates are particularly high among groups with low income, low education, and low socioeconomic status.

As adult obesity has increased so too has childhood obesity. This has occurred in most industrialized countries, with the exception of Russia and Poland, and in several low-income countries, particularly in urbanized areas [9]. Globally, it is estimated that up to 200 million school-aged children are either overweight or obese; of those 40–50 million are classified as obese. In the USA, 17% of children and adolescents aged 2–19-year old were classified as overweight in the National Health and Nutrition Examination Survey (NHANES) carried out in 2003–2004 [10]. Prevalence rates in children doubled or trebled between the early 1970s and late 1990s in Australia, Brazil, Canada, Chile, Finland, France, Germany, Greece, Japan, the UK, and the USA [9]. Some recent evidence suggests that the increase in childhood obesity might be abating in the USA, UK, and Sweden [11, 12], but this remains to be confirmed.

8.1.3   Health Consequences of Obesity

It has been known for centuries that obesity is the cause of many serious conditions and diseases. In more recent years, however, the full spectrum of diseases which are strongly associated with obesity has become clear. The increase in obesity prevalence has had a significant impact on the global incidence of cardiovascular disease, type 2 diabetes, some cancers, dyslipidemia, infertility, and pregnancy complications, among other conditions (Table 8.1). BMI is also a strong predictor of overall mortality, with risk being lowest at 22.5–25 in both males and females of all ages. With each increase of 5 Units of BMI above this range, risk of all-cause mortality is 30% higher, whereas the increased mortality observed below 22.5 is mostly secondary to smoking-related diseases [13].

In addition, individuals who are overweight or obese may suffer from psychological issues. This can particularly be the case for women, due the media perception that being slender is ideal, and can result in feeling increasingly unacceptable, leading to low self-esteem, anxiety, and depression. Individuals who are obese tend to be discriminated against in many ways, including being less acceptable as marriage partners, and also in the workplace in relation to promotions and salary. One study reported that obesity increased the risk of major depression by 37% in American women, although obese men had a 37% lower risk of depression than normal weight men. Moreover, two eating disorders have been linked to depression and obesity: binge eating disorder and night eating syndrome [14].

8.2   Causes of Obesity

In the steady state, the energy balance equation indicates that energy input (EI) equals energy output (EO), so that $EI - EO = 0$. However, both EI and EO are modifiable, in that EI is represented by food intake and EO is represented by energy expenditure. Obesity is caused by a state where EI exceeds EO. This positive energy imbalance leads to an increase in body fat stores. This can be due to several reasons: (1) an increase in energy intake with no compensatory change in energy expenditure; (2) a decrease in energy expenditure with no compensatory change in energy intake; or (3) an increase in energy intake and a decrease in energy expenditure. The following sections focus on factors likely to contribute to an increased energy intake.

8.2.1   Energy Density

Energy density reflects the energy content of foods, and is expressed per unit weight. Foods with a low energy density provide less energy relative to their weight than foods with a higher energy density. Therefore, when eating food with a low energy density, a larger and more satiating portion can be consumed for the same amount of energy. The energy density of foods consumed is dependent on water and macronutrient content; water has a considerable influence on energy density as it adds substantial weight without adding energy. Dietary fat is the most energy-dense macronutrient: 9 kcal/g compared with 4 kcal/g in carbohydrate and protein. Foods high in fat also have a high palatability. For these reasons it is thought that consuming foods with a high energy density leads to an overconsumption of energy. This overeating effect has been referred to as passive overconsumption [15].

Nowadays, there is a higher consumption of readily-available processed convenience foods, which tend to have a high energy density. This is particularly the case in those from lower socioeconomic backgrounds, as a diet based on added fats, sugars, and refined grains is more affordable than the recommended healthy diet of lean meats, fish, fresh fruit, and vegetables [16]. Moreover, foods which are marketed as "low fat" can be high in sugar, and therefore still have a high energy density. The marketing of low-fat food therefore may imply—falsely—that they can benefit weight control.

8.2.2 Portion Size

A strong environmental factor influencing energy intake is food portion size. Studies have demonstrated that the portion sizes of many foods and beverages have increased since the 1970s, both inside and outside the home, in supermarkets and restaurants [17–20]. This has been suggested as a contributing factor to the obesity epidemic [20].

There is a substantial body of evidence showing that when larger portions are provided, significantly greater amounts of food are consumed [21]. In one study, energy intake was increased with the increasing size of a sandwich (from 6 to 8 in. and 12 in.) [22]. No significant differences in hunger or fullness were reported. Research indicates that on average subjects provided with larger portions consume 30% more energy [23]. One might assume that an individual might compensate for this by reducing energy intake in subsequent meals, but, on the contrary, the opposite has been observed. One study which provided varied portion sizes of potato chips observed that when served the largest compared to the smallest portion, subjects consumed an additional 143 kcal at snack and dinner time combined [24]. This additional energy intake was sustained over a period of 2 days [25]. In a longer-term study, participants were provided with 11 days of food and beverages on two occasions, one of standard portions and the other increased by 50%. During the period with the larger portion sizes, a mean increase in intake of 423 kcal/day was observed, compared to the days when a standard portion was provided. This evidence strongly supports increased portion sizes contributing to overconsumption [26].

What is not clear is at which stage of life people become responsive to portion size. Even though portion sizes have increased over the past number of decades, together with changes in our eating environment, surveys reveal that the average portion sizes consumed by very young children (1–2 years) remained relatively stable from the late 1970s to the late 1990s [27]. However, for children older than 2 years, increases in portion sizes were observed for many foods, particularly beverages, over the same time period [17, 18]. For children to establish what is an adequate portion size to satisfy their energy requirements, one potential strategy to help children regulate appetite and food intake is to allow them to serve their own portions, rather than be served by an adult. One study observed that children consumed 25% less of a main meal when allowed to serve themselves compared to when served by an adult [28]. Such a strategy means they are relying on their own biological cues of hunger rather than being influenced by the external environment.

8.2.3 Soft Drinks

In recent years the consumption of sugar-sweetened beverages (SSBs), which includes all soda, fruit drinks (not fruit juices), and energy and vitamin drinks, has increased globally. The largest increase has been documented in the USA with per capita consumption increasing from 65 to 142 kcal/day between the late 1970s and 2006 [29], with SSBs being a major source of added sugar in the American diet [30]. Large prospective studies and meta-analyses have shown that increased consumption of SSBs is associated with weight gain and obesity [31–33], as well as the development of the metabolic syndrome and type 2 diabetes [33–35]. The additional energy intake from SSBs does not appear to be compensated for at mealtimes [31]. Some research suggests that SSBs have a different mechanistic effect on hunger and satiety compared to solid foods, or liquid foods such as milk and fruit juices [36].

8.2.4   Eating Out and Fast Food

The frequency of eating outside the home has increased greatly in Westernized countries in recent years. This shift has occurred in parallel with increasing obesity prevalence rates [37, 38]. There is some evidence that food prepared at home may be generally of higher nutritional quality, with food purchased out of the home being higher in total energy, total fat, saturated fat, cholesterol, and sodium, and lower in fiber and calcium. In addition, the fat content of foods prepared at home has decreased over time, but there has been no change to food prepared outside the home [39].

There has also been a dramatic growth over time in the number of fast-food outlets. This has been classified as the most rapidly expanding sector in US food distribution. Observational studies have shown a positive relationship between fast-food consumption and weight gain and obesity [40]. The American population study CARDIA observed that those who consumed fast food meals twice a week at baseline and at follow-up gained 4.5 kg more weight, and had a twofold greater increase in insulin resistance when compared to those who ate fast food less than once a week [41]. Together, increasing portion size and eating outside the home, in particular from fast-food outlets, are thought to be major contributing factors to the obesity epidemic [40].

A worrying side-effect of the increase in fast food consumption, apart from the possible direct effects on weight gain and obesity, concerns the *trans* fat content of fast food. Foods commonly sold by fast-food outlets, such as French fries, fried meat, and donuts, and baked goods including pastries, pie crusts, cookies, and pizza dough tend to contain large amounts of industrially produced *trans* fat. These are produced by the industrial hardening of vegetable or marine oils, which results in more stable products with a longer shelf life. Observational studies have shown that a high intake of *trans* fats in the diet increases the risk of weight gain and abdominal fat [42].

Currently, the American Heart Association's recommendation is to limit the amount of *trans* fats eaten to less than 1% of total energy intake; however, survey data from the USA have indicated that double this amount is being consumed [43]. In comparison the National Diet and Nutrition Survey (NDNS) rolling program has shown that *trans* fat intake has decreased over time in the UK and is currently within the recommended levels for all age groups [44]. A worldwide investigation of the content of *trans* fat in commonly consumed fast food, cookies, and snacks, found *trans* fats to contribute up to 50% of the fat in these products. This could potentially equate to 36 g of *trans* fat being ingested in one meal in the USA [45]. In addition, fast food from major chains in most countries has been found to still contain unacceptably high levels of *trans* fats [45]. This is alarming, given that research has shown that 5 g/day of *trans* fat (corresponding to approximately 2% energy intake) is associated with a 30% increase in the risk of coronary heart disease [46]. There has been a response from both the food industry and governments to the potential public health implications of this, but this has varied by country [47, 48].

8.2.5 Snacking

Traditionally, people consume three meals every day. However, in recent times this has changed, and it is now common to have more frequent eating episodes per day with variable sizes; this is commonly referred to as snacking. Findings from large dietary surveys suggests that snacking prevalence is increasing, and the energy density of snack foods is also increasing; it therefore follows that the contribution of snacking to total energy is also increasing [49–51]. Snacking prevalence increased significantly from 71 to 97% in surveys conducted between 1989–1994 and 2003–2006 [50]. Foods sources which increased progressively were desserts, salty snacks, candies, and sweetened beverages [50]. However, there is little direct evidence to link snacking behavior directly to obesity.

8.3   Diet and the Prevention and Treatment of Obesity

Prevention is always preferable to treatment and is universally viewed as the best approach to reverse the rising global prevalence of obesity, particularly because body weight tracks from childhood to adulthood [52]. The recommendations for the dietary prevention of obesity are similar to those for the treatment of obesity, and therefore the two topics will be discussed together.

The dietary treatment of obesity is now a major focus for nutritional research. Dietary weight-loss strategies have changed somewhat over time with a number of variations in macronutrient and fiber composition suggested. However, there is still extensive debate as to which dietary strategy, particularly concerning macronutrient composition, is most effective for weight loss and for the long-term maintenance of a healthy body weight.

8.3.1 Level of Energy Deficit

The key aim for overweight and obese individuals is to reduce energy intake whilst maintaining a nutritionally adequate diet. The deficit required for weight loss depends on the person's level of excess weight and his or her energy requirements.

This energy deficit can be anywhere in the region of 300–1,000 kcal/day with the ideal weight loss for most people being 1–2 lb (0.5–1 kg)/week. Clinical guidance for obesity recommends setting an ideal weight-loss target of 5–10% of original weight [53]. Depending on the level of obesity, it may be the case that further weight loss is required and, in that case, progressive weight-loss targets can be set; a series of smaller targets will be more motivating for the individual rather than one large target which may be unattainable.

8.3.1.1   Very-Low-Energy Diet

Traditionally, the recommendation for the treatment of obesity was to follow a very-low-energy diet (VLED) (450–800 kcal/day), but with all the essential nutrients. However, this strategy has proved difficult to construct with natural foods. VLED has proven to be very effective for weight loss, at least in the short term [54], but there are concerns over its safety and long-term efficacy. Most individuals find compliance with VLED particularly difficult because of increased hunger and lack of variety [55].

8.3.1.2   The 600-kcal-Deficit Diet

One of the key strategies in dietary treatment is to reduce energy intake below expenditure in a way that is acceptable and feasible for the patient. Research has suggested that adherence to an energy-deficient diet may be influenced by the extent of the energy deficit suggested [56]. A more modest energy restriction may therefore be more successful. The 600-kcal-deficit diet has been demonstrated to be an effective approach for some individuals. A systematic review reported that it produces an 11.7 lb (5.3 kg) weight loss in 12 months [57]. The diet is in line with the dietary recommendations for good health [58]. Moreover, a systematic review has shown that the diet is an effective long-term lifestyle intervention for the prevention of weight gain after a follow-up period of 2 years [59].

8.3.1.3   Meal-Replacement Products

In recent years, there has been increasing use of commercial replacement meals that are supplemented with the recommended daily amounts of vitamins and minerals. Meal replacement may simplify weight-loss treatment by allowing the replacement of one or two meals a day with a product which has a defined calorific and nutritional content [60]. These diets have proven to be effective, but, like VLED, compliance is an issue due to limited variety and hunger. Furthermore, such diets do nothing to address the suboptimal eating patterns that contributed to obesity and thus do not help to establish lifelong healthy eating habit. As a result, this type of diet is generally unsustainable in the long term [61, 62].

8.3.2   Macronutrient Profile, Weight Gain, and Weight Loss

Much evidence suggests that macronutrients have different effects on satiety and possibly on metabolism. Different proportions of fat, protein, and carbohydrate within the diet may have different effects on weight gain and loss. The main approaches that have been studied are discussed below.

8.3.2.1   Low-Fat Diets

As fat is the most energy-dense nutrient, many studies have investigated whether the manipulation of the fat content of the diet can help in the prevention or treatment of obesity. Population studies have often reported a positive association between dietary fat and obesity or weight gain, particularly in countries with a lower overall intake of fat [63, 64]. Longitudinal studies have shown positive associations between changes in dietary fat intake and obesity in countries which are undergoing the nutrition transition, whereas inverse relationships have been observed in many Westernized countries [64, 65]. The large observational European Prospective Investigation into Cancer and Nutrition (EPIC) study found no significant association between the amount and type of fat consumed and subsequent weight change over time in men or women [66]. However, epidemiological studies are prone to multiple sources of bias, such as different methods of dietary assessment and under-reporting of fat and overall energy intake (particularly from individuals who are overweight or obese); such studies are therefore unable to identify causal relationships. Results from randomized-controlled trials (RCTs) may provide a better insight into this issue.

A recent RCT was the Women's Health Initiative Dietary Modification Trial. Findings from a subset of this study showed that a low-fat diet (compared a to a usual diet) was associated with modest decreases in percentage body fat and fat mass after 1 and 3 year [67]. The evidence from earlier RCTs has been subjected to a number of robust systematic reviews and meta-analyses. One such systematic review compared the effects of low-carbohydrate/high protein diets and low-fat/high-carbohydrate diets on weight loss. It was reported that low-carbohydrate/high-protein diets were more effective at 6 months, and as effective, if not more so, for up to 1 year. However, this review consisted of only 13 studies, some of which allowed the reintroduction of carbohydrates in the low-carbohydrate/high-protein diet; therefore the long-term efficacy of these diets in unknown [68]. In another systematic review of RCTs that investigated long-term diets ($\geq$1 year) in obese adults, low-fat diets produced a mean weight loss of 8 lb (3.6 kg) up to 3 years when compared to control diets. Improvements in blood pressure, lipid profile, and fasting plasma glucose levels were also reported. The authors concluded that very little evidence supports the use of other diets for weight reduction [57]. A further meta-analysis which examined the effectiveness of long-term lifestyle interventions for the prevention of weight gain in normal weight, overweight, and obese adults found a 600-kcal-deficit/low-fat diet an effective strategy to prevent weight gain [59].

Overall, reducing the fat content of the diet consistently results in weight loss. However, it is likely that other macronutrients are also being reduced when fat intake is reduced, such as staple carbohydrates, and such a reduction would also be a contributing factor to the resulting weight loss [69]. One criticism of low-fat diets is that studies show the tendency to return to pre-intervention weight. There are a number of speculated reasons for this including: a reduced compliance, perhaps due to boredom; an environment which is not supportive of healthier eating, such as family members having a diet higher in fat; and overeating of foods which are low in fat [70].

8.3.2.2   Low-Carbohydrate Diets

Low-carbohydrate diets have become increasingly popular in recent years. This dietary regime is often viewed as an alternative to a low-fat diet for weight loss, particularly in the short term [71–73]. In addition to weight loss, beneficial metabolic changes have been documented, such as improvements in lipid profile and glycemic control [71, 74, 75], even in the absence of weight loss and with increases in saturated fat intake [76]. However, some adverse effects have also been reported such as fatigue, constipation, accumulation of ketone bodies, headache, impaired liver and kidney function, and difficulties in maintaining the weight loss after the diet [55, 77]. Although low-carbohydrate diets for weight loss have been around for many years, the most widely-known one is the "Atkins Diet" named after Dr. Atkins who has sold millions of books detailing this diet. This book portrays the low-carbohydrate eating plan as "lifetime nutritional philosophy" rather than simply a diet.

There are numerous short-term studies detailing the effectiveness of a low-carbohydrate diet [71–73], including the systematic review mentioned previously which reported that low-carbohydrate/high-protein diets are more effective at 6 months and 1 year when compared to low-fat diets [68]. In contrast, another systematic review reported that the weight-loss observed with low-carbohydrate diets was associated with a reduced energy intake and the duration of the diet but not with reduced carbohydrate intake [77]. Longer-term (>1 year) investigations are more limited and the results are equivocal. One long-term study reported successful weight loss in healthy men and women. Moreover, the diet did not result in any deleterious metabolic effects and did not increase cardiovascular risk factors [74]. Likewise, in a comparison of three diets after 2 years larger weight decreases were observed in those following a Mediterranean diet and a low-carbohydrate diets than in a low-fat, low-carbohydrate diet. More favorable changes in lipid profile were observed in those following the low-carbohydrate diet [78]. However, in another trial, even though more weight loss was reported at 6 months with a diet low in carbohydrate and high in protein and fat, compared to a conventional diet, the differences were not significant after 1 year [72]. Moreover, little is known about the clinical impact of adhering to a low-carbohydrate diet for weight maintenance: more research is warranted. One study observed higher total cholesterol and LDL cholesterol, and that the increased consumption of saturated fat was inversely associated with endothelial function, therefore demonstrating an overall negative effect on cardiovascular health over time with a low-carbohydrate diet [79].

Therefore, it is still unresolved whether ingestion of a low-carbohydrate diet is more successful than a low-fat diet for long-term weight loss and whether adherence to such a diet has any adverse health outcomes. It is also worth noting that very-low-carbohydrate diets are relatively unpalatable, and may be low in important food components such as fiber [80]. Until the long-term efficacy and safety has been established, caution should be applied when using low-carbohydrate diets for weight loss. Of particular concern is the increased intake of saturated fat and whether long-term adherence to this particular diet will consequently have an adverse effect on cardiovascular health [81].

8.3.2.3   Moderate-Protein Diets

Alongside manipulation of fat and carbohydrate there has also been interest in manipulating the protein content of the diet. It is suggested that a modest increase in protein (30–35% energy), in substitution for other macronutrients, may promote satiety and facilitate weight loss through reduced energy consumption [82]. A higher proportion of dietary protein has also been associated with greater diet-induced thermogenesis than other macronutrients [83], enhanced energy expenditure, protein balance, fat oxidation, and greater maintenance of lean muscle mass [84, 85].

The number of trials in this area is currently limited, but they do provide some support for the effectiveness of moderate-protein diets in relation to weight loss. Reviews of the literature have concluded that a diet moderately increased in protein with modestly restricted consumption of carbohydrates and fat, particularly saturated fats, can have a beneficial effect on body composition and weight [86, 87].

This approach was illustrated in a randomized, crossover intervention study in women who were fed a diet that was either adequate or moderately high in protein for 4 days. Protein and carbohydrate were either 10% and 60% of energy, or 30% and 40%, respectively. Both diets provided 30% of energy as fat. The higher-protein diet produced greater satiety, thermogenesis, sleeping metabolic rate, protein balance, and fat oxidation. This suggests more positive results for weight loss and weight maintenance than a diet with usual protein content [84]. Another multi-center study compared the effectiveness of a diet moderately high in protein for weight maintenance with a conventional high-carbohydrate diet [88]. A 4-month period of active weight loss was followed by an 8-month weight-maintenance diet. Protein and carbohydrate were either 30% and 40% of energy, or 15% and 55%, respectively. Both diets provided 30% of energy as fat. Although similar weight loss was observed in both groups, the moderate-protein diet was more effective for fat-mass loss and body composition improvement in both the initial weight loss and maintenance periods. In addition, the completion rate was better in the moderate-protein group (64% vs. 45%). This study suggests that a moderate-protein diet may be effective for weight loss and also for weight maintenance, and these effects may be more sustainable than those attained with conventional high-carbohydrate diets, but this requires further confirmation. Given this evidence, moderate-protein diets (30–35%) may be an effective method for weight loss and maintenance, particularly as reducing fat mass whilst maintaining lean mass is important for long-term health and the prevention of weight regain. However, studies examining the effect of moderate-protein diets are, to date, limited in number and therefore further work.

8.3.3   Other Nutritional Factors Contributing to Prevention and Treatment of Obesity

8.3.3.1   Fatty Acids

Most current healthy eating strategies emphasize the importance of increasing the consumption of "good fats" such as monounsaturated fatty acids (MUFA) and *n*-3 fatty acids, whilst reducing intake of saturated and *trans* fats. The effect of MUFA and *n*-3 fatty acids on weight loss and weight management has been investigated in recent years. In one study overweight and obese diabetics were randomized to a high-MUFA diet (45% carbohydrate and 40% fat, of which 20% was MUFA) or a high-carbohydrate/low-fat diet (60%/25%). Both diets provided 15% of energy as protein. After 1 year there was similar weight loss in both groups suggesting that a high-MUFA diet may be a suitable alternative to common high-carbohydrate/low-fat diets [89]. Another randomized trial aimed to assess the best method of weight-loss maintenance in overweight and obese men and women after an initial loss of ≥8% of body weight. Three ad libitum diets were assessed: (1) a diet of moderate fat intake (35–45%) and greater than 20% of fat from MUFA; (2) a low-fat diet (20–30%); (3) or a control diet (35% of energy as fat). Over the 6-month assessment period, participants following all three diets regained weight, but weight regain was slower in the low-fat diet group and the MUFA diet group [90]. Overall evidence to date suggests that high-MUFA diets may be beneficial for weight loss and could perhaps be an alternative to conventional low-fat diets; however, research to date is limited.

Fish and fish oil are considered important components of a healthy diet as they are rich in *n*-3 fatty acids. Generally, the benefits of fish and fish oil consumption have been reflected in improved cardiovascular health [91]. In addition it has been suggested that *n*-3 fatty acids may be beneficial for successful weight reduction. A 3.7 lb (1.7 kg) greater weight loss was reported in men adhering to a weight-reduction diet (30% energy restriction) which included 150 g of cod, five times/ week, than those following a similar diet but with no seafood [92]. This may, how-ever, suggest a benefit of fish in general rather than *n*-3 fatty acids, as cod is rela-tively low in *n*-3 fatty acids in comparison with other fish. *n*-3 Fatty acids from fish consumption have been suggested to influence postprandial satiety in overweight and obese subjects, which is an important factor when adhering to a weight-loss diet [93]. Overall, the evidence suggests that fish, fish oils, and foods rich in *n*-3 fatty acids may contribute to weight loss and maintenance but to date the research in this area is limited. Moreover, fish consumption in Western societies is generally low, and major dietary changes would need to take place for this to have any impact on the current obesity rates.

8.3.3.2   Glycemic Index

The glycemic index (GI) is a system for the classification of carbohydrate-containing foods based on their potential to raise the blood glucose level [94]. It was discussed in Chap. 7. The GI of a food varies depending on its rate of digestion, with the higher GI (>70) value meaning a faster rate of digestion. Foods such as potatoes, white bread, and refined cereals are digested and absorbed rapidly and thus have a high GI and induce increased insulin secretion. However, the GI concept is not intuitive and can be confusing. For example, the GI of potatoes is dependent on how they are cooked, and some seemingly less healthy foods, such as chocolate, can have a low GI.

The GI of foods may have an impact on body weight [95]. It is thought that low-GI foods may benefit weight control by promoting satiety, minimizing postprandial insulin secretion, maintaining insulin sensitivity, and by promoting fat oxidation at the expense of carbohydrate oxidation [96]. In a recent Cochrane review of the literature it was reported that overweight and obese individuals adhering to a low-GI diet had significantly greater decreases in BMI, total fat mass, total cholesterol, and LDL-cholesterol over time, when compared to those on high-GI diets or conventional energy-restricted diets [97]. In addition, a low-GI diet may be easier to adhere to than conventional weight-loss diets as there is less need to restrict the actual intake of food [97]. In contrast, in a systematic review of intervention studies to clarify the role of GI in body weight regulation, no clear pattern was observed. The authors concluded that it is still advisable to recommend a low-fat diet with increased amounts of carbohydrates or protein and increased fiber content for weight loss [98]. These systematic reviews differed in their inclusion criteria and therefore in the number of RCTs included, and therefore comparisons between them are difficult. In general the current evidence supports a potential relationship of GI with weight loss and body composition; however studies are needed to investigate compliance and effectiveness in the long term before firm conclusions can be drawn.

8.3.3.3   Portion Size and Energy Density

Approaches to reduce energy intake have primarily focused on reducing portion sizes and cutting the amount of food energy eaten. However, to move from a completely ad libitum diet to one which is restrictive can be associated with low satiety, hunger, boredom, and dissatisfaction; these problems can lead to little weight loss and poor weight loss maintenance [99]. Major reductions in portion size may therefore not be an optimal strategy. Several studies have demonstrated that eating foods with a low energy density, such as fruit, vegetables, and soups, especially those with a high water content, such as tomatoes, apples, and melon, maintains satiety while reducing energy intake [100]. In addition, increasing fruit and vegetable consumption not only lowers the energy density of the diet but also increases the fiber content, which may help curb hunger [101, 102].

Findings from a clinical trial revealed that advising subjects to eat portions of foods with a low energy density was a more successful weight-loss strategy than fat reduction coupled with restriction of portion sizes. Therefore, eating satisfying portions of foods with a low energy density can help to enhance satiety and control hunger while restricting energy intake for weight management [21]. Thus, for the same energy intake, individuals adhering to a diet with a low energy density can eat a greater weight of food and may experience less hunger than those following a diet which has smaller portion sizes. This may help achieve long-term adherence to a weight-loss diet and maintenance of weight loss.

Clinical studies support the effectiveness of this strategy. In a weight-loss intervention conducted on overweight and obese subjects who were following an energy-restricted diet, each of the participants were randomized to consume either one or two servings of soup daily with a low energy density, two servings of energy-dense snack foods daily, or no special food (control group). Even though all groups showed significant weight loss after 6 months, which was maintained after 12 months, consuming soup led to 50% more weight loss than consuming the same amount of energy as energy-dense snack food [103]. Another 1-year trial in obese women tested two strategies; one group was counseled to reduce fat intake and restrict portions, and the other to reduce dietary intake and increase consumption of water-rich foods, particularly fruit and vegetables. Both strategies were effective for weight loss; however the group who were counseled to eat more fruit and vegetables were found to have a greater reduction in the energy density of their diet and had greater weight loss [21]. In a 2-year follow-up of a study which promoted foods with a low energy density for weight loss, adhering to the diet was associated with maintenance of the weight loss. In addition, when the weight-loss maintainers did consume foods with a high energy density, these tended to be in smaller portions [104]. This evidence suggests that incorporation of foods with a low energy density into a reduced-energy diet can increase the weight loss and help with maintenance. However, there is a need for more studies of this dietary strategy with long-term follow-up.

8.3.4   Weight Maintenance

Weight regain after weight loss, as well as weight cycling, is common in overweight and obese individuals; long-term maintenance remains a critical challenge. Moreover, there is the possibility that weight regain and weight cycling may be linked to increased mortality [105], although this was not confirmed in a recent large study [106]. The beginning stages of weight loss are typically the most intensive phase, with subjects being highly motivated. However, at the maintenance stage, motivation can wane as people are no longer receiving gratification from watching their weight reduce week by week, dropping dress sizes, and receiving positive comments from friends and family. Research suggests that only 20% of overweight individuals can lose 10% of their body weight and maintain this at 12 months [107].

The National Weight Control Registry is a US-based registry that records data from a self-selected population of more than 4,000 adults who have lost >30 lb (13.6 kg) and maintained that loss for at least 1 year [108]. Members lost an average of 72.8 lb (33 kg) and maintained the minimum weight loss for an average of >5 year. Analysis of data available from the registry indicates that several factors predict long-term weight maintenance: (1) participation in high levels of physical activity; (2) following a low-calorie, low-fat diet; (3) eating breakfast; (4) self-monitoring weight on a regular basis; (5) maintaining a consistent eating pattern; and (6) identifying minor lapses before they result in larger weight regains [107]. Moreover, weight loss maintenance may get easier over time; after individuals have successfully maintained their weight loss for 2–5 years, the chance of longer-term success greatly increases [107].

8.3.5   Other Non-nutritional Factors Contributing to Prevention and Treatment of Obesity

8.3.5.1   Physical Activity

Physical activity has many health benefits and is an important component in the prevention and treatment of obesity. Evidence from a longitudinal study showed that every extra 30 min/day of walking was associated with an annual reduced weight gain of 1 lb (0.54 kg) or 15 lb over 15 years for women who were heaviest at baseline [109]. A similar trend, albeit smaller, was seen in men. Similar findings were reported in another large prospective study [110]. However, exercise alone is relatively ineffective for weight loss and maintenance, probably owing to the limited ability of physical activity to induce large enough energy deficits. Evidence indicates that diet combined with physical activity has a greater impact on weight reduction than either of these alone [111].

Over time, populations have become more sedentary, with estimates suggesting that over half of Americans and Canadians do not meet the 30 min of recommended physical activity on most days of the week [112]. Those who are overweight and obese should be encouraged to become more physically active. Evidence shows that 150–250 min/week of moderate-intensity physical activity can be effective for preventing weight gain but will only bring about modest weight loss. However, weight loss can be improved if this frequency of physical activity is combined with moderate dietary restriction. Greater amounts of physical activity (>250 min/week) have been associated with clinically significant weight loss [112].

It is important to identify the barriers an individual may encounter as this will have an effect on the adoption of this recommendation. Also, it is important to stress to the individual that increasing physical activity can be achieved with small changes in everyday life, such as walking to work, taking the stairs, and in household tasks, and that the 30 min/day recommendation can be split into smaller bouts of activity, for example two 15 min walks. In the long term, this approach is more likely to be successful and maintained.

8.3.5.2 Breastfeeding

Breastfeeding has been suggested as a protective factor against weight gain in childhood and adolescence, which is important as obesity tends to track into adulthood [113, 114]. A dose-dependent relationship has also been reported between the duration of breastfeeding and the risk of obesity [115]. Thus, infants who are breastfed or who are breastfed for longer periods (and exclusively) have a lower risk of being overweight during older childhood and adolescence [115, 116]. However, there is much variability between the studies and the findings are likely to be affected by residual confounding.

8.3.5.3 Weight-Loss Drugs

The first line of treatment of obesity should always be through diet and physical activity, but many patients also need pharmacological intervention to assist with weight loss or the maintenance of weight loss. Pharmaceuticals will, however, not "cure" obesity, and it is important that patients prescribed weight-control drugs understand the potential side-effects of these mediations and are also aware that they should endeavor to make life-long changes to their lifestyle. Approximately two-thirds of patients can achieve a 5–10% weight-loss in 3–6 months with drug treatment and adequate lifestyle modification. An inadequate response to weight-loss drugs would be a 1–2 kg loss after 6 weeks, except in patients who have already lost weight by lifestyle modifications [117]. Despite promising results on weight loss and some cardiovascular risk factors, the majority of obesity drugs developed to date has not been approved or have been withdrawn due to adverse side effects. Currently, Orlistat is the only drug licensed for long-term use in the USA, Europe, and Australia [118]; a systematic review has shown that a third of obese people taking the standard therapeutic dosage (120 mg three times daily) lost at least 10% of their initial weight [119]. This is the threshold value that is generally assumed to confer clinically important reductions in the metabolic and cardiovascular risks associated with obesity [120]. Other drugs have shown promising results but their long-term safety and efficacy still has to established [121].

8.3.5.4 Surgery

There has been an increasing use in recent years of bariatric surgery for the treatment of obesity. Although this is an extreme and expensive option, the National Institutes of Health (NIH) have proposed that surgery should be considered in patients with a BMI of >40, or of >35 in those with coexisting illnesses, but only after all other treatments have failed [122]. Bariatric surgery leads to weight loss of around 20–30 kg [123, 124] as well as reduced mortality rates [124]. Similarly, in a retrospective study of patients who had undergone surgery, deaths from all causes were reduced by 40%, from diabetes by 92%, from coronary heart disease by 56%, and from cancer by 60% [125]. Obesity is a well-established risk factor for diabetes; research shows that weight loss in patients with prediabetic conditions delays or prevents the development of diabetes [126]. Findings from a clinical trial showed that as a result of weight loss type 2 diabetic patients randomized to surgical treatment were far more likely to achieve remission from diabetes than those given conventional weight-loss therapy (73% vs. 13%) [127].

Surgery should, however, be considered with care by the health professional, and only after the patient has been fully informed of the long-term impact it will have and that the risks are explained [123].

8.3.6 Government-Led Population Strategies to Prevent and Treat Obesity

Earlier sections of this chapter discussed the major causes of obesity. As countries adopt an increasingly Westernized lifestyle and move further along the nutrition transition, their environments, with easy exposure to cheap, unhealthy diets and increasingly sedentary lifestyles, have been termed obesogenic. Any attempt to tackle obesity will have to be multifaceted and target this obesogenic environment. The obesity epidemic is of so much concern, in terms of its potential health-care and economic impact, that governments are exploring various measures to promote weight loss and reduce or prevent weight gain.

8.3.6.1 Fat Tax

Studies have suggested that as the price of a food increases, consumption decreases. It has been estimated that for every 10% increase in price, consumption decreases by 7.8%. Such evidence suggests that a "fat tax" on such foods as SSBs would encourage consumers to switch to healthier beverages. Some governments are considering taxing SSBs and other foods that are thought to contribute to obesity, such as high-sugar/high-fat snacks. This could potentially offset the economic costs of obesity, with the extra revenue generated used to subsidize healthy foods, such as fruit and vegetables.

One of the current population strategies adopted in the USA is the taxation of SSBs. About 40 states have already applied a small tax (mean tax rate 5.2%) on SSBs and snack foods such as chips, cookies, and chocolate; although the level required to elicit the desired effect on consumer behavior is still being debated [128]. Moreover, the extra revenue generated is not being used to fund public-health interventions. If the revenue generated was used to support nutrition and obesity prevention programs, a national tax of one cent per ounce of SSB could raise $14.9 billion in the first year alone [128]. However, this is a very controversial issue and many argue that the government should not interfere with the food market, and that price should continue to be based on costs and demand. This topic is discussed in more detail by Temple in Chap. 23.

8.3.6.2 Nutritional Information Panels and Food Claims

Nutrition information panels have now been introduced on many food and beverage labels across the world, with approximately two-thirds of Americans using labels to help make food purchases [37]. In comparison, about half of the population in the UK read and use nutrition information on food labels [129]. However, their usefulness is still under debate, and the extent to which they actually influence food selection [129]. The number of foods displaying nutritional claims, such as "low in calories" or "fat free," has also increased. Nutrition information on food labels may assist selection of a healthier diet, and ultimately have an impact on obesity rates, but this is reliant on individuals taking time to check labels and carefully consider their food choices. However, nutrition claims are not straightforward; for example, many foods have been modified to reduce the fat content so that a low-fat claim can be made, but these foods may still have a high energy density as sugar is added. The issue of food labels is further discussed in Chap. 17.

8.3.6.3 Food Marketing to Children

It has long been debated whether the volume of marketing directed at children and adolescents should be reduced, particularly television advertising [130]. Food marketing for children is dominated by television advertisements for sugared breakfast cereals, soft drinks, confectionary, savory snacks, and fast food. Such advertising is antithetical to a healthy diet. However, promotion is just one part of the complex process of marketing; actually measuring its effects on consumer behavior is notoriously difficult.

In a systematic review it was found that children engage with food advertising and that it has a significant effect on children's preferences, purchasing behavior, and consumption [131]. Further evidence to support the negative effect of food marketing on children's health was shown in a review of surveys from the USA, Europe, and Australia. This review observed a significant association between the number of overweight children and the number of advertisements per hour on children's television, particularly those which encouraged the consumption of foods with a high energy density [132].

Reducing food marketing to children has been proposed as a central part of any successful anti-obesity strategy [133], but significant social, legal, financial, and public perception barriers stand in the way. Food marketing to children is extensive and has the potential to expand further through such mediums as the internet and video games. However, this is a contentious issue and policies differ from country to country, therefore leaving the effect of different levels of such marketing on obesity rates difficult to evaluate [134].

8.3.7   Tackling Obesity Across a Broad Front

It is clear from the information in this chapter than many factors contribute to the obesity epidemic. Reversing, or at least slowing, the epidemic requires action at all levels, including government policy initiatives, and education of the general public, including action in schools and workplaces. A broad-based but local approach with community support has been shown to be successful [135]. The need for a broad strategy was recognized in a 2007 report by senior scientists working for the UK government [136]:

> The obesity epidemic cannot be prevented by individual action alone and demands a societal approach. Tackling obesity requires far greater change than anything tried so far, and at multiple levels: personal, family, community, and national. Preventing obesity is a societal challenge, similar to climate change. It requires partnership between government, science, business, and civil society.

This broad strategy is explored in Chap. 18.

8.4   Conclusions

The causes of obesity and the dramatic increases documented over time are multifactorial. There has been a steady increase in the availability of foods and beverages with a high energy density at affordable prices; this, together with growing portion sizes, has provided populations with easy access to excessive energy intakes.

There has been much debate as to which weight-loss strategy is most effective, and this has proven to be a lucrative market for industry. A diet which is successful for one person may not be effective for everyone. Ultimately, individuals must weigh up the likelihood of weight-loss success against the potential for adverse effects and long-term maintenance of weight loss when they are making their choice about which weight-loss regime to follow.

The strongest evidence is that adherence to a low-fat diet is successful for the prevention and treatment of obesity. Moreover, a diet with reduced saturated and *trans* fats may have beneficial effects on many other aspects of health, such as reducing risk of cardiovascular disease. However, portion sizes and the energy density of foods consumed also need to be controlled. Consuming more foods with a low energy density, such as fruit, vegetables, and soups, may help to create a negative energy balance whilst still maintaining satiety.

Governments need to develop multilevel strategies that will help to make people aware of the need to monitor their food consumption, including portion sizes, adopt a balanced diet, and increase physical activity. The food industry also needs to consider the portion size issue as well as focusing on the availability and promotion of healthier food choices.

Given that obesity is a chronic disease, and is also related to many other debilitating conditions, there is an urgent need for cost-effective strategies to help achieve the above goals.

References

1. Flegal KM, Carroll MD, Ogden CL, Curtin LR. Prevalence and trends in obesity among US adults, 1999–2008. JAMA. 2010;303:235–41.
2. Flegal KM, Carroll MD, Ogden CL, Johnson CL. Prevalence and trends in obesity among US adults, 1999–2000. JAMA. 2002;288:1723–7.
3. Thompson D, Edelsberg J, Colditz GA, Bird AP, Oster G. Lifetime health and economic consequences of obesity. Arch Intern Med. 1999;159:2177–83.
4. WHO Expert Consultation. Appropriate body-mass index for Asian populations and its implications for policy and intervention strategies. Lancet. 2004;363:157–63.
5. Expert Panel on the Identification, Evaluation, and Treatment of Overweight and Obesity in Adults. Executive summary of the clinical guidelines on the identification, evaluation, and treatment of overweight and obesity in adults. Arch Intern Med. 1998;158:1855–67.
6. International Association for the Study of Obesity (IASO). Overweight and obesity in the EU27. 2008. http://www.iaso.org. Accessed 9 Sept 2011.
7. Berghofer A, Pischon T, Reinhold T, Apovian CM, Sharma AM, Willich SN. Obesity prevalence from a European perspective: a systematic review. BMC Public Health. 2008;8:200.
8. Popkin BM. The nutrition transition and obesity in the developing world. J Nutr. 2001;131:871S–3.
9. Wang Y, Lobstein T. Worldwide trends in childhood overweight and obesity. Int J Pediatr Obes. 2006;1:11–25.
10. Ogden CL, Carroll MD, Curtin LR, McDowell MA, Tabak CJ, Flegal KM. Prevalence of overweight and obesity in the United States, 1999–2004. JAMA. 2006;295:1549–55.
11. Kipping RR, Jago R, Lawlor DA. Obesity in children. Part 1: epidemiology, measurement, risk factors, and screening. BMJ. 2008;337:a1824.
12. Ogden CL, Carroll MD, Flegal KM. High body mass index for age among US children and adolescents, 2003–2006. JAMA. 2008;299:2401–5.
13. Prospective Studies Collaboration, Whitlock G, Lewington S, et al. Body-mass index and cause-specific mortality in 900 000 adults: collaborative analyses of 57 prospective studies. Lancet. 2009;373:1083–96.
14. Haslam DW, James WP. Obesity. Lancet. 2005;366:1197–209.
15. Blundell JE, MacDiarmid JI. Fat as a risk factor for overconsumption: satiation, satiety, and patterns of eating. J Am Diet Assoc. 1997;97:S63–9.
16. Drewnowski A. Obesity and the food environment: dietary energy density and diet costs. Am J Prev Med. 2004;27:154–62.

17. Nielsen SJ, Popkin BM. Patterns and trends in food portion sizes, 1977–1998. JAMA. 2003;289:450–3.
18. Smiciklas-Wright H, Mitchell DC, Mickle SJ, Goldman JD, Cook A. Foods commonly eaten in the United States, 1989–1991 and 1994–1996: are portion sizes changing? J Am Diet Assoc. 2003;103:41–7.
19. Young LR, Nestle M. Portion sizes and obesity: responses of fast-food companies. J Public Health Policy. 2007;28:238–48.
20. Young LR, Nestle M. The contribution of expanding portion sizes to the US obesity epidemic. Am J Public Health. 2002;92:246–9.
21. Ello-Martin JA, Roe LS, Ledikwe JH, Beach AM, Rolls BJ. Dietary energy density in the treatment of obesity: a year-long trial comparing 2 weight-loss diets. Am J Clin Nutr. 2007;85:1465–77.
22. Rolls BJ, Roe LS, Meengs JS, Wall DE. Increasing the portion size of a sandwich increases energy intake. J Am Diet Assoc. 2004;104:367–72.
23. Rolls BJ, Morris EL, Roe LS. Portion size of food affects energy intake in normal-weight and overweight men and women. Am J Clin Nutr. 2002;76:1207–13.
24. Rolls BJ, Roe LS, Kral TV, Meengs JS, Wall DE. Increasing the portion size of a packaged snack increases energy intake in men and women. Appetite. 2004;42:63–9.
25. Rolls BJ, Roe LS, Meengs JS. Larger portion sizes lead to a sustained increase in energy intake over 2 days. J Am Diet Assoc. 2006;106:543–9.
26. Rolls BJ, Roe LS, Meengs JS. The effect of large portion sizes on energy intake is sustained for 11 days. Obesity (Silver Spring). 2007;15:1535–43.
27. McConahy KL, Smiciklas-Wright H, Birch LL, Mitchell DC, Picciano MF. Food portions are positively related to energy intake and body weight in early childhood. J Pediatr. 2002;140: 340–7.
28. Orlet Fisher J, Rolls BJ, Birch LL. Children's bite size and intake of an entree are greater with large portions than with age-appropriate or self-selected portions. Am J Clin Nutr. 2003; 77:1164–70.
29. Popkin BM. Patterns of beverage use across the lifecycle. Physiol Behav. 2010;100:4–9.
30. Block G. Foods contributing to energy intake in the US: data from NHANES III and NHANES 1999–2000. J Food Compos Anal. 2004;17:439–47.
31. Malik VS, Schulze MB, Hu FB. Intake of sugar-sweetened beverages and weight gain: a systematic review. Am J Clin Nutr. 2006;84:274–88.
32. Malik VS, Willett WC, Hu FB. Sugar-sweetened beverages and BMI in children and adolescents: reanalyses of a meta-analysis. Am J Clin Nutr. 2009; 89:438–9, author reply 439–40.
33. Vartanian LR, Schwartz MB, Brownell KD. Effects of soft drink consumption on nutrition and health: a systematic review and meta-analysis. Am J Public Health. 2007;97:667–75.
34. Malik VS, Popkin BM, Bray GA, Despres JP, Willett WC, Hu FB. Sugar-sweetened beverages and risk of metabolic syndrome and type 2 diabetes: a meta-analysis. Diabetes Care. 2010;33:2477–83.
35. Malik VS, Popkin BM, Bray GA, Despres JP, Hu FB. Sugar-sweetened beverages, obesity, type 2 diabetes mellitus, and cardiovascular disease risk. Circulation. 2010;121:1356–64.
36. Almiron-Roig E, Chen Y, Drewnowski A. Liquid calories and the failure of satiety: how good is the evidence? Obes Rev. 2003;4:201–12.
37. French SA, Story M, Jeffery RW. Environmental influences on eating and physical activity. Annu Rev Public Health. 2001;22:309–35.
38. Jeffery RW, French SA. Epidemic obesity in the United States: are fast foods and television viewing contributing? Am J Public Health. 1998;88:277–80.
39. Lin BH, Frazzao E, Guthrie J. Away-from-home foods increasingly important to quality of American Diet. Washington, DC: US Dept of Agriculture; 1999. Economic Research Service no. 749.
40. Binkley JK, Eales J, Jekanowski M. The relation between dietary change and rising US obesity. Int J Obes Relat Metab Disord. 2000;24:1032–9.
41. Pereira MA, Kartashov AI, Ebbeling CB, et al. Fast-food habits, weight gain, and insulin resistance (the CARDIA study): 15-year prospective analysis. Lancet. 2005;365:36–42.

42. Koh-Banerjee P, Chu NF, Spiegelman D, et al. Prospective study of the association of changes in dietary intake, physical activity, alcohol consumption, and smoking with 9-y gain in waist circumference among 16,587 US men. Am J Clin Nutr. 2003;78:719–27.
43. Harnack L, Lee S, Schakel SF, Duval S, Luepker RV, Arnett DK. Trends in the trans-fatty acid composition of the diet in a metropolitan area: the Minnesota Heart Survey. J Am Diet Assoc. 2003;103:1160–6.
44. Pot GK, Prynne CJ, Roberts C, et al. National Diet and Nutrition Survey: fat and fatty acid intake from the first year of the rolling programme and comparison with previous surveys. Br J Nutr. 2012;107:405–15.
45. Stender S, Dyerberg J, Astrup A. High levels of industrially produced trans fat in popular fast foods. N Engl J Med. 2006;354:1650–2.
46. Mozaffarian D, Katan MB, Ascherio A, Stampfer MJ, Willett WC. Trans fatty acids and cardiovascular disease. N Engl J Med. 2006;354:1601–13.
47. Woodside JV, McKinley MC, Young IS. Saturated and trans fatty acids and coronary heart disease. Curr Atheroscler Rep. 2008;10:460–6.
48. Gebauer SK, Psota TL, Kris-Etherton PM. The diversity of health effects of individual trans fatty acid isomers. Lipids. 2007;42:787–99.
49. Zizza C, Siega-Riz AM, Popkin BM. Significant increase in young adults' snacking between 1977–1978 and 1994–1996 represents a cause for concern! Prev Med. 2001;32:303–10.
50. Piernas C, Popkin BM. Snacking increased among U.S. adults between 1977 and 2006. J Nutr. 2010;140:325–32.
51. Popkin BM, Duffey KJ. Does hunger and satiety drive eating anymore? Increasing eating occasions and decreasing time between eating occasions in the United States. Am J Clin Nutr. 2010;91:1342–7.
52. Johnson DB, Gerstein DE, Evans AE, Woodward-Lopez G. Preventing obesity: a life cycle perspective. J Am Diet Assoc. 2006;106:97–102.
53. Dietitians in Obesity Management UK. The dietetic weight management intervention for adults in the one-to-one setting. Is it time for a radical rethink? UK: DOM Committee UK; 2010. DOM no. 2.
54. Franz MJ, VanWormer JJ, Crain AL, et al. Weight-loss outcomes: a systematic review and meta-analysis of weight-loss clinical trials with a minimum 1-year follow-up. J Am Diet Assoc. 2007;107:1755–67.
55. Abete I, Astrup A, Martinez JA, Thorsdottir I, Zulet MA. Obesity and the metabolic syndrome: role of different dietary macronutrient distribution patterns and specific nutritional components on weight loss and maintenance. Nutr Rev. 2010;68:214–31.
56. Frost G, Masters K, King C, et al. A new method of energy prescription to improve weight loss. J Hum Nutr Diet. 2007;20:152–6.
57. Avenell A, Broom J, Brown TJ, et al. Systematic review of the long-term effects and economic consequences of treatments for obesity and implications for health improvement. Health Technol Assess. 2004;8(21):1–182.
58. Department of Health. Dietary reference values for food, energy and nutrients for the United Kingdom. London: The Stationery Office; 1991. DoH no. 41.
59. Brown T, Avenell A, Edmunds LD, et al. Systematic review of long-term lifestyle interventions to prevent weight gain and morbidity in adults. Obes Rev. 2009;10:627–38.
60. Cheskin LJ, Mitchell AM, Jhaveri AD, et al. Efficacy of meal replacements versus a standard food-based diet for weight loss in type 2 diabetes: a controlled clinical trial. Diabetes Educ. 2008;34:118–27.
61. Heymsfield SB. Meal replacements and energy balance. Physiol Behav. 2010;100:90–4.
62. Heymsfield SB, van Mierlo CA, van der Knaap HC, Heo M, Frier HI. Weight management using a meal replacement strategy: meta and pooling analysis from six studies. Int J Obes Relat Metab Disord. 2003;27:537–49.
63. Bray GA, Popkin BM. Dietary fat intake does affect obesity! Am J Clin Nutr. 1998;68: 1157–73.
64. Seidell JC. Dietary fat and obesity: an epidemiologic perspective. Am J Clin Nutr. 1998;67: 546S–50.

65. Lissner L, Heitmann BL. Dietary fat and obesity: evidence from epidemiology. Eur J Clin Nutr. 1995;49:79–90.
66. Forouhi NG, Sharp SJ, Du H, et al. Dietary fat intake and subsequent weight change in adults: results from the European Prospective Investigation into Cancer and Nutrition cohorts. Am J Clin Nutr. 2009;90:1632–41.
67. Carty CL, Kooperberg C, Neuhouser ML, et al. Low-fat dietary pattern and change in body-composition traits in the Women's Health Initiative Dietary Modification Trial. Am J Clin Nutr. 2011;93:516–24.
68. Hession M, Rolland C, Kulkarni U, Wise A, Broom J. Systematic review of randomized controlled trials of low-carbohydrate vs. low-fat/low-calorie diets in the management of obesity and its comorbidities. Obes Rev. 2009;10:36–50.
69. Swinburn BA, Caterson I, Seidell JC, James WP. Diet, nutrition and the prevention of excess weight gain and obesity. Public Health Nutr. 2004;7:123–46.
70. Swinburn BA, Metcalf PA, Ley SJ. Long-term (5-year) effects of a reduced-fat diet intervention in individuals with glucose intolerance. Diabetes Care. 2001;24:619–24.
71. Brehm BJ, Seeley RJ, Daniels SR, D'Alessio DA. A randomized trial comparing a very low carbohydrate diet and a calorie-restricted low fat diet on body weight and cardiovascular risk factors in healthy women. J Clin Endocrinol Metab. 2003;88:1617–23.
72. Foster GD, Wyatt HR, Hill JO, et al. A randomized trial of a low-carbohydrate diet for obesity. N Engl J Med. 2003;348:2082–90.
73. Samaha FF, Iqbal N, Seshadri P, et al. A low-carbohydrate as compared with a low-fat diet in severe obesity. N Engl J Med. 2003;348:2074–81.
74. Grieb P, Klapcinska B, Smol E, et al. Long-term consumption of a carbohydrate-restricted diet does not induce deleterious metabolic effects. Nutr Res. 2008;28:825–33.
75. Nordmann AJ, Nordmann A, Briel M, et al. Effects of low-carbohydrate vs low-fat diets on weight loss and cardiovascular risk factors: a meta-analysis of randomized controlled trials. Arch Intern Med. 2006;166:285–93.
76. Feinman RD, Volek JS. Low carbohydrate diets improve atherogenic dyslipidemia even in the absence of weight loss. Nutr Metab (Lond). 2006;3:24.
77. Bravata DM, Sanders L, Huang J, et al. Efficacy and safety of low-carbohydrate diets: a systematic review. JAMA. 2003;289:1837–50.
78. Shai I, Schwarzfuchs D, Henkin Y, et al. Weight loss with a low-carbohydrate, Mediterranean, or low-fat diet. N Engl J Med. 2008;359:229–41.
79. Miller M, Beach V, Sorkin JD, et al. Comparative effects of three popular diets on lipids, endothelial function, and C-reactive protein during weight maintenance. J Am Diet Assoc. 2009;109:713–7.
80. Astrup A. Dietary management of obesity. J Parenter Enteral Nutr. 2008;32:575–7.
81. Astrup A, Meinert Larsen T, Harper A. Atkins and other low-carbohydrate diets: hoax or an effective tool for weight loss? Lancet. 2004;364:897–9.
82. Astrup A. The satiating power of protein—a key to obesity prevention? Am J Clin Nutr. 2005;82:1–2.
83. Westerterp KR, Wilson SA, Rolland V. Diet induced thermogenesis measured over 24 h in a respiration chamber: effect of diet composition. Int J Obes Relat Metab Disord. 1999;23:287–92.
84. Lejeune MP, Westerterp KR, Adam TC, Luscombe-Marsh ND, Westerterp-Plantenga MS. Ghrelin and glucagon-like peptide 1 concentrations, 24-h satiety, and energy and substrate metabolism during a high-protein diet and measured in a respiration chamber. Am J Clin Nutr. 2006;83:89–94.
85. Paddon-Jones D, Westman E, Mattes RD, Wolfe RR, Astrup A, Westerterp-Plantenga M. Protein, weight management, and satiety. Am J Clin Nutr. 2008;87:1558S–61.
86. Brehm BJ, D'Alessio DA. Benefits of high-protein weight loss diets: enough evidence for practice? Curr Opin Endocrinol Diabetes Obes. 2008;15:416–21.
87. Layman DK. Protein quantity and quality at levels above the RDA improves adult weight loss. J Am Coll Nutr. 2004;23:631S–6.

88. Layman DK, Evans EM, Erickson D, et al. A moderate-protein diet produces sustained weight loss and long-term changes in body composition and blood lipids in obese adults. J Nutr. 2009;139:514–21.
89. Brehm BJ, Lattin BL, Summer SS, et al. One-year comparison of a high-monounsaturated fat diet with a high-carbohydrate diet in type 2 diabetes. Diabetes Care. 2009;32:215–20.
90. Due A, Larsen TM, Mu H, Hermansen K, Stender S, Astrup A. Comparison of 3 ad libitum diets for weight-loss maintenance, risk of cardiovascular disease, and diabetes: a 6-month randomized, controlled trial. Am J Clin Nutr. 2008;88:1232–41.
91. Riediger ND, Othman RA, Suh M, Moghadasian MH. A systemic review of the roles of n-3 fatty acids in health and disease. J Am Diet Assoc. 2009;109:668–79.
92. Ramel A, Jonsdottir MT, Thorsdottir I. Consumption of cod and weight loss in young overweight and obese adults on an energy reduced diet for 8-weeks. Nutr Metab Cardiovasc Dis. 2009;19:690–6.
93. Parra D, Ramel A, Bandarra N, Kiely M, Martinez JA, Thorsdottir I. A diet rich in long chain omega-3 fatty acids modulates satiety in overweight and obese volunteers during weight loss. Appetite. 2008;51:676–80.
94. Jenkins DJ, Wolever TM, Taylor RH, et al. Glycemic index of foods: a physiological basis for carbohydrate exchange. Am J Clin Nutr. 1981;34:362–6.
95. Ludwig DS, Majzoub JA, Al-Zahrani A, Dallal GE, Blanco I, Roberts SB. High glycemic index foods, overeating, and obesity. Pediatrics. 1999;103:E26.
96. Brand-Miller J, McMillan-Price J, Steinbeck K, Caterson I. Carbohydrates—the good, the bad and the whole grain. Asia Pac J Clin Nutr. 2008;17 Suppl 1:16–9.
97. Thomas DE, Elliott EJ, Baur L. Low glycaemic index or low glycaemic load diets for overweight and obesity. Cochrane Database Syst Rev. 2007;(3):CD005105.
98. Raben A. Should obese patients be counselled to follow a low-glycaemic index diet? No. Obes Rev. 2002;3:245–56.
99. Elfhag K, Rossner S. Who succeeds in maintaining weight loss? A conceptual review of factors associated with weight loss maintenance and weight regain. Obes Rev. 2005;6:67–85.
100. Ello-Martin JA, Ledikwe JH, Rolls BJ. The influence of food portion size and energy density on energy intake: implications for weight management. Am J Clin Nutr. 2005;82:236S–41.
101. Tucker LA, Thomas KS. Increasing total fiber intake reduces risk of weight and fat gains in women. J Nutr. 2009;139:576–81.
102. Du H, Du H, Van Der ADL. Dietary fiber and subsequent changes in body weight and waist circumference in European men and women. Am J Clin Nutr. 2010;91:329–36.
103. Rolls BJ, Roe LS, Beach AM, Kris-Etherton PM. Provision of foods differing in energy density affects long-term weight loss. Obes Res. 2005;13:1052–60.
104. Greene LF, Malpede CZ, Henson CS, Hubbert KA, Heimburger DC, Ard JD. Weight maintenance 2 years after participation in a weight loss program promoting low-energy density foods. Obesity (Silver Spring). 2006;14:1795–801.
105. Brownell KD, Rodin J. Medical, metabolic, and psychological effects of weight cycling. Arch Intern Med. 1994;154:1325–30.
106. Field AE, Malspeis S, Willett WC. Weight cycling and mortality among middle-aged or older women. Arch Intern Med. 2009;169:881–6.
107. Wing RR, Phelan S. Long-term weight loss maintenance. Am J Clin Nutr. 2005;82:222S–5.
108. Klem ML, Wing RR, McGuire MT, Seagle HM, Hill JO. A descriptive study of individuals successful at long-term maintenance of substantial weight loss. Am J Clin Nutr. 1997;66: 239–46.
109. Gordon-Larsen P, Hou N, Sidney S, et al. Fifteen-year longitudinal trends in walking patterns and their impact on weight change. Am J Clin Nutr. 2009;89:19–26.
110. Mekary RA, Feskanich D, Malspeis S, Hu FB, Willett WC, Field AE. Physical activity patterns and prevention of weight gain in premenopausal women. Int J Obes (Lond). 2009;33:1039–47.
111. Shaw K, Gennat H, O'Rourke P, Del Mar C. Exercise for overweight or obesity. Cochrane Database Syst Rev. 2006;(4):CD003817.

112. Donnelly JE, Blair SN, Jakicic JM, et al. American College of Sports Medicine Position Stand. Appropriate physical activity intervention strategies for weight loss and prevention of weight regain for adults. Med Sci Sports Exerc. 2009;41:459–71.
113. Dietz WH. Breastfeeding may help prevent childhood overweight. JAMA. 2001;285: 2506–7.
114. Stark O, Atkins E, Wolff OH, Douglas JW. Longitudinal study of obesity in the National Survey of Health and Development. Br Med J (Clin Res Ed). 1981;283:13–7.
115. Harder T, Bergmann R, Kallischnigg G, Plagemann A. Duration of breastfeeding and risk of overweight: a meta-analysis. Am J Epidemiol. 2005;162:397–403.
116. Gillman MW, Rifas-Shiman SL, Camargo Jr CA, et al. Risk of overweight among adolescents who were breastfed as infants. JAMA. 2001;285:2461–7.
117. Lean M, Finer N. ABC of obesity. Management: part II—drugs. BMJ. 2006;333:794–7.
118. Williams G. Orlistat over the counter. BMJ. 2007;335:1163–4.
119. Rucker D, Padwal R, Li SK, Curioni C, Lau DC. Long term pharmacotherapy for obesity and overweight: updated meta-analysis. BMJ. 2007;335:1194–9.
120. Pi-Sunyer FX. A review of long-term studies evaluating the efficacy of weight loss in ameliorating disorders associated with obesity. Clin Ther. 1996;18:1006–35, discussion 1005.
121. Li MF, Cheung BM. Rise and fall of anti-obesity drugs. World J Diabetes. 2011;2:19–23.
122. Consensus Development Conference Panel. NIH conference: gastrointestinal surgery for severe obesity. Ann Intern Med. 1991;115:956–61.
123. Maggard MA, Shugarman LR, Suttorp M, et al. Meta-analysis: surgical treatment of obesity. Ann Intern Med. 2005;142:547–59.
124. Sjostrom L, Narbro K, Sjostrom CD, et al. Effects of bariatric surgery on mortality in Swedish obese subjects. N Engl J Med. 2007;357:741–52.
125. Adams TD, Gress RE, Smith SC, et al. Long-term mortality after gastric bypass surgery. N Engl J Med. 2007;357:753–61.
126. Knowler WC, Barrett-Connor E, Fowler SE, et al. Reduction in the incidence of type 2 diabetes with lifestyle intervention or metformin. N Engl J Med. 2002;346:393–403.
127. Dixon JB, O'Brien PE, Playfair J, et al. Adjustable gastric banding and conventional therapy for type 2 diabetes: a randomized controlled trial. JAMA. 2008;299:316–23.
128. Brownell KD, Farley T, Willett WC, et al. The public health and economic benefits of taxing sugar-sweetened beverages. N Engl J Med. 2009;361:1599–605.
129. Social Science Research Unit. Evidence review of public attitudes towards, and use of, general food labelling. Food Standards Agency; 2010. FSA no. 4.
130. Buttriss J, Deakin k, Smith E. Promotion of foods to children- to ban or not to ban? Nutr Bull. 2003;28:43–6.
131. Hastings G, Stead M, McDermott L, et al. Review of research on the effects of food promotion to children. Food Standards Agency; 2003. FSA no. 1.
132. Lobstein T, Dibb S. Evidence of a possible link between obesogenic food advertising and child overweight. Obes Rev. 2005;6:203–8.
133. Schor JB, Ford M. From tastes great to cool: children's food marketing and the rise of the symbolic. J Law Med Ethics. 2007;35:10–21.
134. Harris JL, Pomeranz JL, Lobstein T, Brownell KD. A crisis in the marketplace: how food marketing contributes to childhood obesity and what can be done. Annu Rev Public Health. 2009;30:211–25.
135. Romon M, Lommez A, Tafflet M, et al. Downward trends in the prevalence of childhood overweight in the setting of 12-year school- and community-based programmes. Public Health Nutr. 2009;12:1735–42.
136. Foresight. Tackling obesities: future choices. 2007. http://www.bis.gov.uk/foresight. Accessed 15 Sept 2011.
137. Haslam D, Sattar N, Lean M. ABC of obesity. Obesity—time to wake up. BMJ. 2006;333: 640–2.

Chapter 9
Diet, the Control of Blood Lipids, and the Prevention of Heart Disease

Michael R. Flock and Penny M. Kris-Etherton

Keywords Cardiovascular disease • Coronary heart disease • Saturated fatty acids • *Trans* fatty acids • Monounsaturated fatty acids • Polyunsaturated fatty acids • Low-density lipoprotein cholesterol • High-density lipoprotein cholesterol • Triglycerides

Key Points

- Lipids and lipoproteins play a key role in the development of coronary heart disease (CHD).
- Many nutrients, especially fatty acids, affect the risk of developing CHD by modifying lipids and lipoproteins.
- Certain foods are effective in improving the lipid profile.
- A variety of different dietary patterns to manage lipids and lipoproteins provide options to target specific CHD risk factors.
- Dietary strategies that affect cardiovascular health are evolving continually, resulting in a better understanding of effective dietary practices to reduce CHD risk.

9.1 Introduction

Lipids and lipoproteins play an important role in modulating risk of coronary heart disease (CHD). It is well established that elevated levels of total cholesterol (TC), low-density lipoprotein cholesterol (LDL-C), and triglycerides (TG) increase CHD

M.R. Flock, BS (✉) • P.M. Kris-Etherton, PhD, RD
Department of Nutritional Sciences, 317 Chandlee Lab, The Pennsylvania State University, University Park, PA 16802, USA
e-mail: mif5098@psu.edu

N.J. Temple et al. (eds.), *Nutritional Health: Strategies for Disease Prevention*, Nutrition and Health, DOI 10.1007/978-1-61779-894-8_9, © Springer Science+Business Media, LLC 2012

risk. In contrast, elevated high-density lipoprotein cholesterol (HDL-C) exerts a cardioprotective effect. Thus, a greater ratio of TC to HDL-C (TC/HDL-C) indicates an increased CHD risk. Many epidemiologic and controlled clinical studies have demonstrated effects of single nutrients, specific foods, and dietary patterns, on lipids and lipoproteins. Diet can increase or decrease CHD risk via changes in the blood lipid profile as well as other risk factors (e.g., elevated blood pressure, inflammation, oxidative stress). This research has led to dietary recommendations that can markedly lower the risk of CHD. Consequently, a healthy diet is important in the prevention of CHD.

This chapter will review both epidemiologic and clinical studies that have evaluated single nutrients, specific foods, and dietary patterns on lipid and lipoprotein CHD risk factors. Identifying the role that dietary factors play in affecting CHD lipids and lipoprotein is important for implementation of diet strategies that maximally reduce CHD risk. New dietary interventions can be implemented to control blood lipids and provide a variety of options for individualizing diets to enhance CHD risk reduction and promote overall adherence.

9.2 Food Components That Modify Lipids and Affect Risk of Heart Disease

The nutrients that modify blood lipids have been the most extensively studied diet components to date. The emphasis has been on examining the effects of different fatty acids and cholesterol on lipids and lipoproteins. Certain saturated fatty acids (SFA), *trans* fatty acids (TFA), conjugated linoleic acids (CLA), and cholesterol adversely affect blood lipid levels, whereas viscous fiber, unsaturated fatty acids (monounsaturated [MUFA] and polyunsaturated fatty acids [PUFA]), plant sterols/stanols, and to a certain extent, polyphenols have favorable effects. Alcohol in moderation positively influences the lipid profile, whereas adverse effects have been associated with excessive consumption.

9.2.1 Saturated Fatty Acids

The role of dietary SFA in relation to CHD has been evaluated extensively in epidemiologic investigations, controlled clinical studies/trials, animal studies, and different in vitro models. There is a large body of clinical evidence showing that SFA (specifically certain SFA) increase blood lipid levels. Early studies by Keys et al. [1] and Hegsted et al. [2] in the 1950s and 1960s culminated in predictive equations evaluating the effect of fatty acids on blood cholesterol levels in humans, using regression analysis of data from many clinical studies. The studies found that SFA raises TC levels compared to carbohydrate (CHO) and MUFA (which both had

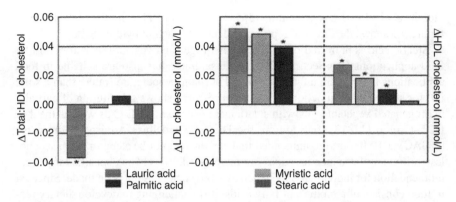

Fig. 9.1 Predicted changes in the ratio of TC/HDL-C and in LDL-C and HDL-C concentrations when carbohydrates are replaced with lauric acid (12:0), myristic acid (14:0), palmitic acid (16:0), or stearic acid (18:0). $^*p<0.05$. Reprinted from Mensink et al. [9] with permission from the *American Journal of Clinical Nutrition*

neutral effects) whereas PUFA lowers TC levels. The effect of SFA is twice as potent in raising TC as PUFA was in lowering TC.

Since the development of the blood cholesterol prediction equations by Keys et al. [1] and Hegsted et al. [2] for TC, several equations have been published that predict the effects of SFA, MUFA, and PUFA on TC, LDL-C, and HDL-C levels [3–5]. Regression analyses demonstrate that for every 1% increase in energy from SFA, LDL-C levels increase approximately 0.033–0.045 mmol/L [3, 6, 7]. In addition to raising TC and LDL-C, SFAs also increase HDL-C. Studies have shown that for every 1% increase in SFA, HDL-C levels increase by 0.011–0.013 mmol/L [3, 6, 7].

9.2.1.1 Individual Saturated Fatty Acids

It is evident that individual SFAs have different effects on lipids and lipoproteins [8]. These results are based on blood cholesterol prediction equations that have been developed for individual fatty acids. Regression analyses have demonstrated that stearic acid (18:0) has a neutral effect on TC, LDL-C, and HDL-C [5], while myristic acid (14:0) is more hypercholesterolemic than lauric acid (12:0) and palmitic acid (16:0) [4].

A meta-analysis of 60 controlled trials determined the effects of different SFA relative to CHO on LDL-C, HDL-C, and TC/HDL-C ratio [9]. All individual SFA with the exception of stearic acid increased LDL-C and HDL-C (Fig. 9.1). Although lauric acid (12:0) was found to have the greatest LDL-C raising effect, it decreases the ratio of TC/HDL-C as a result of causing the greatest increase in HDL-C compared to other SFA. Myristic acid (14:0) and palmitic acid (16:0) have little effect on the ratio due to comparable increases in both TC and HDL-C [10].

Stearic acid (18:0) has a neutral effect on the clinical markers of CHD risk. Figure 9.1 shows the effect of replacing CHO with each individual SFA.

Stearic acid is preferred over partially hydrogenated vegetable oils for solid fat food applications (see Sect. 9.2.2) [11]. Unsaturated fats also are suitable in food applications that require a fat that is liquid at room temperature (rather than a solid fat) for structure and functionality. Current research shows substitution of partially hydrogenated vegetable oils with stearic acid decreases LDL-C as well as the TC/HDL-C ratio [11–13]. In addition, the Dietary Guidelines Advisory Committee (DGAC) 2010 Report recommended that stearic acid not be categorized as a "cholesterol-raising fat" since it is does not increase LDL-C [14]. A specific target recommendation for individual SFAs, such as stearic acid, has not been made. Since fat in food consists of a mixture of fatty acids, it is challenging to develop dietary recommendations for individual SFAs. Therefore, current recommendations focus on limiting total SFA intake. The DGAC 2010 Report and the American Heart Association (AHA) 2020 Impact Goals both recommend that SFA intake be <7% of calories [14, 15].

9.2.1.2 Early Epidemiologic Associations

Early epidemiologic evidence revealed an adverse effect of SFA intake on the lipid profile. The landmark Seven Countries Study reported a significant association between total SFA intake and TC among different populations [16]. Subsequent epidemiologic studies also reported positive correlations between SFA intake and TC levels, as well as the incidence of CHD [17, 18]. An additional analysis of the Seven Countries Study data reported associations between individual SFA and TC, as well as with SFA and CHD mortality. Intakes of lauric acid (12:0) and myristic acid (14:0) were most strongly associated with TC ($r=0.84$, 0.81, respectively) [19].

In the 1960s, men in Finland had the highest rates of CHD mortality in the world. However, during the following decades, community-based interventions aimed at reducing cholesterol levels, blood pressure, and smoking led to remarkable declines in CHD mortality. From 1972 to 2007, CHD mortality decreased by 80%, with risk factor changes explaining 60% of the reduction [20]. Dramatic changes in diet as well as individual fat intakes likely explained a large part of the decrease in CHD. The total intake of SFA in Finland decreased from 22% of total energy in 1972 to 13% in 2007 [20]. Butter and whole-milk consumption declined during this period, while the use of vegetable oil increased. Collectively, the dramatic changes in CHD mortality in a Finnish population demonstrates that decreasing SFA intake (together with reducing other CHD risk factors) was associated with the reduction in CHD risk.

9.2.1.3 Replacement Nutrient for SFA

Inherent to decreasing dietary SFA in a calorie-controlled diet is that an alternative nutrient increases to maintain an equal caloric intake. Consequently, the nutrient(s)

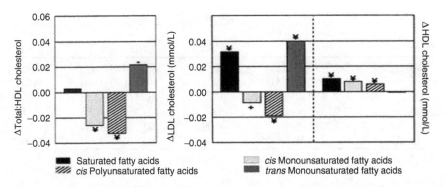

Fig. 9.2   Predicted changes in the ratio of TC/HDL-C and in LDL-C and HDL-C concentrations when carbohydrates are replaced (1% of energy isocalorically) with SFA, MUFA, PUFA, and TFA. $^*p < 0.05$; $^+p < 0.01$; $^\forall p < 0.001$. Reprinted from Mensink et al. [9] with permission from the *American Journal of Clinical Nutrition*

substituted for SFA will impart a unique effect on the blood lipid response, which, in turn, affects overall CHD risk. The DGAC 2010 Report concluded that a 5% energy decrease in SFA replaced by MUFA or PUFA results in a meaningful decrease in the risk of CHD and type 2 diabetes [14].

Replacing SFA with MUFA [21–23] and PUFA [22, 24, 25] consistently lowers LDL-C as does replacing SFA with CHO (Fig. 9.2) [21]. However, isoenergetically replacing SFA with CHO has been associated with a decrease in HDL-C and an increase in TG compared to replacement with MUFA and PUFA [9, 21]. In addition, replacing SFA with refined CHO may increase the number of small LDL particles, another CHD risk factor, while SFA intake has been associated with larger LDL particle size [9, 26]. Some studies have shown that small dense LDL particles are more atherogenic, possibly due to their greater ability to penetrate the endothelium [27–29]. Although LDL and HDL particle size may provide a more detailed assessment of CHD risk, additional research is needed to determine the effect that particle size has on CHD.

Clinical evidence also demonstrates a lower CHD risk when SFA is replaced with MUFA or PUFA (see Sects. 9.2.3 and 9.2.4). Sacks and Katan [30] predicted that diets high in unsaturated fat decrease coronary events by 16% in women and 19% in men. Mozaffarian et al. [31] reported that each 5% energy increase in PUFA as a replacement for SFA decreases CHD events by 10%. Collectively, clinical studies have consistently demonstrated that replacing SFA with MUFA or PUFA improves the lipid profile and decreases CHD risk.

9.2.1.4   Recent Epidemiologic Associations

Recent epidemiologic studies have reported that when SFA is replaced by CHO (which typically represents refined carbohydrates), CHD risk is unchanged; yet,

Table 9.1 Estimates of relative risk of fatty acids on CHD death and event (cohort studies)

Type of fat	Relative risk (95% CI)			
	CHD death	p Value	CHD events	p Value
Total fat	0.94 (0.74–1.18)	0.583	0.93 (0.84–1.03)	0.177
TFA	1.32 (1.08–1.61)	0.006	1.25 (1.07–1.46)	0.007
SFA	1.14 (0.82–1.60)	0.431	0.93 (0.83–1.05)	0.269
MUFA	0.85 (0.60–1.20)	0.356	0.87 (0.74–1.03)	0.110
PUFA	1.25 (1.06–1.47)	0.009	0.97 (0.74–1.27)	0.825
n-3 LCPUFA	0.82 (0.71–0.94)	0.006	0.87 (0.71–1.10)	0.066

Adapted from ref. [32]

when MUFA or PUFA replace SFA, CHD risk is decreased. Thus, the results reflect the replacement nutrient for SFA.

A recent meta-analysis conducted by Skeaff and Miller [32] using results from prospective cohort evaluated fatty acid associations on CHD risk (Table 9.1). While some epidemiologic studies have reported that SFA intake is positively associated with CHD risk, SFA intake was not significantly associated with CHD death (RR 1.14 [95% CI: 0.82–1.60]) or CHD events (RR 0.93 [CI: 0.83–1.05]) in this analysis. The authors concluded that the available evidence was unsatisfactory and unreliable due to various limitations [32, 33]. One problem has been an inadequate number of subjects. Another problem is regression dilution bias, which can be described as the underestimation of associations with disease outcomes due to measurement errors and within-individual variability in the exposure variable. A third problem is dietary assessment methods. Frequency of food consumption in epidemiologic studies is self-reported; therefore, reporting bias can potentially influence the observations. Although many studies used food frequency questionnaires, several studies used a 24-h recall to assess dietary intake, a particularly imprecise method to estimate an individual's long-term dietary habits [34]. Siri-Tarino et al. [35] also conducted a recent meta-analysis summarizing evidence from epidemiologic studies associating SFA with CHD risk. Only six of the 16 studies included that examined SFA intake found a positive association with CHD. Overall, SFA did not have a statistically significant relationship with CHD (RR 1.07 [95% CI: 0.96, 1.19]). However, the authors did not define CHD nor did they evaluate substitutions for SFA. A decrease in dietary SFA intake is associated with an increased intake of another nutrient, which would be expected to affect CHD risk. In addition, more than half of the studies used 24-h recalls or some other unvalidated dietary assessment method, which would question the reliability of the results and potentially reduce the strength of association (through regression dilution bias) [36].

A recent pooled-analysis by Jakobsen et al. [37] of 11 cohort studies from both Europe and the United States reported that substituting SFA with PUFA reduces the risk of CHD death (HR 0.74 [95% CI: 0.61, 0.89]) and CHD events (HR 0.87 [CI: 0.77,0.97]). However, replacing MUFA or CHO for SFA was associated with a greater risk of CHD events (HR 1.19 [CI: 1.00, 1.42] and HR 1.07 [CI: 1.01, 1.14]), but not CHD death. These findings suggest that in order to prevent CHD SFA should be replaced with PUFA rather than MUFA or CHO. Although most of the studies included did adjust for TFA intake in estimating HRs for MUFA intake, the adjustments may have been incomplete due to industrial modifications of the TFA in the foods consumed during the follow-up period [37]. Also, no differentiation was made between CHO sources of different glycemic indexes, which significantly influence CHD risk [37]. In contrast, a recent systematic review by Mente et al. [38] also pooled prospective cohort studies and found strong evidence indicating a positive association between TFA and CHD risk (RR 1.32 [95% CI: 1.16–1.48]) and an inverse association between MUFA and CHD risk (RR 0.80 [CI: 0.67, 0.93]). Only weak evidence was found for SFA (RR 1.06 [CI: 0.96, 1.15]) and PUFA (RR 1.02 [CI: 0.81, 1.23]). It is significant that the Mediterranean dietary pattern, typically low in SFA and high in MUFA, did report a strong inverse association with CHD risk in prospective studies (RR 0.66 [CI: 0.53, 0.72]).

Clearly, the nutrient substituted for SFA is important in assessing the role of SFA on CHD risk. Further research is needed to distinguish the benefits associated with decreasing SFA or increasing MUFA or PUFA as well as complex CHO intake in order to clarify inconsistencies reported among epidemiologic studies and to identify why some epidemiologic studies differ from the consistent body of clinical evidence. Replacement of SFA with MUFA has not yet been tested by randomized controlled trials (RCTs) that have evaluated clinical endpoints; therefore, the strongest and most consistent evidence for reducing CHD risk has been associated with replacing SFA with PUFA [31, 32].

9.2.2   Industrial Trans-Fatty Acids

Industrially synthesized TFAs behave similarly to SFA relative to increasing LDL-C, but also adversely affect (i.e., lower) HDL-C levels. Mensink and Katan [39] measured the effects of TFA on HDL-C by placing subjects on three diets identical in nutrient composition except that 10% of total energy was from oleic acid, *trans*-isomers of oleic acid, or SFA. The mean HDL-C level was the same on the SFA and oleic acid diets, but was 0.17 mmol/L lower on the TFA diet ($p < 0.0001$).

In a controlled clinical trial, Lichtenstein et al. [40] evaluated the effects of different hydrogenated fats on lipids and lipoproteins. The experimental diets provided 30% energy from total fat and were identical with the exception of the test fats fed. Two-thirds of the fat was provided by either soybean oil (<0.5 g TFA per 100 g of fat), semiliquid margarine (<0.5 g/100 g), soft margarine (7.4 g/100 g), shortening (9.9 g/100 g), stick margarine (20.1 g/100 g), or butter (1.25 g/100 g). Compared with the butter diet, the vegetable fat diets elicited the following reductions in TC, LDL-C, and HDL-C:

- Soybean-oil diet: 10%, 12%, and 3%, respectively
- Semiliquid margarine diet: 10%, 11%, and 4%, respectively
- Stick margarine diet: 3%, 5%, and 6%, respectively

Although every vegetable fat diet resulted in lower TC, LDL-C, and HDL-C compared with the butter diet, stick margarine (containing the highest amount of TFA) decreased LDL-C the least and decreased HDL-C the most compared with the other vegetable fats, resulting in a 4% increase in the TC/HDL-C ratio. By comparison, the other vegetable fats slightly decreased the ratio. TFA also decreased LDL particle size in a dose-dependent manner ($p<0.001$). On balance, the soybean oil diet elicited the most favorable effects on the lipid profile with the greatest reduction in TC and LDL-C and the smallest reduction in HDL-C. The soybean oil diet decreased the TC/HDL-C ratio 6%, and the semiliquid margarine diet elicited a 5% reduction. This study demonstrates that increasing industrially synthesized TFA results in a dose-dependent increase in LDL-C and a decrease in HDL-C at high levels (higher than typical consumption, which is about 2.6% of calories), thereby increasing the TC/HDL-C ratio. Consequently, industrially synthesized TFA increases risk for CHD death and CHD events more than other fatty acids, including SFA (Table 9.1).

Judd et al. [41] conducted a clinical trial to assess the change in LDL-C that occurred when CHO was replaced with TFA. Subjects were fed experimental diets that provided approximately 15% of total calories from protein, 39% from total fat, and 46% from CHO. TC and LDL-C increased 5.8% and 10.1%, respectively, when TFA replaced 8% of the calories from CHO. When 8% of calories were replaced with a combination of 4% TFA and 4% stearic acid, TC and LDL-C were increased by 5.6% and 8.7%, respectively. There were no significant differences in HDL-C with either replacement compared to the CHO control diet. Ascherio and Willett [42] demonstrated a linear dose-dependent relationship between TFA intake and the LDL-C/HDL-C ratio from intakes of 0.5–10% of total calories. The magnitude of this LDL-C/HDL-C raising effect is greater for TFA than it is for SFA.

9.2.2.1 Epidemiologic Associations

Through regression analysis, the effects of TFA on TC, LDL-C, HDL-C, and TG levels have been compared to other fatty acids using predictive equations. TFA has been shown to increase TC and LDL-C similarly to SFA; however, it lowers HDL-C vs. SFA. A recent meta-analysis by Mozaffarian et al. [43] found TFA consumption to consistently increase LDL-C and the TC/HDL-C ratio, and consistently decrease HDL-C in both experimental and observational studies. These effects were most prominent when TFA was compared with MUFA and PUFA.

A strong body of epidemiologic evidence supports the link between industrially synthesized TFA and CHD [38]. In a follow-up of the Seven Countries Study, elaidic acid, the predominant TFA in hydrogenated vegetable oil, was significantly associated with TC ($r=0.70$, $p<0.01$) and 25-year mortality rates from CHD ($r=0.78$, $p<0.001$) [19]. Therefore, reducing TFA, specifically elaidic acid, often is recommended to lower CHD risk; although the effects of ruminant TFA on lipids and lipoproteins may be less understood.

9.2.2.2   Conjugated Linoleic Acid and Vaccenic Acid

CLA, an isomer of linoleic acid (n-6 fatty acid), is found in ruminant fats in meat and dairy products. Although CLA does contain a trans bond, an adjacent cis bond may allow CLA to possess different characteristics compared to the TFA found in partially hydrogenated oils [44]. The cis-9, trans-11 CLA isomer in addition with vaccenic acid, a precursor to CLA, combines to form the majority of ruminant TFA in beef and dairy. However, research on naturally occurring TFA has been limited compared to industrial TFA (hydrogenated oils). In addition, most of the conclusive research on CLA has been done in animal models and not humans. In animals, CLA isomers have been shown to decrease body fat, improve insulin sensitivity, improve lipid profiles, and, consequently, decrease risk of atherosclerosis [45]. Human studies, however, have produced conflicting results. Attention has focused recently on the possible benefits of CLA intake, particularly as an aide for achieving an improved body composition. Intervention studies in humans usually involve diets enriched with CLA, often using CLA supplements. One double-blind study found that CLA supplementation (0.6 g three times/day) promoted a greater loss of body fat [46]; yet numerous other double-blind studies reported no effect [47–50]. The evidence of a CLA effect on blood levels of lipids and lipoproteins has also been minimal and inconsistent [51–53].

Wanders et al. [53] recently reported the results of a RCT that evaluated the effects of CLA intake on lipoprotein levels in healthy adults. For 3 weeks subjects received 7% of total calories from oleic acid, industrial TFA, or a mixture of 80% cis-9, trans-11 and 20% trans-10, cis-12 CLA. LDL-C levels were 11.6% ($p < 0.001$) higher after the industrial TFA diet and 8.7% ($p < 0.001$) higher after the CLA diet than after the oleic acid diet. Both the TFA and CLA diets also decreased HDL-C levels comparably. The authors concluded that "an intake of 3 g/day of CLA, as recommended by manufacturers of CLA supplements, could theoretically increase the risk of CHD by 3 to 12%" [53]. Tholstrup et al. [54] conducted a RCT evaluating the effects of vaccenic acid on blood lipids and lipoproteins in healthy young men. Subjects were either on a high (3.6 g/day) or low (0.4 g/day) vaccenic acid diet. The high vaccenic acid diet resulted in a 6% ($p = 0.05$) decrease in TC as well as a 9% ($p = 0.002$) decrease in HDL-C. Consequently, the TC/HDL-C ratio remained unchanged.

Jakobsen et al. [55] evaluated the evidence on ruminant TFA and CHD outcomes and reported no statistically significant association. Similarly, Field et al. [56] recently reviewed current research on vaccenic acid and CHD risk. Evidence did not show convincingly an association between vaccenic acid and MI. More well-controlled clinical trials are needed to determine benefits and/or risks associated with ruminant TFA consumption.

9.2.2.3   Ruminant vs. Industrial TFA

Ruminant and industrial TFAs are structurally similar, while the fat sources of these TFAs have a different fatty acid profile. Ruminant fat contains significant amounts of CLA, while partially hydrogenated oils contain very small amounts [57]. Thus, there may be metabolic differences between ruminant and industrial TFAs.

Brouwer et al. [57] recently conducted a quantitative review to determine the effects of both ruminant TFA and industrial TFA on HDL-C and LDL-C in humans. The analysis of 39 randomized trials reported that all classes of TFA (0.4–10.9% of energy) increase the LDL-C/HDL-C ratio. Most studies used supplements with an equal ratio of cis-9, trans-11 and trans-10, cis-12 CLAs (1.8 and 6.8 g/day). Although the difference was not significant, the effect of ruminant TFA was less than that of industrial TFA. However, the daily intake of ruminant TFA is a fairly small amount, approximately 1.5 g for men and 0.9 g for women (less than 2% of total daily caloric intake) [14]. The DGAC 2010 Report concluded that total TFA intake should be the focus for dietary change rather than individual TFAs due to insufficient evidence suggesting differences among types of TFA on CHD risk [14]. Clearly, further clinical research is needed that compares ruminant and industrial TFAs, particularly with respect to CHD risk.

9.2.3 Monounsaturated Fatty Acids

Evidence has demonstrated that substituting MUFA for SFA in the diet improves the lipid profile. In particular, replacing SFA with MUFA decreases LDL-C (Fig. 9.2) [21–23]. HDL-C levels are reduced less or unchanged when SFA is replaced with MUFA rather than PUFA [58]. However, replacement of SFA with CHO has been standard advice for reducing CHD risk, even though low-fat, high-CHO diets may unfavorably affect lipid and lipoprotein risk factors. There has been much debate about whether high-MUFA diets or high-CHO diets are more effective in reducing obesity and CHD risk factors [59]. Rodriguez-Villar et al. [59] conducted a study reporting that a high MUFA diet (25% kcal from MUFA) compared to a high-CHO diet (12% kcal from MUFA) significantly lowered VLDL-C by 35% ($p=0.023$) and VLDL-TG by 16% ($p=0.016$) in subjects with type 2 diabetes. As a result, particles were believed to be less atherogenic (due to greater lipid depletion) although TC, LDL-C, HDL-C, and TG levels were similar after both diets [59]. Substantial evidence also suggests that MUFA is protective against oxidative stress and smooth muscle cell proliferation, two factors that promote atherosclerosis [60]. In addition, compared to PUFAs, MUFAs are less likely to be converted to disruptive bioactive compounds, due to a lower susceptibility to oxidation [61–63].

Clinical studies have shown that MUFA—and oleic acid in particular—have demonstrated either a neutral or slight lowering effect on TC levels, particularly when SFA intake is low [8]. A meta-analysis conducted by Garg [64] evaluated the effects of high-MUFA diets (22–33% kcal) vs. high-CHO diets (49–60% kcal). High-MUFA diets led to a decrease in VLDL-C (22%) and TG levels (19%) as well as a slight decrease in TC (3%), when compared with high-CHO diets. High-MUFA diets also caused a slight increase in HDL-C (4%), although inconsistent among studies with either a small increase or no change, depending on the comparison diet.

No change in LDL-C was observed in the high-MUFA diets compared to high-CHO diets; however, high-MUFA diets were noted as having a lower susceptibility of LDL oxidation, potentially reducing the risk of atherosclerosis [64].

Mensink et al. [9] conducted a meta-analysis of dietary fatty acids and CHO effects on lipids and lipoproteins. MUFAs increased HDL-C levels vs. CHO while both significantly lowered LDL-C levels (Fig. 9.2). In addition, MUFA decreased apolipoprotein B levels, the primary apolipoprotein in LDL-C. The TC/HDL-C ratio did not change when SFA was replaced with CHO; however, the ratio decreased when SFA was replaced with MUFA. More recently, Cao et al. [65] conducted a meta-analysis of 30 controlled trials evaluating moderate vs. low-fat diets (MF and LF) in subjects with and without diabetes. MF diets contained on average 23.6% of total calories as MUFA, whereas LF had 11.4%. Both diets lowered TC and LDL-C similarly; however, the MF diets resulted in more favorable HDL-C and TG levels compared to the LF diets. HDL-C increased 0.06 mmol/L ($p < 0.0001$) and TG decreased 0.11 mmol/L ($p < 0.00001$) after MF diets vs. LF diets. Consequently, the reduction in the TC/HDL-C ratio was significant after MF diets (-0.36, $p < 0.0001$) with a -0.30 ($p < 0.001$) difference between MF and LF diets. Predicted changes in CHD risk were calculated based on these changes in blood lipids, estimating a greater reduction in CHD risk after MF diets compared to LF diets for both men and women (-6.4% and -9.3%, respectively). Therefore, it was concluded that a diet high in MUFA and unsaturated fatty acids reduces CHD risk greater than a LF diet, typically higher in refined CHO [65].

9.2.3.1   Epidemiologic Associations

Epidemiologic evidence for replacing SFA with MUFA has been mixed with regard to CHD risk. A strong prospective cohort study consisting of 5,672 diabetic women from the Nurses' Health Study [66] reported that replacement of 5% of calories from SFA with equivalent calories from MUFA was associated with a 37% lower CVD risk, greater than the 22% decrease in CVD risk associated with CHO replacement. In contrast, a recent pooled analysis of prospective cohort studies reported a direct association between substitution of MUFA for SFA and CHD events (HR: 1.19 [95% CI: 1.00, 1.42]). It was speculated that the increased CHD risk associated with higher MUFA intake may be explained in part by intake of TFAs, which were included in MUFAs. Most of the studies included did adjust for TFA intake in estimating HRs for MUFA intake; however, the adjustments may have been incomplete due to industrial modifications of the TFA in the foods consumed during the follow-up period [37].

The DGAC 2010 Report stated a 5% energy replacement of SFA with MUFA improves lipids and lipoproteins as well as decreases the risk of CHD and type 2 diabetes [14]. In addition, MUFA intake in place of high CHO intake improves glucose tolerance and insulin and diabetic control [14]. Therefore, MUFA can serve as a suitable replacement for SFA, TFA, and/or CHO to improve the lipid profile as well as other biomarkers of CHD, although further research is needed to clarify inconsistencies reported in epidemiologic studies.

9.2.4 Polyunsaturated Fatty Acids

There is strong evidence that PUFA improves the lipid profile, particularly when replacing SFA or TFA. Some of the earliest clinical trials evaluated the effects of consuming a diet high in PUFA (13–21% of energy) and low in SFA on TC and CHD events [67–70]. Three of the studies reported a 13–15% decrease in TC that was accompanied by a 25–43% decrease in CHD events [67, 69, 70]. These studies established that consuming a diet high in PUFA substantially decreases TC. In addition, data from many controlled clinical studies have been used to develop predictive equations demonstrating that PUFA lowers both TC and LDL-C. These equations predict that for each 1% increase in PUFA, TC is reduced by 0.024 mmol/L; this cholesterol-lowering effect is approximately half that of the cholesterol-raising effect of SFA [1, 2]. In addition, replacement of SFA with PUFA may increase LDL receptor mediated uptake of LDL-C from the blood [14]. A blunted receptor mediated uptake of LDL-C results in increased levels of LDL-C in circulation. Therefore, replacing SFA with PUFA lowers LDL-C by increasing LDL receptor mediated removal [71].

9.2.4.1   n-6 Fatty Acids

Predictive equations for individual fatty acids demonstrate that n-6 fatty acids, specifically linoleic acid (18:2, LA), have potent TC and LDL-C lowering effects (Fig. 9.2). In addition, LA raises HDL-C when compared with stearic acid (18:0) [8]. Mattson and Grundy [72] reported an HDL-C lowering effect (−0.13 mmol/L, $p < 0.02$) of PUFA intake at very high levels (28% of energy). However, other studies reported no significant change in HDL-C at a high PUFA intake (15.9% energy, with a range of 10–30%) [73]. A recent review by Czernichow et al. [74] reported that the decrease in HDL-C related to n-6 fatty acid consumption, although not observed in all intervention studies, could be challenged considering the increased CHD risk associated with low HDL-C levels. However, the TC/HDL-C ratio is considered a better predictor of CHD risk [75] and the authors conclude that eating a diet enriched with n-6 PUFA or replacing SFA with n-6 PUFA significantly reduces TC, LDL-C, and the TC/HDL-C, which may reduce the risk of CHD [74]. LA is also more susceptible to lipid peroxidation compared to MUFA or SFA. This may increase the risk of oxidizing LDL-C particles in the blood, potentially exacerbating the development of atherosclerosis if consumed in excessive amounts. However, LA clearly imparts beneficial effects on the lipid profile. A recent AHA Scientific Statement based on a review of clinical and epidemiologic evidence recommended that n-6 PUFA should provide at least 5–10% of total calories and that any reduction in n-6 PUFA would increase CHD risk [76]. The AHA review reported that higher intakes of n-6 PUFA appear to be safe and may even provide additional benefits as part of a low-SFA, low-cholesterol diet. Based on clinical evidence, n-6 PUFA intakes of 10–21% of energy, compared with lower intakes, might reduce

CHD risk without evidence of harmful effects [76]. In addition, epidemiologic evidence shows no indication that higher n-6 PUFA intakes increase CHD risk [31, 76]. Mozaffarian et al. [31], in a recent systematic review and meta-analysis, concluded that increasing PUFA consumption as a replacement for SFA reduces the occurrence of CHD by 19%; each 5% energy increase in PUFA consumption decreased CHD events by 10%. No increase in CHD risk was associated with very high levels of PUFA consumption in long-term studies (mean of 14.9% total energy; range 8–20%). Therefore, the authors concluded that current recommendations of <10% energy from PUFA need to be reevaluated [31]. More research, particularly long-term studies, is needed to determine the effect and safety of high intakes of n-6 PUFA.

9.2.5   n-3 Fatty Acids

Much research has been conducted on marine long-chain n-3 fatty acids: eicosapentaenoic acid (EPA) and docosahexaenoic acid (DHA). A strong body of evidence demonstrates that these fats significantly decrease TG levels, although LDL-C may increase in proportion to the TG lowering effect. For this reason, n-3 consumption combined with therapies that lower LDL-C would be expected to concurrently reduce both LDL-C and TG levels [77]. The consistent TG-lowering effect of marine n-3 fatty acids generally has not been found at physiologically relevant concentrations of plant sources of n-3 fatty acids (alpha-linolenic acid, ALA). Although dietary ALA is a source of EPA and DPA, the conversion efficiency is very limited, converting between 0.01 and 8% of ALA to EPA and less to DHA [78, 79].

Numerous clinical trials have shown that consumption of EPA and DHA from fish or from fish oil supplements reduces all-cause mortality and/or various CHD outcomes [80–82]. A review by Mozaffarian and Rimm [83] demonstrated that the effects of EPA + DHA intake on CHD depend on the dose and time responses. The authors reported that antiarrhythmic effects are strongest in reducing risk of CHD death and sudden death, having an effect within weeks of a modest intake (<750 mg/day EPA + DHA). However, at higher intakes (>750 mg/day EPA + DHA) maximum antiarrhythmic effects are achieved, while TG lowering begins to have a clinically relevant effect within months to years. The major mechanism responsible for the cardioprotective effect of n-3 fatty acids does not appear to be associated with changes in lipids. Although TG lowering provides some benefit in reducing CHD risk, it is likely a minor factor. Rather, the reduced CHD risk associated with n-3 fatty acid consumption is thought to be the result of stabilizing atherosclerotic plaques, decreasing production of chemoattractants, growth factors, adhesion molecules, inflammatory eicosanoids and cytokines, increasing endothelial relaxation and vascular compliance, and reducing sudden death [80, 82–84]. This is discussed in greater detail in the following chapter.

9.2.6   Dietary Cholesterol

9.2.6.1   Effect on Blood Lipids

The effect of dietary cholesterol on CHD risk varies from person to person. The effect of dietary cholesterol on lipids and lipoproteins depends largely on whether an individual's cholesterol synthesis is increased or decreased by cholesterol consumption [85]. Hyperresponders are more inclined to experience a greater increase in LDL-C following intake of dietary cholesterol than hypo-responders. Dietary cholesterol is absorbed with the aid of bile acids, packaged in chylomicrons, then transported to the liver, where it can suppress LDL receptor activity, leading to an increase in LDL-C as a result of decreased cholesterol clearance from plasma [86].

Jones et al. [87] conducted a study to determine whether cholesterol synthesis is suppressed by cholesterol intake in subjects with normal TC or elevated TC. They were randomized to consume varying amounts of cholesterol (50, 350, or 650 mg cholesterol/2,800 kcal). Cholesterol was provided as egg yolk and whole egg added to baked or cooked products. In all subjects, cholesterol synthesis was significantly lower ($p < 0.05$) following the transition from low to medium and low to high cholesterol diets. The authors concluded that dietary cholesterol causes a slight feedback inhibition of cholesterol biosynthesis, independent of blood cholesterol levels [87].

Clinical studies evaluating the effect of dietary cholesterol on blood lipids have been somewhat inconsistent. Green et al. [88] conducted a randomized crossover trial where subjects 29–60 years of age consumed diets consisting of 3 eggs/day or a fat-free egg substitute. Consuming dietary cholesterol from 3 eggs/day (~640 mg dietary cholesterol) significantly increased LDL-C ($p < 0.05$) and HDL-C ($p < 0.001$) in both men and women, with no significant change in the LDL-C/HDL-C or TC/HDL-C ratios. Lecithin cholesteryl acyl transferase (LCAT) activity also was greater in hyperresponders. The increase in LCAT activity in addition to the elevation of lipoprotein cholesterol levels suggests that the reverse cholesterol transport pathway actively mobilizes excess dietary cholesterol in order to maintain cholesterol homeostasis [88]. Some studies have shown that dietary cholesterol (2–3 eggs/day) may increase the size of LDL-C particles by 0.4–1.5 nm [88, 89]. However, not all clinical trials have found that egg consumption increases cholesterol levels [90, 91]. For example, Goodrow et al. [90] saw no significant increase in LDL-C, HDL-C, or TG when elderly adults (mean age of 79 year) consumed 1 egg/day.

Based on a meta-analysis of 27 controlled feeding studies, each 100 mg/day increase in dietary cholesterol (up to 400 mg/day) results in a corresponding increase of TC from 0.16 to 0.51 mmol/L, when baseline intake of dietary cholesterol is zero [92]. However, with a higher baseline cholesterol intake (300 mg/day), the increase in TC induced by an additional 100 mg/day of dietary cholesterol is only 0.05–0.16 mmol/L. Therefore, additional dietary cholesterol has a greater impact on blood lipids when baseline dietary cholesterol consumption is low. An increase in LDL-C accounts for 80% of the increase predicted in TC [92]. Calculations from a similar analysis reported that a 100 mg/day increase in dietary cholesterol increases the LDL-C/HDL-C ratio from 3.00 to 3.01 in subjects with elevated blood cholesterol. It is not clear how this small change in the LDL-C/HDL-C ratio would affect CHD risk [93].

9.2.6.2   Epidemiologic Associations

Moderate evidence from epidemiologic studies indicates a positive relationship between dietary cholesterol and TC and LDL-C, yet many recent prospective studies have reported no association [14]. The Harvard Egg Study [94] concluded that adverse effects of egg yolk cholesterol on LDL-C levels are often counterbalanced by benefits on HDL-C, TG, and other nutrients related to the risk of CHD. Kritchevsky and Kritchevsky [95] reviewed epidemiologic studies that evaluated egg consumption and reported that there was no association between eating one egg a day and risk of CHD. The Oxford Vegetarian Health Study [96] and the Seventh-day Adventist Study [97] both found no association between egg consumption and CHD risk when they followed subjects over several years (13 and 6 years, respectively). However, several studies have reported that egg consumption in type 2 diabetics was associated with an increase in CHD risk [66, 94, 98]. The Health Professionals Study and the Nurses' Health Study both reported no association between egg consumption and CHD [94]. However, when considering individuals with diabetes separately, 1 egg/day was found to significantly increase CHD risk twofold compared to an intake of less than 1 egg/week [94, 98]. Consequently, the DGAC 2010 Report concluded that consuming 1 egg/day does not lead to negative changes in lipid and lipoprotein levels, although, a greater risk is associated with individuals having type 2 diabetes [14]. Therefore, limiting cholesterol to <200 mg/day is recommended for individuals with or at high risk for type 2 diabetes or CHD, while <300 mg/day remains the recommendation for persons at low risk for CHD [14].

9.2.7   Dietary Fiber

9.2.7.1   Effect on Blood Lipids

Some fibers that form viscous solutions in water have been shown to interfere with the reabsorption of bile in the intestines, resulting in an increased bile acid production and/or enhanced LDL receptor status [99]. These viscous fibers include pectins, beta-glucans, some gums, and mucilages (ex. psyllium). Numerous intervention studies have demonstrated that viscous fiber reduces TC and LDL-C [92, 100–106] and may prevent TG increases associated with greater CHO consumption. The National Cholesterol Education Program (NCEP) Adult Treatment Panel (ATP) III Report [107] stated that an increase in viscous fiber of 5–10 g/day would be expected to lower LDL-C by approximately 3–5%. The NCEP recommends 10–25 g/day of viscous fiber as an additional option of the Therapeutic Lifestyle Changes (TLC) diet to lower LDL-C and reduce CHD risk.

Although LDL-C lowering effects of viscous fiber have been well established in clinical trials, the changes in TG appear to be mixed. In the Dietary Approaches to Stop Hypertension (DASH) Trial [103], individuals with elevated blood pressure were randomly assigned to the DASH diet (58% total calories from CHO, 27% from total fat, 7% SFA) that provided 30 g of fiber/day had significantly lower TC (7.3%), LDL-C (9.0%), and HDL-C (7.5%), compared to subjects fed a control diet (50% calories from CHO, 37% from total fat, 14% SFA) with intakes of only 11 g of fiber/day (all $p < 0.0001$). TG levels were not significantly affected by the DASH diet. However, a study of individuals with type 2 diabetes following a diet that provided 50 g of fiber/day (25 g viscous fiber) reported a 10.2% decrease in TG levels in addition to a 6.7 and 12.5% reduction in LDL-C and TC, respectively (all $p < 0.05$), compared to a diet that contained 24 g of fiber/day (8 g viscous fiber) [100]. In a double-blind study conducted by Naumann et al. [106], healthy subjects were randomly assigned to consume either a fruit drink daily enriched with 5 g of beta-glucan from oats or a rice starch-based placebo. After 5 weeks, the beta-glucan enriched drink significantly decreased TC by 4.8% and LDL-C by 7.7%, respectively (both $p < 0.05$), when compared to the placebo, but no significant difference in TG was reported between groups. Therefore, the effects of fiber on TG levels may vary depending on the subject population, the amount of fiber, and the type of fiber. In the case of diabetes, a decrease in TG following high fiber consumption might be attributed to the improvement in glycemic control [100].

A study involving 36 overweight men with normal blood glucose levels found that consuming about 30 g/day of oat fiber for 12 weeks lowered LDL-C by 2.5% ($p = 0.02$) and reduced the concentration of small, dense LDL-C particles by 17.3% ($p = 0.01$). By contrast, subjects who consumed 30 g/day of wheat fiber for 12 weeks increased LDL-C by 8.0% ($p = 0.02$) and increased the concentration of small, dense LDL-C particles by 60.4% ($p = 0.02$), with a significant time-by-treatment interaction ($p < 0.05$). TG levels tended to be lower in subjects consuming oat fiber but higher in subjects consuming the wheat fiber ($p < 0.07$) [101]. This lipid-lowering effect of oat fiber is likely due to the action of viscous fiber, specifically beta-glucan, which is not present in wheat fiber. The FDA states that intakes of beta-glucan ≥ 3 g/day can effectively lower LDL-C [108], although the changes in TG appear to be mixed.

A meta-analysis of 20 controlled trials using viscous fiber from oat sources demonstrated that the reduction in TC ranged from 0.1 to 2.5% per gram intake of viscous fiber [109]. Larger reductions were seen in studies conducted with subjects who had higher baseline blood cholesterol levels (≥ 5.9 mmol/L), specifically supplemental viscous fiber intakes of at least 3 g/day. A meta-analysis of 67 controlled trials evaluating different sources of viscous fiber (oats, pectin, guar, and psyllium) reported that 2–10 g/day significantly reduced LDL-C (-0.057 mmol/L, $p < 0.001$) [110]. Not all studies reported a cholesterol-lowering effect, perhaps due to variation in viscosity as a result of different molecular weights and solubility of beta-glucan sources. Beta-glucan, with a higher molecular weight, has been shown to reduce LDL-C by about 5%, whereas lower molecular weight beta-glucan does not appear to have a significant effect [111]. The molecular weight of beta-glucan often is reduced during extraction and occasionally decreased to improve the functional properties of food products [111]. Therefore, it is likely that beta-glucan effects vary among products, indicating that the molecular weight and solubility are fundamental for establishing the effects of fiber on increasing intestinal viscosity and lowering cholesterol levels.

9.2.7.2 Epidemiologic Associations

Observational data have demonstrated an inverse relationship between dietary fiber and TC and LDL-C. However, the cardiovascular benefits associated with fiber depend largely on the particular food source. Jacobs et al. [112] demonstrated that fiber from whole grains but not refined grains was associated with reduced risk of mortality. The authors concluded that other constituents (e.g., vitamins, minerals, and phytonutrients) found in whole grains may provide benefits beyond the individual effects of fiber (see Sect. 9.3.2) [112].

Overall, evidence suggests that dietary fiber improves blood lipids, lowers blood pressure, and reduces markers of inflammation [113]. These benefits may occur with intakes as low as 12–33 g/day of fiber from whole foods (or up to 43 g from supplements) [113]. Americans on average only consume 15 g/day of fiber, below the recommended intake of 25 g for a 2,000 kcal diet [14]. In addition, white flour and processed potatoes provide the majority of fiber consumed in the average American diet [18]. Manufacturers are allowed to label a food a "good source of fiber" if it contains 10% of the recommended amount (2.5 g/serving) and an "excellent source of fiber" if it contains 20% of the recommended amount (5 g/serving) [14]. However, a consistent definition of dietary fiber continues to be a challenge.

In the 2002 DRIs, total fiber was defined as dietary fiber and added fiber (functional fiber); specific definitions were issued for each type of fiber. Dietary fiber was defined as nondigestible CHOs and lignin that are intrinsic and intact in plants [92]. Added fiber, or functional fiber, consists of isolated nondigestible CHOs that have beneficial effects in humans [92]. The Codex Alimentarius Commission (Codex), the food standards program of the United Nations Food and Agriculture Organization (FAO) and the World Health Organization (WHO), recently established a similar dietary fiber definition after years of opposition and debate. Codex defined dietary fiber as CHO polymers with ten or more monomeric units, which are not hydrolyzed by the endogenous enzymes in the small intestine [114]. However, several countries continue to use more restrictive definitions. For example, Canada requires that "novel fibers" can only be included in the definition after approval from Health Canada and cannot include certain chemically processed ingredients [115]. Therefore, it may be possible that the dietary fiber definition and analytical method used by one organization varies from that of another. Consistency in the measurement and description of dietary fiber is needed.

More information on the bioactive components contained in fiber and the mechanisms of action is also needed to understand the effects of fiber consumption on CHD risk, specifically lipids and lipoproteins. In addition, research addressing the effect of different dietary fibers as well as added fibers on lipids and lipoproteins will play a role in updating current dietary fiber recommendations.

9.2.8   Plant Sterols/Stanols

9.2.8.1   Effect on Blood Lipids

Evidence from several randomized studies has demonstrated that plant sterols/ stanols, similar to cholesterol in structure, decrease TC and LDL-C in subjects with both normal and elevated cholesterol levels [113, 116–119]. Sterols/stanols may reduce dietary and biliary cholesterol absorption by displacing cholesterol from micelles in the intestine, resulting in lower cholesterol solubility. Both sterols and stanols modify lipids and lipoproteins similarly [120].

Clinical studies involving normocholesteromic subjects have demonstrated that sterols may reduce TC by 4–9% and LDL-C by 6–15%, respectively [118, 119]. Similar studies have evaluated the effectiveness of sterols in hypercholesterolemic adults. Devaraj et al. [117] randomly assigned 72 free-living adults with mild hypercholesterolemia to consume a placebo orange juice or an orange juice fortified with plant sterols (2 g/day) for 8 weeks while following an average American diet. The sterols significantly decreased TC (7.2%) and LDL-C (12.4%) compared with baseline and placebo ($p < 0.01$). Other studies have reported similar reductions in TC (2.8–11.9%) and LDL-C (1.7–13.4%) when consuming 1.6–3.2 g/day of sterol-enriched products [121–123]. Christiansen et al. [116] measured the lipid lowering effect of sterols at different doses. Subjects in the control group consumed a placebo spread while test groups consumed a spread containing either 1.5 or 3.0 g/day of plant sterols for 6 months. TC and LDL-C levels were significantly lower in groups consuming the sterols compared to the control, i.e., 7.5–11.6% (0.46–0.62 mmol/L). However, no significant difference was reported between 1.5 and 3.0 g/day of sterols on TC and LDL levels [116]. Moruisi et al. [124] conducted a systematic review and meta-analysis to evaluate the cholesterol-lowering effects of sterols in subjects with hypercholesterolemia for 4 weeks to 3 months of intervention. An intake of 2.3 g/day of sterols/stanols was associated with a reduction in TC of 7–11% with a mean decrease of 0.65 mmol/L and LDL-C by 10–15% with a mean decrease of 0.64 mmol/L (both $p < 0.00001$).

9.2.8.2   Epidemiologic Associations

Epidemiologic evidence of an association between plant sterols and CHD is limited. Data from population studies have been mixed, reporting inconsistent associations between plasma sterol concentration and CHD risk [125–127]. The Prospective Cardiovascular Munster study [125] found that the upper quartile of plasma sterol concentration (>2.18 mg/L) was associated with a greater CHD risk (1.8-fold increase) compared with the lower three quartiles ($p < 0.05$), while the European Prospective Investigation into Cancer-Norfolk Study [127] and the Longitudinal Aging Study Amsterdam [126] both reported that high plasma sterol concentrations were associated with lower CHD risk (OR 0.79 and 0.78, $p < 0.05$). Plant sterols can interfere with the absorption of fat-soluble vitamins and carotenoids [116, 122, 124], which may explain part of the discrepancy in epidemiologic evidence. However, a diet rich in carotenoids and fat-soluble vitamins has been shown to maintain nutrient adequacy when consuming sterols/stanols [116, 124]. Although sterols/stanols occur naturally in some vegetable oils, nuts/seeds, fruits, vegetables, legumes, and whole grains, many commercial foods, such as juices, breads, and dairy products, are fortified with sterol/stanols. Additional research is still needed though to evaluate the long-term safety of plant sterols/stanols.

9.2.9   Other Bioactive Compounds

A large body of evidence has reported that numerous bioactive compounds reduce CHD risk through a variety of mechanisms, many of which are nonlipid related. Bioactive compounds, found in small quantities in plant-based foods, may exhibit antioxidant, antithrombotic, and/or anti-inflammatory properties [128]. Some compounds may also protect against several forms of cancer. Though the lipid-lowering effects of bioactive compounds are minimal at best, the cardioprotective benefits are well established. Therefore, incorporating foods that contain these bioactive compounds may be beneficial in the prevention of CHD.

9.2.9.1   Polyphenols

Data from experimental and clinical studies suggest polyphenols, particularly flavonoid-rich foods, can reduce CHD risk by improving endothelial function and decreasing blood pressure (BP) [129]. In addition, flavonoids have a high antioxidant capacity capable of inactivating reactive oxygen species (ROS) associated with oxidizing LDL-C in blood. Flavanols, a subgroup of flavonoids, are present in high concentrations in plant foods, specifically fruits and vegetables. Green and black tea, chocolate, red wine or grapes, apples, pears, berries, and cocoa are all rich sources of flavanols that have been associated with a lower CHD mortality (Table 9.2) [130]. Although flavanols occur as monomers catechin and epicatechin, they often are assembled as procyanidins (also known as condensed tannins) [131]. Resveratrol is a nonflavonoid antioxidant found in grapes, red wine, peanuts, and some berries. It also has been associated with reducing CHD risk by inhibiting LDL-C oxidation, improving endothelial function, and providing antiplatelet and anti-inflammatory effects [132–135].

Table 9.2 Catechin/epicatechin content found in food

Source	Mean content[a] (mg/kg FW or mg/L)
Cocoa power	2,711
Dark chocolate	960
Green tea	657
Black tea	495
Milk chocolate	192
Peaches	135
Grapes (black)	124
Red wine	115
Apricots	77
Pistachios	69
Strawberries	68
Raspberries	56
Beans	53
Almonds	49
Pears	41
Lentils	37

Adapted from ref. [214]

[a]Values determined by chromatography; *FW* fresh weight for solid foods

9.2.9.2 Cocoa

A moderate amount of evidence demonstrates that modest consumption of chocolate, specifically cocoa, has cardiovascular benefits capable of reducing CHD risk [14]. These benefits have been accredited specifically to the flavanol compounds contained in cocoa. The flavanol content and the antioxidant capacity in blood have been shown to increase following cocoa consumption [136]. Both the manufacturing process as well as the type of cocoa bean (along with the geographic region of agricultural production) determines the amount of flavanols contained in chocolate. Cocoa beans are often treated with alkali to reduce the bitterness and increase the dispersability of cocoa; however, this process also modifies or destroys the polyphenols and flavanols (Table 9.2) [137]. As a result milk chocolate generally contains a lower flavanol content than cocoa powder or dark chocolate [136]. It is noteworthy that the predominant fat in chocolate is cocoa butter which is high in stearic acid, a fatty acid that does not significantly affect blood lipids [9]. However, the DGAC 2010 Report advised that since chocolate is energy dense, the potential benefits need to be balanced with caloric intake [14].

In a crossover study by Grassi et al. [138], 20 hypertensive patients were randomly assigned to consume 100 g/day of dark chocolate (containing 88 mg of flavanols) or 90 g/day of flavanol-free white chocolate for 2 weeks. The dark chocolate was found to increase insulin sensitivity and flow-mediated vasodilation (FMD) with a 12% reduction in LDL-C ($p < 0.05$) as well as a decrease in systolic and diastolic BP ($-11.9/-8.5$ mmHg, both $p < 0.0001$). Other studies also have reported beneficial effects of cocoa on lipids and lipoproteins [139–141]; however, neutral effects on lipids and biomarkers of inflammation have also been reported as well [142, 143]. In a clinical study by Taubert et al. [143], subjects in good health except for upper-range prehypertension (BP between 130/85 and 139/89 mmHg) or stage 1 hypertension (BP between 140/90 and 160/100 mmHg) were randomly assigned to consume either 6.3 g/day dark chocolate (30 mg of polyphenols) or polyphenol-free white chocolate for 18 weeks. Following the intervention, dark chocolate decreased mean systolic and diastolic BP (−2.9/−1.9 mmHg, both $p < 0.001$) with no significant changes in blood lipids. White chocolate did not change BP or plasma biomarkers of CVD risk. Taubert et al. [143] provided subjects with much less chocolate than Grassi et al. [138] (6.3 vs. 100 g/day) which may explain the smaller changes in BP. The American Dietetic Association (ADA) Evidence Analysis Library concluded that limited evidence demonstrates positive effects of cocoa or chocolate on BP when consumed for more than 2 weeks [144].

Flavanols and procyanidins in cocoa also provide strong antioxidant effects that may counteract lipid peroxidation. The improvement in FMD following chocolate consumption suggests induction of endothelial NO synthase, resulting in elevated NO levels and enhanced endothelial function [131]. A randomized study involving cardiac transplant patients evaluated the effect of flavonoid-rich dark chocolate (40 g) vs. a cocoa-free control chocolate on coronary vasomotion, the change in diameter of coronary artery [145]. Coronary vasomotion was analyzed with quantitative coronary angiography and cold pressor testing (immersion of the right hand in ice-cold water for 2 min) before and 2 h after chocolate consumption. After 2 h, the dark chocolate consumption induced coronary vasodilatation and decreased platelet adhesion. Both effects were proportional to the amount of serum epicatechin, suggesting that the flavanol epicatechin may be responsible for improving coronary vasomotion [145].

A meta-analysis of 135 RCTs by Hooper et al. [129] evaluated the effects of various flavonoid subclasses as well as foods rich in flavonoids, including cocoa and chocolate, on CHD risk factors. The heterogeneity was significant for LDL-C ($p = 0.002$ for heterogeneity), which likely reflects the diverse effects of each flavonoid subclass. Only soy protein isolate (SPI), a source of isoflavones, and green tea had a significant LDL-C lowering effect (−0.19 mmol/L, $p = 0.03$ for heterogeneity and −0.23 mmol/L, $p = 0.62$). It must be noted, however, that it requires a rather large amount of these food products to achieve significant reductions in LDL-C: 20–56 g of SPI and 2–5 mugs/day (~480–1,200 mL) of green tea. Chocolate intake did not significantly affect levels of LDL-C or HDL-C ($p = 0.03$ for heterogeneity). Chocolate consumption did, however, increase FMD after acute and chronic intake (4.0 and 1.45%) as well as decrease systolic and diastolic BP (−5.9/−3.3 mmHg, $p < 0.001$). Many of the chocolate and cocoa studies used formulations that are not readily available to the public. Therefore, additional studies using commonly consumed formulations would allow for more relevant implications. Also, consistent flavonoid content is important to evaluate the individual effects of each flavonoid subclass on CHD risk.

Evidence from epidemiologic studies suggests an inverse relationship between flavonoid-rich foods and CHD risk. A meta-analysis of seven prospective cohort studies with 105,000 subjects reported that a high intake of flavonoids from fruits and vegetables, red wine, and tea was inversely associated with CHD risk [146]. Individuals in the top tertile of dietary flavonoid intake had a 20% lower risk of CHD mortality compared with those in the lower tertile; this was independent of other dietary components and CHD risk factors ($p < 0.001$).

Research is needed to address dose–response effects of polyphenols, particularly flavonoids, on risk factors associated with CHD. In addition, the effectiveness of foods rich in flavonoids may be in part due to a variety of other confounding factors within the food, such as vitamins, minerals, and other bioactive compounds. There is sufficient evidence to recommend consuming foods rich in bioactive compounds, including polyphenols. More research is needed to understand the effects of individual polyphenols and the cardiovascular benefits associated with each polyphenol [128]; however, a complex interaction likely exists among food substances. Chapter 14 discusses food synergy in greater detail.

9.3   Foods That Modify Lipids and Affect Risk of Heart Disease

The Dietary Guidelines (DG) and AHA provide food-based dietary recommendations for healthy Americans to decrease risk of CHD and other chronic diseases. These guidelines recommend an overall healthy eating pattern to achieve and maintain a healthy body weight, a desirable cholesterol profile, and a desirable BP. The recommendations provide flexibility in selecting foods to meet nutritional needs and reduce CHD risk. A healthy eating pattern includes a variety of fruits, vegetables, grain products (of which whole grains should comprise at least one half of the total servings), low-fat or nonfat dairy products, fish, legumes, poultry and lean meats, and liquid vegetable oil. To achieve a desirable lipid profile, the DG and AHA recommend limiting foods high in SFA, TFA, and cholesterol that are provided by full-fat dairy products, fatty meats, partially hydrogenated vegetable oils, and egg yolks [147]. In addition, SFA should be decreased and replaced with calories from either nutrient-dense CHO (from whole grain products, fruits, vegetables, and fat-free dairy products) or unsaturated fat (from vegetables, fish, legumes, and nuts).

The following is an overview of the specific food groups that are recommended as part of a healthy diet and their apparent beneficial effects on blood lipids and on risk of CHD.

9.3.1   Fruits and Vegetables

A large epidemiologic database has found that an increased intake of fruits and vegetables is associated with a significant reduction in at least one CHD outcome, including total mortality [14]. The mechanisms of action are thought to be mediated, in part, through CHD risk factor reduction including beneficial effects on lipids and lipoproteins. Tande et al. [148] used NHANES III data to evaluate the relationship between the Food Guide Pyramid recommendations (i.e., food groups) and serum lipids in 9,111 participants. Fruit intake was found to be inversely related to TC, LDL-C, and HDL-C ($p=0.012, 0.014, 0.001$, respectively) and positively related to TG levels ($p=0.003$). High intake of viscous fiber may have accounted for part of the lower TC and LDL-C levels. Antioxidants and phytonutrients, including poly- phenols, may provide additional benefits. The National Heart, Lung, and Blood Institute Family Heart Study [149] reported similar findings. Consumption of fruit and vegetables was inversely related to LDL-C. Subjects consuming more fruit and vegetables (>4 vs. <1.9 servings/day) had lower LDL-C concentrations (3.14 vs. 3.36 mmol/L; $p<0.0001$) [149]. In a recent cross-sectional study, individuals with the highest fruit and vegetable consumption (compared to those with lower con- sumption) had lower LDL-C levels (2.46 vs. 3.36 mmol/L, $p<0.01$) [150].

A meta-analysis by Dauchet et al. [151] found that each additional serving per day of fruit and vegetables resulted in a 4% ($p=0.0027$) decrease in CHD risk and for each additional serving per day of fruit, a 7% ($p<0.0001$) decrease in CHD risk. The DGAC 2010 Report found that the lowest risk of CHD was associated with a fruit and vegetable intake >5 servings/day, and highest risk with an intake <3 servings/day, suggesting a linear relationship between fruit and vegetable con- sumption and CHD risk [14].

9.3.2   Whole Grain Breads and Cereals

A moderate amount of evidence suggests that whole grain consumption may protect against CHD [14]. In the United States, grains are typically overconsumed com- pared to current recommendations, although consumption of whole grains is inad- equate [14]. White flour makes up a large portion of the grain consumed in the average American diet [18].

Distinguishing whole grains from terms such as "whole wheat" or "multigrain" can be difficult. The FDA and the American Association of Cereal Chemists estab- lished that to be considered a "whole grain," the main components of the grain (endosperm, germ, and bran) need to be provided in the same relative proportion as it would be in the natural seed [152]. Therefore, a product containing 100% whole wheat flour may not technically be a whole grain since the germ is often removed. In order for a product to be labeled as a whole grain, it must contain whole grain as the first ingredient, making up 51% of the products total weight [153].

Clinical evidence regarding the effect of whole grain intake on lipids and lipo- proteins has been mixed, depending on the type of whole grain consumed. Oats and barley demonstrate a greater lipid lowering effect than wheat or brown rice, likely due to their viscous fiber content (see Sect. 9.2.7) [154]. A systematic review of 10 intervention studies found that whole grain intake, specifically oats, was associated with lower TC (−0.20 mmol/L, $p=0.0001$) and LDL-C (−0.18 mmol/L, $p<0.0001$)

Table 9.3 Fatty acid composition and caloric value in 1 oz of nuts

	Amount							
	1 oz (28.35 g)	Energy (kcal)	Total fat (g)	SFA (g)	MUFA (g)	PUFA (g)	n-3 (g)	n-6 (g)
Almonds	22 nuts	169	15.0	1.2	9.5	3.6	0.00	3.59
Brazil nuts	6–8 nuts	186	18.8	4.3	7.0	5.8	0.01	5.82
Cashews	12 nuts	163	13.1	2.6	7.7	2.2	0.05	2.17
Hazelnuts	12 nuts	183	17.7	1.3	13.2	2.4	0.02	2.38
Macadamia nuts	10–12 nuts	204	21.6	3.4	16.8	0.4	0.06	0.37
Peanuts	35 pieces	168	14.6	2.0	8.9	3.1	0.05	2.99
Pecans	15 halves	201	21.1	1.8	12.5	5.8	0.28	5.55
Pistachios	49 kernels	162	13.9	1.6	6.9	3.9	0.07	3.87
Walnuts	14 halves	175	16.7	1.0	4.3	9.9	0.57	9.38

Adapted from ref. [215]
All values given are for dried or dry roasted nuts

compared to diets based on refined grains; there were no difference in HDL-C or TG [155]. However, the studies included in the analysis only evaluated short-term effects. Long-term controlled trials are needed to determine the sustained effects of whole grains on blood lipids.

Epidemiologic evidence has also suggested an inverse relationship between whole grain intake and CHD risk. Jacobs and Gallaher [156] reviewed 13 prospective epidemiologic studies and reported a 20–40% reduction in risk of CHD when comparing habitual consumers of whole grains vs. those who seldom consumed them. Most whole grains are eaten as whole wheat which has only a small lowering effect on blood lipids. This might be expected to reduce CHD by a few percent, but not by 20–40%. This suggests that other mechanisms are largely responsible for the lower CHD risk associated with whole grains.

9.3.3 Nuts

Both clinical and epidemiologic evidence have firmly established the benefit of nut consumption in lowering CHD risk across a variety of population groups, including elderly persons. Nuts are low in SFA and high in MUFAs and PUFAs (Table 9.3), which favorably affect blood lipids and lipoproteins [157]. In addition, they are rich sources of other nutrients including folate, vitamin E, and polyphenols. Furthermore, nuts contain plant sterols and dietary fiber, which would also contribute to their cholesterol-lowering effect.

Numerous clinical studies have consistently reported a TC and LDL-C lowering effect of diets low in SFA and cholesterol and high in unsaturated fat provided by a variety of nuts. Based on clinical studies, an intake of 1–2 oz/day of nuts reduces LDL-C by 2–19% [158, 159]. A systematic review by Mukuddem-Petersen et al. [159] showed that consuming 50–100 g (1.7–3.5 oz) of nuts ≥5 times/

week as part of a moderate-fat diet (35% of total calories) results in greater reductions of TC (2–16%) and LDL-C (2–19%) compared to a low-fat, cholesterol-lowering diet.

In a recent pooled analysis of 25 intervention studies ranging from 3 to 8 weeks in length, a mean daily nut consumption of 67 g reduced TC by 5.1%, LDL-C by 7.4%, the LDL-C/HDL-C ratio by 8.3%, and the TC/HDL-C ratio by 5.6% ($p < 0.001$ for all) [160]. In addition, increasing amounts of nuts resulted in a dose–response effect when compared to nut-free control diets. Subjects with a high baseline LDL-C or low BMI experienced the greatest cholesterol-lowering effect. This indicates that nuts were less effective in lowering blood cholesterol among obese subjects. Obesity is associated with decreased intestinal cholesterol absorption, due in part to the development of insulin resistance that results from adipose-induced inflammation. Insulin resistance causes impaired receptor mediated removal of LDL-C and decreases chylomicron clearance [161], thereby diminishing reductions in LDL-C associated with plant sterols.

Large cohort studies consistently reveal that frequent nut consumption is associated with a 30–50% decrease in risk of CHD [162]. In the Nurses' Health Study, women who reported eating more than 5 oz/week of nuts had a significantly lower relative risk of total CHD (0.65; $p = 0.0009$) compared to women who ate less than 1 oz/month, after adjusting for various risk factors for CHD [163]. Similarly, the Seventh-day Adventist Study reported that men and women consuming ≥5 oz nuts/week had 50% less risk of fatal and nonfatal CHD compared to individuals eating ≤1 oz/week (RR of 0.49, 0.52, respectively) [162].

Other biomarkers of CHD have also been associated with nut consumption. A review by Ros [164] found that nuts, high in MUFAs, moderately improve oxidative status by reducing the susceptibility of LDL-C to oxidation. In addition, anti-inflammatory effects and improved endothelial function were associated with nuts and their bioactive components (antioxidants and polyphenols) [164]. Therefore, nuts appear to reduce CHD risk through a variety of mechanisms. Additional studies are needed to better understand all of the cardioprotective effects of nut consumption beyond changes in the lipid profile.

9.3.4 Vegetable Oils

Vegetable oils are excellent sources of unsaturated fatty acids—MUFA and PUFA—depending upon the seed oil (Table 9.4). Beneficial effects of unsaturated fatty acids have been studied extensively in both epidemiologic and clinical studies. The TC/HDL-C ratio has been shown to decrease the most when CHO (mostly refined) is replaced with MUFA or PUFA in the diet [9]. The Nurses' Health Study found that a 5% increment in energy from MUFA was associated with a decrease in CHD risk (RR = 0.81, $p = 0.05$) compared to an equivalent energy from mostly refined CHO [17]. In addition, a 5% increment in energy from PUFA was associated with lower CHD risk (0.62, $p = 0.003$) compared with CHO. This is

Table 9.4 Fatty acid composition and caloric value of vegetable oils in 1 tbsp

Oil	Energy (kcal)	Total fat (g)	SFA (g)	MUFA (g)	PUFA (g)	n-3 (g)	n-6 (g)
Canola	124	14.0	1.0	8.2	4.1	1.30	2.84
Canola (high oleic)	124	14.0	0.9	10.1	2.4	0.36	2.03
Corn	120	13.6	1.7	3.3	8.0	0.16	7.28
Cottonseed	120	13.6	3.5	2.4	7.1	0.03	7.00
Flaxseed oil	120	13.6	1.3	2.8	9.0	7.25	1.73
Olive	119	13.5	1.8	10.0	1.4	0.11	1.24
Palm	120	13.6	6.7	5.0	1.3	0.03	1.24
Palm-kernel	117	13.6	11.1	1.6	0.2	0.00	0.22
Peanut	119	13.5	2.3	6.2	4.3	0.00	4.32
Safflower	120	13.6	0.8	2.0	10.2	0.00	10.15
Safflower (high oleic)	120	13.6	0.8	10.2	2.0	0.00	1.95
Sesame	120	13.6	1.9	5.4	5.7	0.04	5.62
Soybean	120	13.6	2.0	3.2	7.9	0.93	6.94
Sunflower	120	13.6	1.4	2.7	8.9	0.00	8.94
Sunflower (high oleic)	124	14.0	1.4	11.7	0.5	0.03	0.51
Walnut oil	120	13.6	1.2	3.1	8.6	1.41	7.19

Adapted from ref. [215]

SFA saturated fatty acids; *MUFA* monounsaturated fatty acids; *PUFA* polyunsaturated fatty acids

similar to the lower risk associated with replacing SFA with PUFA or MUFA (see Sect. 9.2.1.3).

Evidence from clinical trials convincingly affirms the beneficial effects reported in the epidemiologic literature of MUFA and PUFA is partially due to favorable effects on lipids and lipoproteins (see Sects. 9.2.3 and 9.2.4). The contrasting fatty acid profile of different vegetable oils has enabled several clinical studies to be done that have manipulated dietary fatty acid composition in a variety of ways. These studies have shown consistently that n-6 PUFA lowers TC and LDL-C and that MUFA has either a slight or neutral cholesterol-lowering effect, when substituted isoenergetically for CHO [3].

Some vegetable oils are solid at room temperature. This is especially the case with tropical oils, notably palm and palm kernel oil; they contain higher amounts of SFA and are associated with adverse effects on blood lipids. Many organizations, including the WHO, encourage that palm oil consumption be reduced to decrease CHD risk [165]. Coconut oil is another tropical oil rich in SFA.

Unsaturated oils typically lack the chemical stability necessary for certain processed foods; therefore, food manufacturers frequently use partially hydrogenated oils as an alternative to achieve functionality in some food products (see Sects. 9.2.1.1 and 9.2.2). A current challenge for many food manufactures has been to decrease the use of saturated and partially hydrogenated oils while maintaining the chemical stability and functionality required for their products. The removal of partially hydrogenated oils from foods would confer health benefits, especially when replaced with oils rich in MUFA and/or PUFA [43].

New unhydrogenated oils, particularly high-oleic oils, recently entered the marketplace to provide a healthier alterative to partially hydrogenated oils and SFA. Through selective breeding or genetic modification, several oils, including canola, sunflower, safflower, and soybean, have been developed with an increasing proportion of oleic acid. These new high-oleic oils deliver a higher resistance to oxidation, allowing for greater use in food applications. Lichtenstein et al. [22] analyzed several oils from genetically modified soybeans and compared them with common soybean and partially hydrogenated soybean oils on CVD risk. Subjects consumed five experimental diets for 35 day, each containing the same foods and providing 30% of energy from fat. Two-thirds of the fat energy was soybean oil, low-SFA soybean oil, high-oleic soybean oil, low-ALA soybean oil, or partially hydrogenated soybean oil. All varieties of soybean oils resulted in more favorable lipoprotein profiles except the partially hydrogenated oil. LDL-C was highest following the partially hydrogenated and low-ALA oil diets. HDL-C was not significantly different between diets, although for men the high-oleic diet resulted in significantly greater HDL-C levels compared to the common soybean oil diet. All unhydrogenated soybean oil diets resulted in lower TC/HDL-C ratios compared to the partially hydrogenated form, with high-oleic oil diets resulting in the lowest TC/HDL-C ratio. The findings from this study indicate, therefore, that several unhydrogenated oils, including high-oleic oils, are viable substitutes for TFA and/or SFA. Additional studies are needed to evaluate the effect of novel vegetable oils on lipids and lipoproteins as well as other CHD risk factors. The effect of different vegetable oils on cholesterol lowering will reflect the type and amount of oil(s) used in the diet. Furthermore, the fatty acid profile of the total diet is the defining determinant of the blood cholesterol response.

9.3.5   Dairy Products

Multiple systematic reviews and meta-analyses have reported that low-fat/fat-free dairy intake is protective against CHD. In contrast, Tande et al. [148] analyzed NHANES III data and found that dairy consumption was associated with an increase in LDL-C. This is likely due to the higher SFA intake associated with fat-containing dairy products. Therefore, consumption of low-fat/fat-free dairy rather than higher fat dairy would reduce SFA intake. Other dairy constituents, such as calcium, magnesium, potassium, and vitamin D, may decrease CHD risk via other mechanisms. CLA also is found in dairy products; however, the amount is probably too low to have any effect on CHD risk [166].

Tholstrup et al. [167] investigated the effect of the different physical states of fat on blood lipids and lipoproteins. Fourteen healthy men were randomized to one of three groups (whole milk, butter, or hard cheese) for 3 weeks, each in a controlled crossover study. All diets resulted in a slight increase of LDL-C. The most noteworthy findings was that butter diet resulted in 0.21 mmol/L higher LDL-C than cheese ($p = 0.037$). More research is needed to understand how the physical state of fat in dairy foods affects blood lipids.

A model community-based CHD intervention program, the North Karelia Project in eastern Finland, reported that a decrease in SFA intake (principally from dairy fat) from 20–21% of energy in 1972 to 14–15% of energy in 1997 was associated with a decrease in TC levels of 15–19% among men and women [168]. The primary changes in diet were decreased use of butter and increased use of low-fat and skim milk instead of whole milk. In the North Karelia Project, CHD mortality declined 73% from 1971 to 1995 [169]. The reduced mortality was associated with a decrease in blood cholesterol and also beneficial changes in BP and cigarette smoking.

More recently, Elwood et al. [170] conducted a systematic review of 15 prospective cohort studies evaluating the association between dairy consumption and CHD risk. The highest consumption of milk was associated with a significant reduction in the risk of CHD, stroke, and MI. The RR of CHD risk in subjects with high milk or dairy consumption was 0.84 and 0.79, respectively, compared to those with low consumption. In addition, several prospective studies have reported that consumption of milk and dairy products is inversely associated with the prevalence of metabolic syndrome, a significant contributor of CHD [170, 171]. Therefore, a cardioprotective effect appears to be associated with dairy consumption. Clinical evidence has demonstrated a consistent BP-lowering effect of dairy consumption, possibly due in part to bioactive dairy peptides and the micronutrient profile of dairy products (i.e., calcium, magnesium, and potassium). Hilpert et al. [172] found that stage I hypertensive patients fed a diet rich in dairy foods correlated with a decrease in diastolic BP ($r = 0.52$, $p < 0.05$). The increase in calcium associated with dairy intake presumably caused the decrease in intracellular calcium by depressing the levels of calcitropic hormones (parathyroid hormone and 1,25-hydroxy vitamin D) responsible for regulating intracellular calcium. This reduction in intracellular calcium also has been shown to inhibit lipogenesis and stimulate lipolysis [173]; therefore, dairy products containing calcium may influence whether fat is stored or broken down in the body.

Although not entirely consistent, epidemiologic evidence also suggests an inverse relationship between dairy consumption and elevated BP. The Coronary Artery Risk Development in Young Adults (CARDIA) study found that the 15-year incidence of elevated BP was inversely related to dairy consumption [174]. However, there is still a need to evaluate dairy components, including milk proteins, to understand the underlying mechanisms associated with a reduced risk of CHD in response to dairy consumption. Further research also should be conducted to investigate the effects of dairy consumption on both established and novel CHD risk factors.

9.3.6   Meat

The Third National Health and Nutrition Examination Survey [148] found that increasing consumption of higher fat red meats was associated with an increase in LDL-C, which could be explained by the increase in SFA intake. However, controlled clinical studies have shown consistently that including lean red meat (beef, veal, and pork) in a NCEP Step I diet in hypercholesterolemic subjects decreases LDL-C and increases HDL-C similar to that of lean white meat (poultry and fish) [175–177].

Studies of the relationship between meat consumption and CHD have generated inconsistent findings. The Nurses' Health Study [178] and the Iowa Women's Health Study [179] both found no relationship between intake of meat and CHD. However, in a prospective study of 84,136 women in the Nurses' Health Study followed for 26 years, Bernstein et al. [180] found that higher intakes of both red meat and red meat excluding processed meats were significantly associated with increased risk of CHD. Consumption of these foods was associated with higher rates of smoking, TFA consumption, lower physical activity, and decreased intake of various vitamins. It is therefore essential to adjust for these confounding variables. After doing so the associations were still significant for both red meat and red meat excluding processed meats (RR = 1.16 per serving/day, p for trend <0.001, and RR = 1.19, p = 0.02, respectively). Serving size was not defined, but according to the USDA 3 oz is the recommended serving size. Similarly, the National Institutes of Health-AARP Diet and Health Study [181], a prospective study of over half a million people, reported that CVD mortality risk was elevated for men and women in the highest intake of red (HR = 1.27 and 1.50, respectively, p for trend <0.001) and processed meat (HR = 1.09 and 1.38, p for trend <0.001) after adjusting for confounding variables.

A review by McAfee et al. [182] reported that many studies evaluating red meat consumption and CHD risk were not consistent in measuring meat intake, with several studies including processed meat in the red meat category. In the absence of a clear definition about what can be classified as processed or red meat, results from future prospective studies will most likely continue to vary. However, the review did not find any increase in CHD risk associated with consuming moderate amounts of lean red meat.

Micha et al. [183] recently conducted a systematic review and meta-analysis. The analysis included 17 prospective cohort studies and 3 case-control studies conducted in the United States (n = 11), Europe (n = 6), Asia (n = 2), and Australia (n = 1), consisting of over one million subjects. Red meat intake was not associated with CHD risk (RR = 1.00 per 50 g serving/day); however, processed meat was associated with a 42% higher risk of CHD. The authors concluded that consumption of processed meats, but not red meats, was associated with an increase in CHD risk; therefore, differentiating between processed and unprocessed meats is necessary to assess health effects and underlying mechanisms [183]. In addition, no RCTs of red or processed meat consumption and risk of CHD were identified in this analysis; therefore, a controlled clinical study is needed to validate these findings.

Lean red meat can be included in a cholesterol-lowering diet as long as the quantity is consistent with the daily amount recommended (5 oz/day for the TLC diet). In addition, the total diet must meet current recommendations for cholesterol-znutrients. SFA and cholesterol content of selected cuts/species of red meat, poultry, and fish are shown in Table 9.5. A recommendation of less than 7% energy from SFA translates to 15.6 g/day for a 2,000 kcal diet. A 6-oz serving of lean beef

Table 9.5 Fatty acid composition, cholesterol, and caloric value of raw meat, fish, poultry, dairy, and soy products in 100 g portions

Source	Energy (kcal)	Total fat (g)	SFA (g)	MUFA (g)	PUFA (g)	n-3 (g)	n-6 (g)	Cholesterol (mg)
Ground beef, 95% lean	137	6.0	2.2	2.1	0.2	0.04	0.22	62
Beef, bottom round, fat trimmed	187	7.7	2.8	3.3	0.3	0.04	0.22	86
Beef, top sirloin, fat trimmed	183	5.8	2.2	2.3	0.2	0.02	0.17	58
Pork tenderloin	120	3.5	1.2	1.4	0.6	0.03	1.02	65
Ham, lean	195	8.3	2.8	3.8	1.0	0.09	0.88	70
Chicken, light meat, no skin	114	1.7	0.4	0.4	0.4	0.05	0.28	58
Chicken, dark meat, no skin	125	4.3	1.1	1.3	1.1	0.11	0.82	80
Turkey, light meat, no skin	108	0.5	0.2	0.1	0.1	0.01	0.12	66
Salmon, farm raised	183	10.9	2.2	3.9	3.9	2.16	1.74	59
Tuna, light, canned in water	116	0.8	0.2	0.2	0.3	0.29	0.04	30
Tofu, firm	145	8.7	1.3	1.9	4.9	0.58	4.34	0
Yogurt, low fat, plain	63	1.6	1.0	0.4	<0.1	0.01	0.03	0.6
Cottage cheese, low fat, 1%	72	1.0	0.6	0.3	<0.1	0.01	0.02	4

Adapted from ref. [215]

SFA saturated fatty acids; *MUFA* monounsaturated fatty acids; *PUFA* polyunsaturated fatty acids

provides about 8 g of SFA. Lean red meats include bottom round and sirloin cuts, whereas lean poultry includes skinless white meat.

9.3.7 Seafood

A moderate amount of evidence suggests that consuming two servings of seafood per week (4 oz/serving) reduces cardiac mortality from CHD [14]. However, not all fish or seafood provide equal benefits given that certain sources contain substantially more EPA+DHA than others. The AHA recommends consuming two servings of fish per week, preferably fatty fish, to decrease CHD risk [184]. Furthermore, choosing fish with a low content of methylmercury is recommended so as to maximize cardiovascular benefits. Children, pregnant women, and women who may become pregnant are advised to avoid eating fish with higher levels of mercury. Good sources of fatty fish lower in mercury include salmon, light tuna, trout, herring, and sardines. Albacore ("white") tuna is a commonly eaten fish, yet it contains a moderately high level of mercury compared to canned light tuna, intermediate to the high and low mercury seafood (Fig. 9.3).

Many other marine sources of *n*-3 fatty acids, high in EPA and DHA, have gained recent attention, especially krill, a zooplankton crustacean found in cold ocean waters. Maki et al. [185] examined the effects of krill oil supplementation on plasma EPA and DHA concentrations compared to fish oil in overweight and obese patients. Subjects were randomly assigned to one of three treatment groups and consumed 2 g/day of krill oil (216 mg/day EPA and 90 mg/day DHA), fish (menhaden) oil

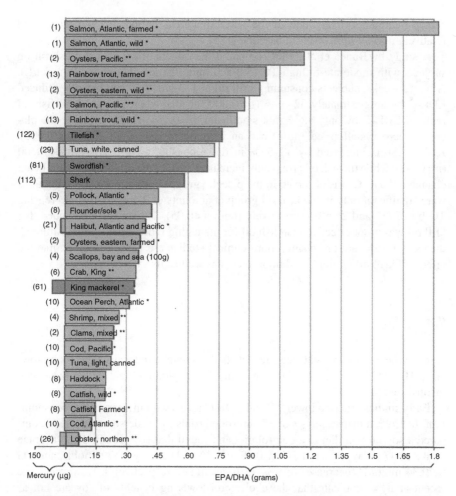

Fig. 9.3 EPA/DHA intake and methyl mercury intake exposure from one 3-oz portion of seafood. Reprinted from ref. [216] with permission by the *National Academy of Sciences*. *Asterisk* cooked, dry heat; *double asterisk* cooked, moist heat; *triple asterisk* The EPA and DHA content in Pacific salmon is a composite from chum, coho, and sockeye

(212 mg/day EPA and 178 mg/day DHA), or control (olive oil). At the end of the 4-week treatment, plasma EPA and DHA concentrations increased significantly more in the krill oil (178 and 90 μmol/L) and fish oil (132 and 150 μmol/L) groups compared to the control (3 and −1 μmol/L) ($p < 0.001$). Mean plasma EPA was greater in the krill oil group compared to the fish oil group (377 vs. 293 μmol/L). Although the krill oil supplement provided only half the amount of DHA as the fish oil supplement, plasma DHA was comparable at the end the trial (476 vs. 478 μmol/L). Therefore, EPA + DHA in krill oil appear to be absorbed equally as well if not better than that from fish oil. Blood lipid levels did not differ significantly

across treatments, likely due to the short treatment period, lower dosage, and relatively normal lipid levels at baseline [185].

A study by Bunea et al. [186] compared the effects of krill oil and fish oil on subjects with moderate-to-high TC (5.0–9.0 mmol/L) and TG (2.3–4.0 mmol/L). After 12 weeks, subjects consuming krill oil at 1–3 g/day all significantly reduced LDL-C by approximately 32–39% ($p=0.000$ for all), whereas 3 g/day of fish oil reduced LDL-C by only 4.6% (not significant, $p=0.14$). Those receiving the placebo (microcrystalline cellulose) had an increase in LDL-C of 13% ($p=0.000$). HDL-C levels increased by 43–59% in those consuming 1–3 g/day of krill oil ($p=0.000$) while the fish oil group only increased by 4.2% ($p=0.000$). No significant change in HDL-C was observed in the placebo group. Furthermore, TG reductions were significant only in the krill oil group at amounts of 2 and 3 g/day, decreasing TG by 27.6% and 26.5%, respectively (both $p<0.05$). The authors concluded that krill oil was more effective than fish oil for managing blood lipids [186]. Although the early results are promising, more clinical studies are needed to determine the effects of krill oil on lipids and lipoproteins as well as its safety.

9.3.8 Soy

A moderate amount of evidence suggests that soy intake reduces CHD risk by lowering TC and LDL-C levels, although results from epidemiologic studies have been inconsistent [14].

Early studies reported lower TC and LDL-C levels with increasing soy consumption. In 1995, a meta-analysis of 38 controlled trials by Anderson et al. [187] compared a soy protein diet to a control diet and found that soy protein diets (average intake of 47 g/day) lowered TC by 0.59 mmol/L, LDL-C by 0.57 mmol/L, and TG by 0.34 mmol/L (decreases of 9%, 13%, and 11%, respectively). However, more recent analyses indicate that the cholesterol-lowering benefits of soy are not as significant.

In 1999, the FDA issued a health claim based largely on the meta-analysis by Anderson et al. stating that 25 g/day of soy protein was associated with a decrease in CHD risk [188]. A food must contain at least 6.25 g of soy protein in order for the label to claim a cholesterol-lowering effect. Three recent meta-analyses of RCTs supported the claim, finding that soy protein lowered TC, LDL-C, and TG in adults with or without hyperlipidemia, although the lowering effect sizes for LDL-C (0.11, 0.21, and 0.23 mmol/L) were substantially less than reported in the original meta-analysis by Anderson et al. (a decrease of 0.57 mmol/L) [189–191]. Perhaps the difference in effect size was due in part to the inconsistency of inclusion criteria, particularly the use of nonrandomized trials only in the initial meta-analysis, as well as the variety of soy protein products comprising total soy intake.

The AHA Soy Advisory published a review of 22 clinical studies that found a mean LDL-C reduction of 4.3% ($p=0.003$), much lower than the 13% reported by Anderson et al. [192]. Both the AHA and Anderson et al. meta-analyses included controlled studies evaluating the inherent effect of soy consumption and did not assess the effect of substituting soy for other foods in the diet. In other words, the macronutrient compositions among diet groups were matched. However, Jenkins et al. [193] recently discovered that six studies reporting a large LDL-C lowering effect (20.9%) in the Anderson et al. meta-analysis used a particular type of texturized vegetable protein (Cholsoy). Removing these studies from the analysis lowered the LDL-C reduction from 13 to 9% (0.57–0.36 mmol/L). Therefore, the particular types of soy products included in each analysis likely explained in part the different effect sizes reported.

The Agency for Healthcare Research and Quality (AHRQ) Report summarized the clinical evidence for soy protein and its isoflavones on CVD risk and found that soy protein (14–113 g/day, with a median of 36 g) yielded a small lowering effect on LDL-C (−3%) and TG (−6%), despite a wide range of treatment effects [194]. However, based on limited evidence and poor quality studies, no beneficial or harmful effects of soy protein on CVD risk factors were demonstrated. A considerable variety of soy products explained a large part of the heterogeneity among studies [194]. The ADA Evidence Analysis Library also reported that studies vary greatly in estimating the effect of soy protein on lipids and lipoproteins. Individuals with normal and elevated cholesterol have shown 0–20% lower TC, 0–22% lower TG, and 4–24% lower LDL-C after consuming 26–50 g/day of soy protein [144].

The DGAC 2010 Report stated that soy protein intake may have small effects on TC and LDL-C, while the relationship with CHD appears to be inconsistent [14]. Mechanisms other than lipid and lipoprotein changes, perhaps improved vascular function following soy intake, may be responsible for part of the cardiovascular benefit [195]. Replacing foods high in SFA and cholesterol with soy protein may provide some cardiovascular benefit as a low-fat alternative. Some authors indicate that soy protein lacking isoflavones may not provide the same cholesterol-lowering effect, suggesting that the benefit of soy protein may be attributed to a combination of amino acids with isoflavones, saponins, or bioactive components [196–198]. Further research is needed to understand the effect of soy protein and isoflavones on blood lipids and lipoproteins, and the amount necessary to achieve modest improvements. In addition, long-term studies are needed to determine how soy intake may influence CHD risk.

9.4 Dietary Patterns to Prevent Heart Disease

Many dietary patterns consist of varying nutrient compositions and specific foods can be implemented to beneficially affect lipids and lipoproteins. The Step I and Step II diets were formerly considered the "gold standard" dietary approaches for lipid management. A Step I diet is defined as: total fat <30% of calories; saturated fat <10% of calories; and dietary cholesterol <300 mg/day. A Step II diet is defined as: total fat <30% of calories; saturated fat <7% of calories; and dietary cholesterol <200 mg/day. However, the AHA recently dropped the "Step I" and "Step II"

Table 9.6 Therapeutic lifestyle changes diet recommendations

Nutrient	Recommended intake	For a 2,000-kcal diet
Total fat	25–35% of total energy (TE)	55–78 g/day
Saturated fat	<7% TE	16 g or less/day
Monounsaturated fat	Up to 20% TE	Up to 44 g/day
Polyunsaturated fat	Up to 10% TE	Up to 22 g/day
Carbohydrate[a]	50–60% TE	250–300 g/day
Viscous fiber (g/day)	10–25	10–25
Protein	~15% TE	~75 g/day
Cholesterol (mg/day)	<200	<200
Plant sterols (g/day)	2	2

Adapted from ref. [107]
[a]Carbohydrate intake should be derived predominately from foods rich in complex carbohydrates, including grains, especially whole grains, fruits, and vegetables

Table 9.7 Descriptions of dietary patterns that beneficially affect blood lipids

Diet	Featured food components
TLC	Step II with food sources of plant stanol/sterols, such as margarine, orange juice, yogurt, etc., and of viscous fiber, such as fruits, vegetables, whole grains, and legumes
DASH	Fruits, vegetables, whole grains, low-fat dairy
Portfolio	Vegetarian. Oats, barley, psyllium, soy protein, almonds, eggplant, okra, and sterol-enriched margarine
Mediterranean	Bread, root vegetables, green vegetables, fish, olive oil, canola oil, mustard oil, and/or soybean oil

TLC therapeutic lifestyle change; *DASH* dietary approaches to stop hypertension

designations. Although still emphasizing the same principles, the new focus is on foods rather than individual food components and percentages of macronutrients. The TLC diet, a contemporary version of a Step II diet, currently serves as the new standard (Table 9.6).

In addition to the TLC diet, there are other dietary patterns recommended for the management of blood lipids (Table 9.7). Several clinical studies have evaluated the effects of these diets on lipids and lipoproteins. Typically, these diets lower TC and LDL-C, while the effects on TG and HDL-C often are diet specific. A variety of different dietary patterns to manage lipids and lipoproteins provide options to target specific CHD risk factors as well as to achieve maximal adherence.

9.4.1 Therapeutic Lifestyle Changes Diet

The NCEP and AHA currently recommend the TLC diet in place of Step I and II diets for individuals with CHD or those at risk of developing CHD. The TLC diet is designed to maximally lower LDL-C and recommends 25–35% of calories from fat, <7% calories from saturated fat, and <200 mg/day of cholesterol (Table 9.6). Additional options for LDL-C lowering in the TLC diet include incorporation of 2 g/day of plant sterols/stanols, 10–25 g/day of viscous fiber, as well as emphasis on weight loss and physical activity. The TLC diet is expected to decrease LDL-C by approximately 20–30% [107]. The decrease in LDL-C induced by each of these changes is as follows: approximately 8–10% by the reduction in SFA to <7% of calories, 3–5% by the reduction in dietary cholesterol to <200 mg, up to 5% by the addition of 5–10 g/day of viscous fiber, about 5–8% by a weight loss of about 10 lbs, and 6–15% by the inclusion of 2 g/day of plant stanol/sterol esters [107].

9.4.2 Dietary Approaches to Stop Hypertension Diet

The DASH Trial was conducted to compare the effects of three dietary patterns on BP [103]. In a secondary analysis, the effect of the DASH Diet on blood lipids was also assessed. In the DASH Trial, 459 subjects consumed a 3-week run-in control diet followed by 8 weeks on one of three experimental diets:

1. Control diet: high in saturated fat, low in fiber (14% of calories from protein, 50% CHO, 37% fat [14% SFA, 13% MUFA, and 7% PUFA], 246 mg/day of cholesterol, 10.8 g/day of fiber).
2. Fruit and vegetable diet, comparable to control diet in macronutrients (15% protein, 52% CHO, 37% fat [13% SFA, 14% MUFA, and 7% PUFA], 188 mg/day of cholesterol, 29.9 g/day of fiber).
3. DASH diet: high in fruits (5.2 servings/day), vegetables (4.4 servings/day), and low-fat dairy products (2 servings/day and 0.7 servings of high-fat dairy/day). Lower in fat and saturated fat but high in fiber (18% protein, 58% CHO, 27% fat [7% SFA, 10% MUFA, and 8% PUFA], 141 mg/day of cholesterol, 29.7 g/day of fiber).

TC, LDL-C, and HDL-C significantly decreased by 7.3%, 9.0%, and 7.5%, respectively, in the DASH group vs. the control group ($p < 0.0001$). TC, LDL-C, and HDL-C decreased nonsignificantly by 1.9%, 1.5%, and 0.4%, respectively, in the fruit and vegetable group vs. the control group. Although TG levels decreased 8.4% in the fruit and vegetable group ($p = 0.076$), they were not significantly affected by the DASH diet [103]. Thus, a high-fiber diet may prevent an increase in TG levels on a lower fat diet.

A prospective cohort study by Fung et al. [199] evaluated the association between a DASH-style diet adherence score and CHD risk in women. Diet was assessed seven times during 24 years of follow-up with food frequency questionnaires. A DASH score was calculated based on eight food and nutrient components (fruits, vegetables, whole grains, nuts, and legumes, low-fat dairy, red and processed meats, sweetened beverages, and sodium). Women with a high DASH score had a lower risk of CHD: those in the top quintile of the DASH score had a RR of 0.76, compared with those in the bottom quintile ($p < 0.001$ for trend). The risk reduction was significant for both fatal and nonfatal CHD ($p = 0.002$, $p < 0.001$).

A similar study by Parikh et al. [200] used NHANES III data in order to evaluate the association between the DASH diet and mortality in 5,532 hypertensive adults. Dietary intakes were estimated using a single 24-h dietary recall. NHANES participants consuming a typical western diet were compared to participants consuming a DASH-like diet (7.1% of subjects). The DASH-like diet was associated with lower mortality from all causes (HR = 0.69, $p = 0.01$); however, the mortality risk due to CVD and CHD did not differ in the two dietary groups. Limitations associated with dietary recall data may explain in part the inconsistencies reported among prospective studies.

Clinical evidence has consistently demonstrated that the DASH diet reduces BP, TC, and LDL-C, but also reduces HDL-C with no effect on TG. The unfavorable effect on HDL-C requires further evaluation. Although evidence suggests that the DASH diet can lower the relative risk of CHD, more epidemiologic studies are needed to determine the efficacy of a DASH-like diet in reducing CHD risk across different populations.

9.4.3 Portfolio Diet

The goal of the Portfolio diet is to achieve a maximum LDL-C lowering effect. It is a vegetarian diet that includes plant sterols and viscous fibers, primarily from oat, barley, and psyllium, in addition to soy protein (21 g/1,000 kcal) and almonds (14 g/1,000 kcal) [201, 202]. It meets the recommendations of the TLC diet.

Jenkins et al. [201] reported that subjects following a Portfolio diet (22.4% of calories as protein [97% as vegetable protein], 50.6% CHO, 27.0% fat [4.3% SFA, 11.8% MUFA, and 9.9% PUFA], 10 mg cholesterol per 1,000 kcal, 30.7 g fiber per 1,000 kcal) for 4 weeks had marked reductions in TC and LDL-C (22.4% and 29.0%, respectively; $p < 0.001$). HDL-C and TG were unaffected. In a follow-up study, subjects consumed either a Portfolio diet (20.0% of calories as protein [almost entirely as vegetable protein], 56.6% CHO, 23.2% fat [4.9% SFA, 9.5% MUFA, and 7.9% PUFA], 48 mg cholesterol per 1,000 kcal, 37.2 g fiber per 1,000 kcal) or a Step II diet (19.6% of calories as protein [less than one third as vegetable protein], 58.8% CHO, 21.6% fat [4.4% SFA, 8.5% MUFA, and 7.5% PUFA], 34 mg cholesterol per 1,000 kcal, 26.6 g fiber per 1,000 kcal) for 4 weeks [202]. TC and LDL-C significantly decreased 9.9% and 12.1%, respectively, ($p < 0.001$) on the Step II diet, but by a much larger amount on the Portfolio diet: 26.6 and 35.0% ($p < 0.001$). In addition, nonsignificant reductions in HDL-C occurred on both diets, although TG levels rose 4.9% following a Step II diet and decreased $6.3 \pm 7.8\%$ following the intervention diet ($p = 0.25$). This study demonstrates the benefits of a higher fiber intake in preventing an increase in TG on a reduced fat diet.

Jenkins et al. [203] more recently conducted a year-long study on hypercholesterolemic subjects to determine long-term effects of the Portfolio diet. Subjects were prescribed diets high in plant sterols (1.0 g/1,000 kcal), soy protein (22.5 g/1,000 kcal), viscous fibers (10 g/1,000 kcal), and almonds (23 g/1,000 kcal) for 1 year. Maximum reductions in LDL-C were evident after 12 weeks (14.0% ($p < 0.001$)) and were sustained for 1 year (12.8% ($p < 0.001$)). However, this is far less than the 29% LDL-C reduction reported in the previous Portfolio diet study when subjects were fed the Portfolio diet for 4 weeks in a controlled setting. Self-reported dietary compliance over 1 year was significantly correlated with LDL-C reduction ($r = -0.42$, $p < 0.001$). This likely explains part of the variability in LDL-C change. More than 30% of the subjects lowered LDL-C levels >20%, similar to the response of statin treatment under metabolically controlled conditions; however, only two of the 26 subjects with a compliance <55% lowered LDL-C levels >20% [203]. This suggests that sustaining long-term dietary compliance is necessary to achieve significant effects. Therefore, the Portfolio diet could be an effective primary option to control lipid levels if individuals are motivated and compliant throughout the duration of the dietary intervention.

9.4.4 Mediterranean Diet

The Mediterranean region encompasses a broad geographical area; hence, there is no one common Mediterranean diet. Despite this, there is a dietary pattern that characterizes a Mediterranean-style diet; it emphasizes fruits, vegetables, bread, cereals, potatoes, beans, nuts, and seeds. In addition, it includes olive oil as an important fat source; dairy products, fish, and poultry are consumed in low to moderate amounts; eggs are consumed zero to four times weekly; while little red meat is included. In addition, wine is consumed in low to moderate amounts. Historically, there has been great interest in the Mediterranean diet because populations that consume this dietary pattern have low rates of coronary disease.

The Seven Countries Study spurred interest in the Mediterranean diet when it reported that the 15-year mortality rate from cardiovascular disease in Southern Europe was two to three times lower than in Northern Europe or the United States [204]. In the Ustica Project conducted by Barbagallo et al. [205], the 576 participants of a small Mediterranean island, >14 years of age, had mean TC and LDL-C levels of 5.4 and 3.7 mmol/L, respectively. Overall, 22.8% of the Ustica population was hypercholesterolemic (TC >5.17 mmol/L) and 22.5% had low HDL-C levels (<1.03 mmol/L). In the ATTICA Study, 43% of the 2,282 participants (>18 years of age) living in the greater area of Athens had TC >5.17 mmol/L, while 49% had LDL-C levels >3.36 mmol/L. HDL-C levels were <0.91 mmol/L in 14% of the participants [206]. Collectively, these studies suggest that the cardioprotective effect of the Mediterranean diet may be mediated, in part, by factors other than blood lipids.

The Prevención con Dieta Mediterránea (PREDIMED) Study [207] is a large multicenter clinical trial that evaluates the efficacy of a Mediterranean diet on the primary prevention of CVD in Spain. First launched in 2003, PREDIMED is an ongoing study that randomly assigns participants 55–80 years of age with diabetes or ≥3 major CVD risk factors (hypertension, hypercholesterolemia, family history, tobacco use, or overweight/obese) to consume a low-fat diet, a Mediterranean diet with virgin olive oil (1 L/week), or a Mediterranean diet with tree nuts (30 g/day). Early results showed that after 3 months, subjects in the Mediterranean diet groups had lower TC/HDL-C ratios than those in the low-fat group, −0.38 ($p < 0.001$) with olive oil and −0.26 ($p = 0.002$) with nuts. Salas-Salvado et al. [208] recruited participants from the PREDIMED Study to determine if the same Mediterranean diets also could reduce the prevalence of metabolic syndrome in older adults (55–80 years old). After 1 year, the Mediterranean diet with 30 g/day of tree nuts reduced the overall prevalence of metabolic syndrome (−13.7%, $p = 0.01$) compared to a low-fat diet; however, the reduction using the Mediterranean diet with olive oil (1 L/week) was not significant (−6.7%, $p = 0.18$). This study shows that the Mediterranean diet with nuts can improve the metabolic syndrome status independent of weight loss or increased energy expenditure. Other dietary factors also have been addressed using PREDIMED participants, including eating competence and dietary adherence [209, 210]. Studies of a longer duration are ongoing and will clarify the clinical CVD outcomes of the Mediterranean diet [208].

The Lyon Diet Heart Study [211] was a randomized, single-blind secondary prevention trial that tested the effects of a western diet vs. a Mediterranean diet on the reoccurrence of MI. Subjects consumed a control diet (32.7% fat [11.7% SFA, 10.3% 18:1, 5.3% 18:2, 0.27% 18:3], 318 mg cholesterol) or an experimental diet rich in ALA (30.5% fat [8.3% SFA, 12.9% 18:1, 3.6% 18:2, 0.81% 18:3], 217 mg cholesterol). After 104 weeks on the experimental diet, TC, LDL-C, and TG levels were unchanged. Despite a similar coronary risk factor profile (plasma lipids and lipoproteins, systolic and diastolic BP, BMI, and smoking status), subjects following the Mediterranean-style diet had a 50–70% lower risk of recurrent heart disease as measured by three different combinations of outcome measures including: (1) cardiac death and nonfatal heart attacks; (2) the preceding plus other measures (unstable angina, stroke, heart failure, pulmonary or peripheral embolism); and (3) all of these measures plus events that required hospitalization.

More recent evidence supports these findings. A meta-analysis by Mente et al. [38] included 146 prospective cohort studies and 43 RCTs in reviewing the evidence supporting a causal link between dietary factors and CHD. The Mediterranean dietary pattern was found to have strong evidence of a causal relationship for protective factors, compatible with the evidence from the Lyon Diet Heart Study. In addition, only the Mediterranean dietary pattern was related to CHD in the RCTs. This suggests that modification of the entire dietary pattern may be more effective than individual nutrient or food modifications in lowering CHD risk.

Sofi et al. [212] systematically reviewed prospective cohort studies analyzing the relationship between adherence to a Mediterranean diet, mortality, and incidence of chronic diseases. The analysis included 12 studies with a total of almost 1.6 million subjects followed for 3–18 years. Greater adherence to a Mediterranean diet was associated with a significant reduction in overall mortality (9%) and mortality

Table 9.8   AHA diet and lifestyle recommendations

Balance calorie intake and physical activity to achieve or maintain a healthy body weight

Consume a diet rich in vegetables and fruits

Choose whole-grain, high fiber foods

Consume fish, especially oily fish, at least twice a week

Limit intake of saturated fat to <7% of calories, *trans* fat to <1% of calories, and cholesterol to <300 mg/day by

   Choosing lean meats and vegetable alternatives

   Selecting fat-free (skim), 1%-fat, and low-fat dairy products

   Minimizing intake of partially hydrogenated fats

Minimize your intake of beverages and foods with added sugars

Choose and prepare foods with little or no salt

If you consume alcohol, do so in moderation

When you eat food that is prepared outside of the home, follow the AHA Diet and Lifestyle Recommendations

Adapted from ref. [147]

from CVD (9%). However, the authors noted that even with the benefits of the Mediterranean diet, a shift to a non-Mediterranean dietary pattern has gradually developed, including countries bordering the Mediterranean Sea [212]. Both the AHA and the DGAC 2010 promote the consumption of a Mediterranean-like dietary pattern to reduce CHD risk. Further studies are needed to identify the underlying mechanisms that account for this benefit. While there is some evidence that it is mediated by way of blood lipids, it is clear that other mechanisms also account for the effects observed.

9.4.5   Guidelines and Recommendations

9.4.5.1   AHA Revised Diet and Lifestyle Goals

In 2006, the AHA published a revised version of their diet and lifestyle recommendations for CVD risk reduction (Table 9.8). These recommendations, established for adults and children >2 years old, allow for flexibility in implementation [147].

The dietary pattern recommended is consistent with the previous AHA recommendations, emphasizing fruits and vegetables, legumes, whole grains, fat-free and low-fat dairy, poultry, lean meats, and fatty fish at least twice a week. Dietary cholesterol should be limited to <300 mg/day for healthy individuals and <200 mg/day for high-risk individuals. The previous recommendation combined SFA and TFA into one category, suggesting that they be <10% of total daily calories; however, the revised diet now recommends that SFA consumption be <7% total daily calories and TFA <1% total daily calories. Sodium intake should be <2,300 mg/day, although in 2010 the AHA lowered the recommended intake to <1,500 mg/day for all adults, excluding individuals at risk for losing large amounts of sodium in sweat [213]. Also, the revised diet recommends reducing consumption of foods and beverages containing added sugars to lower calorie intake and promote nutrient adequacy [147]. Fish oil supplements are recommended only for individuals with CHD (1 g/day) or hypertriglyceridemia (2–4 g/day), both under physician's supervision, and plant sterols/stanols (up to 2 g/day) are considered helpful for individuals with elevated LDL-C. In addition to diet, the AHA recommends being physically active, avoiding exposure to tobacco products, and achieving a healthy body weight [147].

9.4.5.2 Dietary Guidelines for Americans 2010

The DGAC 2010 Report [14] recommends several strategies to promote good health across the life span. The aim is to summarize and provide nutrition recommendations that can lower the risk of developing chronic disease while meeting individual nutrient requirements. DGAC 2010 supports a total diet approach, recommending the following components:

- Energy balance, limited calories, and be portion controlled
- Nutrient-dense and include:

  - Vegetables, fruits, high-fiber whole grains
  - Fat-free or low-fat fluid milk and milk products
  - Seafood, lean meat and poultry, eggs, soy products, nuts, seeds, and liquid vegetable oils

- Very low in solid fats and added sugars (SoFAS)
- Reduced in sodium

DGAC 2010 recommends that Americans consume nutrient-dense foods that meet but do not exceed their energy needs. Vegetables, fruits, whole grains high in fiber, fish, eggs, low-fat dairy, nuts, lean meat, and poultry—all prepared without SoFAS—are considered "nutrient-dense foods" [14]. Plant foods should be the main feature of the diet with fat intake consisting predominately of MUFA and PUFA, including n-3 fatty acids from fish sources. Intake of CHO should come from a majority of complex sources with minimal processing, such as beans, peas, and whole grain products. SoFAS, a significant portion of the American diet (approximately 35% of calories), should be no more than 5–15% of total calories. DGAC 2005 recommended that sodium intake be <2,300 mg/day for the general population and <1,500 mg/day for African-Americans, hypertensive individuals, middle-aged adults, and older adults; however, considering almost 70% of the population meets the criteria for the lower amount, DGAC 2010 now recommends that the current sodium goal be <1,500 mg/day for all adults [14], similar to the recent recommendation by the AHA. DGAC 2010 also notes that there is no one specific dietary pattern for the population. Rather, individuals have a great deal of flexibility in selecting a diet that incorporates foods they enjoy, meets their nutrient needs, helps control their weight, and reduces the risk of disease [14].

9.5   Summary

CHD is the leading cause of death in the United States. While much progress has been made in various medical therapies to reduce CHD, it is apparent that a focus should be on prevention. As discussed in this chapter, diet plays a key role in modulating lipids and lipoproteins, critical risk factors for CHD. Dietary patterns, which reflect a composite of foods and nutrients consumed, can be modified in numerous ways to reduce CHD risk. SFA, TFA, and dietary cholesterol increase LDL-C levels, whereas PUFA, viscous fiber, and plant stanol/sterol esters reduce LDL-C. Since LDL-C is a primary target for CHD intervention efforts, dietary strategies that lower LDL-C can consequently reduce CHD risk. In addition, some dietary strategies can have significant effects on CHD risk with little or no effect on LDL-C. Increasing HDL-C can lower the TC/HDL-C ratio, thereby reducing CHD risk. Contemporary efforts to reduce CHD risk include strategies that target not only blood lipid risk factors but also involve dietary patterns that target the wide array of other CHD risk factors. Evidence from intervention studies has identified foods and dietary patterns that merit making recommendations for implementation in practice to reduce CHD. Cardioprotective foods include fruits and vegetables, whole grains, nuts, vegetable oils, lean meats, fish (preferably fatty and low in mercury content), and fat-free and low-fat dairy products. Recommendations by the AHA and DGAC 2010 encourage individuals to include these foods as part of their daily diet to prevent CHD. Soy, plant sterols/stanols, and numerous bioactive compounds also may be effective in preventing in CHD. However, the challenge that remains is to identify effective population-based strategies for implementing healthy diets that reduce CHD risk. Successful adoption of healthy dietary practices on a population basis will favorably affect lipids and lipoproteins and, consequently, play a critical role in the prevention of CHD.

References

1. Keys A, Anderson JT, Grande F. Serum cholesterol response to changes in the diet. IV. Particular saturated fatty acids in the diet. Metabolism. 1965;14:776–87.
2. Hegsted DM, McGandy RB, Myers ML, et al. Quantitative effects of dietary fat on serum cholesterol in man. Am J Clin Nutr. 1965;17:281–95.
3. Mensink RP, Katan MB. Effect of dietary fatty acids on serum lipids and lipoproteins. A meta-analysis of 27 trials. Arterioscler Thromb. 1992;12:911–9.
4. Muller H, Kirkhus B, Pedersen JI. Serum cholesterol predictive equations with special emphasis on trans and saturated fatty acids. An analysis from designed controlled studies. Lipids. 2001;36:783–91.
5. Yu S, Derr J, Etherton TD, Kris-Etherton PM. Plasma cholesterol-predictive equations demonstrate that stearic acid is neutral and monounsaturated fatty acids are hypocholesterolemic. Am J Clin Nutr. 1995;61:1129–39.
6. Clarke R, Frost C, Collins R, et al. Dietary lipids and blood cholesterol: quantitative meta-analysis of metabolic ward studies. BMJ. 1997;314:112–7.
7. Hegsted DM, Ausman LM, Johnson JA, et al. Dietary fat and serum lipids: an evaluation of the experimental data. Am J Clin Nutr. 1993;57:875–83.

8. Kris-Etherton PM, Yu S. Individual fatty acid effects on plasma lipids and lipoproteins: human studies. Am J Clin Nutr. 1997;65(5 Suppl):1628S–44.

9. Mensink RP, Zock PL, Kester AD, et al. Effects of dietary fatty acids and carbohydrates on the ratio of serum total to HDL cholesterol and on serum lipids and apolipoproteins: a meta-analysis of 60 controlled trials. Am J Clin Nutr. 2003;77:1146–55.

10. Katan MB, Zock PL, Mensink RP. Effects of fats and fatty acids on blood lipids in humans: an overview. Am J Clin Nutr. 1994;60(6 Suppl):1017S–22.

11. Aro A, Jauhiainen M, Partanen R, et al. Stearic acid, trans fatty acids, and dairy fat: effects on serum and lipoprotein lipids, apolipoproteins, lipoprotein(a), and lipid transfer proteins in healthy subjects. Am J Clin Nutr. 1997;65:1419–26.

12. Sundram K, Karupaiah T, Hayes KC. Stearic acid-rich interesterified fat and trans-rich fat raise the LDL/HDL ratio and plasma glucose relative to palm olein in humans. Nutr Metab (Lond). 2007;4:3.

13. Zock PL, Katan MB. Hydrogenation alternatives: effects of trans fatty acids and stearic acid versus linoleic acid on serum lipids and lipoproteins in humans. J Lipid Res. 1992;33:399–410.

14. Diet Guidelines. The executive summary. 2010. http://www.cnpp.usda.gov/Publications/DietaryGuidelines/2010/PolicyDoc/ExecSumm.pdf. The full report is available at: http://www.health.gov/dietaryguidelines/dga2010/DietaryGuidelines2010.pdf. Accessed 1 Feb 2011.

15. Lloyd-Jones DM, Hong Y, Labarthe D, et al. Defining and setting national goals for cardiovascular health promotion and disease reduction: the American Heart Association's Strategic Impact Goal through 2020 and beyond. Circulation. 2010;121:586–613.

16. Keys A. Coronary heart disease in seven countries. Circulation. 1970;41:I1–211.

17. Hu FB, Stampfer MJ, Manson JE, et al. Dietary fat intake and the risk of coronary heart disease in women. N Engl J Med. 1997;337:1491–9.

18. Posner BM, Cobb JL, Belanger AJ, et al. Dietary lipid predictors of coronary heart disease in men. The Framingham Study. Arch Intern Med. 1991;151:1181–7.

19. Kromhout D, Menotti A, Bloemberg B, et al. Dietary saturated and trans fatty acids and cholesterol and 25-year mortality from coronary heart disease: the Seven Countries Study. Prev Med. 1995;24:308–15.

20. Vartiainen E, Laatikainen T, Peltonen M, et al. Thirty-five-year trends in cardiovascular risk factors in Finland. Int J Epidemiol. 2010;39:504–18.

21. Berglund L, Lefevre M, Ginsberg HN, et al. Comparison of monounsaturated fat with carbohydrates as a replacement for saturated fat in subjects with a high metabolic risk profile: studies in the fasting and postprandial states. Am J Clin Nutr. 2007;86:1611–20.

22. Lichtenstein AH, Matthan NR, Jalbert SM, et al. Novel soybean oils with different fatty acid profiles alter cardiovascular disease risk factors in moderately hyperlipidemic subjects. Am J Clin Nutr. 2006;84:497–504.

23. Yu-Poth S, Etherton TD, Reddy CC, et al. Lowering dietary saturated fat and total fat reduces the oxidative susceptibility of LDL in healthy men and women. J Nutr. 2000;130:2228–37.

24. Chung BH, Cho BH, Liang P, et al. Contribution of postprandial lipemia to the dietary fat-mediated changes in endogenous lipoprotein-cholesterol concentrations in humans. Am J Clin Nutr. 2004;80:1145–58.

25. Kralova Lesna I, Suchanek P, Kovar J, et al. Replacement of dietary saturated FAs by PUFAs in diet and reverse cholesterol transport. J Lipid Res. 2008;49:2414–8.

26. Siri-Tarino PW, Sun Q, Hu FB, et al. Saturated fat, carbohydrate, and cardiovascular disease. Am J Clin Nutr. 2010;91:502–9.

27. Gardner CD, Fortmann SP, Krauss RM. Association of small low-density lipoprotein particles with the incidence of coronary artery disease in men and women. JAMA. 1996;276:875–81.

28. Lamarche B, Tchernof A, Moorjani S, et al. Small, dense low-density lipoprotein particles as a predictor of the risk of ischemic heart disease in men. Prospective results from the Quebec Cardiovascular Study. Circulation. 1997;95:69–75.

29. Stampfer MJ, Krauss RM, Ma J, et al. A prospective study of triglyceride level, low-density lipoprotein particle diameter, and risk of myocardial infarction. JAMA. 1996;276:882–8.
30. Sacks FM, Katan M. Randomized clinical trials on the effects of dietary fat and carbohydrate on plasma lipoproteins and cardiovascular disease. Am J Med. 2002;113(Suppl 9B):13S–24.
31. Mozaffarian D, Micha R, Wallace S. Effects on coronary heart disease of increasing polyun-saturated fat in place of saturated fat: a systematic review and meta-analysis of randomized controlled trials. PLoS Med. 2010;7:e1000252.
32. Skeaff CM, Miller J. Dietary fat and coronary heart disease: summary of evidence from pro-spective cohort and randomised controlled trials. Ann Nutr Metab. 2009;55:173–201.
33. Clarke R, Shipley M, Lewington S, et al. Underestimation of risk associations due to regres-sion dilution in long-term follow-up of prospective studies. Am J Epidemiol. 1999;150: 341–53.
34. Beaton GH, Milner J, Corey P, et al. Sources of variance in 24-hour dietary recall data: impli-cations for nutrition study design and interpretation. Am J Clin Nutr. 1979;32:2546–59.
35. Siri-Tarino PW, Sun Q, Hu FB, et al. Meta-analysis of prospective cohort studies evaluating the association of saturated fat with cardiovascular disease. Am J Clin Nutr. 2010;91:535–46.
36. Katan MB, Brouwer IA, Clarke R, et al. Saturated fat and heart disease. Am J Clin Nutr. 2010;92:459–60; author reply 60–1.
37. Jakobsen MU, O'Reilly EJ, Heitmann BL, et al. Major types of dietary fat and risk of coronary heart disease: a pooled analysis of 11 cohort studies. Am J Clin Nutr. 2009;89: 1425–32.
38. Mente A, de Koning L, Shannon HS, et al. A systematic review of the evidence supporting a causal link between dietary factors and coronary heart disease. Arch Intern Med. 2009;169:659–69.
39. Mensink RP, Katan MB. Effect of dietary trans fatty acids on high-density and low-density lipoprotein cholesterol levels in healthy subjects. N Engl J Med. 1990;323:439–45.
40. Lichtenstein AH, Ausman LM, Jalbert SM, et al. Effects of different forms of dietary hydro-genated fats on serum lipoprotein cholesterol levels. N Engl J Med. 1999;340:1933–40.
41. Judd JT, Baer DJ, Clevidence BA, et al. Dietary cis and trans monounsaturated and saturated FA and plasma lipids and lipoproteins in men. Lipids. 2002;37:123–31.
42. Ascherio A, Willett WC. Health effects of trans fatty acids. Am J Clin Nutr. 1997;66(4 Suppl):1006S–10.
43. Mozaffarian D, Aro A, Willett WC. Health effects of trans-fatty acids: experimental and observational evidence. Eur J Clin Nutr. 2009;63 Suppl 2:S5–21.
44. Lawrence G. The fats of life: essential fatty acids in health and disease. New Brunswick: Rutgers University Press; 2010.
45. Mitchell PL, McLeod RS. Conjugated linoleic acid and atherosclerosis: studies in animal models. Biochem Cell Biol. 2008;86:293–301.
46. Thom E, Wadstein J, Gudmundsen O. Conjugated linoleic acid reduces body fat in healthy exercising humans. J Int Med Res. 2001;29:392–6.
47. Blankson H, Stakkestad JA, Fagertun H, et al. Conjugated linoleic acid reduces body fat mass in overweight and obese humans. J Nutr. 2000;130:2943–8.
48. Mougios V, Matsakas A, Petridou A, et al. Effect of supplementation with conjugated linoleic acid on human serum lipids and body fat. J Nutr Biochem. 2001;12:585–94.
49. Smedman A, Vessby B. Conjugated linoleic acid supplementation in humans–metabolic effects. Lipids. 2001;36:773–81.
50. Zambell KL, Keim NL, Van Loan MD, et al. Conjugated linoleic acid supplementation in humans: effects on body composition and energy expenditure. Lipids. 2000;35:777–82.
51. Salas-Salvado J, Marquez-Sandoval F, Bullo M. Conjugated linoleic acid intake in humans: a systematic review focusing on its effect on body composition, glucose, and lipid metabo-lism. Crit Rev Food Sci Nutr. 2006;46:479–88.
52. Tricon S, Burdge GC, Jones EL, et al. Effects of dairy products naturally enriched with cis-9, trans-11 conjugated linoleic acid on the blood lipid profile in healthy middle-aged men. Am J Clin Nutr. 2006;83:744–53.

53. Wanders AJ, Brouwer IA, Siebelink E, et al. Effect of a high intake of conjugated linoleic acid on lipoprotein levels in healthy human subjects. PLoS One. 2010;5:e9000.
54. Tholstrup T, Raff M, Basu S, et al. Effects of butter high in ruminant trans and monounsaturated fatty acids on lipoproteins, incorporation of fatty acids into lipid classes, plasma C-reactive protein, oxidative stress, hemostatic variables, and insulin in healthy young men. Am J Clin Nutr. 2006;83:237–43.
55. Jakobsen MU, Bysted A, Andersen NL, et al. Intake of ruminant trans fatty acids and risk of coronary heart disease-an overview. Atheroscler Suppl. 2006;7:9–11.
56. Field CJ, Blewett HH, Proctor S, et al. Human health benefits of vaccenic acid. Appl Physiol Nutr Metab. 2009;34:979–91.
57. Brouwer IA, Wanders AJ, Katan MB. Effect of animal and industrial trans fatty acids on HDL and LDL cholesterol levels in humans–a quantitative review. PLoS One. 2010;5:e9434.
58. Rudel LL, Parks JS, Hedrick CC, et al. Lipoprotein and cholesterol metabolism in diet-induced coronary artery atherosclerosis in primates. Role of cholesterol and fatty acids. Prog Lipid Res. 1998;37:353–70.
59. Rodriguez-Villar C, Perez-Heras A, Mercade I, et al. Comparison of a high-carbohydrate and a high-monounsaturated fat, olive oil-rich diet on the susceptibility of LDL to oxidative modification in subjects with type 2 diabetes mellitus. Diabet Med. 2004;21:142–9.
60. Colette C, Percheron C, Pares-Herbute N, et al. Exchanging carbohydrates for monounsaturated fats in energy-restricted diets: effects on metabolic profile and other cardiovascular risk factors. Int J Obes Relat Metab Disord. 2003;27:648–56.
61. Ahuja KD, Ashton EL, Ball MJ. Effects of two lipid-lowering, carotenoid-controlled diets on the oxidative modification of low-density lipoproteins in free-living humans. Clin Sci (Lond). 2003;105:355–61.
62. Ashton EL, Best JD, Ball MJ. Effects of monounsaturated enriched sunflower oil on CHD risk factors including LDL size and copper-induced LDL oxidation. J Am Coll Nutr. 2001;20:320–6.
63. Hargrove RL, Etherton TD, Pearson TA, et al. Low fat and high monounsaturated fat diets decrease human low density lipoprotein oxidative susceptibility in vitro. J Nutr. 2001;131:1758–63.
64. Garg A. High-monounsaturated-fat diets for patients with diabetes mellitus: a meta-analysis. Am J Clin Nutr. 1998;67(3 Suppl):577S–82.
65. Cao Y, Mauger DT, Pelkman CL, et al. Effects of moderate (MF) versus lower fat (LF) diets on lipids and lipoproteins: a meta-analysis of clinical trials in subjects with and without diabetes. J Clin Lipidol. 2009;3:19–32.
66. Tanasescu M, Cho E, Manson JE, et al. Dietary fat and cholesterol and the risk of cardiovascular disease among women with type 2 diabetes. Am J Clin Nutr. 2004;79:999–1005.
67. Dayton S, Pearce ML, Goldman H, et al. Controlled trial of a diet high in unsaturated fat for prevention of atherosclerotic complications. Lancet. 1968;2:1060–2.
68. Frantz Jr ID, Dawson EA, Ashman PL, et al. Test of effect of lipid lowering by diet on cardiovascular risk. The Minnesota Coronary Survey. Arteriosclerosis. 1989;9:129–35.
69. Leren P. The Oslo diet-heart study. Eleven-year report. Circulation. 1970;42:935–42.
70. Turpeinen O, Karvonen MJ, Pekkarinen M, et al. Dietary prevention of coronary heart disease: the Finnish Mental Hospital Study. Int J Epidemiol. 1979;8:99–118.
71. Mustad VA, Etherton TD, Cooper AD, et al. Reducing saturated fat intake is associated with increased levels of LDL receptors on mononuclear cells in healthy men and women. J Lipid Res. 1997;38:459–68.
72. Mattson FH, Grundy SM. Comparison of effects of dietary saturated, monounsaturated, and polyunsaturated fatty acids on plasma lipids and lipoproteins in man. J Lipid Res. 1985;26:194–202.
73. Gardner CD, Kraemer HC. Monounsaturated versus polyunsaturated dietary fat and serum lipids. A meta-analysis. Arterioscler Thromb Vasc Biol. 1995;15:1917–27.
74. Czernichow S, Thomas D, Bruckert E. n-6 Fatty acids and cardiovascular health: a review of the evidence for dietary intake recommendations. Br J Nutr. 2010;104:788–96.

75. Yusuf S, Hawken S, Ounpuu S, et al. Effect of potentially modifiable risk factors associated with myocardial infarction in 52 countries (the INTERHEART study): case-control study. Lancet. 2004;364:937–52.
76. Harris WS, Mozaffarian D, Rimm E, et al. Omega-6 fatty acids and risk for cardiovascular disease: a science advisory from the American Heart Association Nutrition Subcommittee of the Council on Nutrition, Physical Activity, and Metabolism; Council on Cardiovascular Nursing; and Council on Epidemiology and Prevention. Circulation. 2009;119:902–7.
77. Skulas-Ray AC, West SG, Davidson MH, et al. Omega-3 fatty acid concentrates in the treatment of moderate hypertriglyceridemia. Expert Opin Pharmacother. 2008;9:1237–48.
78. Brenna JT. Efficiency of conversion of alpha-linolenic acid to long chain n-3 fatty acids in man. Curr Opin Clin Nutr Metab Care. 2002;5:127–32.
79. Plourde M, Cunnane SC. Extremely limited synthesis of long chain polyunsaturates in adults: implications for their dietary essentiality and use as supplements. Appl Physiol Nutr Metab. 2007;32:619–34.
80. GISSI-Prevenzione Investigators. Dietary supplementation with n-3 polyunsaturated fatty acids and vitamin E after myocardial infarction: results of the GISSI-Prevenzione trial. Gruppo Italiano per lo Studio della Sopravvivenza nell'Infarto miocardico. Lancet. 1999; 354:447–55.
81. Burr ML, Fehily AM, Gilbert JF, et al. Effects of changes in fat, fish, and fibre intakes on death and myocardial reinfarction: diet and reinfarction trial (DART). Lancet. 1989;2: 757–61.
82. Yokoyama M, Origasa H, Matsuzaki M, et al. Effects of eicosapentaenoic acid on major coronary events in hypercholesterolaemic patients (JELIS): a randomised open-label, blinded endpoint analysis. Lancet. 2007;369:1090–8.
83. Mozaffarian D, Rimm EB. Fish intake, contaminants, and human health: evaluating the risks and the benefits. JAMA. 2006;296:1885–99.
84. Calder PC. n-3 Fatty acids and cardiovascular disease: evidence explained and mechanisms explored. Clin Sci (Lond). 2004;107:1–11.
85. Jones PJ. Dietary cholesterol and the risk of cardiovascular disease in patients: a review of the Harvard Egg Study and other data. Int J Clin Pract Suppl. 2009;163(1–8):28–36.
86. Katan MB. The response of lipoproteins to dietary fat and cholesterol in lean and obese persons. Curr Cardiol Rep. 2006;8:446–51.
87. Jones PJ, Pappu AS, Hatcher L, et al. Dietary cholesterol feeding suppresses human cholesterol synthesis measured by deuterium incorporation and urinary mevalonic acid levels. Arterioscler Thromb Vasc Biol. 1996;16:1222–8.
88. Greene CM, Zern TL, Wood RJ, et al. Maintenance of the LDL cholesterol:HDL cholesterol ratio in an elderly population given a dietary cholesterol challenge. J Nutr. 2005;135: 2793–8.
89. Ballesteros MN, Cabrera RM, Saucedo Mdel S, et al. Dietary cholesterol does not increase biomarkers for chronic disease in a pediatric population from northern Mexico. Am J Clin Nutr. 2004;80:855–61.
90. Goodrow EF, Wilson TA, Houde SC, et al. Consumption of one egg per day increases serum lutein and zeaxanthin concentrations in older adults without altering serum lipid and lipoprotein cholesterol concentrations. J Nutr. 2006;136:2519–24.
91. Harman NL, Leeds AR, Griffin BA. Increased dietary cholesterol does not increase plasma low density lipoprotein when accompanied by an energy-restricted diet and weight loss. Eur J Nutr. 2008;47:287–93.
92. Trumbo P, Schlicker S, Yates AA, et al. Dietary reference intakes for energy, carbohydrate, fiber, fat, fatty acids, cholesterol, protein and amino acids. J Am Diet Assoc. 2002;102:1621–30.
93. McNamara DJ. Dietary cholesterol and atherosclerosis. Biochim Biophys Acta. 2000;1529: 310–20.
94. Hu FB, Stampfer MJ, Rimm EB, et al. A prospective study of egg consumption and risk of cardiovascular disease in men and women. JAMA. 1999;281:1387–94.

95. Kritchevsky SB, Kritchevsky D. Egg consumption and coronary heart disease: an epidemiologic overview. J Am Coll Nutr. 2000;19(5 Suppl):549S–55.
96. Mann JI, Appleby PN, Key TJ, et al. Dietary determinants of ischaemic heart disease in health conscious individuals. Heart. 1997;78:450–5.
97. Fraser GE. Associations between diet and cancer, ischemic heart disease, and all-cause mortality in non-Hispanic white California Seventh-day Adventists. Am J Clin Nutr. 1999;70(3 Suppl):532S–8.
98. Djousse L, Gaziano JM. Egg consumption in relation to cardiovascular disease and mortality: the Physicians' Health Study. Am J Clin Nutr. 2008;87:964–9.
99. Van Horn L, McCoin M, Kris-Etherton PM, et al. Evidence for dietary prevention and treatment of cardiovascular disease. J Am Diet Assoc. 2008;108:287–331.
100. Chandalia M, Garg A, Lutjohann D, et al. Beneficial effects of high dietary fiber intake in patients with type 2 diabetes mellitus. N Engl J Med. 2000;342:1392–8.
101. Davy BM, Davy KP, Ho RC, et al. High-fiber oat cereal compared with wheat cereal consumption favorably alters LDL-cholesterol subclass and particle numbers in middle-aged and older men. Am J Clin Nutr. 2002;76:351–8.
102. Jenkins DJ, Kendall CW, Vuksan V, et al. Soluble fiber intake at a dose approved by the US Food and Drug Administration for a claim of health benefits: serum lipid risk factors for cardiovascular disease assessed in a randomized controlled crossover trial. Am J Clin Nutr. 2002;75:834–9.
103. Obarzanek E, Sacks FM, Vollmer WM, et al. Effects on blood lipids of a blood pressure-lowering diet: the Dietary Approaches to Stop Hypertension (DASH) Trial. Am J Clin Nutr. 2001;74:80–9.
104. Saltzman E, Das SK, Lichtenstein AH, et al. An oat-containing hypocaloric diet reduces systolic blood pressure and improves lipid profile beyond effects of weight loss in men and women. J Nutr. 2001;131:1465–70.
105. Van Horn L, Liu K, Gerber J, et al. Oats and soy in lipid-lowering diets for women with hypercholesterolemia: is there synergy? J Am Diet Assoc. 2001;101:1319–25.
106. Naumann E, van Rees AB, Onning G, et al. Beta-glucan incorporated into a fruit drink effectively lowers serum LDL-cholesterol concentrations. Am J Clin Nutr. 2006;83:601–5.
107. Expert Panel on Detection, Evaluation, and Treatment of High Blood Cholesterol in Adults. Executive summary of the Third Report of The National Cholesterol Education Program (NCEP) Expert Panel on Detection, Evaluation, and Treatment of High Blood Cholesterol in Adults (Adult Treatment Panel III). JAMA. 2001;285:2486–97.
108. Food and Drug Administraton, HHS. Food labeling: health claims; soluble dietary fiber from certain foods and coronary heart disease. Interim final rule. Fed Regist. 2002;67(191):61773–83.
109. Ripsin CM, Keenan JM, Jacobs DR, et al. Oat products and lipid lowering. A meta-analysis. JAMA. 1992;24(267):3317–25.
110. Brown L, Rosner B, Willett WW, et al. Cholesterol-lowering effects of dietary fiber: a meta-analysis. Am J Clin Nutr. 1999;69:30–42.
111. Wolever TM, Tosh SM, Gibbs AL, et al. Physicochemical properties of oat beta-glucan influence its ability to reduce serum LDL cholesterol in humans: a randomized clinical trial. Am J Clin Nutr. 2010;92:723–32.
112. Jacobs DR, Pereira MA, Meyer KA, et al. Fiber from whole grains, but not refined grains, is inversely associated with all-cause mortality in older women: the Iowa women's health study. J Am Coll Nutr. 2000;19(3 Suppl):326S–30.
113. Slavin JL. Position of the American Dietetic Association: health implications of dietary fiber. J Am Diet Assoc. 2008;108:1716–31.
114. Report of the 31st session of the Codex Committee on Nutrition and Foods for Special Dietary Uses. Germany: Codex Alimentarius Commission, Joint FAO/WHO Food Standards Programme; 2009.
115. The elements within the nutrition facts table: section 6.8. Guide to food labelling and advertising, Canadian Food Inspection Agency, 2010;32–3.

116. Christiansen LI, Lahteenmaki PL, Mannelin MR, et al. Cholesterol-lowering effect of spreads enriched with microcrystalline plant sterols in hypercholesterolemic subjects. Eur J Nutr. 2001;40:66–73.

117. Devaraj S, Jialal I, Vega-Lopez S. Plant sterol-fortified orange juice effectively lowers cholesterol levels in mildly hypercholesterolemic healthy individuals. Arterioscler Thromb Vasc Biol. 2004;24:e25–8.

118. Hendriks HF, Brink EJ, Meijer GW, et al. Safety of long-term consumption of plant sterol esters-enriched spread. Eur J Clin Nutr. 2003;57:681–92.

119. Quilez J, Rafecas M, Brufau G, et al. Bakery products enriched with phytosterol esters, alpha-tocopherol and beta-carotene decrease plasma LDL-cholesterol and maintain plasma beta-carotene concentrations in normocholesterolemic men and women. J Nutr. 2003;133:3103–9.

120. Talati R, Sobieraj DM, Makanji SS, et al. The comparative efficacy of plant sterols and stanols on serum lipids: a systematic review and meta-analysis. J Am Diet Assoc. 2010; 110:719–26.

121. Hallikainen MA, Sarkkinen ES, Uusitupa MI. Plant stanol esters affect serum cholesterol concentrations of hypercholesterolemic men and women in a dose-dependent manner. J Nutr. 2000;130:767–76.

122. Nestle M. Genetically engineered "golden" rice unlikely to overcome vitamin A deficiency. J Am Diet Assoc. 2001;101:289–90.

123. Vanstone CA, Raeini-Sarjaz M, Parsons WE, et al. Unesterified plant sterols and stanols lower LDL-cholesterol concentrations equivalently in hypercholesterolemic persons. Am J Clin Nutr. 2002;76:1272–8.

124. Moruisi KG, Oosthuizen W, Opperman AM. Phytosterols/stanols lower cholesterol concentrations in familial hypercholesterolemic subjects: a systematic review with meta-analysis. J Am Coll Nutr. 2006;25:41–8.

125. Assmann G, Cullen P, Erbey J, et al. Plasma sitosterol elevations are associated with an increased incidence of coronary events in men: results of a nested case-control analysis of the Prospective Cardiovascular Munster (PROCAM) study. Nutr Metab Cardiovasc Dis. 2006;16:13–21.

126. Fassbender K, Lutjohann D, Dik MG, et al. Moderately elevated plant sterol levels are associated with reduced cardiovascular risk–the LASA study. Atherosclerosis. 2008;196: 283–8.

127. Pinedo S, Vissers MN, von Bergmann K, et al. Plasma levels of plant sterols and the risk of coronary artery disease: the prospective EPIC-Norfolk Population Study. J Lipid Res. 2007;48:139–44.

128. Kris-Etherton PM, Hecker KD, Bonanome A, et al. Bioactive compounds in foods: their role in the prevention of cardiovascular disease and cancer. Am J Med. 2002;113(Suppl 9B):71S–88.

129. Hooper L, Kroon PA, Rimm EB, et al. Flavonoids, flavonoid-rich foods, and cardiovascular risk: a meta-analysis of randomized controlled trials. Am J Clin Nutr. 2008;88:38–50.

130. Mink PJ, Scrafford CG, Barraj LM, et al. Flavonoid intake and cardiovascular disease mortality: a prospective study in postmenopausal women. Am J Clin Nutr. 2007;85:895–909.

131. Corti R, Flammer AJ, Hollenberg NK, et al. Cocoa and cardiovascular health. Circulation. 2009;119:1433–41.

132. Frankel EN, Waterhouse AL, Kinsella JE. Inhibition of human LDL oxidation by resveratrol. Lancet. 1993;341:1103–4.

133. Miller NJ, Rice-Evans CA. Antioxidant activity of resveratrol in red wine. Clin Chem. 1995;41:1789.

134. Olas B, Wachowicz B, Szewczuk J, et al. The effect of resveratrol on the platelet secretory process induced by endotoxin and thrombin. Microbios. 2001;105:7–13.

135. Orsini F, Pelizzoni F, Verotta L, et al. Isolation, synthesis, and antiplatelet aggregation activity of resveratrol 3-O-beta-D-glucopyranoside and related compounds. J Nat Prod. 1997;60:1082–7.

136. Miller KB, Stuart DA, Smith NL, et al. Antioxidant activity and polyphenol and procyanidin contents of selected commercially available cocoa-containing and chocolate products in the United States. J Agric Food Chem. 2006;54:4062–8.

137. Miller KB, Hurst WJ, Payne MJ, et al. Impact of alkalization on the antioxidant and flavanol content of commercial cocoa powders. J Agric Food Chem. 2008;56:8527–33.

138. Grassi D, Necozione S, Lippi C, et al. Cocoa reduces blood pressure and insulin resistance and improves endothelium-dependent vasodilation in hypertensives. Hypertension. 2005;46:398–405.

139. Baba S, Osakabe N, Kato Y, et al. Continuous intake of polyphenolic compounds containing cocoa powder reduces LDL oxidative susceptibility and has beneficial effects on plasma HDL-cholesterol concentrations in humans. Am J Clin Nutr. 2007;85:709–17.

140. Davies MJ, Judd JT, Baer DJ, et al. Black tea consumption reduces total and LDL cholesterol in mildly hypercholesterolemic adults. J Nutr. 2003;133:3298S–302.

141. Fraga CG, Actis-Goretta L, Ottaviani JI, et al. Regular consumption of a flavanol-rich chocolate can improve oxidant stress in young soccer players. Clin Dev Immunol. 2005;12:11–7.

142. Kris-Etherton PM, Derr JA, Mustad VA, et al. Effects of a milk chocolate bar per day substituted for a high-carbohydrate snack in young men on an NCEP/AHA Step 1 Diet. Am J Clin Nutr. 1994;60(6 Suppl):1037S–42.

143. Taubert D, Roesen R, Lehmann C, et al. Effects of low habitual cocoa intake on blood pressure and bioactive nitric oxide: a randomized controlled trial. JAMA. 2007;298:49–60.

144. Myers EF. ADA evidence analysis library. J Am Diet Assoc. 2005;105(5 Suppl 1):S79.

145. Flammer AJ, Hermann F, Sudano I, et al. Dark chocolate improves coronary vasomotion and reduces platelet reactivity. Circulation. 2007;116:2376–82.

146. Huxley RR, Neil HA. The relation between dietary flavonol intake and coronary heart disease mortality: a meta-analysis of prospective cohort studies. Eur J Clin Nutr. 2003;57:904–8.

147. Lichtenstein AH, Appel LJ, Brands M, et al. Summary of American Heart Association Diet and Lifestyle Recommendations revision 2006. Arterioscler Thromb Vasc Biol. 2006;26:2186–91.

148. Tande DL, Hotchkiss L, Cotugna N. The associations between blood lipids and the Food Guide Pyramid: findings from the Third National Health and Nutrition Examination Survey. Prev Med. 2004;38:452–7.

149. Djousse L, Arnett DK, Coon H, et al. Fruit and vegetable consumption and LDL cholesterol: the National Heart, Lung, and Blood Institute Family Heart Study. Am J Clin Nutr. 2004;79:213–7.

150. Mirmiran P, Noori N, Zavareh MB, et al. Fruit and vegetable consumption and risk factors for cardiovascular disease. Metabolism. 2009;58:460–8.

151. Dauchet L, Amouyel P, Hercberg S, et al. Fruit and vegetable consumption and risk of coronary heart disease: a meta-analysis of cohort studies. J Nutr. 2006;136:2588–93.

152. Administration USFaD. Draft guidance: whole grains label statements, guidance for industry and FDA staff. College Park, MD: Administration UFaD; 2006.

153. Administration USFaD. Health claims notification for whole grain foods. College Park, MD: Administration UFaD; 1999.

154. Harris K, Kris-Etherton P. Effects of whole grains on coronary heart disease risk. Curr Atheroscler Rep. 2010;12:368–76.

155. Kelly SA, Summerbell CD, Brynes A, et al. Wholegrain cereals for coronary heart disease. Cochrane Database Syst Rev. 2007;(2):CD005051.

156. Jacobs Jr DR, Gallaher DD. Whole grain intake and cardiovascular disease: a review. Curr Atheroscler Rep. 2004;6:415–23.

157. Kris-Etherton PM, Zhao G, Binkoski AE, et al. The effects of nuts on coronary heart disease risk. Nutr Rev. 2001;59:103–11.

158. Griel AE, Kris-Etherton PM. Tree nuts and the lipid profile: a review of clinical studies. Br J Nutr. 2006;96 Suppl 2:S68–78.

159. Mukuddem-Petersen J, Oosthuizen W, Jerling JC. A systematic review of the effects of nuts on blood lipid profiles in humans. J Nutr. 2005;135:2082–9.

160. Sabate J, Oda K, Ros E. Nut consumption and blood lipid levels: a pooled analysis of 25 intervention trials. Arch Intern Med. 2010;170:821–7.

161. Lefevre M, Champagne CM, Tulley RT, et al. Individual variability in cardiovascular disease risk factor responses to low-fat and low-saturated-fat diets in men: body mass index, adiposity, and insulin resistance predict changes in LDL cholesterol. Am J Clin Nutr. 2005;82:957–63.

162. Fraser GE. Nut consumption, lipids, and risk of a coronary event. Clin Cardiol. 1999;22(7 Suppl):III11–5.

163. Hu FB, Stampfer MJ, Manson JE, et al. Frequent nut consumption and risk of coronary heart disease in women: prospective cohort study. BMJ. 1998;317:1341–5.

164. Ros E. Nuts and novel biomarkers of cardiovascular disease. Am J Clin Nutr. 2009;89:1649S–56.

165. Diet, nutrition and the prevention of chronic diseases. World Health Organ Tech Rep Ser 2003;916:i–viii; 1–149; backcover.

166. van Meijl LE, Vrolix R, Mensink RP. Dairy product consumption and the metabolic syndrome. Nutr Res Rev. 2008;21:148–57.

167. Tholstrup T, Hoy CE, Andersen LN, et al. Does fat in milk, butter and cheese affect blood lipids and cholesterol differently? J Am Coll Nutr. 2004;23:169–76.

168. Vartiainen E, Jousilahti P, Alfthan G, et al. Cardiovascular risk factor changes in Finland, 1972-1997. Int J Epidemiol. 2000;29:49–56.

169. Pekka P, Pirjo P, Ulla U. Influencing public nutrition for non-communicable disease prevention: from community intervention to national programme–experiences from Finland. Public Health Nutr. 2002;5:245–51.

170. Elwood PC, Pickering JE, Fehily AM. Milk and dairy consumption, diabetes and the metabolic syndrome: the Caerphilly prospective study. J Epidemiol Community Health. 2007;61:695–8.

171. Lutsey PL, Steffen LM, Stevens J. Dietary intake and the development of the metabolic syndrome: the Atherosclerosis Risk in Communities study. Circulation. 2008;117:754–61.

172. Hilpert KF, West SG, Bagshaw DM, et al. Effects of dairy products on intracellular calcium and blood pressure in adults with essential hypertension. J Am Coll Nutr. 2009;28:142–9.

173. Zemel MB. Mechanisms of dairy modulation of adiposity. J Nutr. 2003;133:252S–6.

174. Steffen LM, Kroenke CH, Yu X, et al. Associations of plant food, dairy product, and meat intakes with 15-y incidence of elevated blood pressure in young black and white adults: the Coronary Artery Risk Development in Young Adults (CARDIA) Study. Am J Clin Nutr. 2005;82:1169–77.

175. Beauchesne-Rondeau E, Gascon A, Bergeron J, et al. Plasma lipids and lipoproteins in hypercholesterolemic men fed a lipid-lowering diet containing lean beef, lean fish, or poultry. Am J Clin Nutr. 2003;77:587–93.

176. Davidson MH, Hunninghake D, Maki KC, et al. Comparison of the effects of lean red meat vs lean white meat on serum lipid levels among free-living persons with hypercholesterolemia: a long-term, randomized clinical trial. Arch Intern Med. 1999;159:1331–8.

177. Scott LW, Dunn JK, Pownall HJ, et al. Effects of beef and chicken consumption on plasma lipid levels in hypercholesterolemic men. Arch Intern Med. 1994;154:1261–7.

178. Halton TL, Willett WC, Liu S, et al. Low-carbohydrate-diet score and the risk of coronary heart disease in women. N Engl J Med. 2006;355:1991–2002.

179. Kelemen LE, Kushi LH, Jacobs Jr DR, et al. Associations of dietary protein with disease and mortality in a prospective study of postmenopausal women. Am J Epidemiol. 2005;161:239–49.

180. Bernstein AM, Sun Q, Hu FB, et al. Major dietary protein sources and risk of coronary heart disease in women. Circulation. 2010;122:876–83.

181. Sinha R, Cross AJ, Graubard BI, et al. Meat intake and mortality: a prospective study of over half a million people. Arch Intern Med. 2009;169:562–71.

182. McAfee AJ, McSorley EM, Cuskelly GJ, et al. Red meat consumption: an overview of the risks and benefits. Meat Sci. 2010;84:1–13.

183. Micha R, Wallace SK, Mozaffarian D. Red and processed meat consumption and risk of incident coronary heart disease, stroke, and diabetes mellitus: a systematic review and meta-analysis. Circulation. 2010;121:2271–83.

184. Kris-Etherton PM, Harris WS, Appel LJ. Fish consumption, fish oil, omega-3 fatty acids, and cardiovascular disease. Circulation. 2002;106:2747–57.

185. Maki KC, Reeves MS, Farmer M, et al. Krill oil supplementation increases plasma concentrations of eicosapentaenoic and docosahexaenoic acids in overweight and obese men and women. Nutr Res. 2009;29:609–15.

186. Bunea R, El Farrah K, Deutsch L. Evaluation of the effects of Neptune Krill Oil on the clinical course of hyperlipidemia. Altern Med Rev. 2004;9:420–8.

187. Anderson JW, Johnstone BM, Cook-Newell ME. Meta-analysis of the effects of soy protein intake on serum lipids. N Engl J Med. 1995;333:276–82.

188. Food and Drug Administration, HHS. Food labeling: health claims; soy protein and coronary heart disease. Final rule. Fed Regist. 1999;64(206):57700–33.

189. Harland JI, Haffner TA. Systematic review, meta-analysis and regression of randomised controlled trials reporting an association between an intake of circa 25 g soya protein per day and blood cholesterol. Atherosclerosis. 2008;200:13–27.

190. Reynolds K, Chin A, Lees KA, et al. A meta-analysis of the effect of soy protein supplementation on serum lipids. Am J Cardiol. 2006;98:633–40.

191. Zhan S, Ho SC. Meta-analysis of the effects of soy protein containing isoflavones on the lipid profile. Am J Clin Nutr. 2005;81:397–408.

192. Sacks FM, Lichtenstein A, Van Horn L, et al. Soy protein, isoflavones, and cardiovascular health: an American Heart Association Science Advisory for professionals from the Nutrition Committee. Circulation. 2006;113:1034–44.

193. Jenkins DJ, Mirrahimi A, Srichaikul K, et al. Soy protein reduces serum cholesterol by both intrinsic and food displacement mechanisms. J Nutr. 2010;140:2302S–11.

194. Balk E, Chung M, Chew P, et al. Effects of soy on health outcomes. Evid Rep Technol Assess (Summ). 2005;(126):1–8.

195. Cuevas AM, Irribarra VL, Castillo OA, et al. Isolated soy protein improves endothelial function in postmenopausal hypercholesterolemic women. Eur J Clin Nutr. 2003;57:889–94.

196. Lichtenstein AH. Got soy? Am J Clin Nutr. 2001;73:667–8.

197. Vitolins MZ, Anthony M, Burke GL. Soy protein isoflavones, lipids and arterial disease. Curr Opin Lipidol. 2001;12:433–7.

198. Weggemans RM, Trautwein EA. Relation between soy-associated isoflavones and LDL and HDL cholesterol concentrations in humans: a meta-analysis. Eur J Clin Nutr. 2003; 57:940–6.

199. Fung TT, Chiuve SE, McCullough ML, et al. Adherence to a DASH-style diet and risk of coronary heart disease and stroke in women. Arch Intern Med. 2008;168:713–20.

200. Parikh A, Lipsitz SR, Natarajan S. Association between a DASH-like diet and mortality in adults with hypertension: findings from a population-based follow-up study. Am J Hypertens. 2009;22:409–16.

201. Jenkins DJ, Kendall CW, Faulkner D, et al. A dietary portfolio approach to cholesterol reduction: combined effects of plant sterols, vegetable proteins, and viscous fibers in hypercholesterolemia. Metabolism. 2002;51:1596–604.

202. Jenkins DJ, Kendall CW, Marchie A, et al. The effect of combining plant sterols, soy protein, viscous fibers, and almonds in treating hypercholesterolemia. Metabolism. 2003;52:1478–83.

203. Jenkins DJ, Kendall CW, Faulkner DA, et al. Assessment of the longer-term effects of a dietary portfolio of cholesterol-lowering foods in hypercholesterolemia. Am J Clin Nutr. 2006;83:582–91.

204. Keys A, Menotti A, Karvonen MJ, et al. The diet and 15-year death rate in the seven countries study. Am J Epidemiol. 1986;124:903–15.

205. Barbagallo CM, Polizzi F, Severino M, et al. Distribution of risk factors, plasma lipids, lipoproteins and dyslipidemias in a small Mediterranean island: the Ustica Project. Nutr Metab Cardiovasc Dis. 2002;12:267–74.

206. Panagiotakos DB, Pitsavos C, Chrysohoou C, et al. Status and management of blood lipids in Greek adults and their relation to socio-demographic, lifestyle and dietary factors: the ATTICA Study. Blood lipids distribution in Greece. Atherosclerosis. 2004;173:353–61.

207. Estruch R, Martinez-Gonzalez MA, Corella D, et al. Effects of a Mediterranean-style diet on cardiovascular risk factors: a randomized trial. Ann Intern Med. 2006;145:1–11.

208. Salas-Salvado J, Fernandez-Ballart J, Ros E, et al. Effect of a Mediterranean diet supplemented with nuts on metabolic syndrome status: one-year results of the PREDIMED randomized trial. Arch Intern Med. 2008;168:2449–58.

209. Lohse B, Psota T, Estruch R, et al. Eating competence of elderly Spanish adults is associated with a healthy diet and a favorable cardiovascular disease risk profile. J Nutr. 2010;140:1322–7.

210. Zazpe I, Sanchez-Tainta A, Estruch R, et al. A large randomized individual and group intervention conducted by registered dietitians increased adherence to Mediterranean-type diets: the PREDIMED study. J Am Diet Assoc. 2008;108:1134–44; discussion 45.

211. de Lorgeril M, Renaud S, Mamelle N, et al. Mediterranean alpha-linolenic acid-rich diet in secondary prevention of coronary heart disease. Lancet. 1994;343:1454–9.

212. Sofi F, Cesari F, Abbate R, et al. Adherence to Mediterranean diet and health status: meta-analysis. BMJ. 2008;337:a1344.

213. Lloyd-Jones DM. Cardiovascular risk prediction: basic concepts, current status, and future directions. Circulation. 2010;121:1768–77.

214. Neveu V, Perez-Jimenez J, Vos F, et al. Phenol-Explorer: an online comprehensive database on polyphenol contents in foods. Database (Oxford). 2010;2010:bap024.

215. U.S. Department of Agriculture, ARS. USDA National nutrient database for standard reference, release 22. Washington, DC: USDA; 2009.

216. Institutes of Medicine. Seafood choices: balancing benefits and risks. Washington, DC: National Academy of Sciences; 2006.

Chapter 10
Fish, *n*-3 Polyunsaturated Fatty Acids, and Cardiovascular Disease

Claire McEvoy, Ian S. Young, and Jayne V. Woodside

Keywords Fish • Fish oil • *n*-3 Polyunsaturated fatty acids • Eicosapentaenioc acid • Docosahexaenoic acid • α-Linolenic acid • Cardiovascular disease • Coronary heart disease

Key Points

- *n*-3 Fatty acids are polyunsaturated fatty acids (PUFA). Those present in fish and fish oil consist of eicosapentaenioc acid (EPA, C20:5 *n*-3) and docosahexaenoic acid (DHA, C22:6 *n*-3).
- Observational studies suggest that fish intake and/or *n*-3 PUFA protect against coronary heart disease (CHD), heart failure, and sudden cardiac death. Clinical trial data for CHD are less consistent, and the results of further trials are awaited.
- Both US and UK health agencies recommend an increase in consumption of fish and *n*-3 PUFA.
- In the general population the benefits of fish consumption within recommended amounts outweigh the risk posed by environmental contaminants such as methylmercury.
- Evidence indicates that *n*-3 PUFA may also protect against the development of a number of other diseases such as some cancers, rheumatoid arthritis, and dementia.

C. McEvoy, MPhil (✉) • I.S. Young, MD • J.V. Woodside, PhD
Nutrition and Metabolism Research Group, School of Medicine, Dentistry and Biomedical Sciences, Centre for Public Health, Institute of Clinical Science, Queen's University Belfast, Block B, Grosvenor Road, Belfast BT12 6BJ, UK
e-mail: c.mcevoy@qub.ac.uk; i.young@qub.ac.uk; j.woodside@qub.ac.uk

N.J. Temple et al. (eds.), *Nutritional Health: Strategies for Disease Prevention*,
Nutrition and Health, DOI 10.1007/978-1-61779-894-8_10,
© Springer Science+Business Media, LLC 2012

10.1   Introduction

Cardiovascular disease (CVD) is a major cause of morbidity and mortality in the western world [1]. There are a number of well-established risk factors for CVD including smoking, hypertension, and family history [2]. In terms of nutrition, a diet high in fat, particularly saturated fat, has been shown to be associated with CVD incidence [3]. The observation that Greenland Eskimos (Inuit) have a low incidence of CVD despite a high saturated fat intake [4] has led to much scientific and public interest in the role of n-3 fatty acids found in fish and fish oils in the prevention and treatment of disease, and particularly CVD. In this chapter, the biochemistry and normal dietary intake of these compounds will be discussed, and the evidence linking them and their food sources with CVD reviewed. The safety of both fish oil supplements and fish will be assessed, and finally the potential effect of fish and fish oil consumption on other diseases considered.

10.2   Biochemistry of n-3 Fatty Acids

n-3 Fatty acids are long-chain polyunsaturated fatty acids (PUFA). They have 18–22 carbon atoms with the first of two or more double bonds beginning with the third carbon atom (when counting from the methyl end). n-3 PUFA from fish and fish oil consist of eicosapentaenioc acid (EPA, C20:5 n-3) and DHA (C22:6 n-3). Dietary α-linolenic acid (ALA, C18:3 n-3) can also be converted into EPA and DHA (e.g., in the brain, liver, and testes). However, the extent of this conversion is likely to be modest, and remains under debate. For example, Emken et al. [5] reported a 15% conversion, whereas more recently Pawlosky et al. [6] reported only a 0.2% conversion. Both reported that conversion to DHA was much lower than to EPA. The metabolism of n-3 PUFA is shown schematically in Fig. 10.1.

10.3   Food Sources of n-3 PUFA

The major food sources of ALA acid are vegetable oils, principally canola and soybean oils [7]. Other rich sources include flaxseed and walnuts [7]. Fish are the main source of EPA and DHA [7]. All fish contain EPA and DHA; however, content can differ dramatically by species, and also within species, with factors such as the diet of the fish and wild vs. farm-raising influencing EPA and DHA content. Fatty fish, such as mackerel, salmon, herring, sardines, and trout where fat is stored in muscle, contain more EPA and DHA than white fish such as cod, hake, and haddock, where fat is stored in the liver. Examples of food sources of ALA and of EPA and DHA are shown in Table 10.1.

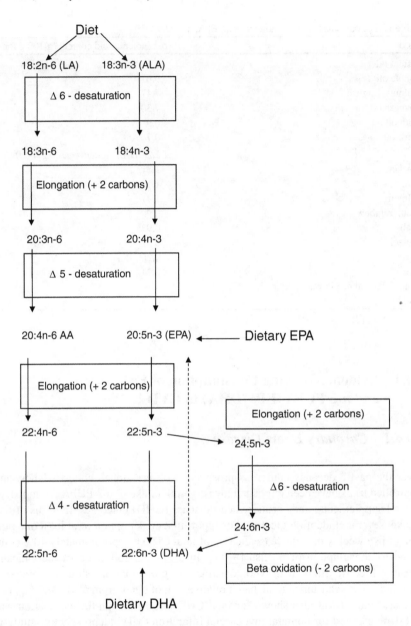

Fig. 10.1 Metabolism of *n*-3 PUFA (*LA* linoleic acid; *ALA* α-linolenic acid; *AA* arachidonic acid)

Table 10.1 Food sources of α-linolenic acid, EPA, and DHA

Food	α-Linolenic acid content (g/100 g food)
Walnut oil	11.5
Canola oil	9.6
Walnuts, English	7.47
Soybean oil	7.30
Corn oil	0.9
Olive oil	0.7
	EPA + DHA content (g/100 g food)
Mackerel	1.81
Sardines	1.71
Salmon	1.41
Trout, rainbow	0.97
Crab	0.92
Mussels	0.57
Plaice	0.26
Cod	0.24
Tuna (light, canned in water, drained)	0.16
Haddock	0.15

Source: Ref. [7]

10.4 Evidence Linking Consumption of Fish and *n*-3 PUFA (EPA/DHA) to CVD

10.4.1 Coronary Heart Disease

Over the past 3 decades numerous prospective observational, cohort studies, and controlled trials have been published investigating intake of *n*-3 PUFA (from fish or fish oil supplements) and coronary heart disease (CHD) outcomes. In terms of fish intake, several epidemiological studies report that consumption of at least one portion of fish weekly may decrease the risk of fatal CHD by approximately 40% compared to consumption of no fish [8–11]. In a 30-year follow-up of the Chicago Western Electric Study, men who consumed 35 g or more of fish daily compared with those who consumed none had a relative risk of death from CHD of 0.62 [12]. The association was also shown for fatal CHD [13], although the association was not demonstrated for nonfatal myocardial infarction (MI) in a prospective study in the elderly. Several meta-analyses of prospective studies have supported an inverse association of fish intake and fatal CHD outcome. Whelton et al. [14] in a meta-analysis of observational studies found a protective effect of fish consumption on CHD (RR = 0.86; 95% CI 0.81–0.92, $p < 0.005$) [14]. Additionally, a recent systematic review of cohort studies reported similar summary estimates for fish intake and CHD risk (RR = 0.81; 95% CI, 0.70–0.92) [15]. In this review, associations between fish intake and CHD risk were stronger for populations at high risk of CHD than initially healthy populations [15]. A further meta-analysis confirmed a strong positive association of fish intake and CHD within high-risk groups [16]. He et al. [17], in a meta-analysis of cohort studies, suggested a dose–response effect of fish intake on CHD mortality where a 20 g increased intake of fish was associated with a 7% lower risk of CHD mortality. The inverse association between fish consumption and mortality from CHD has been shown to be consistent internationally in an ecological study of 36 countries [18].

Some notable studies have not reported a significant association between fish consumption and CHD risk. In the Health Professionals Follow-up Study, there was no significant association between fish intake (and *n*-3 PUFA intake) and risk of CHD [19]. Similarly, the Physicians Health Study failed to show an association between fish consumption or *n*-3 PUFA intake and risk of MI, nonsudden cardiac death, or total CVD mortality, although there was a reduced risk of total mortality [20]. The Seven Countries data also showed a lack of association between fish consumption and both CHD incidence and mortality [21]. The Alpha-Tocopherol, Beta-Carotene Cancer Prevention Study in fact found that estimated *n*-3 PUFA intake from fish was associated with a trend towards increased risk of coronary death (after adjustment for *trans*, saturated, and *cis*-monounsaturated fatty acids) [22].

The authors of the American Heart Association (AHA) scientific statement on fish consumption, fish oil, *n*-3 fatty acids, and CVD have summarized the possible reasons for the conflicting data from the epidemiological studies [23]. There have been suggestions that the conflicting data reflect variability in endpoints studied, experimental design, dietary assessment of fish intake, or study populations [24, 25]. Albert et al. [20] attempted to explain the lack of association in their study because only a small fraction of their population reported little or no fish consumption, whereas the studies reporting an inverse association between fish consumption and coronary mortality have had a much higher proportion of the study population that were noneaters of fish. This suggests that *n*-3 PUFA from fish may have its greatest benefit for CHD risk when fish consumption is increased in habitually low consumers of fish. Another explanation, based on a summary of 11 prospective studies, is that fish is only protective in populations at high risk of CHD (RR = 0.4–0.6), but not in populations at low risk [16]. Another consideration is the type of fish consumed and how it is prepared: Oomen et al. [26] reported a lower CHD mortality (RR = 0.66) only in those eating fatty fish, and not in those eating lean fish. An emerging explanation relates to levels of methylmercury in fish (see later) with several, although not all, studies showing an association between methylmercury exposure and CHD risk [27, 28]. Thus methylmercury in fish may mask the beneficial effects of *n*-3 PUFA on CHD risk.

Several studies have measured tissue concentrations of fatty acids in relation to CHD risk. Tissue concentrations of DHA and EPA have been postulated to be a more reliable biomarker of fish consumption than is dietary intake assessment. A recent meta-analysis evaluating the association between tissue *n*-3 PUFA and risk of major CHD events in 25 prospective and case–control studies found an inverse association between the concentration of EPA and DHA in plasma and cellular phospholipids and major CHD events [29]. Additional research should determine the prognostic value of blood levels of n-3 PUFA.

To date, therefore, cumulative epidemiological evidence appears to support the hypothesis that consumption of n-3 PUFA reduces the risk of CHD, including cardiac death. However, data from intervention studies is less consistent and often conflicting. This is considered in more detail below.

The first randomized controlled trial using n-3 PUFA was the DART (Diet and Reinfarction Trial) study, which examined the effects of increased fatty fish intake on secondary prevention of CHD [30]. There was a 29% reduction in all-cause mortality over a 2-year period in male MI survivors advised to increase intake of fatty fish by 200–400 g/week (providing an extra 500–800 mg/day of n-3 PUFA). Analysis of a subset of patients who received fish oil capsules (900 mg/day EPA + DHA) suggests the effect was due to these fatty acids [31].

The Italian GISSI-Prevention Study [32], a secondary prevention study, randomized 11,324 patients with pre-existing CHD, to either 300 mg vitamin E, fish oil (850 mg of EPA + DHA), both, or neither. After 3.5-year follow-up, those given n-3 PUFA alone had a 15% reduction in the primary endpoint of death, nonfatal MI, and nonfatal stroke ($p < 0.02$), a 20% reduction in all-cause mortality ($p = 0.01$), and a 45% reduction in sudden death ($p < 0.001$) compared to the control group. Vitamin E had no apparent effect on the primary endpoint, whether given alone or when added to n-3 PUFA, although p-values approached significance. This trial was large and carried out in a relatively usual care setting (in that subjects were receiving conventional cardiac therapy). However, it was not placebo controlled (the control group received no intervention) and therefore is methodologically weaker than if a placebo had been utilized. Dropout rates were also high ($> 25\%$).

A smaller secondary prevention study compared corn oil with 3.5 g/day of fish oil that was concentrated in DHA + EPA in 300 post-MI patients in Norway. No effect was observed in rates of cardiac events after 1.5 years of intervention [33]. The authors speculated that this lack of observed benefit may be due to the high habitual fish intake in western Norway. A follow up of the DART study (DART II), comprising over 3,000 patients with angina, failed to demonstrate any benefit of fish or fish oil supplements on cardiac death over a follow-up period of 9 years [34]. In fact, the results indicate a significant increase in sudden cardiac death (SCD) in the group given specific advice to increase fish intake. The conduct of the DART II trial has been criticized as a result of methodological limitations. For example, there were no objective markers of dietary compliance, and the authors postulate that compliance with the intervention may have been low. Results from this trial should therefore be interpreted with caution [35].

Data from the OMEGA secondary prevention trial [36] assessing a 1-year supplementation with n-3 PUFA (460 mg EPA + 380 mg DHA) vs. an olive oil placebo in 3,851 post-MI patients in Germany were recently presented. The patients were aggressively managed with modern secondary prevention therapies, including statins, ACE inhibitors, beta blockers, aspirin, and clopidogrel. Preliminary results suggest no significant difference between the groups in the primary endpoint (SCD) or in secondary endpoints (major cardiac events). However, the overall mortality rate was much lower than expected in this study and, coupled with the relatively short follow-up period, the study may have been seriously underpowered to detect significant differences in the primary endpoint.

The results of a similar Dutch multicenter secondary prevention trial, The Alpha-Omega Trial, have recently been published [37]. A total of 4,837 post-MI patients were randomized to placebo, approximately 400 mg/day of EPA–DHA, 2 g/day of ALA, or a combination of EPA–DHA and ALA for an average of 3.7 years. As in the previous study, the patients were aggressively managed with lipid-lowering medications, antihypertensives, and antithrombotic agents. There was no significant difference between the groups in terms of major CVD events (fatal and nonfatal CVD and the cardiac interventions). In this instance, the supplementation dose of EPA–DHA was small (especially in comparison to the GISSI-P trial) and the study design may have been underpowered to detect meaningful differences between the four groups.

Overall, this recent evidence questions whether *n*-3 PUFA supplementation offers additional benefit over and above current optimal conventional treatment regimens in CHD secondary prevention. Further studies are warranted in this area.

No intervention studies have as yet shown a benefit of *n*-3 PUFA in primary prevention of CVD. The largest RCT published to date using *n*-3 PUFA supplementation is the JELIS trial (Japan EPA Lipid Intervention Study) [38]. In this trial, 18,645 patients (14,981 in primary prevention and 3,664 in secondary prevention) with hypercholesterolemia were randomized to receive statin alone or statin combined with a mean dose of 1,800 mg/day of purified EPA. At the end of the 4.6-year follow-up, the EPA group had a 19% reduction in major CVD events [38], but this benefit did not reach statistical significance in the primary prevention group when analyzed separately. It is worth noting that overall reported incidence of major cardiac events in this Japanese population was much lower (1%) than that reported in other populations [39]. Additionally, in contrast to the GISSI-P trial, there was a virtual absence of SCD. The disparity in these results may be partly explained by the high baseline fish intake in this Japanese population. Furthermore, the JELIS cohort consisted of approximately 70% women which may have contributed to an observed low incidence of major cardiac events in this cohort.

In summary, epidemiological studies provide convincing evidence that fish and supplementation with EPA and/or DHA are associated with a reduced risk of coronary disease death in both men and women. Consumption of fish appears to provide cardio-protection in a dose–response effect in persons with or without pre-existing coronary disease, although the cardioprotective benefits of fish may be greater in populations with low habitual fish consumption and those at greater risk of developing CHD. There is also some evidence supporting the use of *n*-3 PUFA supplements for secondary prevention of coronary events, particularly after MI, although questions are beginning to emerge regarding the efficacy of *n*-3 PUFA supplementation in post-MI patients receiving optimal drug treatments. Current data from randomized controlled trials using *n*-3 PUFA for primary prevention of CVD are limited. A further primary prevention randomized controlled trial is currently underway in the UK (NCT00135226) and is due for completion in 2011.

10.4.2 Sudden Cardiac Death

A number of studies have shown an association between fish consumption and a reduced risk of SCD. In the Physicians' Health Study, men who consumed fish >1/ week had a relative risk of sudden death of 0.48 ($p = 0.04$) compared to men who consumed fish <1/month [20]. A similar association was reported when blood levels of n-3 PUFA were examined, with the relative risk of SCD being significantly lower in men in the third and fourth quartiles compared with those in the first quartile [40]. In another nested case–control study, fish intake was also associated with a reduced risk of SCD: intake of 5.5 g/month of n-3 fatty acids were associated with a 50% reduced risk of primary cardiac arrest [41].

The GISSI-P trial, which was mentioned above, tested the effect of fish oil supplementation on CVD. Although not a stated primary endpoint, there was a 45% reduction in SCD [32]. The results from this study were reanalyzed and it was determined that the reduction in risk of SCD approached significance after 3 months of consumption, accounting for 57% of the overall mortality benefit. The reduction became significant at 4 months and was highly significant ($p = 0.0006$) at 3.5 years, the end of the study, when it accounted for 59% of the n-3 PUFA advantage in mortality [42].

The evidence linking n-3 fatty acids with SCD has recently been reviewed by Leaf et al. [43]. Overall, evidence relating n-3 PUFA to disease is greater for SCD than CHD events, and it would also appear that the effect is stronger for fatal CHD than for nonfatal MI. A meta-analysis of all RCTs examining n-3 PUFA (including ALA) has confirmed this, with the risk ratios for nonfatal MI, fatal CHD, and sudden death being 0.8, 0.7, and 0.7, respectively [44]. The cardioprotective benefit of n-3 PUFA on SCD has been consistent in the majority of meta-analyses to date [45–47].

10.4.3 Heart Failure

Recent evidence suggests a cardioprotective role for n-3 PUFA in the prevention and treatment of heart failure (HF). Data from large prospective cohort studies in the USA, Sweden, and Japan suggest an inverse association of total fish and/or n-3 PUFA intake and incident HF [48–51]. This was recently confirmed in the GISSI-HF trial [52], a large factorial randomized controlled trial. Almost 7,000 adults with class II to IV HF were randomized to 1 g of n-3 PUFA, a statin, both, or dual placebo. Significant benefits were observed for those receiving n-3 PUFA, including a reduction in mortality (-9%, $p < 0.05$) and significantly fewer hospital admissions for arrhythmias. Importantly, the cardioprotective benefits appeared to be additive to conventional secondary prevention treatments in these patients diagnosed with HF. It has been suggested that pharmacological doses of n-3 PUFA may be required for treatment in HF patients, particularly for those in the end stage of disease [35]. In a recent double-blind, randomized, controlled pilot trial, 43 patients with severe HF received 1 g/day n-3-PUFA ($n = 14$), 4 g/day n-3-PUFA ($n = 13$), or placebo ($n = 16$) for 3 months [53]. The study showed a dose-dependent benefit of n-3 supplementation on left ventricular ejection fraction, a known prognostic indictor in HF. Further intervention studies are required in HF patients to ascertain the optimal dose for cardioprotection.

10.4.4   Stroke

Compared with those focusing on CHD, relatively few studies have examined the effects of *n*-3 PUFA on stroke. Ecologic, cross-sectional, and case–control studies generally show an inverse association between consumption of fish and fish oils and stroke risk [54]. Results from prospective studies have been less consistent [55–60]. In the largest of these, the Nurses' Health Study, the relative risk of total stroke was lower among women who regularly ate fish than among those who did not, although this was not statistically significant. A significant decrease in the risk of ischemic stroke (RR = 0.49) was observed among women who ate fish at least two times per week compared with women who ate fish <1/month, after adjustment for age, smoking, and other cardiovascular risk factors. The results also showed a nonsignificant decrease among women in the highest quintile of *n*-3 PUFA intake. No association was observed between fish or fish oil consumption and hemorrhagic stroke [57]. Similarly, in the Health Professionals Follow-up Study, a protective effect of eating fish at least once per month or more on ischemic stroke after 12-year follow-up was demonstrated, but no effect on hemorrhagic stroke [60]. Recent data from the Cardiovascular Health Study in 4,775 adults (>65 years old) showed a 30% lower risk of ischemic stroke with a fish consumption five or more times per week and a 27% lower risk with a fish consumption of 1–4 times per week compared to consumption of <1/month [61].

In a review, Skerrett and Hennekens [54] concluded that the data currently support the hypothesis that consumption of fish several times per week reduces the risk of ischemic stroke (the leading cause of stroke), but not hemorrhagic stroke. However, not all studies are consistent; for example, a recent prospective study of a British population (*n* = 24,312) found no association between total fish consumption and reduced risk of subsequent stroke, although the results indicate that there appeared to be an inverse association between oily fish intake and stroke risk in women [62]. Similarly, a recent Finnish prospective cohort study found no association between total fish intake and risk of overall stroke, although consumption of salted fish was associated with a higher risk of hemorrhagic stroke [63]. In both the Lyon Diet Heart Study [64] and the GISSI-P study [32] there was no significant effect of increased fish intake or *n*-3 PUFA supplementation on total stroke. A subanalysis of the JELIS findings found no significant difference in stroke incidence between the control and EPA-supplemented group. However, a 20% risk reduction of recurrent stroke in the EPA-treated group was reported (hazard ratio, 0.80; 95% CI, 0.64–1.00) [65]. More intervention studies are required to examine the effect of oily fish or *n*-3 PUFA supplementation for type-specific stroke risk rather than total stroke in high-risk populations.

The evidence detailed here suggests that regular consumption of fish (and *n*-3 PUFA) may be associated with reduced risk of the major forms of CVD: CHD, SCD, stroke, and HF. The potential mechanisms by which this reduction in risk occurs are explored below.

10.4.5   Mechanisms

There are a number of possible mechanisms of action for *n*-3 PUFA.

- Decreased risk of arrhythmias
  SCD is a major cause of death in industrialized countries. Mortality statistics from the USA indicate that up to 80% of SCDs are due to ventricular fibrillation [66]. Studies in cell cultures and animal models, observational studies, and human intervention trials all suggest that *n*-3 PUFA may protect against fatal arrhythmia [67–69]. This mechanism may well be the most important as indicated by the strong protective association observed between *n*-3 PUFA and SCD and the fact that *n*-3 PUFA appear to exert their protective effect early in an intervention [42].

  The proposed antiarrhythmic mechanism involves a stabilizing effect on the myocardium itself [23, 35, 68, 70]. In contrast, three recent trials using *n*-3 PUFA supplementation in patients with implantable cardiac defibrillators (ICD) have produced heterogeneous results. One trial suggested proarrhythmic effects of *n*-3 PUFA supplements in certain patients with ICD [70]. However, two further trials have not confirmed these findings [71, 72], and a recent systematic review by Jenkins et al. [73], including these trials, showed no cardioprotective benefit of fish oil supplements in patients with ICD.
- Reduced triglycerides (TG)
  The reduction of TG with *n*-3 PUFA supplementation is well established [23]. In a review of human studies, Harris et al. [74] reported that around 4 g/day of *n*-3 PUFA from fish oil decreases serum TG by 25–30%, with an accompanying increase in LDL-cholesterol (see later) and HDL-cholesterol of 1–3%. The lowering effect increases as the supplement dose increases [74]. Therefore, fish oil can have a therapeutic role in hypertriglyceridemia at doses of 3–5 g/day, levels only consistently obtainable by supplementation [23]. Both EPA and DHA appear to be able to reduce TG [75]. The exact mechanism is not completely understood; however, EPA/DHA can modulate gene transcription to reduce hepatic very-low-density lipoprotein (VLDL)–TG synthesis and secretion and enhance TG clearance from chylomicrons and VLDL particles [76].
- Lowered blood pressure
  n-3 PUFA seem to have a small, dose-dependent, hypotensive effect [77]. A meta-analysis of 36 trials indicated a significant reduction in systolic/diastolic blood pressure of −2/−1.6 mmHg with an average consumption of 3.7 g/day of *n*-3 PUFA [78]. The effect was greater in older persons (>45 year, −3.5/−2.4 mmHg)

and hypertensive individuals (>140/90 mmHg, −4.0/−2.5 mmHg) [78]. The hypotensive effect of fish oil is thought to be due to improved arterial compliance and vasodilation [79]. Fish oil may regulate ACE activity, angiotensin II formation, tumor growth factor-beta expression, nitric oxide generation, vasoactive prostanoid release, and vasomotor tone [79]. Other biological actions of fish oil include altering the physicochemical structure of the phospholipid cell membrane leading to changes in permeability and function [80]. This can also play a role in blood pressure modulation through enhanced nitric oxide production.

- Reduced thrombosis and hemostasis

  n-3 PUFA have been shown to reduce platelet aggregation [81, 82], thereby reducing hemostasis. Fish oil has been shown to favorably influence the production of inflammatory and platelet aggregation intermediates including prostaglandin E_3, thromboxane A_4, and leukotriene B_4 [83]. Their effect on thrombosis has yet to be clearly determined. A negative association between n-3 PUFA intake and levels of fibrinogen, factor VIII, and von Willebrand factor has been demonstrated [84], but the more recent CARDIA study found no such association between either fish or n-3 PUFA intake and these coagulation factors [85]. There have been both positive [86] and negative [87] studies of n-3 PUFA supplementation and coagulation factors. Therefore, while it seems clear that n-3 PUFA reduce platelet aggregation, thereby affecting hemostasis, their effects on thrombosis are uncertain and further well-designed intervention studies are required.

- Reduced atherosclerotic plaque growth and increased plaque stability

  n-3 PUFA appear to have an antiatherogenic action in that they can inhibit new plaque development [35]. For example, EPA and DHA may alter expression of adhesion molecules. Abe et al. [88] reported a 9% reduction in ICAM-1 and a 16% reduction in E-selectin but no change in VCAM-1 in hypertriglyceridemic subjects receiving n-3 PUFA supplementation over 7–12 months. However, other studies have failed to confirm these results [86, 89]. Fish oil has also been postulated to stabilize existing plaque formation. Thies et al. [90] demonstrated absence of inflammation and fewer macrophages in carotid plaques of patients treated with fish oil supplements (1.4 g EPA and DHA) compared to the placebo group.

- Improved endothelial function

  Fish oil has been shown to improve endothelial function at a supplemental dose of >3 g/day (EPA/DHA) [91]. Putative mechanisms include hypotensive effects of fish oil (see above) and, particularly, enhanced nitric oxide production. However, a small randomized, double-blind, placebo-controlled crossover trial, designed to examine the dose–response effect of 0.85 g/day vs. 3.4 g/day of n-3 PUFA, in 26 healthy adults with elevated TG over an 8-week treatment period, found no significant effect on endothelial function [92]. Clearly, further large-scale randomized trials are required.

- Reduced inflammatory response

  Experimental data have provided some evidence that fish oil is anti-inflammatory. In vitro data suggest that DHA reduces the expression of IL-6 and IL-8 in stimulated cells (and endothelial expression of adhesion molecules) [93, 94].

Epidemiological evidence has suggested an inverse relationship between consumption of fish and/or n-3 PUFA and inflammatory biomarkers, including CRP, IL-6, and soluble intercellular adhesion molecule-1 [95, 96]. Fish oil also affects other inflammatory mediators such as TNF-α, and these molecules may affect atherogenesis and plaque stability [97]. A recent study by Trebble et al. [98] demonstrated that dietary fish oil supplementation is associated with decreases in prostaglandin E_2 production and simultaneous increases in lymphocyte proliferation and IFN-γ production, and a trend towards increased IL-4 production by peripheral blood mononuclear cells [98]. Other studies have not demonstrated a favorable effect of fish oil on inflammatory mediators [99, 100]. It has been proposed that n-3 PUFA (particularly EPA) compete with n-6 PUFA at a cellular level to modulate the eicosanoid derivatives from arachidonic acid substrate resulting in a less pro-atherogenic/inflammatory prostaglandin release [100, 101]. In this regard, n-3 PUFA have been proposed to play a role in the treatment of other chronic diseases, such as rheumatoid arthritis, inflammatory bowel disease, and psoriasis, perhaps through a suppression of immune and inflammation responses. This is considered in more detail later.

10.4.6 Effects on CVD of Plant-Derived Versus Marine-Derived n-3 Fatty Acids

ALA, in contrast to EPA and DHA, is found in plant foods and not marine sources (Table 10.1). Evidence from epidemiological studies of ALA and CVD indicates that ALA is associated with a lower risk of both MI and fatal CHD in both women and men [23]. A meta-analysis of five prospective cohort studies demonstrated an inverse relationship between ALA intake and CHD death, indicating a 20% risk reduction which was considered borderline significant [102]. In addition, in an observational study, higher consumption of ALA was associated with a lower prevalence of carotid plaques, and with lesser thickness of segment-specific carotid intima-media thickness [103]. A more recent cohort study showed an inverse association between ALA intake and SCD in women [104]. Also, in a post-MI population in Costa Rica, ALA intake and blood levels predicted a better prognosis in terms of survival, independent of fish intake and EPA/DHA levels [105].

The effect of ALA supplementation in CHD prevention has been examined in four trials. In the Lyon Diet Heart Study, a randomized, controlled trial of an ALA-rich Mediterranean diet with free-living subjects, those in the intervention group had a 50–70% reduction of cardiac endpoints [64]. Although plasma levels of ALA were significantly associated with the endpoints, it is, however, impossible to ascribe the benefit unambiguously to ALA because of the many other dietary variables present. The Alpha-Omega randomized controlled trial (outlined above), comprising 4,837 post-MI patients, observed a nonsignificant 9% reduction in major cardiac events with ALA supplementation in the whole population, when compared with placebo and EPA–DHA supplementation [37]. Subgroup analyses demonstrated a 27% reduction in major cardiac events among women in the ALA group, but again, this only approached statistical significance [37].

These two positive or borderline positive studies of ALA are balanced by two negative studies. In the Norwegian Vegetable Oil Experiment, 13,000 men aged 50–59 years with no history of MI were randomly assigned to consume either 5.5 g/ day of ALA (from 10 mL linseed oil) or 10 mL sunflower oil for 1 year. There were no differences in sudden death, death from CHD, or all deaths between the groups [106]. Similarly, the MARGARIN (Mediterranean Alpha-Linolenic Enriched Groningen Dietary Intervention) examined 282 subjects with multiple CVD risk factors and randomized them to receive margarines rich in either ALA or linoleic acid. Follow-up was for 2 years. There was no difference between the groups in CHD risk, although there was a trend towards reduced CVD events in the ALA group ($p = 0.20$) [107].

These contradictory studies indicate that further well-designed trials must be carried out to determine the role of ALA acid in CVD etiology. Currently, the effects of plant vs. marine *n*-3 PUFA are difficult to determine as few studies have set out to test this. For example, a meta-analysis of the available RCTs examined all intervention trials including marine and/or plant sources of *n*-3 PUFA and found significant reductions in risk of nonfatal MI, fatal MI, and sudden death, but did not distinguish between the two sources of *n*-3 PUFA [44].

10.5   Fish and Fish Oil Intake: Recommendations for CVD Health

The mean current daily intake of EPA and DHA combined in a typical North American diet (which includes about one fish serving every 10 days) approaches 130 mg/day, which is about 0.15% of total dietary fat intake. This is markedly lower than Japanese intakes and only a small fraction of the EPA and DHA consumed by the Greenland Inuit. Fish consumed 2.5–3 times/week would provide an EPA + DHA intake of about 500 mg/day, an intake about four times that currently consumed in North America. Epidemiological data from the MRFIT study in the USA have indicated that progressively higher intakes of fish-derived *n*-3 PUFA (up to about 665 mg/day) over 10.5 years were associated with a reduction in total mortality as well as in CHD mortality.

The AHA guidelines include the following recommendations with respect to *n*-3 PUFA in patients with coronary disease [23]: Consumption of one fatty fish meal per day (or alternatively, a fish oil supplement) could result in an *n*-3 PUFA intake of around 1 g/day, an amount shown to beneficially affect CHD mortality rates in patients with coronary disease. Individuals without CVD should consume at least 0.25–0.5 g/day *n*-3 PUFA. Dietary oily fish intake is generally advocated as fish contains nutrients—vitamin D, selenium, and antioxidants—usually absent from fish oil supplements [108].

Table 10.2 Summary of AHA recommendations for *n*-3 PUFA intake in different population and patient groups

Population	Recommendation
Patients without documented CHD	Eat a variety of (preferably oily) fish at least twice a week. Include oils and foods rich in α-linolenic acid (flaxseed, canola, and soybean oils; flaxseed and walnuts)
Patients with documented CHD	Consume ≈ 1 g/day of EPA + DHA, preferably from oily fish. EPA + DHA supplements could be considered in consultation with the physician
Hypertriglyceridemic patients	2–4 g/day of EPA + DHA provided as capsules under a physician's care

The FDA recently approved a 4 g daily dose *n*-3 PUFA (containing both EPA and DHA) formulation available by prescription for patients with very high TG. These recommendations are summarized in Table 10.2.

Current UK guidelines recommend consuming two portions of fish per week, of which one should be oily fish [109]. This will provide approximately 0.45 g/day of *n*-3 PUFA. There are no specific recommendations made for individuals with pre-existing CVD.

10.6   Other Health Effects of Fish and *n*-3 PUFA

10.6.1   Pregnancy

Adequate DHA intake is important, particularly in the third trimester during pregnancy, for optimal fetal brain and retina development [110]. Observational evidence has shown that frequent consumption of fish is associated with improved infant neuro-cognitive development and function [110]. DHA has been shown to improve infant visual acuity, IQ, and behavioral scores in a dose-dependent manner [111]. Furthermore, recent meta-analyses of randomized controlled trials, assessing the efficacy of *n*-3 supplementation in pregnancy, suggest a small but significant benefit for *n*-3 PUFA on gestation length (increase of approximately 2.5 days) and lower risk of preterm birth before 34-week gestation [112, 113]. However, observational data, involving a cohort of 657 Spanish women with high habitual fish consumption, found that total seafood intake, in particular, crustaceans and canned tuna (at least once per week), but not fatty fish intake, was significantly associated with small-for-gestational-age births (SGA), even after adjustment for serum persistent organic pollutants (POPs) [114]. Some species of fish and seafood contain POPs as well as methylmercury (see later) and moderate prenatal mercury exposure can have severe adverse effects on infant development [115]. Therefore, current recommendations for *n*-3 PUFA intake for pregnant women or women planning pregnancy should aim to provide optimal DHA for fetal development while minimizing overall methylmercury exposure. The FDA has provided specific advice regarding fish intake in pregnancy and childhood, as discussed later. Even so, a prospective UK study of almost 12,000 pregnant women found that those exceeding the FDA recommendations for fish intake had offspring with better cognitive and behavioral development than offspring from women consuming less fish during pregnancy. This suggests that the benefit of consuming extra nutrients (presumably *n*-3 PUFA, which is essential for brain development) exceeds the risk of harm from exposure to trace contaminants [116].

10.6.2   Cancer

EPA and DHA have consistently been shown to inhibit the proliferation of breast and prostate cancer cell lines in vitro and to reduce both the risk and progression of these tumors in animal models [117]. There is some evidence that increased fish intake may contribute to lower risk of colorectal, prostate, and breast cancer [117, 118]. This is not supported by data from a recent meta-analysis of case–control and cohort studies, which found no evidence of a protective effect of fish consumption on prostate cancer risk [119]. However, in this analysis, a significant association was observed between increased fish intake and reduced prostate cancer mortality (RR 0.37; 95% CI: 0.18, 0.74). In addition, a recent systematic review found no significant association between *n*-3 PUFA and cancer incidence [120]. However, further well-designed epidemiological studies are required before definitive conclusions can be reached. Multiple mechanisms would appear to be involved in any chemopreventive activity of *n*-3 PUFA, including suppression of neoplastic transformation, cell growth inhibition, enhanced apoptosis, reduced oxidative stress, and anti-angiogenesis [118]. Recent trials have suggested a role for *n*-3 PUFA, particularly EPA, for enhanced nutritional support in patients with cancer cachexia [121]. However, a recent Cochrane review of the literature failed to find sufficient evidence to support the use of oral EPA in treatment of cancer cachexia [122].

10.6.3   Alzheimer's Disease

Epidemiologic investigation of dietary *n*-3 PUFA and Alzheimer's disease (AD) provides some evidence of a beneficial effect [123]. One case–control study reported that *n*-3 PUFA levels in plasma phospholipids of AD patients were 60–70% of those found in age-matched control subjects [124]. The Framingham study, involving 899 dementia-free adults (mean age 76 years) over approximately 9 years, found that those in the highest quintile range of phosphatidylcholine DHA levels had a significant 47% risk reduction in developing all-cause dementia [125]. Furthermore, a prospective study reported that subjects who ate fish once a week or more, had a 60% lower risk of developing AD compared with those who never or rarely ate fish, in a model adjusted for age and other risk factors [126]. Total intake of *n*-3 PUFA was also associated with reduced risk of AD [126]. However, a recent randomized, double-blind, placebo-controlled trial in 295 patients with established mild to moderate AD, assessed supplementation of 2 g/day of DHA vs. a placebo control for 18 months and found no effect of supplementation on cognitive decline or brain atrophy [127]. Emerging evidence suggests n-3 PUFA may exert its greatest protection in the development of late-onset AD in individuals without pre-existing genetic risk. The presence of the Apo-E e4 allele is associated with an increased risk of late-onset AD [128]. The recent Three City cohort study in France examined 8,085 dementia-free adults (age >65 years) over a 4-year period. Weekly consumption of fish was associated with a significant reduction in risk of AD (HR = 0.65; 95% CI 0.43–0.99) and all-cause dementia (HR 0.60; 95% CI 0.40–0.90) among those without the Apo-E genotype [129]. The Cardiovascular Health Cognition Study in 2,233 adults found that consumption of oily fish twice per week was associated with a significant reduction in dementia in noncarriers of Apo-E e4 [130]. This suggests that Apo-E status may have a modulating effect on n-3 PUFA and requires further study. While data from prospective and cohort studies is suggestive of a beneficial effect of n-3 PUFA in AD, there is a need for well-designed large-scale randomized controlled trials.

10.6.4   Rheumatoid Arthritis

Rheumatoid arthritis (RA) is a debilitating disease and is associated with increased risk of CVD and osteoporosis. Poor nutrient status in RA patients has been reported, and some drug therapies, such as nonsteroidal anti-inflammatory drugs (NSAIDs), prescribed to alleviate RA symptoms, may increase the requirement for nutrients because of reduced absorption [131]. As mentioned previously, n-3 PUFAs are precursors of anti-inflammatory eicosanoids and may therefore attenuate inflammatory activity. Supplementation with n-3 PUFA has consistently been shown to improve RA symptoms, delay the progression of established RA, and lead to a reduction in NSAID usage [131–133].

10.6.5   Diabetes

A high consumption of fish and fish oils has been associated with a lower CHD incidence and total mortality among diabetic women [134]. However, a recent prospective study, involving 36,328 women over approximately 12.4 years, found that increased intake of marine n-3 PUFA (from fish and/or fish oil) was positively associated with an increased risk of type 2 diabetes, particularly in those with higher intakes >0.2 g/day n-3 PUFA or >2 servings/week of fish [135]. One study suggested an adverse effect of n-3 PUFA on glycemic control in established diabetes [136] which is in contrast with findings from a recent Cochrane review. This systematic review examined the effect of fish oil supplementation on people with type 2 diabetes mellitus and included 23 randomized placebo-controlled trials involving 1,075 participants [137]. The reviewers concluded that fish oil supplementation in type 2 diabetes lowers TG, may raise LDL-cholesterol (especially in hypertriglyceridemic patients on higher doses of fish oil), and has no statistically significant effect on glycemic control (as above). The reviewers also stated that trials with vascular events or mortality-defined endpoints are needed before definite conclusions can be made. A meta-analysis of 26 trials of subjects with type 1 or type 2 diabetes confirmed that fish oil has no effect on HbA1c, although fasting glucose levels rose slightly in the type 2 diabetics [138].

10.7   Safety and Side-Effects of *n*-3 PUFA Supplements

Whereas the ratio of *n*-6 to *n*-3 fatty acid intake of early humans is estimated to have been about 1:1 [139], the ratio in the American diet is now around 10:1. This is due to both the reduction in intake of *n*-3 PUFA and the increase in use of vegetable oils rich in linoleic acid [23]. The FDA has ruled that intakes of up to 3 g/day of marine *n*-3 PUFA are Generally Recognized as Safe (GRAS) [23]. The FDA has also approved a qualified health claim for EPA and DHA in dietary supplements [23].

Some side effects of *n*-3 PUFA supplementation can occur [140], with the most common complaint being a fishy aftertaste. In the GISSI-P intervention study, which provided 0.85 g of *n*-3 PUFA daily for 3.5 years, 3.8% of subjects in the *n*-3 PUFA arm reported discontinuing their supplements because of side effects, compared to 2.1% in the vitamin E group. The most commonly reported side effects were gastrointestinal upset and nausea [32].

High doses of *n*-3 PUFA seem to modestly prolong skin bleeding time, and may also increase the tendency for nosebleeds [141]. However, a daily intake of ≤3 g/day EPA + DHA does not appear to appreciably increase bleeding time [141]. Similarly, in the clinical trials using >3 g/day, no clinically significant increase in bleeding has been reported [142].

While *n*-3 PUFA are known to reduce TG, they can also increase LDL-cholesterol. Harris [143] reported that around 4 g/day of *n*-3 PUFA from fish oil increases LDL-cholesterol by 5–10%. There is also a suggestion that consumption of *n*-3 PUFA may be associated with a higher susceptibility to in vitro oxidation of LDL [144, 145], although this remains to be confirmed [146–148]. However, it is likely that the established beneficial effects of *n*-3 PUFA on other cardiovascular risk factors far outweigh their effects on either LDL-cholesterol or its oxidizability.

More recently, concerns suggesting a positive association between ALA intake and prostate cancer in men have been published [102]. A meta-analysis of nine cohort and case–control studies reported an increased risk of prostate cancer in men with a high intake or blood level of ALA (combined RR = 1.70; 95% CI 1.12–2.58) [102]. It is widely accepted that the current evidence is weak and further studies are presently underway to elucidate the relationship (if any) between ALA and prostate cancer.

10.8   Fish: Contamination Safety Issues

There have been recent safety concerns about the levels of environmental contaminants in fish. Some species of fish may contain significant levels of methylmercury, polychlorinated biphenyls (PCBs), dioxins, and other potentially toxic chemicals. While these compounds are present at low levels in both fresh and sea waters, they bioconcentrate in the aquatic food chain such that levels may be significant in larger predatory fish and marine mammals, such as swordfish and shark (see Fig. 9.3). Eating these fish on a regular basis is not advised [109]. Despite this recommendation, a recent meta-analysis of five published studies demonstrated no significant relationship between increased mercury levels and CHD risk (RR = 1.12; 95% CI 0.71–1.75, $p = 0.62$) [111].

10.9   Fish and Fish Oil Intake: Recommendations

In the USA, the Environmental Protection Agency (EPA) regulates sport-caught fish, whereas the FDA regulates all commercial fish. The EPA recommends that women who are pregnant or may become pregnant, and nursing mothers, should limit their consumption of fish to one 6-oz/week (170 g/week) meal [23]. They also recommend that young children consume ≤2 oz/week (60 g/week) of fish. In order to minimize methylmercury exposure the FDA recommends that pregnant women, nursing mothers, and young children eliminate shark, swordfish, king mackerel, and tilefish (also known as golden bass or golden snapper) from their diet completely and limit other fish consumption to 12 oz/week (340 g or 3–4 servings/week) [23]. The FDA also recommends that for adults other than pregnant women or those who may become pregnant, the maximum intake should be <7 oz/week (200 g) of fish with very high (around 1 ppm) methylmercury levels (as listed above) and <14 oz/week (400 g) of fish with high (around 0.5 ppm) methylmercury levels (fresh tuna, marlin, red snapper).

In the UK, the Food Standards Agency (FSA) has advised pregnant and breast-feeding women, and women who intend to become pregnant, to limit their consumption of tuna to no more than four medium-size cans or two fresh tuna steak per week. These women are also advised to avoid eating shark, swordfish, and marlin. For children, the FSA advise that children under 16 should avoid shark, marlin, and swordfish, but they can still eat tuna [149].

In summary, pregnant women or those considering pregnancy should aim to consume a wide variety of fish, avoiding species with known high methylmercury concentrations in accordance with current public health fish/seafood intake recommendations. Adherence to current recommendations will provide DHA for fetal brain and eye development and minimize risk of prenatal mercury exposure.

The sustainability of fish stocks is also of some concern, with fishing controls being implemented to attempt conserve and develop stocks. Information is also being made available to consumers to enable them to make informed decisions about sustainability when purchasing fish and other seafood [150].

10.10   Conclusions

Evidence from epidemiological studies and some clinical trials suggests that *n*-3 PUFA protects against CHD, SCD, and HF. However, further trials are required to confirm this. In particular, evidence from clinical trial data for a cardioprotective benefit in both primary and secondary CVD prevention is not entirely consistent, particularly for primary prevention. The disparity in clinical trial results is probably due to several reasons, but the ones most likely to be important are variation in sample sizes (and therefore power), study duration, doses of EPA and DHA, the background fish intake of the populations studied, and the endpoints used. There have been relatively few studies that investigated the efficacy of *n*-3 PUFA in primary prevention of CVD, while secondary prevention studies should aim to fully elucidate the role of *n*-3 PUFA supplementation in patients who are optimally treated with pharmacological regimens. Large clinical trials are also still required in high-risk groups, such as patients with type 2 diabetes, dyslipidemia, and hypertension, and those with congestive HF, who are at high risk of sudden death.

Fish oil supplements would appear to be safe, while the benefits of fish consumption within recommended amounts seem to outweigh the risk attributed to environmental contaminants. Fish intake during pregnancy and childhood is currently advised to provide adequate intake of DHA although specific recommendations apply to minimize mercury exposure. There are inconsistent results from epidemiological data examining fish intake, POP exposure, and infant outcomes. In part, results are limited by current dietary assessment methods, which provide imprecise estimates of short-term fish intake and/or POP exposure and can result in misclassification error. Therefore, further research in this area should focus on direct measurement of serum POPs in pregnant women to determine lifelong exposure in relation to pregnancy outcomes [151]. In addition, it has been suggested that low-exposure populations should also be studied in order to provide valid risk estimates for POP levels and pregnancy outcomes [151]. The competing adverse effect of chemical pollutants should be considered when examining the association between fish consumption and other endpoints such as CVD and diabetes.

Fish and fish oil consumption may also protect against a variety of other diseases including cancer at various sites, and Alzheimer's disease. Further intervention studies investigating ALA supplementation and disease outcomes are warranted.

References

1. Ross R. The pathogenesis of atherosclerosis in a perspective for the 1990's. Nature. 1993;362:801–9.
2. Heller RF, Chinn S, Tunstall-Pedoe H, Rose G. How well can we predict coronary heart disease? Findings in the United Kingdom Heart Disease Prevention Project. Br Med J. 1984;288:1409–11.
3. Mann JI. Diet and risk of coronary heart disease and type 2 diabetes. Lancet. 2002;360: 783–9.
4. Dyerberg J, Bang HO. Haemostatic function and platelet polyunsaturated fatty acids in Eskimos. Lancet. 1979;2:433–5.
5. Emken EA, Adlof RO, Gulley RM. Dietary linoleic acid influences desaturation and acylation of deuterium-labelled linoleic and linolenic acids in young adult males. Biochim Biophys Acta. 1994;1213:277–88.
6. Pawlosky RJ, Hibbeln JR, Novotny JA, Salem Jr N. Physiological compartmental analysis of alpha-linolenic acid metabolism in adult humans. J Lipid Res. 2001;42:1257–65.
7. Ministry of Agriculture, Fisheries and Food. Fatty acids. Seventh supplement to the fifth edition of McCance and Widdowson's The Composition of Foods. London: HMSO; 1998.
8. Kromhout D, Bosschieter EB, de Lezenne Coulander C. The inverse relation between fish consumption and 20-year mortality from coronary heart disease. N Engl J Med. 1985; 312:1205–9.
9. Kromhout D, Feskens EJ, Bowles CH. The protective effect of a small amount of fish on coronary heart disease mortality in an elderly population. Int J Epidemiol. 1995;24:340–5.
10. Shekelle RB, Missell L, Paul O, Shryock AM, Stamler J. Fish consumption and mortality from coronary heart disease. N Engl J Med. 1985;313:820–4.
11. Dolecek TA, Granditis G. Dietary polyunsaturated fatty acids and mortality in the Multiple Risk Factor Intervention Trial (MRFIT). World Rev Nutr Diet. 1991;66:205–16.
12. Daviglus ML, Stamler J, Orencia AJ, et al. Fish consumption and the 30-year risk of fatal myocardial infarction. N Engl J Med. 1997;336:1046–53.
13. Lemaitre RN, King IB, Mozaffarian D, Kuller LH, Tracy RP, Siscovick DS. N-3 polyunsaturated fatty acids, fatal ischemic disease, and nonfatal myocardial infarction in older adults: the Cardiovascular Health Study. Am J Clin Nutr. 2003;77:319–25.
14. Whelton SP, He J, Whelton PK, Muntner P. Meta-analysis of observational studies on fish intake and coronary heart disease. Am J Cardiol. 2004;93:1119–23.
15. Mente A, de Koning L, Shannon HS, Anand SS. A systematic review of the evidence supporting a causal link between dietary factors and coronary heart disease. Arch Intern Med. 2009;169:659–69.
16. Marckmann P, Gronbaek M. Fish consumption and coronary heart disease mortality: a systematic review of prospective cohort studies. Eur J Clin Nutr. 1999;53:585–90.
17. He K, Song Y, Daviglus ML, et al. Accumulated evidence on fish consumption and coronary heart disease mortality: a meta-analysis of cohort studies. Circulation. 2004;109:2705–11.
18. Zhang J, Sasaki S, Amano K, Kestelot H. Fish consumption and mortality from all causes, ischemic heart disease, and stroke: an ecological study. Prev Med. 1999;28:520–9.
19. Ascherio A, Rimm EB, Stampfer MJ, Giovannucci EL, Willett WC. Dietary intake of marine n-3 fatty acids, fish intake, and the risk of coronary heart disease among men. N Engl J Med. 1995;332:977–82.
20. Albert CM, Hennekens CH, O'Donnell CJ, et al. Fish consumption and risk of sudden cardiac death. JAMA. 1998;279:23–8.
21. Kromhout D, Bloemberg BP, Feskens EJ, Hertog MG, Menotti A, Blackburn H. Alcohol, fish, fibre and antioxidant vitamin intake do not explain population differences in coronary heart disease mortality. Int J Epidemiol. 1996;25:753–9.
22. Pietinen P, Ascherio A, Korhonen P, et al. Intake of fatty acids and risk of coronary heart disease in a cohort of Finnish men: the Alpha-Tocopherol, Beta-Carotene Cancer Prevention Study. Am J Epidemiol. 1997;145:876–87.

23. Kris-Etherton PM, Harris WS, Appel LJ, for the Nutrition Committee. Fish consumption, fish oil, omega-3 fatty acids, and cardiovascular disease. Circulation. 2002;106:2747–57.

24. Kromhout D. Fish consumption and sudden cardiac death. JAMA. 1998;279:65–6.

25. Sheard NF. Fish consumption and risk of sudden cardiac death. Nutr Rev. 1998;56:177–9.

26. Oomen CM, Feskens EJ, Rasanen L, et al. Fish consumption and coronary heart disease mortality in Finland, Italy and the Netherlands. Am J Epidemiol. 2000;151:999–1006.

27. Salonen JT, Seppanen K, Lakka TA, Salonen R, Kaplan GA. Mercury accumulation and accelerated progression of carotid atherosclerosis: a population-based prospective 4-year follow-up study in men in eastern Finland. Atherosclerosis. 2000;148:265–73.

28. Ahlqwist M, Bengtsson C, Lapidus L, Gergdahl IA, Schutz A. Serum mercury concentration in relation to survival, symptoms, and diseases: results from the prospective population study of women in Gothenburg, Sweden. Acta Odontol Scand. 1999;57:168–74.

29. Harris WS, Poston WC, Haddock CK. Tissue n-3 and n-6 fatty acids and risk for coronary heart disease events. Atherosclerosis. 2007;193:1–10.

30. Burr ML, Fehily AM, Gilbert JF, et al. Effects of changes in fat, fish, and fibre intakes on death and myocardial reinfarction: Diet and Reinfarction Trial (DART). Lancet. 1989;2: 757–61.

31. Burr ML, Sweetnam PM, Fehily AM. Diet and reinfarction. Eur Heart J. 1994;15:1152–3.

32. GISSI-Prevenzione Investigators. Dietary supplementation with n-3 polyunsaturated fatty acids and vitamin E after myocardial infarction: results of the GISSI-Prevenzione trial. Gruppo Italiano per lo Studio della Sopravvivenza nell'Infarcto miocardico. Lancet. 1999; 354:447–55.

33. Nilsen DW, Albrektsen G, Landmark K, Moen S, Aarsland T, Woie L. Effects of a high dose concentrate of n-3 fatty acids or corn oil introduced early after an acute myocardial infarction on serum triacylglycerol and HDL cholesterol. Am J Clin Nutr. 2001;74:50–6.

34. Burr ML, Ashfield-Watt PAL, Dunstan FDJ, et al. Lack of benefit of dietary advice to men with angina: results of a controlled trial. Eur J Clin Nutr. 2003;57:193–200.

35. Lavie CJ, Milani RV, Mehra MR, Ventura HO. Omega-3 polyunsaturated fatty acids and cardiovascular diseases. J Am Coll Cardiol. 2009;54:585–94.

36. Rauch B, Schiele R, Schneider S, et al. OMEGA, a randomized, placebo-controlled trial to test the effect of highly purified omega-3 fatty acids on top of modern guideline-adjusted therapy after myocardial infarction. Circulation. 2010;122:2152–9.

37. Kromhout D, Giltay EJ, Geleijnse J, for the Alpha Omega Trial Group. n-3 Fatty acids and cardiovascular events after myocardial infarction. N Engl J Med. 2010;363:2015–26.

38. Yokoyama M, Origasa H, Matsuzaki M, et al. Effects of eicosapentaenoic acid on major coronary events in hypercholesterolaemic patients (JELIS): a randomised open-label, blinded endpoint analysis. Lancet. 2007;369:1090–8.

39. von Schacky C, Harris WS. Cardiovascular benefits of omega-3 fatty acids. Cardiovasc Res. 2007;73:310–5.

40. Albert CM, Campos H, Stampfer MJ, et al. Blood levels of long-chain n-3 fatty acids and the risk of sudden death. N Engl J Med. 2002;346:1113–8.

41. Siscovick DS, Raghunathan TE, King I, et al. Dietary intake and cell membrane levels of long-chain n-3 polyunsaturated fatty acids and the risk of primary cardiac arrest. JAMA. 1995;274:1363–7.

42. Marchioli R, Barzi F, Bomba E, et al. Early protection against sudden death by n-3 polyunsaturated fatty acids after myocardial infarction: time course analysis of the results of the GISSI-Prevenzione. Circulation. 2002;105:1897–903.

43. Leaf A, Kang JX, Xiao Y-F, Billman GE. Clinical prevention of sudden cardiac death by n-3 polyunsaturated fatty acids and mechanism of prevention of arrhythmias by n-3 fish oils. Circulation. 2003;107:2646–52.

44. Bucher HC, Hengstler P, Schindler C, Meier G. N-3 polyunsaturated fatty acids in coronary heart disease: a meta-analysis of randomised controlled trials. Am J Med. 2002;112: 298–304.

45. Lee JH, O'Keefe JH, Lavie CJ, Marchioli R, Harris WS. Omega-3 fatty acids for cardioprotection. Mayo Clin Proc. 2008;83:324–32.

46. Marik PE, Varon J. Omega-3 dietary supplements and the risk of cardiovascular events: a systematic review. Clin Cardiol. 2009;32:365–72.

47. Wang C, Harris WS, Cheung M, et al. n-3 Fatty acids from fish or fish-oil supplements, but not α-linolenic acid, benefit cardiovascular disease outcomes in primary- and secondary-prevention studies: a systematic review. Am J Clin Nutr. 2006;84:5–17.

48. Mozaffarian D, Bryson CL, Lemaitre RN, et al. Fish intake and risk of incident heart failure. J Am Coll Cardiol. 2005;45:2015–21.

49. Levitan EB, Wolk A, Mittleman MA. Fish consumption, marine omega-3 fatty acids, and incidence of heart failure: a population-based prospective study of middle-aged and elderly men. Eur Heart J. 2009;30:1495–500.

50. Yamagishi J, Nettleton JA, Folsom AR. Plasma fatty acid composition and incident heart failure in middle-aged adults: the Atherosclerosis Risk in Communities (ARIC) study. Am Heart J. 2008;156:965–74.

51. Yamagishi K, Iso H, Date C, et al. Fish, ω-3 polyunsaturated fatty acids, and mortality from cardiovascular diseases in a nationwide community-based cohort of Japanese men and women. J Am Coll Cardiol. 2008;52:988–96.

52. GISSI-HF Investigators, Tavazzi L, Maggioni AP, Marchioli R, et al. Effect of n-3 polyunsaturated fatty acids in patients with chronic heart failure (the GISSI-HF trial): a randomized, double-blind, placebo-controlled trial. Lancet. 2008;372:1223–30.

53. Moertl D, Hammer A, Steiner S, et al. Dose-dependent effects of omega-3 polyunsaturated fatty acids on systolic left ventricular function, endothelial function, and markers of inflammation in chronic heart failure of nonischemic origin: a double-blind, placebo-controlled, 3-arm study. Am Heart J. 2011;161:915.e–9.

54. Skerrett PJ, Hennekens CH. Consumption of fish and fish oils and decreased risk of stroke. Prev Cardiol. 2003;6:38–41.

55. Orencia AJ, Daviglus ML, Dyer AR, Shekelle RB, Stamler J. Fish consumption and stroke in men: 30-year findings of the Chicago Western Electric Study. Stroke. 1996;27:204–9.

56. Morris MC, Manson JE, Rosner B, Buring JE, Willett WC, Hennekens CH. Fish consumption and cardiovascular disease in the Physicians' Health Study: a prospective study. Am J Epidemiol. 1995;142:166–75.

57. Iso H, Rexrode KM, Stampfer MJ, et al. Intake of fish and omega-3 fatty acids and risk of stroke in women. JAMA. 2001;285:304–12.

58. Keli SO, Feskens EJ, Kromhout D. Fish consumption and risk of stroke: the Zutphen Study. Stroke. 1994;25:328–32.

59. Gillum RF, Mussolino ME, Madans JH. The relationship between fish consumption and stroke incidence: the NHANES I Epidemiologic Follow-up Study. Arch Intern Med. 1996;156:537–42.

60. He K, Rimm EB, Merchant A, et al. Fish consumption and risk of stroke in men. JAMA. 2002;288:3130–6.

61. Mozaffarian D, Longstreth Jr WT, Lemaitre RN, et al. Fish consumption and stroke risk in elderly individuals: the cardiovascular health study. Arch Intern Med. 2005;165:200–6.

62. Myint P, Welch AA, Bingham SA, et al. Habitual fish consumption and risk of incident stroke: the European Prospective Investigation into Cancer (EPIC)-Norfolk prospective population study. Public Health Nutr. 2006;9:882–8.

63. Montonen J, Järvinen R, Reunanen A, Knekt P. Fish consumption and the incidence of cerebrovascular disease. Br J Nutr. 2009;102:750–6.

64. de Lorgeril M, Salen P, Martin JL, Monjaud I, Delaye J, Mamelle N. Mediterranean diet, traditional risk factors, and the rate of cardiovascular complications after myocardial infarction. Final report of the Lyon Diet Heart Study. Circulation. 1999;99:779–85.

65. Tanaka K, Ishikawa Y, Yokoyama M, et al. Reduction in the recurrence of stroke by eicosapentaenoic acid for hypercholesterolemic patients. Stroke. 2008;39:2052–8.

66. Charnock JS. Lipids and cardiac arrhythmia. Prog Lipid Res. 1994;33:355–85.

67. Nair SSD, Leitch JW, Falconer J, Garg ML. Prevention of cardiac arrhythmia by dietary (n-3) polyunsaturated fatty acids and their mechanism of action. J Nutr. 1997;127:383–93.

68. Anand RG, Alkadri M, Lavie CJ, Milani RV. The role of fish oil in arrhythmia prevention. J Cardiopulm Rehabil Prev. 2008;28:2–8.
69. Lee KW, Lip GYH. The role of omega-3 fatty acids in the secondary prevention of cardiovascular disease. Q J Med. 2003;96:465–80.
70. Raitt MH, Connor WE, Morris C, et al. Fish oil supplementation and risk of ventricular tachycardia and ventricular fibrillation in patients with implantable defibrillators: a randomized controlled trial. JAMA. 2005;293:2884–91.
71. Leaf A, Albert CM, Josephson M, et al. Prevention of fatal arrhythmias in high-risk subjects by fish oil n-3 fatty acid intake. Circulation. 2005;112:2762–8.
72. Brouwer IA, Zock PL, Camm AJ, et al. Effect of fish oil on ventricular tachyarrhythmia and death in patients with implantable cardioverter defibrillators: the Study on Omega-3 Fatty Acids and Ventricular Arrhythmia (SOFA) randomized trial. JAMA. 2006;295:2613–9.
73. Jenkins DJ, Josse AR, Beyene J, et al. Fish-oil supplementation with implantable cardioverter defibrillators: a meta-analysis. CMAJ. 2008;178:157–64.
74. Harris WS, Connor WE, Alam N, Illingworth DR. Reduction of postprandial triglyceridaemia in humans by dietary n-3 fatty acids. J Lipid Res. 1988;29:1451–60.
75. Grimsgaard S, Bonaa KH, Hansen JB, Nordoy A. Highly purified eicosapentaenoic acid and docosahexaenoic acid in humans have similar triacylglycerol-lowering effects but divergent effects on serum fatty acids. Am J Clin Nutr. 1997;66:649–59.
76. Harris WS, Miller M, Tighe AP, Davidson MH, Schaefer EJ. Omega-3 fatty acids and coronary heart disease risk: clinical and mechanistic perspectives. Atherosclerosis. 2008;197:12–24.
77. Howe PR. Dietary fats and hypertension: focus on fish oil. Ann N Y Acad Sci. 1997;827:339–52.
78. Geleijnse JM, Giltay EJ, Grobbee DE, Donders AR, Kok FJ. Blood pressure response to fish oil supplementation: metaregression analysis of randomized trials. J Hypertens. 2002;20:1493–9.
79. Cicero AF, Ertek S, Borghi C. Omega-3 polyunsaturated fatty acids: their potential role in blood pressure prevention and treatment. Curr Vasc Pharmacol. 2009;7:330–7.
80. Mori TA. Omega-3 fatty acids and hypertension in humans. Clin Exp Pharmacol Physiol. 2006;33:842–6.
81. Agren JJ, Vaisanen S, Hanninen O, Muller AD, Hornstra G. Hemostatic factors and platelet aggregation after a fish-enriched diet or fish oil or docosahexaenoic acid supplementation. Prostaglandins Leukot Essent Fatty Acids. 1997;57:419–21.
82. Mori TA, Beilin LJ, Burke V, Morris J, Ritchie J. Interactions between dietary fat, fish and fish oils and their effects on platelet function in men at risk of cardiovascular disease. Arterioscler Thromb Vasc Biol. 1997;17:279–86.
83. He K. Fish, long-chain omega-3 polyunsaturated fatty acids and prevention of cardiovascular disease—eat fish or take fish oil supplement? Prog Cardiovasc Dis. 2009;52:95–114.
84. Shahar E, Folsom AR, Wu KK, et al. Associations of fish intake and dietary n-3 polyunsaturated fatty acids with a hypocoagulable profile. The Atherosclerosis Risk in Communities (ARIC) study. Arterioscler Thromb. 1993;13:1205–12.
85. Archer SL, Green D, Chamberlain M, Dyer AR, Liu K. Association of dietary fish and n-3 fatty acid intake with hemostatic factors in the coronary artery risk development in young adults (CARDIA) study. Arterioscler Thromb Vasc Biol. 1998;18:1119–23.
86. Johansen O, Brekke M, Seljeflot I, Abdelnoor M, Arnesen H. N-3 fatty acids do not prevent restenosis after coronary angioplasty: results from the CART study. Coronary Angioplasty Restenosis Trial. J Am Coll Cardiol. 1999;33:1619–26.
87. Marckmann P, Bladbjerg EM, Jespersen J. Dietary fish oil (4 g daily) and cardiovascular risk markers in healthy men. Arterioscler Thromb Vasc Biol. 1997;17:3384–91.
88. Abe Y, El-Masri B, Kimball KT, et al. Soluble cell adhesion molecules in hypertriglyceridaemia and potential significance on monocyte adhesion. Arterioscler Thromb Vasc Biol. 1998;18:723–31.
89. Seljeflot I, Arnesen H, Brude IR, Nenseter MS, Drevon CA, Hjermann I. Effects of omega-3 fatty acids and/or antioxidants on endothelial cell markers. Eur J Clin Invest. 1998;28:629–35.

90. Thies F, Garry JMC, Yaqoob P, et al. Association of n-3 polyunsaturated fatty acids with stability of atherosclerotic plaques: a randomized controlled trial. Lancet. 2003;361:477–85.
91. Brown A, Hu F. Dietary modulation of endothelial function: implications for cardiovascular disease. Am J Clin Nutr. 2001;73:673–86.
92. Skulas-Ray AC, Kris-Etherton PM, Harris WS, et al. Dose-response effects of omega-3 fatty acids on triglycerides, inflammation, and endothelial function in healthy persons with moderate hypertriglyceridemia. Am J Clin Nutr. 2011;93:243–52.
93. De Caterina R, Libby P. Control of endothelial leukocyte adhesion molecules by fatty acids. Lipids. 1996;31:S57–63.
94. De Caterina R, Liao JK, Libby P. Fatty acid modulation of endothelial activation. Am J Clin Nutr. 2000;71:213S–23.
95. Lopez-Garcia E, Schulze MB, Manson JE, et al. Consumption of (n-3) fatty acids is related to plasma biomarkers of inflammation and endothelial activation in women. J Nutr. 2004;134:1806–11.
96. He K, Liu K, Daviglus ML, et al. Associations of dietary long-chain n-3 polyunsaturated fatty acids and fish with biomarkers of inflammation and endothelial activation (from the Multi-Ethnic Study of Atherosclerosis [MESA]). Am J Cardiol. 2009;103:1238–43.
97. Endres S, von Schacky C. n-3 Polyunsaturated fatty acids and human cytokine synthesis. Curr Opin Lipidol. 1996;7:48–52.
98. Trebble TM, Wootton SA, Miles EA, et al. Prostaglandin E_2 production and T cell function after fish-oil supplementation: response to antioxidant cosupplementation. Am J Clin Nutr. 2003;78:376–82.
99. Hjerkinn EM, Seljeflot I, Ellingsen I, et al. Influence of long-term intervention with dietary counseling, long-chain n-3 fatty acid supplements, or both on circulating markers of endothelial activation in men with long-standing hyperlipidemia. Am J Clin Nutr. 2005;81: 583–9.
100. Wall R, Ross RP, Fitzgerald GF, Stanton C. Fatty acids from fish: the anti-inflammatory potential of long chain omega-3 fatty acids. Nutr Rev. 2010;68:280–9.
101. Simopoulos AP. Omega-3 fatty acids in inflammation and autoimmune disease. J Am Coll Nutr. 2002;21:495–505.
102. Brouwer IA, Katan MB, Zock PL. Dietary alpha-linolenic acid is associated with reduced risk of fatal coronary heart disease, but increased prostate cancer risk: a meta-analysis. J Nutr. 2004;134:919–22.
103. Djousse L, Folsom AR, Province MA, Hunt SC, Ellison RC. Dietary linolenic acid and carotid atherosclerosis: the National Heart, Lung and Blood Institute Family Heart Study. Am J Clin Nutr. 2003;77:819–25.
104. Albert CM, Oh K, Whang W, et al. Dietary-linolenic acid intake and risk of sudden cardiac death and coronary heart disease. Circulation. 2005;112:3232–8.
105. Campos H, Baylin A, Willett WC. α-Linolenic acid and risk of non-fatal acute myocardial infarction. Circulation. 2008;118:339–45.
106. Natvig H, Borchgrevink CF, Dedichen J, Owren PA, Schiotz EH, Westlund K. A controlled trial of the effect of linolenic acid on the incidence of coronary heart disease. Scand J Clin Lab Med. 1968;105:S1–20.
107. Bemelmans WJ, Broer J, Feskens EJ, et al. Effect of an increased intake of alpha-linolenic acid and group nutritional education on cardiovascular risk factors: the Mediterranean Alpha-linolenic Enriched Groningen Dietary Intervention (MARGARIN) study. Am J Clin Nutr. 2002;75:221–7.
108. Lee JH, O'Keefe JH, Lavie CJ, Harris WS. Omega-3 fatty acids: cardiovascular benefits, sources and sustainability. Nat Rev Cardiol. 2009;6:753–8.
109. Scientific Advisory Committee on Nutrition. Advice on fish consumption: benefits and risks. London: Stationary Office; 2004. www.food.gov.uk/multimedia/pdfs/fishreport200401.pdf. Accessed 13 Sept 2010.
110. Oken E, Belfort MB. Fish, fish oil, and pregnancy. JAMA. 2010;304:1717–8.
111. Mozaffarian D, Rimm EB. Fish intake, contaminants, and human health: evaluating the risks and benefits. JAMA. 2006;296:1885–99.

112. Horvath A, Koletzko B, Szajewska H. Effect of supplementation of women in high-risk pregnancies with long-chain polyunsaturated fatty acids on pregnancy outcomes and growth measures at birth: a meta-analysis of randomized controlled trials. Br J Nutr. 2007;98:253–9.
113. Makrides M, Duley L, Olsen SF. Marine oil, and other prostaglandin precursor, supplementation for pregnancy uncomplicated by preeclampsia or intrauterine growth restriction. Cochrane Database Syst Rev. 2006;3:CD003402.
114. Mendez MA, Plana E, Guxens M, et al. Seafood consumption in pregnancy and infant size at birth: results from a prospective Spanish cohort. J Epidemiol Community Health. 2010;64:216–22.
115. Oken E, Bellinger DC. Fish consumption, methylmercury and child neurodevelopment. Curr Opin Pediatr. 2008;20:178–83.
116. Hibbeln JR, Davis JM, Steer C, et al. Maternal seafood consumption in pregnancy and neurodevelopmental outcomes in childhood (ALSPAC study): an observational cohort study. Lancet. 2007;369:578–85.
117. Terry PD, Rohan TE, Wolk A. Intakes of fish and marine fatty acids and the risks of cancers of the breast and prostate and of other hormone-related cancers: a review of the epidemiologic evidence. Am J Clin Nutr. 2003;77:532–43.
118. Rose DP, Connolly JM. Omega-3 fatty acids as cancer chemopreventive agents. Pharmacol Ther. 1999;83:217–44.
119. Szymanski KM, Wheeler DC, Mucci LA. Fish consumption and prostate cancer risk: a review and meta-analysis. Am J Clin Nutr. 2010;92:1223–33.
120. MacLean CH, Newberry SJ, Mojica WA, et al. Effects of omega-3 fatty acids on cancer risk: a systematic review. JAMA. 2006;295:403–15.
121. Fearon KC, Barber MD, Moses AG, et al. Double blind, placebo-controlled, randomized study of eicosapentaenoic acid diester in patients with cancer cachexia. J Clin Oncol. 2006;24:3401–7.
122. Dewey A, Baughan C, Dean TP, Higgins B, Johnson I. Eicosapentaenoic acid (EPA, an omega-3 fatty acid from fish oils) for the treatment of cancer cachexia. Cochrane Database Syst Rev. 2007;1:CD004597.
123. Friedland RP. Fish consumption and the risk of Alzheimer disease. Is it time to make dietary recommendations? Arch Neurol. 2003;60:923–4.
124. Conquer JA, Tierney MC, Zecevic J, Bettger WJ, Fisher RH. Fatty acid analysis of blood plasma of patients with Alzheimer's disease, other types of dementia, and cognitive impairment. Lipids. 2000;35:1305–12.
125. Schaefer EJ, Bongard V, Beiser AS, et al. Plasma phosphatidylcholine docosahexaenoic acid content and risk of dementia and Alzheimer disease: The Framingham Heart Study. Arch Neurol. 2006;63:1545–50.
126. Morris MC, Evans DA, Bienias J, et al. Consumption of fish and n-3 fatty acids and risk of incident Alzheimer disease. Arch Neurol. 2003;60:940–6.
127. Quinn JF, Raman R, Thomas RG, et al. Docosahexaenoic acid supplementation and cognitive decline in Alzheimer disease: a randomized trial. JAMA. 2010;304:1903–11.
128. Coon KD, Myers AJ, Craig DW. A high-density whole-genome association study reveals that APOE is the major susceptibility gene for sporadic late-onset Alzheimer's disease. J Clin Psychiatry. 2007;68:613–8.
129. Barberger-Gateau P, Raffaitin C, Letenneur L, et al. Dietary patterns and risk of dementia: the Three-City cohort study. Neurology. 2007;69:1921–30.
130. Huang TL, Zandi PP, Tucker KL, et al. Benefits of fatty fish on dementia risk are stronger for those without APOE epsilon4. Neurology. 2005;65:1409–14.
131. Rennie KL, Hughes J, Lang R, Jebb SA. Nutritional management of rheumatoid arthritis: a review of the evidence. J Hum Nutr Diet. 2003;16:97–109.
132. Cleland LG, James MJ, Proudman SM. The role of fish oils in the treatment of rheumatoid arthritis. Drugs. 2003;63:845–53.
133. Calder P. Joint Nutrition Society and Irish Nutrition and Dietetic Institute Symposium on 'Nutrition and autoimmune disease' PUFA, inflammatory process and rheumatoid arthritis. Proc Nutr Soc. 2008;67:409–18.

134. Hu FB, Cho E, Rexrode KM, Albert CM, Manson JE. Fish and long-chain omega-3 fatty acid intake and risk of coronary heart disease and total mortality in diabetic women. Circulation. 2003;107:1852–7.
135. Djousse L, Gaziano JM, Buring JE, Lee I-M. Dietary omega-3 fatty acids and fish consumption and risk of type 2 diabetes. Am J Clin Nutr. 2011;93:143–50.
136. Vessby B, Karlsrom B, Boberg M, Lithell H, Berne C. Polyunsaturated fatty acids may impair blood glucose control in type 2 diabetic patients. Diabet Med. 1992;9:126–33.
137. Hartweg J, Perera R, Montori VM, et al. Omega-3 polyunsaturated fatty acids (PUFA) for type 2 diabetes mellitus (review). Cochrane Database Syst Rev. 2008;1:CD003205.
138. Friedberg CE, Janssen MJ, Heine RJ, Grobbee DE. Fish oil and glycemic control in diabetes. A meta-analysis. Diabetes Care. 1998;21:494–500.
139. Simopoulos AP. Evolutionary aspects of omega-3 fatty acids in the food supply. Prostaglandins Leukot Essent Fatty Acids. 1999;60:421–9.
140. Harris WS, Ginsberg HN, Arunakul N, et al. Safety and efficacy of Omacor in severe hypertriglyceridaemia. J Cardiovasc Risk. 1997;4:385–91.
141. De Deckere EAM. Health effects of fish and n-3 polyunsaturated fatty acids from plant and marine origin. In: Wilson T, Temple NJ, editors. Nutritional health: strategies for disease prevention. Totowa, NJ: Humana Press; 2001. p. 195–206.
142. Lox CD. The effects of dietary marine oils (omega-3 fatty acids) on coagulation profiles in men. Gen Pharmacol. 1990;21:241–6.
143. Harris WS. N-3 fatty acids and serum lipoproteins: human studies. Am J Clin Nutr. 1997; 65:1645S–54.
144. Hau MF, Smelt AH, Bindels AJ, et al. Effects of fish oil on oxidation resistance of VLDL in hypertriglyceridemic patients. Arterioscler Thromb Vasc Biol. 1996;16:1197–202.
145. Sorensen NS, Marckmann P, Hoy CE, van Duyvenvoorde W, Princen HM. Effect of fish-oil-enriched margarine on plasma lipids, low-density-lipoprotein particle composition, size, and susceptibility to oxidation. Am J Clin Nutr. 1998;68:235–41.
146. Brude IR, Drevon CA, Hjermann I, et al. Peroxidation of LDL from combined-hyperlipidemic male smokers supplied with omega-3 fatty acids and antioxidants. Arterioscler Thromb Vasc Biol. 1997;17:2576–88.
147. Bonanome A, Biasia F, De Luca M, et al. n-3 Fatty acids do not enhance LDL susceptibility to oxidation in hypertriacylglycerolemic hemodialyzed subjects. Am J Clin Nutr. 1996;63: 261–6.
148. Higdon JV, Du SH, Lee YS, Wu T, Wander RC. Supplementation of postmenopausal women with fish oil does not increase overall oxidation of LDL ex vivo compared to dietary oils rich in oleate and linoleate. J Lipid Res. 2001;42:407–18.
149. Food Standards Agency. Dietary recommendations for fish and shellfish. http://tna.europarchive.org/20110116113217/cot.food.gov.uk/pdfs/fishreport2004full.pdf. Accessed 8 July 2011.
150. SeaChoice. Canada's seafood guide. http://www.seachoice.org. Accessed 15 Sept 2011.
151. Lee DH, Jacobs DR. Inconsistent epidemiological findings on fish consumption may be indirect evidence of harmful contaminants in fish. J Epidemiol Community Health. 2010;64: 190–2.

Chapter 11
Hypertension and Nutrition

Ali Zirakzadeh

Keywords Hypertension • Blood pressure • Sodium • DASH diet

Key Points

- Currently, normal blood pressure (BP) is defined as a systolic and diastolic BP of <120/80 mmHg. Prehypertension is considered a systolic or diastolic BP of 120–139/80–89 mmHg. Hypertension (HTN) is defined as a systolic BP>140 mmHg, a diastolic BP>90 mmHg, or under blood pressure lowering treatment.
- Seventy-four million people living in the United States, including 65% over the age of 65, have HTN.
- HTN is responsible for an estimated 395,000 annual deaths from cardiovascular disease, including coronary artery disease, heart failure, and cerebrovascular accidents.
- Individuals with prehypertension are at increased risk for developing HTN. They also have a higher risk of cardiovascular disease.
- Many cases of HTN are due to subtle renal injury caused by long-term lifestyle choices, such as a high-salt diet and increased weight.
- Lifestyle modifications have been associated with significant BP reductions. These lifestyle changes include salt reduction, weight loss, exercise, limited alcohol consumption, diets high in fruit, vegetables, and grains, but low in sweetened foods.

A. Zirakzadeh, MD, MS (✉)
Denver Health and Hospital Authority, University of Colorado Denver-UCHSC,
777 Bannock St., Denver, CO 80204, USA
e-mail: ali.zirakzadeh@dhha.org

N.J. Temple et al. (eds.), *Nutritional Health: Strategies for Disease Prevention*,
Nutrition and Health, DOI 10.1007/978-1-61779-894-8_11,
© Springer Science+Business Media, LLC 2012

11.1 Introduction

No other chronic medical condition in the United States matches hypertension (HTN) in terms of prevalence and long-term health consequences. HTN will affect the majority of people living in the United States at some point. It results in more morbidity and mortality than any other preventable cause of death, with the exception of cigarette smoking. Yet, despite its high prevalence, our knowledge regarding the mechanisms underlying hypertensive disease remains incomplete. What we have learned, however, through epidemiologic and clinical studies, is that a number of environmental factors influence the development of HTN. Moreover, we have come to understand that specific dietary and lifestyle changes can both prevent and treat high blood pressure (BP).

In this chapter we review the epidemiology of hypertensive disease and discuss the spectrum of dietary factors known to lead to HTN. In addition, we examine the evidence for nonpharmacologic and dietary treatment strategies. Finally, we briefly review the more recently designated risk category, prehypertension, and consider its consequences.

11.2 Epidemiology

The Seventh Report of the Joint National Committee on Prevention, Detection, Evaluation, and Treatment of High Blood Pressure (JNC 7) defines HTN as a BP of greater than 140 mmHg systolic, 90 mmHg diastolic, or under blood pressure lowering treatment [1]. In the United States, approximately 74 million people (one in three) over the age of 20 have HTN and more than half of this group suffers from uncontrolled disease [2].

Importantly, the frequency of HTN has been shown to increase with age and vary by race and ethnicity. While less than 4% of those aged 18–29 in the United States have HTN [3], this figure rises to greater than 65% of all persons over the age of 65 (Fig. 11.1) [4]. Notably, African Americans have the highest rates of HTN (43% in men, 45% in women). Non-Hispanic whites (34% of men, 31% of women) and Mexican Americans (26% of men, 32% of women) have substantially lower rates [2]. Studies have also shown that hypertensive African Americans tend to average higher mean BPs as compared to hypertensive non-Hispanic whites and Mexican Americans (Fig. 11.2). Moreover, African Americans show significantly higher rates of severe HTN (BP >180/110 mmHg) as compared to population controls (8.5% vs. <1%) [5]. Surprisingly, however, economic status appears to have little predictive value in determining rates of HTN. While 35% of Americans living under the poverty line have HTN, Americans who earn at least 200% above the poverty line show a similarly high, 30%, prevalence rate [6].

Research has demonstrated that HTN exacts a high toll on human health resulting in a wide range of adverse outcomes. As a primary example, an estimated

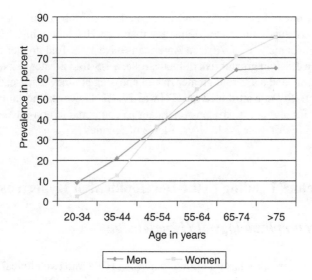

Fig. 11.1 Prevalence of hypertension by age. Adapted from ref. [6]

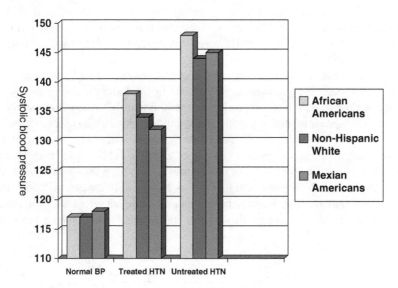

Fig. 11.2 Mean systolic blood pressures in mmHg for normotensives, and treated and untreated hypertensives in the general population. Adapted from ref. [5]

395,000 Americans die annually from the cardiovascular complications of HTN [7]. These complications include coronary artery disease events, heart failure, cerebrovascular accidents (including stroke), peripheral artery disease, and aortic aneurysms. Additionally, HTN is present in 69% of patients experiencing their first heart attack, 77% of patients during their first stroke, and 74% of patients at the time of their first episode of heart failure. Furthermore, these complications appear to affect African Americans with greater frequency than non-Hispanic whites. In comparison to non-Hispanic whites, African Americans have a 1.3-fold greater risk of non-fatal stroke, a 1.8-fold greater risk of fatal stroke, a 1.5-fold greater risk of death due to heart disease, and a 4.2-fold greater risk of end-stage kidney disease. Overall, the 2008 US mortality rate attributed to HTN was 17.8%. However, mortality attributed to HTN was far higher for African Americans as compared to non-Hispanic whites: 51.1% vs. 15.6% of males, 37.7% vs. 14.3% of females [8].

11.3 Factors Leading to the Development of Hypertension

11.3.1 Pathophysiology of Hypertension

The medical community has traditionally divided HTN into two clinical categories: essential and secondary. This chapter focuses on essential HTN, also known as primary HTN. This condition comprises up to 90% of all cases of HTN. In contrast, secondary HTN results from a heterogeneous group of disorders which lead to the development of elevated BP (Table 11.1). Further discussion of secondary HTN is beyond the scope of this chapter. However, Acelajado et al. [9] provides an up-to-date review on the subject.

Scientists have yet to fully describe the pathophysiology of essential HTN. Evidence suggests HTN results from a defect in the kidney's ability to regulate sodium leading to fluid retention and an eventual rise in BP. A small proportion of clinical cases appear related to genetic abnormalities which cause excessive sodium reabsorption by the kidney [10]. Additionally, there is research to suggest that maternal factors influence the development of essential HTN in offspring. Women with HTN, obesity, malnutrition, or preeclampsia have been shown to deliver children with a low birth weight at a higher rate as compared to the general population. These offspring have been found to have impaired kidney development in utero and a decreased number of kidney cells available for sodium regulation after birth [10].

Table 11.1 Secondary causes of hypertension

- Underlying kidney disease
- Obstructive sleep apnea
- Renal artery stenosis
- Primary aldosteronism
- Pheochromocytoma
- Cushing's disease
- Hyperparathyroidism
- Coarctation of the aorta
- Intracranial tumors

Adapted from ref. [9]

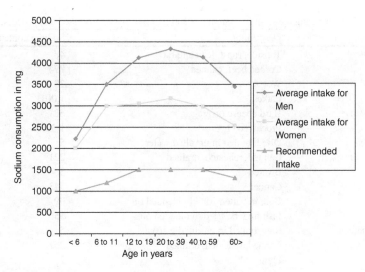

Fig. 11.3 Average daily sodium intake (mg) by age (years). Adapted from ref. [13, 15]

However, most cases of HTN appear to be related to subtle kidney injury occurring over many years, the result of lifestyle and diet. In support of this concept, early autopsy studies have revealed that virtually all kidneys examined from patients with HTN show microscopic damage within vessels and tubules [11]. This type of microscopic damage has been shown to result in sodium and fluid retention which in turn leads to HTN [12].

11.3.2 The Role of Sodium, Kidney Injury, and the Development of Hypertension

Humans require less than 400 mg/day of sodium (1,000 mg of salt) for proper physiologic functioning. The National Academy of Sciences (NAS) recommends that sodium consumption not exceed 1,500 mg/day in adults, and 1,300 mg/day in those over the age of 70. The NAS also sets 2,300 mg/day as the nutritional upper limit of sodium consumption in all adults, with levels below this threshold thought likely to pose no significant risk of adverse events. These recommendations have been adopted by both the United States Department of Agriculture and the JNC 7 [1, 13]. In contrast, however, typical sodium consumption in American adults is, in fact, much higher, ranging from 3,500 to 4,300 mg/day for males and 2,500 to 3,200 mg/day for females [14]. Research has shown that these high-sodium diets are usually adopted early in life and increase steadily with age. Even before the age of 6, children in the United States typically consume twice the recommended level (Fig. 11.3) [15].

Processed or restaurant-prepared food appears to be primarily responsible for the majority of excess sodium consumption in the United States. In fact, processed

Table 11.2 Comparing the sodium content per 100 g (3.5 oz) of unprocessed and processed food

Food item	Description	Sodium content (mg)
Beef	Topside, roast, lean and fat	48
	Corned beef, canned	874
Bran	Wheat	28
	Flakes	1,001
Cheese	Hard cheese (average)	621
	Processed cheese	1,320
Chick peas	Dried, boiled in unsalted water	5
	Canned, reheated, drained	221
Crab	Broiled	384
	Canned	543
Cod	Cod, in batter, fried in blended oil	99
	Fish fingers, fried in blended oil	350
New Potatoes	Raw, boiled in unsalted water	5
	Canned, reheated, drained	251
Peanuts	Plain	2
	Dry roasted	400
Peas	Raw, boiled in unsalted water	Trace
	Canned, reheated, drained	251
French fries	Home-made, fried in blended oil	12
	Oven fries, frozen, baked	53
Salmon	Raw, steamed	110
	Smoked	1,863
Sweet corn	On the cob, boiled in unsalted water	Trace
	Kernels, canned, reheated, drained	269
Tuna	Raw	46
	Canned in oil, drained	290

Dietary sodium measured in mg. Each dietary serving = 100 g (3.2 oz).
Adapted from ref. [14]

foods alone provide an impressive 75% of all sodium consumed in the United States. Studies have shown that processed meals tend to contain significantly more sodium than fresh or home prepared foods. For example, a home-prepared steak and fries has 12 mg of sodium per 100 g (3.2 oz) serving, while similar items at a chain steakhouse may contain as much as 53 mg of sodium per 100 g serving. As a second example, homemade risotto may have as little as 20 mg of sodium per serving, while "ready-made" risotto may contain up to 60 times more sodium. Further examples of similar discrepancies in sodium content between processed and home-prepared meals are listed in Table 11.2 [14].

The link between chronic high sodium ingestion and elevated BP has, in fact, long been established. The 1988 INTERSALT study evaluated the BP effects of sodium intake in over 10,000 participants in 52 international communities. Average sodium intake in this study varied widely, from less than 1,200 mg/day in some communities, up to 10,000 mg/day in others. Significantly, the study concluded that, on average, an increase in sodium intake by 2,400 mg/day over a 30 year period will increase systolic BP by 9 mmHg [16].

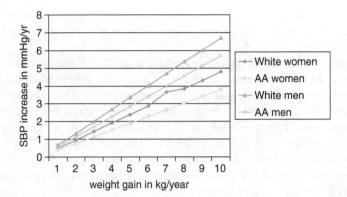

Fig. 11.4 Mean annual systolic increase in blood pressure for every kg gained per year. Adapted from ref. [23]

Sodium intake early in life also appears to influence BP in both the immediate and longer-term follow-up intervals. A study of 476 newborn Dutch infants compared BP effects in infants fed for 6 month on a low-sodium infant diet vs. a usual infant diet. Not only did infants fed the low-sodium diet tend to have lower BPs than the control group at the end of 6 month, but this difference also persisted at the 15-year follow-up [17, 18].

Chronic high sodium intake is now known to lead to kidney injury. These findings have been documented in animal studies. In laboratory experiments with rats, researchers have shown that high-sodium diets result in kidney inflammation. This inflammation leads to damage involving the lining of the renal vasculature, increased extracellular matrix formation, and decreased renal cell survival [19, 20]. Notably, the resulting injury has been shown to lead to a reduction in the ability of the kidney to excrete sodium and results in eventual fluid retention. In this animal model, increased BP eventually becomes necessary to facilitate the excretion of excess sodium and fluid from the damaged kidneys.

11.3.3 The Role of Obesity in the Development of Hypertension

The Centers for Disease Control (CDC) reports that 26.7% of Americans currently meet criteria for the definition of obesity (body mass index [BMI] > 30 kg/m²), with even higher rates (36.8%) seen in the African American population [21]. Significantly, obesity is an important risk factor for the development of HTN. Data from the Framingham study showed that obesity may contribute to 75% of cases of essential HTN in both genders [22]. Additionally, epidemiologic studies from diverse populations show a linear correlation between weight and BP. One study involving 9,309 subjects showed that annual weight gain is directly associated with an elevated risk of HTN. Moreover, this trend persists across race and gender, with Caucasian American men showing the strongest association (Fig. 11.4) [23].

A second large population study reported that for every 4.5 cm increase in waist circumference in males and 2.4 cm in females there is a corresponding 1 mmHg increase in systolic BP [24].

The pathophysiology of obesity-related HTN appears is complex and not fully understood. Multiple potential mechanisms have been proposed to explain the interactions between weight gain and the development of HTN. One line of evidence suggests that obesity increases sodium retention through activation of the renin–angiotensin–aldosterone system [25, 26]. Additionally, weight gain is also thought to activate neural pathways which lead to increased BP [27–29]. Finally, obesity-related insulin resistance may also induce elevations in BP [30, 31]. Further research will be necessary to better understand the important link between weight gain and BP.

11.3.4   Other Factors Contributing to the Development of Hypertension

Current evidence also supports an association between dietary consumption of animal protein, alcohol, sugar, and high BP. Studies have shown that Western diets, with their high content of animal protein, contain large quantities of acidic precursors. The elevation of acidic contents may lead to an acidic metabolic state which, in turn, is thought to influence BP. As evidence, previous animal studies have shown an association between increased serum acidity and BP [32]. More recently, a study of 5,043 nonhypertensive adults showed a link between serum acidosis and increases in BP [33]. To further evaluate the association between dietary animal protein and BP, researchers followed 87,293 adult females with no known history of HTN for 16 year as part of the Nurses Health Study II. By the completion of the study, 15,385 of them were found to have developed HTN. After adjusting for multiple potential confounding factors, intake of high levels of animal protein was found to be an independent risk factor for the development of HTN [34]. However, the mechanism by which acidosis leads to HTN remains unclear.

Alcohol consumption has also been shown to lead to increased BP. Large population studies dating back to the 1980s have shown a linear relationship between alcohol consumption and BP [35]. Importantly, this trend appears to be influenced by gender, race and ethnicity. For example, several studies have shown that the risk of HTN is decreased for adult females who consume one or fewer drinks per day [36–38]. However, a separate study found an association between alcohol consumption and 6-year incidence of HTN, but only for African American men (at all levels of weekly alcohol consumption) and Caucasian women who consume more than 210 g/week of alcohol (approximately 2.5 drinks/day) [39]. To better define this complex association, Taylor et al. [40] conducted a systematic review and meta-analysis of the best available evidence regarding BP and alcohol consumption. The authors showed a linear association between alcohol consumption and incident HTN in men, but a slight J-shaped association in women. This finding was interpreted to be consistent with a small decrease in the prevalence of HTN for adult females who consume less than one alcoholic beverage per day and an increased risk of HTN was observed for females who consume higher levels of alcohol. Interestingly, the strongest association between alcohol consumption and BP was observed for Asian men (Korean and Japanese). The observed gender differences in BP outcomes are thought to reflect behavioral tendencies as adult males exhibit higher rates of binge drinking as compared to females. However, it has been speculated that the stronger association observed for Asian males may reflect genetic differences involving key enzymes important to alcohol metabolism, resulting in increased alcohol metabolite accumulation [40]. Further studies will be necessary to better evaluate the roles of race and alcohol in the development of HTN.

Sugar consumption may also be important to the development of HTN. Notably, Americans consume approximately 18 kg more sugar today than they did 3 decades ago [41]. Multiple studies have linked consumption of fructose (which represents half the sucrose molecule) with increased lipid production and insulin resistance [42–45]. Additionally, numerous animal studies have also shown that increased fructose and sucrose consumption leads to increased BP [46–48]. Four large prospective studies which included multi-year follow-up lend support for a modest association between dietary sugar consumption and BP. The Nurses Health II study, which enrolled 116,671 females, showed that daily sugar-sweetened beverage (SSB) consumption is associated with a small but significant risk of incident HTN. A similar trend (though not reaching statistical significance) was seen in both the Nurses Health I study of 121,700 females and the Framingham study of 6,039 subjects. The recent PREMIER study of adults with a systolic BP of 120–159 mmHg showed a small but still significant drop in systolic BP (0.70 mmHg) for every 10.5-oz decrease in daily SSB intake [49].

11.3.5 Overall Lifestyle Choices and the Development of Hypertension

Multiple studies from diverse communities throughout the world have revealed that lifestyle choices significantly influence the development of HTN. Diets comprised of low levels of fruits and vegetables appear to increase the risk of developing HTN. A study of 549 Portuguese males and females over the age of 40 showed an inverse relationship between fruit/vegetable intake and HTN. The adjusted prevalence rate ratio between the third and first tertiles of fruit/vegetable consumption was 0.61 (95% confidence intervals 0.40–0.93) [50]. Furthermore, a more comprehensive US study followed 4,304 young adults (age 18–40) for 15 year and evaluated the relationship between animal/plant food intake and the development of HTN. Participants in the highest quintile of plant food intake showed a significantly lower risk of developing HTN as compared to those in the lowest quintile (hazard ratio [HR] 0.64 [0.50–0.83]). This association persisted even after multivariate analysis adjusted for several possible confounding variables. Conversely, participants in the highest quintile group of meat, poultry, and seafood consumption showed the highest prevalence of HTN (HR 1.67 [1.21–2.30]), with red and processed meats identified as key contributing factors [51].

A larger population study of 83,882 healthy women aged 27–44 year also evaluated the role of healthy lifestyle choices (diet, exercise, BMI < 25, moderate alcohol intake, minimal analgesic use, and folic acid supplementation) on the development of HTN over a 14-year period. Healthy diet, daily exercise, and BMI < 25 each predicted a significantly decreased risk of developing HTN. The greatest effect was observed for participants who adopted all six healthy lifestyle choices. For these participants, the HR for developing HTN over the next 14 year was only 0.22 (0.10–0.51) [52].

Finally, it has been generally accepted that there is correlation between low levels of physical activity and the development of HTN. A systematic review and meta-analysis of 53 studies performed in 2000 showed an inverse relationship between the intensity of physical activity and BP [53]. However, few of the studies included in this analysis were large, prospective studies. In fact, a review of several such studies revealed a more complex association. There is currently strong evidence to suggest that moderate-to-strenuous exercise may protect against the development of HTN in white males [49, 54–58]. A similar protective role was observed for Japanese males [59]. However, the only large study of African American men and women conducted to date failed to show a protective role for moderate-to-vigorous exercise [54]. Similarly, studies focusing on non-Hispanic white women yielded conflicting results [54, 56, 58, 60].

11.4   Diet in the Treatment of Hypertension

Research has shown that lifestyle choices not only lead to the development of HTN, but, importantly, that changes in lifestyle may effectively decrease BP in hypertensive patients (Table 11.3). As a result, lifestyle modifications are

Table 11.3 Efficacy of various treatments for hypertension

Intervention	Efficacy (measurements 0, ±, +, ++, +++, ++++, +++++)
Usual lifestyle	0
Potassium, calcium supplementation	±
Dairy products	±
Salt restriction	+
Weight reduction	+ to ++
Alcohol reduction	+
Fiber supplementation	+
DASH study diet	++ to +++
DASH study diet plus salt restriction	++ to ++++
Combination lifestyle changes	++ to ++++
Medication	+++ to ++++
Medication plus lifestyle changes	+++ to +++++

recommended by the JNC 7 for all patients with HTN. Recommendations focus on diet, exercise, weight-loss goals, and reductions of alcohol consumption. Weight loss is recommended to achieve a BMI of 18.5–25. Dietary modifications include the following: increased fruits, vegetables, whole grains, and nuts, consumption of low-fat dairy products, and reduction of dietary fats (Dietary Approach to Stop Hypertension [DASH] eating plan). Dietary sodium intake should be reduced to no more than 2,400 mg/day. Patients with HTN are advised to engage in 30 min of aerobic exercise for at least 4 day/week. Finally, alcohol consumption should be restricted to no more than 2 drinks/day for men and one drink/day for women. In clinical studies these interventions have been shown to decrease systolic BP by 2–20 mmHg [1].

11.4.1 Sodium Restriction

Strong evidence supports the adoption of dietary sodium restriction by patients with HTN. In an early influential study examining the effects of sodium restriction, two communities with similar rates of HTN (30%) and sodium intake (~8,500 mg/day) were studied. While one group was treated with a sodium-restricted diet (50% reduction), the other continued with their usual diet. After 2 year the sodium-restricted community experienced an average drop of 5.0/5.1 mmHg in BP while the control group showed an average *increase* of 9 mmHg in systolic pressure, with no change in diastolic pressure [61]. Subsequently, a large number of clinical studies have examined the efficacy of sodium restriction for the treatment of HTN. These studies were analyzed in a meta-analysis of 28 trials of 2,954 individuals randomized to either sodium restriction (decrease from 4,000 to 2,000 mg/day) or usual diet. Participants with HTN experienced an average BP decrease of 5.0/2.8 mmHg with sodium reduction. Furthermore, a BP benefit was also seen in nonhypertensive participants (average drop of 2.0/1.0 mmHg) [62].

Additionally, research has shown that BP reduction appears proportional to sodium reduction. A study of 412 participants showed a 2.1 mmHg drop in systolic BP following the reduction of daily sodium intake from 3,200 to 2,400 mg/day. Further reduction of sodium to 1,600 mg/day resulted in an extra 4.6 mmHg decrease in systolic BP. Corresponding diastolic BP drops were also noted: 1.1 mmHg (from 3,200 to 2,400 mg/day) and 2.4 mmHg (from 2,400 to 1,600 mg/day) [63].

Importantly, large-scale achievement of BP reductions with simple dietary sodium restriction is predicted to have widespread impact on public health. In the United States it is estimated that a 1,200 mg/day reduction of dietary sodium would result in 51,000–81,000 fewer annual deaths. It would also be expected to decrease the annual incidence of coronary artery disease by up to 110,000 and stroke by as much as 40,000. Resulting annual health-care cost savings could reach an estimated 10–24 billion dollars [64].

11.4.2   Weight Reduction

Clinical studies have consistently shown that weight loss resulting from diet and exercise leads to improvements in BP. Weight loss ranging from 2 to 12 kg has been associated with reductions in systolic BP ranging from 4 to 17 mmHg (in a dose-dependent manner). This trend persists regardless of initial BMI (25–30, >30), BP (hypertensive or prehypertensive), or age (adult or child) [65–72].

In cases where dietary interventions appear to fail, gastric bypass surgery has also been shown to lead to significant long-term weight loss and BP reduction. Interestingly, gastric bypass patients with preexisting HTN have been shown to experience large declines in BP, with systolic BP reductions of up to 20 mmHg [73–75].

Additionally, weight loss appears to be a particularly effective tool for the treatment of obese children with HTN. One previous study enrolled 184 obese children ages 6–18 who had a mean starting BP of 130/82 mmHg. After a 6-month weight-loss intervention, which included diet and exercise, participants experienced an average BP reduction of 17/13 mmHg. The degree of BP reduction was directly proportional to weight loss, decrease in BMI, and waist circumference [65]. Bariatric surgery has also been shown to benefit severely obese adolescents with HTN. To date, four clinical studies have shown 50–100% resolution of HTN for obese adolescents treated with either laparoscopic banding or gastric bypass surgery. Participants in these studies experienced a reduction in BMI of up to 23 kg/m^2 [76, 77].

11.4.3   Exercise

Aerobic exercise appears to lower BP in individuals with HTN. In a review of 28 trials involving 492 subjects with a mean age of 53 year, aerobic exercise for 4–52 week was associated with a decrease in BP of 6.9/4.9 mmHg. Interestingly, this drop in BP appeared to be independent of weight loss. The subjects in the reviewed trials were overweight at baseline, with a mean BMI of 28. However, the net loss of weight during the studies was only 1.1 kg. In addition, BP reductions occurred over a wide range of exercise intensity (30–87% of heart rate reserve) [78].

A more recent study directly compared low-intensity to high-intensity aerobic exercise. For this study, 48 middle-aged subjects were randomized to 60 min of aerobic exercise three times/week for 10 week at either 33% or 66% of heart rate reserve. Higher-intensity training resulted in a modest weight loss of 1.1 kg along with a BP drop of 5.9/4.2 mmHg. However, lower-intensity training resulted in a nonsignificant weight loss of 0.2 kg, a decrease in systolic BP that was statistically similar to the higher-intensity group (4.6 mmHg), and a nonsignificant drop in diastolic BP (1.7 mmHg) [79].

The above results indicate that both low- and higher-intensity aerobic exercise lower systolic BP. Higher-intensity exercise appears to confer greater benefits. The decision to pursue a lower- or higher-intensity exercise regimen should thus be based on the patient's cardiovascular risk factors and patient preference [80].

11.4.4 Alcohol

Given the near linear relationship between alcohol consumption and BP, it is not surprising that alcohol reduction has been found to improve high BP [81]. However, moderate alcohol consumption, less than 2–3 drinks/day for men and one or fewer drinks/day for women, confers cardiovascular disease risk protection, even for patients with preexisting HTN or cardiovascular disease [82–85]. As a result, despite alcohol's role in raising BP, the JNC 7 only recommends limiting alcohol consumption to 2 or fewer drinks/day for men, and one or fewer drinks/day for women [1].

11.4.5 Fiber Intake

Treatment with fiber supplementation also appears to have a beneficial BP-lowering effect. The American Heart Association recommends consumption of approximately 30 g/day of fiber. In Western countries, however, intake tends to be much lower, around 15 g/day. The mechanism of action by which fiber reduces BP is unclear. However, it may be related to decreased absorption of sodium from the intestines into the blood. In a systematic review, Streppel et al. [86] analyzed 8 randomized, placebo-controlled trials involving 321 hypertensive participants. Multivariate regression analysis showed a moderate BP drop of 2.4/2.4 mmHg after 2–24 week of treatment with 5.5–20 g/day of fiber supplementation. Interestingly, however, fiber did not appear to have an effect on the BP of normotensive individuals. It must be noted that in these studies fiber intake may be acting as a surrogate for consumption of fiber-rich foods such as vegetables and whole grains.

11.4.6 DASH Study Diet

For many years the primary nonpharmacologic treatments for HTN included salt restriction, weight loss, and alcohol reduction. Then, in the 1990s, observation that vegetarian diets (high in fiber, fruit, and vegetables) were associated with low rates of HTN led to the DASH trial. In this study, 459 adults with elevated BP (systolic <160, diastolic 80–95 mmHg) were randomized to a standard American diet, a standard American diet plus increased fruit and vegetables, or the DASH diet. The DASH diet includes a relatively high daily content of fruit, vegetables, and grain; moderate amounts of low-fat dairy products, fats, and oils; and decreased content of meat, regular-fat dairy products, snacks, and sweets (Tables 11.4 and 11.5). All meals were premade for the participants and had similar sodium content (approximately 3,000 mg/day).

Overall, DASH diet participants achieved better BP reductions as compared to participants randomized to the standard diets. The magnitude of BP reduction was found to vary according to baseline BP, gender, and ethnicity. For hypertensive

Table 11.4 Nutrient components of the DASH study diet

Nutrients	Nutrient targets
Fat (% of total kcal)	27
Saturated	6
Monounsaturated	13
Polyunsaturated	8
Carbohydrates (% of total kcal)	55
Protein (% of total kcal)	18
Cholesterol (mg/day)	150
Fiber (g/day)	31
Potassium (mg/day)	4,700
Calcium (mg/day)	1,240
Magnesium (mg/day)	500
Sodium (mg/day)	3,000[a]

[a]Subsequent studies using DASH have used 2,400 mg/day sodium restriction

Table 11.5 Food group components of the DASH study diet

Food groups	Targets (no. of servings per day)
Fruits and Juices	5.2
Vegetables	4.4
Grains	7.5
Low-fat dairy	2.0
Regular fat dairy	0.7
Nuts, seeds, legumes	0.7
Beef, pork, and ham	0.5
Poultry	0.6
Fish	0.5
Fat, oils, and salad dressing	2.5
Snacks and sweets	0.7

Adapted from ref. [87]

subjects, the DASH diet resulted in an average systolic BP reduction of 11.4 mmHg. Normotensive individuals achieved a less dramatic but still statistically significant drop (3.5 mmHg systolic). Women averaged a higher, 6.2 mmHg, drop in systolic BP with the DASH diet as compared to the 4.9 mmHg reduction averaged by male participants. African American participants experienced more than a twofold decrease in BP with the DASH diet as compared to Caucasians (6.8 vs. 3.0 mmHg) [87]. As a result of this study, DASH diets have become standard first-line treatment for HTN in most primary-care medical settings.

11.4.7 DASH-Sodium Study

The DASH-sodium study followed soon after the initial DASH trial as an effort to better understand the combined effects of DASH dietary changes plus sodium restriction on BP control. This study enrolled 412 participants who, as in the initial study, had elevated BP (120–159/80–95 mmHg) at baseline. Participants were randomized to one of two diets, standard or DASH, and three different levels of daily sodium intake, 1,150, 2,300, or 3,450 mg/day. Both the standard + 1,150 mg/day sodium diet (very low salt) and DASH + 3,450 mg/day sodium diet (regular salt) resulted in similar BP drops (6.7 and 5.9 mmHg systolic, respectively). However, the combination of the DASH diet with sodium restriction (1,150 mg and 2,300 mg daily intake) resulted in even larger drops in BP (systolic BP drop of 8.9 and 7.2 mg Hg, respectively).

Again, the degree of BP reduction was found to vary according to baseline BP, gender, and race. Hypertensive participants on the DASH + 1,150 mg sodium restriction diet had an average reduction of 11.5 mmHg systolic. In addition, women randomized to the DASH + 1,150 mg sodium restriction diet had a greater reduction in BP as compared to men (10.5 vs. 6.8 mmHg). Hypertensive African Americans experienced the largest drops in BP (12.6 mmHg systolic) [63].

11.4.8 Combination Lifestyle Modifications

Because of evidence indicating that several lifestyle modifications successfully achieve BP reduction, subsequent clinical studies were performed to evaluate the additive effects of these interventions. The recent ENCORE trial examined the combined efficacy of weight loss, exercise, and the DASH diet on BP control. In this study, 144 middle-aged adults, mean BMI of 33, and mean BP of 138/86 mmHg were randomized to either: (1) the DASH diet, (2) the DASH diet plus lifestyle intervention (weight loss counseling and three weekly sessions of 45 min of moderate aerobic exercise), or (3) usual dietary and exercise habits. The DASH diet group showed a significant 11.2/7.5 mmHg decrease in BP. Addition of lifestyle intervention resulted in a further drop in BP (4.9/2.4 mmHg). The DASH diet plus lifestyle intervention group also showed a large reduction in weight (8.7 kg, compared with 0.3 kg in the DASH diet group and a 0.9 kg gain in the usual habits group) [88]. Thus, it is currently recommended that nonpharmacologic treatments of HTN should include multiple modalities, especially in overweight and obese individuals.

11.4.9 Substituting Pharmacotherapy with Lifestyle Modifications

Several recent studies have also suggested that carefully selected patients with well-controlled HTN may successfully discontinue antihypertensive medications following the adoption of lifestyle modifications. In one study, 945 participants with medication-controlled BP had their pharmacotherapy withdrawn 90 day after randomization into one of four lifestyle interventions: (1) salt reduction, (2) weight

Table 11.6 Predicted success rates (in %) of hypertensive patients remaining medication free with lifestyle modifications

Years on medication	Baseline systolic blood pressure (mmHg)		
	<120	120–134	≥135
No lifestyle modifications (one medication at screening)			
0–4	48.7	37.6	17.9
5–9	36.7	26.9	11.8
≥10	31	22.2	9.4
No lifestyle modifications (two medications at screening)			
0–4	38.0	28.0	12.4
5–9	27.3	19.2	8.0
≥10	22.5	15.5	6.3
Combined sodium reduction/weight loss (one medication at screening)			
0–4	89.3	84.2	65.9
5–9	83.7	76.5	54.2
≥10	79.9	71.6	47.8
Combined sodium reduction/weight loss (two medications at screening)			
0–4	84.4	77.5	55.6
5–9	76.8	67.8	43.3
≥10	71.9	61.9	37.2

Adapted from ref. [89]

reduction, (3) combination salt and weight reduction, or (4) maintenance of usual lifestyle. After 15–36 month of follow-up, only 6.3–37.6% of the subjects randomized to their usual lifestyles were able to achieve control of BP without medication. Participants found least likely to achieve BP control without medication (6.3%), had a baseline BP > 135 mmHg, were treated with at least two BP medications, and had a diagnosis of HTN for more than 10 year. In contrast, 62–89% of the combined salt and weight reduction group were able achieve BP control without medication. The most successful group (89%) had a baseline BP < 120 mmHg, were using only one medication, and had a diagnosis of HTN for <4 year. The single intervention groups (salt reduction only or weight reduction only) fared better than the usual lifestyle group, but failed to achieve the same degree of BP control as the combined intervention group (Table 11.6) [89].

Two smaller studies have also demonstrated that lifestyle intervention may be successfully substituted for medications in subjects with controlled HTN (BP < 160/95 mmHg). In these studies, subjects were randomized to either continuation of usual medication or replacement of medication with lifestyle modification (diet, exercise, and DASH diet). After 4–9 month follow-up, this study found no significant differences between subjects randomized to the two groups [90, 91]. Findings from this study suggest, therefore, that selected individuals with well-controlled BP may be able to maintain good BP control through a combination of dietary change, exercise, and weight reduction.

11.4.10 The Effect of Lifestyle Modification Plus Pharmacotherapy

Individuals treated with pharmacotherapy have been found to achieve additional BP benefit from lifestyle modification. In one Australian study, 241 overweight hypertensive subjects treated with up to two antihypertensive medications were randomized to either (1) addition of DASH diet, exercise, increased fish intake, weight loss, and decreased alcohol consumption or (2) usual lifestyle. Subjects randomized to the lifestyle intervention achieved an extra 4.2/2.1 mmHg drop in BP [91]. A similar US study involving 45 overweight individuals with medication-controlled HTN showed an even greater drop in BP with the adoption of lifestyle changes (DASH diet, hypocaloric intake, and moderate exercise). Participants in this trial achieved an additional 10.5 mmHg drop in BP as compared to the 1.1 mmHg drop averaged by the control group [92].

11.4.11 Medication vs. Lifestyle Interventions as First-Line Treatment for HTN

Despite extensive research in the field of HTN, few studies have made direct comparisons between antihypertensive medication and dietary modification. A systematic review and meta-analysis found only four studies comparing medication to dietary intervention in previously untreated hypertensive individuals. While one study observed better results with diet, another favored medication, and two trials showed similar BP reductions in both treatment groups. At this time the authors felt that there is insufficient evidence to draw conclusions regarding the superiority of dietary changes over medication for first line treatment of HTN.

11.4.12 Treatment Options Without Proven Benefits

Multiple vitamins and minerals have been BP in experimental settings. Increased potassium and calcium intake have all been shown to decrease BP in laboratory rats. However, in human clinical trials, potassium and calcium supplementation have not consistently improved BP parameters. Moreover, no conclusive evidence yet supports the adoption of diets enriched with either calcium or potassium for either the treatment or prevention of HTN. Studies to date have, in fact, yielded contradictory results [93–98]. In one of the largest studies conducted to evaluate the impact of nutrition on BP, the TOHP-1 trial, neither potassium nor calcium supplementation were associated with BP reduction [99]. The most recent systematic reviews in 2006 revealed that many of the clinical trials conducted to examine the role of calcium and potassium supplementation on the treatment of HTN were of insufficient quality to produce reliable results [100, 101]. The most recent clinical trial of potassium supplementation failed to show any effect on BP control [102]. Similarly, critical assessment of clinical trials involving vitamin D, magnesium, and combined vitamin therapy have all revealed poor and unreliable evidence to support their use in the treatment of HTN [103–105]. Hence, it is currently recommended that vitamin and mineral supplementation should not be used as a primary treatment for HTN.

11.5 Prehypertension

In 2003 the JNC 7 created a newly designated risk category, called prehypertension, defined as BP in the range 120–139/80–89 mmHg [1]. This category identifies individuals at high risk to develop HTN and related adverse health outcomes. In adults 35–64 year of age, the odds of developing HTN are 4.1 times greater with BP values of 120–129/80–84 mmHg and 11.6 times greater for BP values of 130–139/84–89, as compared to those with BP below 120/80 mmHg [106]. Although two previous large population studies have found no increase in overall mortality for individuals with prehypertension, the majority of patients included in those studies were under the age of 50 [107, 108]. Other studies, which enrolled older adults, have shown that prehypertension is associated with an increased risk of cardiovascular disease [34, 109–111]. Furthermore, the risk of cardiovascular mortality appeared increased in both men and women with BP 130–139/85–89 [107, 111]. As a result, the JNC 7 recommends lifestyle modifications and control of cardiovascular risk factors (e.g., smoking, cholesterol) for patients with prehypertension. However, medications are not currently recommended for the treatment of prehypertension.

11.6 Summary

HTN affects approximately one-third of all Americans during their lifetime, including the majority of adults over the age of 65. Moreover, it contributes to 395,000 annual deaths. While researchers continue to work to identify specific pathways leading to disease, most cases of HTN are thought to be related to lifestyle choices which cause microscopic kidney damage. High-salt and high animal protein diets, obesity, and sugar and alcohol consumption all appear to increase the risk for developing HTN. Lifestyle interventions focused on eliminating offending dietary agents have been shown to improve BP in people with HTN. Lifestyle and dietary changes in select patients may also allow for discontinuation of antihypertensive medications. Similar lifestyle modifications are also recommended for those with prehypertension.

References

1. Chobanian AV, Bakris GL, Black HR, et al. Seventh report of the Joint National Committee on prevention, detection, evaluation, and treatment of high blood pressure. Hypertension. 2003;42:1206–52.
2. High Blood Pressure-Statistics. American Heart Association; 2010.
3. National Center for Health Statistics. Health, United States, 2008: with special feature on the health of young adults. Hyattsville, MD: National Center for Health Statistics. Health, United States; 2009.
4. Centers for Disease Control and Prevention. Trends in health status and health care use among older men. Hyattsville, MD: National Center for Health Statistics; 2010.
5. Burt VL, Whelton P, Roccella EJ, et al. Prevalence of hypertension in the US adult population. Results from the Third National Health and Nutrition Examination Survey, 1988–1991. Hypertension. 1995;25:305–13.
6. National Center for Health Statistics. Health, United States, 2009: with special feature on medical technology. Hyattsville, MD: National Center for Health Statistics (US); 2010.
7. Danaei G, Ding EL, Mozaffarian D, et al. The preventable causes of death in the United States: comparative risk assessment of dietary, lifestyle, and metabolic risk factors. PLoS Med. 2009;6:e1000058.
8. Lloyd-Jones D, Adams RJ, Brown TM, et al. Heart disease and stroke statistics—2010 update: a report from the American Heart Association. Circulation. 2010;121:e46–215.
9. Acelajado MC, Calhoun DA. Resistant hypertension, secondary hypertension, and hypertensive crises: diagnostic evaluation and treatment. Cardiol Clin. 2010;28:639–54.
10. Johnson RJ, Rodriguez-Iturbe B, Nakagawa T, Kang DH, Feig DI, Herrera-Acosta J. Subtle renal injury is likely a common mechanism for salt-sensitive essential hypertension. Hypertension. 2005;45:326–30.
11. Sommers SC, Relman AS, Smithwick RH. Histologic studies of kidney biopsy specimens from patients with hypertension. Am J Pathol. 1958;34:685–715.
12. Johnson RJ, Feig DI, Nakagawa T, Sanchez-Lozada LG, Rodriguez-Iturbe B. Pathogenesis of essential hypertension: historical paradigms and modern insights. J Hypertens. 2008;26: 381–91.
13. Food and Nutrition Board, Institute of Medicine. Dietary reference intakes for water, potassium, sodium, chloride, and sulfate. Washington, DC: National Academies Press; 2005.
14. Brown IJ, Tzoulaki I, Candeias V, Elliott P. Salt intakes around the world: implications for public health. Int J Epidemiol. 2009;38:791–813.
15. Wright JD, Wang CY, Kennedy-Stephenson J, Ervin RB. Dietary intake of ten key nutrients for public health, United States: 1999–2000. Adv Data. 2003;334:1–4.
16. Intersalt Cooperative Research Group. Intersalt: an international study of electrolyte excretion and blood pressure. Results for 24 hour urinary sodium and potassium excretion. Br Med J. 1988;297:319–28.
17. Hofman A, Hazebroek A, Valkenburg HA. A randomized trial of sodium intake and blood pressure in newborn infants. JAMA. 1983;250:370–3.
18. Geleijnse JM, Hofman A, Witteman JC, Hazebroek AA, Valkenburg HA, Grobbee DE. Long-term effects of neonatal sodium restriction on blood pressure. Hypertension. 1997;29:913–7.
19. Gu JW, Bailey AP, Tan W, Shparago M, Young E. Long-term high salt diet causes hypertension and decreases renal expression of vascular endothelial growth factor in Sprague-Dawley rats. J Am Soc Hypertens. 2008;2:275–85.
20. Gu JW, Young E, Pan ZJ, et al. Long-term high salt diet causes hypertension and alters renal cytokine gene expression profiles in Sprague-Dawley rats. Beijing Da Xue Xue Bao. 2009;41:505–15.
21. Centers for Disease Control and Prevention (CDC). Vital signs: state-specific obesity prevalence among adults—United States 2009. MMWR Morb Mortal Wkly Rep. 2010;59:1–5.

22. Aneja A, El-Atat F, McFarlane SI, Sowers JR. Hypertension and obesity. Recent Prog Horm Res. 2004;59:169–205.
23. Juhaeri, Stevens J, Chambless LE, et al. Associations between weight gain and incident hypertension in a bi-ethnic cohort: the Atherosclerosis Risk in Communities Study. Int J Obes Relat Metab Disord. 2002;26:58–64.
24. Kissebah AH, Krakower GR. Regional adiposity and morbidity. Physiol Rev. 1994;74: 761–811.
25. Hall JE, da Silva AA, do Carmo JM, et al. Obesity-induced hypertension: role of sympathetic nervous system, leptin, and melanocortins. J Biol Chem. 2010;285:17271–6.
26. Willenberg HS, Schinner S, Ansurudeen I. New mechanisms to control aldosterone synthesis. Horm Metab Res. 2008;40:435–41.
27. Asferg C, Mogelvang R, Flyvbjerg A, et al. Leptin, not adiponectin, predicts hypertension in the Copenhagen City Heart Study. Am J Hypertens. 2010;23:327–33.
28. Galletti F, D'Elia L, Barba G, et al. High-circulating leptin levels are associated with greater risk of hypertension in men independently of body mass and insulin resistance: results of an eight-year follow-up study. J Clin Endocrinol Metab. 2008;93:3922–6.
29. Kotchen TA. Obesity-related hypertension: epidemiology, pathophysiology, and clinical management. Am J Hypertens. 2010;23:1170–8.
30. Tran LT, Yuen VG, McNeill JH. The fructose-fed rat: a review on the mechanisms of fructose-induced insulin resistance and hypertension. Mol Cell Biochem. 2009;332:145–59.
31. Reddy KJ, Singh M, Bangit JR, Batsell RR. The role of insulin resistance in the pathogenesis of atherosclerotic cardiovascular disease: an updated review. J Cardiovasc Med. 2010;11: 633–47.
32. Lucas PA, Lacour B, Comte L, Drueke T. Pathogenesis of abnormal acid-base balance in the young spontaneously hypertensive rat. Clin Sci (Lond). 1988;75:29–34.
33. Taylor EN, Forman JP, Farwell WR. Serum anion gap and blood pressure in the national health and nutrition examination survey. Hypertension. 2007;50:320–4.
34. Zhang L, Curhan GC, Forman JP. Diet-dependent net acid load and risk of incident hypertension in United States women. Hypertension. 2009;54:751–5.
35. Klatsky AL, Friedman GD, Armstrong MA. The relationships between alcoholic beverage use and other traits to blood pressure: a new Kaiser Permanente study. Circulation. 1986;73: 628–36.
36. Halanych JH, Safford MM, Kertesz SG, et al. Alcohol consumption in young adults and incident hypertension: 20-year follow-up from the Coronary Artery Risk Development in Young Adults Study. Am J Epidemiol. 2010;171:532–9.
37. Thadhani R, Camargo Jr CA, Stampfer MJ, Curhan GC, Willett WC, Rimm EB. Prospective study of moderate alcohol consumption and risk of hypertension in young women. Arch Intern Med. 2002;162:569–74.
38. Nanchahal K, Ashton WD, Wood DA. Alcohol consumption, metabolic cardiovascular risk factors and hypertension in women. Int J Epidemiol. 2000;29:57–64.
39. Fuchs FD, Chambless LE, Whelton PK, Nieto FJ, Heiss G. Alcohol consumption and the incidence of hypertension: the Atherosclerosis risk in Communities Study. Hypertension. 2001;37:1242–50.
40. Taylor B, Irving HM, Baliunas D, et al. Alcohol and hypertension: gender differences in dose-response relationships determined through systematic review and meta-analysis. Addiction. 2009;104:1981–90.
41. Samuel VT. Fructose induced lipogenesis: from sugar to fat to insulin resistance. Trends Endocrinol Metab. 2011;22:60–5.
42. Stanhope KL, Schwarz JM, Keim NL, et al. Consuming fructose-sweetened, not glucose-sweetened, beverages increases visceral adiposity and lipids and decreases insulin sensitivity in overweight/obese humans. J Clin Invest. 2009;119:1322–34.
43. Le KA, Ith M, Kreis R, et al. Fructose overconsumption causes dyslipidemia and ectopic lipid deposition in healthy subjects with and without a family history of type 2 diabetes. Am J Clin Nutr. 2009;89:1760–5.

44. Teff KL, Elliott SS, Tschop M, et al. Dietary fructose reduces circulating insulin and leptin, attenuates postprandial suppression of ghrelin, and increases triglycerides in women. J Clin Endocrinol Metab. 2004;89:2963–72.
45. Teff KL, Grudziak J, Townsend RR, et al. Endocrine and metabolic effects of consuming fructose- and glucose-sweetened beverages with meals in obese men and women: influence of insulin resistance on plasma triglyceride responses. J Clin Endocrinol Metab. 2009;94:1562–9.
46. El Hafidi M, Perez I, Carrillo S, et al. Effect of sex hormones on non-esterified fatty acids, intra-abdominal fat accumulation, and hypertension induced by sucrose diet in male rats. Clin Exp Hypertens. 2006;28:669–81.
47. Farah V, Elased KM, Chen Y, et al. Nocturnal hypertension in mice consuming a high fructose diet. Auton Neurosci. 2006;130:41–50.
48. Giani JF, Mayer MA, Munoz MC, et al. Chronic infusion of angiotensin-(1–7) improves insulin resistance and hypertension induced by a high-fructose diet in rats. Am J Physiol Endocrinol Metab. 2009;296:E262–71.
49. Chen L, Caballero B, Mitchell DC, et al. Reducing consumption of sugar-sweetened beverages is associated with reduced blood pressure: a prospective study among United States adults. Circulation. 2010;121:2398–406.
50. Camoes M, Oliveira A, Pereira M, Severo M, Lopes C. Role of physical activity and diet in incidence of hypertension: a population-based study in Portuguese adults. Eur J Clin Nutr. 2010;64:1441–9.
51. Steffen LM, Kroenke CH, Yu X, et al. Associations of plant food, dairy product, and meat intakes with 15-y incidence of elevated blood pressure in young black and white adults: the Coronary Artery Risk Development in Young Adults (CARDIA) Study. Am J Clin Nutr. 2005;82:1169–77.
52. Forman JP, Stampfer MJ, Curhan GC. Diet and lifestyle risk factors associated with incident hypertension in women. JAMA. 2009;302:401–11.
53. Whelton SP, Chin A, Xin X, He J. Effect of aerobic exercise on blood pressure: a meta-analysis of randomized, controlled trials. Ann Intern Med. 2002;136:493–503.
54. Pereira MA, Folsom AR, McGovern PG, et al. Physical activity and incident hypertension in black and white adults: the Atherosclerosis Risk in Communities Study. Prev Med. 1999;28:304–12.
55. Paffenbarger Jr RS, Wing AL, Hyde RT, Jung DL. Physical activity and incidence of hypertension in college alumni. Am J Epidemiol. 1983;117:245–57.
56. Haapanen N, Miilunpalo S, Vuori I, Oja P, Pasanen M. Association of leisure time physical activity with the risk of coronary heart disease, hypertension and diabetes in middle-aged men and women. Int J Epidemiol. 1997;26:739–47.
57. Chase NL, Sui X, Lee DC, Blair SN. The association of cardiorespiratory fitness and physical activity with incidence of hypertension in men. Am J Hypertens. 2009;22:417–24.
58. Hu G, Lindstrom J, Valle TT, et al. Physical activity, body mass index, and risk of type 2 diabetes in patients with normal or impaired glucose regulation. Arch Intern Med. 2004;164:892–6.
59. Hayashi T, Tsumura K, Suematsu C, Okada K, Fujii S, Endo G. Walking to work and the risk for hypertension in men: the Osaka Health Survey. Ann Intern Med. 1999;131:21–6.
60. Blair SN, Goodyear NN, Gibbons LW, Cooper KH. Physical fitness and incidence of hypertension in healthy normotensive men and women. JAMA. 1984;252:487–90.
61. Forte JG, Miguel JM, Miguel MJ, de Padua F, Rose G. Salt and blood pressure: a community trial. J Hum Hypertens. 1989;3:179–84.
62. He FJ, MacGregor GA. Effect of modest salt reduction on blood pressure: a meta-analysis of randomized trials. Implications for public health. J Hum Hypertens. 2002;16:761–70.
63. Sacks FM, Svetkey LP, Vollmer WM, et al. Effects on blood pressure of reduced dietary sodium and the Dietary Approaches to Stop Hypertension (DASH) diet. DASH-Sodium Collaborative Research Group. N Engl J Med. 2001;344:3–10.

64. Bibbins-Domingo K, Chertow GM, Coxson PG, Moran A, Lightwood JM, Pletcher MJ, et al. Projected effect of dietary salt reductions on future cardiovascular disease. N Engl J Med. 2010;362:590–9.

65. Wan YP, Xu RY, Wu YJ, Chen ZQ, Cai W. Diet intervention on obese children with hypertension in China. World J Pediatr. 2009;5:269–74.

66. Goodpaster BH, Delany JP, Otto AD, et al. Effects of diet and physical activity interventions on weight loss and cardiometabolic risk factors in severely obese adults: a randomized trial. JAMA. 2010;304:1795–802.

67. Bennett GG, Herring SJ, Puleo E, Stein EK, Emmons KM, Gillman MW. Web-based weight loss in primary care: a randomized controlled trial. Obesity (Silver Spring). 2010;18: 308–13.

68. Dengo AL, Dennis EA, Orr JS, et al. Arterial destiffening with weight loss in overweight and obese middle-aged and older adults. Hypertension. 2010;55:855–61.

69. Ueki K, Sakurai N, Tochikubo O. Weight loss and blood pressure reduction in obese subjects in response to nutritional guidance using information communication technology. Clin Exp Hypertens. 2009;31:231–40.

70. Croft PR, Brigg D, Smith S, Harrison CB, Branthwaite A, Collins MF. How useful is weight reduction in the management of hypertension? J R Coll Gen Pract. 1986;36:445–8.

71. Fogari R, Zoppi A, Corradi L, et al. Effect of body weight loss and normalization on blood pressure in overweight non-obese patients with stage 1 hypertension. Hypertens Res. 2010;33:236–42.

72. Minami J, Kawano Y, Ishimitsu T, Matsuoka H, Takishita S. Acute and chronic effects of a hypocaloric diet on 24-hour blood pressure, heart rate and heart-rate variability in mildly-to-moderately obese patients with essential hypertension. Clin Exp Hypertens. 1999;21: 1413–27.

73. Hofso D, Nordstrand N, Johnson LK, et al. Obesity-related cardiovascular risk factors after weight loss: a clinical trial comparing gastric bypass surgery and intensive lifestyle intervention. Eur J Endocrinol. 2010;163:735–45.

74. Adams TD, Pendleton RC, Strong MB, et al. Health outcomes of gastric bypass patients compared to nonsurgical, nonintervened severely obese. Obesity (Silver Spring). 2010;18: 121–30.

75. Sjostrom L, Lindroos AK, Peltonen M, et al. Lifestyle, diabetes, and cardiovascular risk factors 10 years after bariatric surgery. N Engl J Med. 2004;351:2683–93.

76. Treadwell JR, Sun F, Schoelles K. Systematic review and meta-analysis of bariatric surgery for pediatric obesity. Ann Surg. 2008;248:763–76.

77. Holterman AX, Browne A, Tussing L, et al. A prospective trial for laparoscopic adjustable gastric banding in morbidly obese adolescents: an interim report of weight loss, metabolic and quality of life outcomes. J Pediatr Surg. 2010;5:74–78; discussion 8–9.

78. Cornelissen VA, Fagard RH. Effects of endurance training on blood pressure, blood pressure-regulating mechanisms, and cardiovascular risk factors. Hypertension. 2005;46:667–75.

79. Cornelissen VA, Arnout J, Holvoet P, Fagard RH. Influence of exercise at lower and higher intensity on blood pressure and cardiovascular risk factors at older age. J Hypertens. 2009;27: 753–62.

80. Fagard RH, Cornelissen VA. Effect of exercise on blood pressure control in hypertensive patients. Eur J Cardiovasc Prev Rehabil. 2007;14:12–7.

81. Xin X, He J, Frontini MG, Ogden LG, Motsamai OI, Whelton PK. Effects of alcohol reduction on blood pressure: a meta-analysis of randomized controlled trials. Hypertension. 2001;38:1112–7.

82. Thun MJ, Peto R, Lopez AD, et al. Alcohol consumption and mortality among middle-aged and elderly U.S. adults. N Engl J Med. 1997;337:1705–14.

83. Freiberg MS, Chang YF, Kraemer KL, Robinson JG, Adams-Campbell LL, Kuller LL. Alcohol consumption, hypertension, and total mortality among women. Am J Hypertens. 2009;22:1212–8.

84. Gaziano JM, Gaziano TA, Glynn RJ, et al. Light-to-moderate alcohol consumption and mortality in the Physicians' Health Study enrollment cohort. J Am Coll Cardiol. 2000;35: 96–105.

85. Djousse L, Lee IM, Buring JE, Gaziano JM. Alcohol consumption and risk of cardiovascular disease and death in women: potential mediating mechanisms. Circulation. 2009;120:237–44.

86. Streppel MT, Arends LR, van't Veer P, Grobbee DE, Geleijnse JM. Dietary fiber and blood pressure: a meta-analysis of randomized placebo-controlled trials. Arch Intern Med. 2005;165:150–6.

87. Appel LJ, Moore TJ, Obarzanek E, et al. A clinical trial of the effects of dietary patterns on blood pressure. DASH Collaborative Research Group. N Engl J Med. 1997;336:1117–24.

88. Blumenthal JA, Babyak MA, Hinderliter A, et al. Effects of the DASH diet alone and in combination with exercise and weight loss on blood pressure and cardiovascular biomarkers in men and women with high blood pressure: the ENCORE study. Arch Intern Med. 2010;170: 126–35.

89. Espeland MA, Whelton PK, Kostis JB, et al. Predictors and mediators of successful long-term withdrawal from antihypertensive medications. TONE Cooperative Research Group. Trial of Nonpharmacologic Interventions in the Elderly. Arch Fam Med. 1999;8:228–36.

90. Reid CM, Maher T, Jennings GL. Substituting lifestyle management for pharmacological control of blood pressure: a pilot study in Australian general practice. Blood Press. 2000;9: 267–74.

91. Burke V, Beilin LJ, Cutt HE, Mansour J, Wilson A, Mori TA. Effects of a lifestyle programme on ambulatory blood pressure and drug dosage in treated hypertensive patients: a randomized controlled trial. J Hypertens. 2005;23:1241–9.

92. Miller III ER, Erlinger TP, Young DR, et al. Results of the Diet, Exercise, and Weight Loss Intervention Trial (DEW-IT). Hypertension. 2002;40:612–8.

93. Skrabal F, Gasser RW, Finkenstedt G, Rhomberg HP, Lochs A. Low-sodium diet versus low-sodium/high-potassium diet for treatment of hypertension. Klin Wochenschr. 1984;62: 124–8.

94. Davis BR, Oberman A, Blaufox MD, et al. Lack of effectiveness of a low-sodium/high-potassium diet in reducing antihypertensive medication requirements in overweight persons with mild hypertension. TAIM Research Group. Trial of Antihypertensive Interventions and Management. Am J Hypertens. 1994;7:926–32.

95. Grossman E, Vald A, Peleg E, Sela B, Rosenthal T. The effects of a combined low-sodium, high-potassium, high-calcium diet on blood pressure in patients with mild hypertension. J Hum Hypertens. 1997;11:789–94.

96. Morgan T, Nowson C. Comparative studies of reduced sodium and high potassium diet in hypertension. Nephron. 1987;47 Suppl 1:21–6.

97. Bompiani GD, Cerasola G, Morici ML, et al. Effects of moderate low sodium/high potassium diet on essential hypertension: results of a comparative study. Int J Clin Pharmacol Ther Toxicol. 1988;26:129–32.

98. Makynen H, Arvola P, Vapaatalo H, Porsti I. High calcium diet effectively opposes the development of deoxycorticosterone-salt hypertension in rats. Am J Hypertens. 1994;7:520–8.

99. Trials of Hypertension Prevention Collaborative Research Group. The effects of nonpharmacologic interventions on blood pressure of persons with high normal levels. Results of the trials of hypertension prevention, Phase I. JAMA. 1992;267:1213–20.

100. Dickinson HO, Nicolson DJ, Cook JV, et al. Calcium supplementation for the management of primary hypertension in adults. Cochrane Database Syst Rev. 2006;19:CD004639.

101. Dickinson HO, Nicolson DJ, Campbell F, Beyer FR, Mason J. Potassium supplementation for the management of primary hypertension in adults. Cochrane Database Syst Rev. 2006;3:CD004641.

102. Berry SE, Mulla UZ, Chowienczyk PJ, Sanders TA. Increased potassium intake from fruit and vegetables or supplements does not lower blood pressure or improve vascular function in UK men and women with early hypertension: a randomised controlled trial. Br J Nutr. 2010;104:1839–47.

103. Dickinson HO, Nicolson DJ, Campbell F, et al. Magnesium supplementation for the management of essential hypertension in adults. Cochrane Database Syst Rev. 2006;3:CD004640.
104. Pilz S, Tomaschitz A, Ritz E, Pieber TR. Vitamin D status and arterial hypertension: a systematic review. Nat Rev Cardiol. 2009;6:621–30.
105. Beyer FR, Dickinson HO, Nicolson DJ, Ford GA, Mason J. Combined calcium, magnesium and potassium supplementation for the management of primary hypertension in adults. Cochrane Database Syst Rev. 2006;3:CD004805.
106. Vasan RS, Larson MG, Leip EP, Kannel WB, Levy D. Assessment of frequency of progression to hypertension in non-hypertensive participants in the Framingham Heart Study: a cohort study. Lancet. 2001;358:1682–6.
107. Mainous III AG, Everett CJ, Liszka H, King DE, Egan BM. Prehypertension and mortality in a nationally representative cohort. Am J Cardiol. 2004;94:1496–500.
108. Lee J, Heng D, Ma S, Chew SK, Hughes K, Tai ES. Influence of pre-hypertension on all-cause and cardiovascular mortality: the Singapore Cardiovascular Cohort Study. Int J Cardiol. 2009;135:331–7.
109. Kshirsagar AV, Carpenter M, Bang H, Wyatt SB, Colindres RE. Blood pressure usually considered normal is associated with an elevated risk of cardiovascular disease. Am J Med. 2006;119:133–41.
110. Qureshi AI, Suri MF, Kirmani JF, Divani AA, Mohammad Y. Is prehypertension a risk factor for cardiovascular diseases? Stroke. 2005;36:1859–63.
111. Hsia J, Margolis KL, Eaton CB, et al. Prehypertension and cardiovascular disease risk in the Women's Health Initiative. Circulation. 2007;115:855–60.

Chapter 12
Diet, Physical Activity, and Cancer Prevention

Cindy D. Davis and John A. Milner

Keywords Diet • Cancer • Obesity • Bioactive food components • Nutrigenomics

Key Points

- Overweight (body mass index [BMI] > 25 kg/m^2) and obesity (BMI > 30 kg/m^2) are associated with increased risk for many, but not all, common cancers.
- One of the most important ways to reduce overall cancer risk is to maintain a healthy weight throughout life.
- Epidemiologic and other (animal and cell culture) studies indicate that increased consumption of fruits, vegetables, and whole grains is associated with reduced cancer risk. Therefore, eat mostly foods of plant origin. This includes at least five portions/servings (at least 400 g or 14 oz) of a variety of vegetables and fruits every day; eating whole grains and/or pulses (legumes) with every meal; and limiting refined starchy foods.
- Alcohol consumption has been associated with an increased risk of some cancers. Although there is no consumption that is not associated with an increased cancer risk, modest amounts of alcohol (two drinks a day for men and one drink a day for women) can protect against coronary heart disease.
- Cancer survivors should follow the recommendations for cancer prevention regarding diet, healthy weight, and physical activity.

C.D. Davis, PhD (✉)
Director of Grants and Extramural Activities Office of Dietary Supplements National Institutes of Health, Bethesda, Maryland
e-mail: davisci@mail.nih.gov

J.A. Milner, PhD
Nutritional Science Research Group, Division of Cancer Prevention, National Cancer Institute, 6130 Executive Boulevard, Suite 3159, Rockville, MD 20892-7328, USA

N.J. Temple et al. (eds.), *Nutritional Health: Strategies for Disease Prevention*, Nutrition and Health, DOI 10.1007/978-1-61779-894-8_12, © Springer Science+Business Media, LLC 2012

• Not all individuals should be expected to respond identically to bioactive food components because of genetic and environmental factors. The "omics" of nutrition can be used to identify responders and nonresponders.

12.1 Introduction

"Cancer" is a general term that represents more than 100 diseases, each with their own etiology. The probability of suffering from cancer is extremely high in the United States; estimates are that approximately 44% of men and 37% of women will develop cancer in their lifetime [1]. It has also been projected that cancer will surpass ischemic heart disease as the leading cause of death worldwide in 2010 [2]. Cancer risk is influenced by both genetic and environmental factors, including dietary habits. While each type of cancer has unique characteristics, cancers share one common feature, namely they begin when a single cell acquires genetic changes and loses control of its normal growth and replication processes [3]. Most cancers develop to the stage of being clinically identifiable only years or decades after the initial cell damage.

Cancer is no longer being perceived as an inevitable consequence of aging. Only about 5–10% of cancers can be classified as resulting from genetics and thus most are associated with multiple environmental factors, including one's eating behaviors. The North American Association of Central Cancer Registries provides evidence that death rates from cancer have been dropping about 2.1% a year in the United States since 2002 [4]. This positive trend is thought not to be a result of miraculous medical breakthroughs but a result of improvements in prevention, early detection, and treatment of some causes of cancer death. In contrast, cancer rates are increasing in less developed and economically transitioning countries because of adoption of unhealthy Western lifestyles, including smoking and physical inactivity and consumption of excess calories [1]. Modifying dietary behaviors and nutritional habits based on current guidelines represents a proactive, practical, and cost-effective approach to cancer prevention that is also likely to promote overall good health. The cancer process includes fundamental, yet diverse, cellular processes that can be influenced by diet, such as carcinogen bioactivation, cellular differentiation, DNA repair, cellular proliferation/signaling, hormonal regulation, inflammation and immunity, angiogenesis, and apoptosis (see Fig. 12.1) [3].

Evidence continues to mount that changing dietary habits can reduce cancer risk and modify the biological behavior of tumors [5]. The importance of diet was emphasized more than a quarter century ago when Doll and Peto [6] suggested that approximately 35% (10–70%) of all cancers in the United States might be attributable to dietary factors. In 2007, similar conclusions were reached by the World Cancer Research Fund/American Institute of Cancer Research (WCRF/AICR) after evaluating over 7,000 studies. Their report concluded that diet and physical activity were major determinants of cancer risk [5]. On a global scale, this could represent over 3–4 million cancer cases that can be prevented each year [5].

Fig. 12.1 Bioactive dietary components can influence many cellular processes associated with carcinogenesis including, but not limited to, carcinogen bioactivation, cellular differentiation, DNA repair, cellular proliferation/signaling, hormonal regulation, inflammation and immunity, angiogenesis, and apoptosis

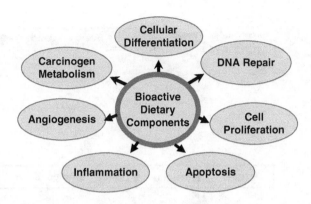

While considerable evidence points to diet as a critical factor in determining cancer risk, there are numerous inconsistencies in the literature. Much of this variation in response may relate to the genetic background of the individual. Recent studies provide important proof that genetic polymorphisms can markedly influence the response to specific foods [7, 8]. By utilizing genetic information, we may be able to identify those individuals who must assure an adequate intake of a nutrient for cancer prevention. For example, dietary calcium can interact with a polymorphism in the vitamin D receptor (the *Fok 1* restriction site) to affect colon cancer risk. Whereas dietary calcium was not important in determining colon cancer risk in individuals that were homozygous for the capital F genotype for the vitamin D receptor, low dietary calcium increased colon cancer risk with increasing copies of the little f allele for the vitamin D receptor [7]. Selected polymorphisms may also be useful as surrogate markers for those who might be placed at risk from excessive exposures [8]. Colorectal adenoma risk is modified by the interplay between polymorphisms in the PPAR delta and fish consumption. In individuals with the variant allele for PPAR delta (<10% of the population), increased fish consumption actually increased the risk of developing colorectal adenomas [8]. However, the existence of about 30,000 genes and many million single nucleotide polymorphisms indicates that understanding individual responses to foods or components will be extremely complicated.

12.2   Body Fatness

Excess body weight is a strong determinant as well as modifiable risk factor for cancer development, and as such carries the potential for primary prevention. Typically, obesity, defined as a body mass index (BMI) > 30, is associated with about a 20% increase in risk for most, but not all, cancers [9]. This lack of a correlation may simply reflect the imprecision in using BMI as a surrogate risk marker for adiposity. The distribution of total body fat, how we measure it, and the ratio of body fat to fat-free mass explain to some degree the inconsistencies in the literature [10]. The use of biomarkers of the metabolic syndrome holds promise for determining which shifts in body energetics are likely contributing to increased cancer risk or changes in the behavior of tumors [11]. Regardless, the WCRF/AICR panel judged

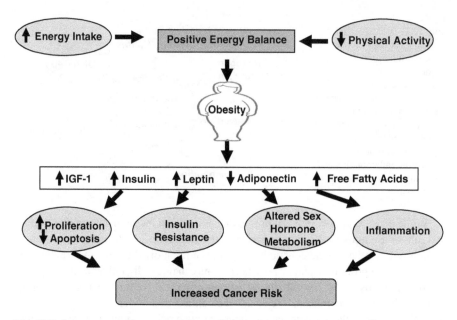

Fig. 12.2 Long-term positive energy balance due to excessive energy intake and/or low levels of energy expenditure can lead to obesity. The metabolic consequences of long-term positive energy balance and the accumulation of excessive body fat include increased IGF-1, insulin, leptin, and free fatty acids, and decreased adiponectin concentrations. These changes can stimulate cellular proliferation, inhibit apoptosis, increase insulin resistance, alter steroid hormone metabolism, and stimulate inflammatory/oxidative stress processes, all of which can contribute to increased cancer risk

the evidence as convincing that greater body fatness is a cause of cancers of the esophagus, pancreas, colorectum, breast (postmenopausal), endometrium, and kidney [5]. Greater body fatness may precipitate cancer of the gallbladder [5]. In terms of weight-related factors, body weight alone does not completely determine an individual's ability to prevent or survive cancer. Location also appears important since abdominal fat accumulation may be more detrimental than peripheral fat accumulation. Thus, a high waist circumference may be especially hazardous.

Scientists have historically used the term "energy balance" to describe the complex interaction among diet and physical activity on body weight over an individual's lifetime. Along with genetics these factors may influence cancer risk (see Fig. 12.2). Long thought of as inert, adipose tissue, particularly visceral fat, is an important metabolic tissue. Excessive adiposity is associated with increased oxidative stress, insulin resistance, inflammation, and changes in hormone and growth factor concentrations that play a role in the pathogenesis of many cancers (Fig. 12.2). High circulating levels of insulin promote the synthesis of IGF-1 and decrease the production of IGF binding protein-1, thus increasing the biologic activity of IGF-1. The role of IGF-1 in the pathogenesis of some human cancers is supported by epidemiologic studies, which have found that high serum concentrations of IGF-1 are associated with increased risk of several human cancers, including those of the prostate, breast, and colon [12]. Evidence in transgenic animals suggests it is not fat per se but the associated metabolic changes that appear to increase cancer in some sites.

The mechanisms by which excess weight may influence cancer risk is unresolved but may relate to homeostasis in the steroid hormones, insulin-like growth factors, insulin resistance and lipid metabolism, and immune function and inflammatory factors, including cytokines and prostaglandins [13]. The identification of which lead to a change in cancer risk and adverse health outcomes in specific populations is fundamental to the development of public health initiatives and to personalized strategies for diet and exercise interventions. The interplay of dietary components with genetic susceptibility can modify the relationship between energy balance and cancer risk.

Caloric restriction is the most effective single intervention for preventing cancer in experimental animals [14]. Recent reports of extended life span and delayed cancer development in response to caloric restriction in rhesus monkeys [15] and observations that caloric restriction during the premenopausal years decreases postmenopausal breast cancer risk in women [16] suggest the cancer protective effects of caloric restriction reported in rodent models extends to primates, including humans. Caloric restriction reduces the levels of several anabolic hormones, growth factors, and inflammatory cytokines; reduces oxidative stress and cell proliferation; and enhances autophagy and several DNA repair processes [17]. At least parts of these anticancer properties associated with caloric restriction likely involve changes in the IGF-1 pathway (Fig. 12.2) [11].

Maintenance of a healthy weight throughout life is one important way to protect against most cancers, as well as a number of other common chronic diseases, including hypertension and stroke, type 2 diabetes, and coronary heart disease.

12.3   Physical Activity

Although reducing excess caloric consumption will help control weight, another key variable in the energy balance equation is to expend energy via physical activity. Despite the circulatory benefits associated with physical activity, Americans are not incorporating enough physical activity into their daily routines. The Centers for Disease Control and Prevention (CDC) estimates that more than 50% of American adults do not get enough physical activity to prove beneficial to their health, and more than 25% are not active even during leisure time [18]. Unfortunately, statistics with children and adolescents are no more encouraging. Overall, industrialization, urbanization, and mechanization have fostered a largely sedentary population in many parts of the world.

Regular, sustained physical activity protects against cancer of some sites independent of its effects on body fatness [5]. The WCRF/AICR panel judged that the evidence that physical activity protects against colon cancer is convincing [5]. Physical activity probably protects against endometrial and postmenopausal breast cancer; however, the evidence suggesting that it protects against premenopausal breast cancer is limited [5]. Moreover, because physical activity promotes healthy weight, it would be anticipated that exercise would also protect against cancers whose risk is increased by obesity.

Physical activity most likely influences the development of cancer through multiple, perhaps overlapping, biological pathways, several of which are mentioned in Fig. 12.2. Many researchers believe physical activity aids in regular bowel movements, which may decrease the time the colon is exposed to potential carcinogens; causes changes in insulin resistance, metabolism, and hormone levels, which may help prevent tumor development; and alters a number of inflammatory and immune factors [19].

Despite considerable epidemiological evidence linking cancer with obesity and with lack of physical activity, the appropriate level and type of physical activity required to bring about protection remains unclear [20]. According to the WCRF/ Report recommendations, individuals should be moderately physically active, equivalent to brisk walking for at least 30 min every day. As fitness improves, individuals should aim for at least 60 min of moderate activity or 30 min of more vigorous physical activity every day [5]. *Healthy People 2010* recommended that adults engage in at least 30 min of moderate-intensity activity, 5 days/week, or 20 min of vigorous-intensity activity, 3 days/week [21]. In contrast, under the *2008 Physical Activity Guidelines for Americans*, the minimum recommended aerobic physical activity required to produce substantial health benefits in adults is 150 min/ week of moderate-intensity activity, or 75 min/week of vigorous-intensity activity, or an equivalent combination of moderate- and vigorous-intensity physical activity [22]. Additional controlled studies are needed to determine if the amount and duration of exercise needed to prevent obesity and to treat obesity are similar and to clarify the importance of the type of physical activity—moderate intensity or vigorous—that is needed to prevent weight gain and improve health [23].

12.4   Fruits and Vegetables

Evidence that vegetables and fruits provide protection against cancer comes principally from epidemiological investigations and from a host of animal and cell culture studies, although the data are not compelling. Vegetables and fruits are usually low in energy density and when consumed in variety provide many vitamins, minerals, and other bioactive compounds (phytochemicals). Recommendations for consumption tend to exclude starchy vegetables such as potato, yam, sweet potato, and cassava. Nonstarchy vegetables probably protect against cancers of the mouth, pharynx, and larynx, and those of the esophagus and stomach [5]. Limited evidence also suggests that they may protect against cancers of the nasopharynx, lung, colorectum, ovary, and endometrium [5]. Fruits probably protect against cancers of the mouth, pharynx, and larynx, and those of the esophagus, lung, and stomach [5]. The possibility that fruits may also protect against cancers of the nasopharynx, pancreas, liver, and colorectum has also surfaced [5]. While these relationships are based upon the epidemiologic literature, it must be pointed out that there are a number of limitations/considerations that are specific to the analysis of dietary intake of fruits and vegetables. These include: most studies of consumption of dietary fruits and vegetables have been conducted in populations with relatively homogeneous diets; smokers consume fewer fruits and vegetables than nonsmokers; fat intake inversely correlates with fruit and vegetable intake in the United States; and studies using self-reporting tend to overreport vegetable and fruit consumption. Thus, it is not surprising that many uncertainties exist about the relationship between plant-based diets and cancer prevention.

Fruits and vegetables contain as many as 100,000 different bioactive food components including both essential micronutrients (e.g., vitamins C, D, E, and folate and the minerals selenium, zinc, iodine, and calcium) and phytochemicals. Phytochemicals, which are discussed below, is a collective term for a variety of plant components that often perform important functions in the plant, such as providing color, flavor, or protection. The phytochemical composition of fruits and vegetables depends both on the species and subtype, as well as the environmental, cultivation, growing, harvesting, and storage conditions.

It is widely believed that many of the health benefits, including cancer prevention, of diets enriched in fruits and vegetables are due, in part, to the presence of multiple phytochemicals. For example, foods containing β-carotene or vitamin C seem to protect against esophageal cancer. There is limited evidence to suggest that foods containing quercetin protect against lung cancer [5]. Food sources of pyridoxine and/or vitamin E may also protect against esophageal and prostate cancers [5]. The relationship between micronutrients, phytochemicals, and risk of different types of cancer is discussed below.

The magnitude of the response to fruit and vegetables is probably influenced by many factors, including the consumer's genetic background and a host of environmental factors, as well as the type, quantity, and duration of consumption of these foods, and interactions among food components. Of 20 effect estimates from 17 cohort studies that reported comparisons of the highest and lowest intake groups of fruits and vegetables and colorectal cancer, 11 were in the direction of reduced risk, 3 of which were statistically significant [5]. The evidence for other cancer sites is not as compelling.

12.5 Micronutrients and Phytochemicals

Both micronutrients and phytochemicals can bring about a multitude of biological responses that may be important in the cancer process. Phytochemicals are classified according to their chemical structure and functional characteristics; they include salicylates, phytosterols, saponins, glucosinolates, polyphenols, protease inhibitors, monoterpenes, phytoestrogens, sulfides, indoles, isothiocyanates, terpenes, and lectins. Examples include plant phenols (from grapes, strawberries, and apples) and

Table 12.1 Examples of dietary phytochemicals that may be protective against cancer

Phytochemical	Dietary sources
Allyl sulfur compounds	Onions, garlic, leeks
Anthocyanadins	Berries, grapes
β-Carotene[a]	Citrus fruit, carrots, squash, pumpkin
Capsaicin	Peppers
Catechins	Tea, berries
Curcumin	Turmeric
Ellagic acid	Grapes, strawberries, raspberries, walnuts
Genistein	Soy
Indoles	Cruciferous vegetables
Isoflavones	Soybeans and other legumes
Isothiocyanates	Cruciferous vegetables
Lycopene	Tomatoes and tomato products, guava, watermelon
Quercetin	Onion, red grapes, citrus fruit, broccoli
Resveratrol	Grapes (skin), red wine
Silymarin	Milk thistle
Terpenes	Citrus fruits
Thioethers	Garlic, onions (*Allium* foods)

[a]In some cases supplemental β-carotene may increase risk in humans [50–52]

phytoestrogens (from soy and soy products). An incomplete list of phytochemicals and some of their sources is given in Table 12.1. Since a comprehensive review of the interactions between micronutrients, phytochemicals, and cancer is beyond the scope of this chapter and has been published elsewhere [24, 25], only a couple of examples are provided to illustrate the principle that these food components are capable of modifying a variety of cancer processes. These examples reveal the magnitude and complexity of the potential interactions. Regardless, it is clear that several factors influence the response to these dietary components including the time of introduction, the quantity and duration of consumption, interactions among food components, and the genetic background of the consumer.

12.5.1 Garlic and Allyl Sulfur Compounds

Food is generally complex as illustrated by the *Allium* family containing about 500 species including garlic, onion, leeks, chives, and scallions. Allyl sulfur compounds arising from these foods are thought to be a primary factor in their anticancer properties. However, it is also clear that they contain many other constituents which may provide protection, including amino acids, carbohydrates, and flavonoids. Similarly, the health benefits of other foods cannot typically be related to a single component.

Epidemiologic findings and preclinical studies provide evidence that garlic and related sulfur constituents can suppress cancer risk and alter the biological behavior of tumors [26]. A total of 2 cohort studies, 27 case–control studies, and 2 ecological studies investigated the relationship between *Allium* vegetables and cancer [5]. Meta-analysis of these data showed a 23% decreased risk per 50 g/day of *Allium* vegetables and a 59% decreased risk per serving of garlic per day [5]. One randomized controlled trial reported a statistically significant 29% reduction in both size and number of colon adenomas in colorectal patients taking aged garlic extract [27]. Other studies have shown that garlic and/or its related organosulfur compounds suppress mammary, colon, skin, uterine, esophagus, lung, renal, forestomach, and liver cancer incidence in animal models [24].

Similar to other foods, garlic and its sulfur constituents appear to exert their cancer protective effects through multiple mechanisms including inhibition of carcinogen metabolism, inhibition of DNA adduct formation, upregulation of antioxidant defenses and DNA repair, suppression of cell proliferation, induction of apoptosis, decreased inflammation, and blocked angiogenesis [28]. It is likely that many of these processes are modified simultaneously. In eukaryotic cells, the nuclear DNA is tightly wrapped around core histone proteins. Acetylation of these proteins is an important posttranslational modification for the regulation of gene transcription. Garlic and its sulfur constituents have been shown to increase acetylation of the core nucleosomal histones in various cell lines in vitro. For example, Druesne et al. reported that diallyl disulfide increased histone H3 acetylation in cultured Caco-2 and HT-29 cells and normal rat colonocytes by reducing histone deacetylase activity [29]. This hyperacetylation was accompanied by an increase in tumor promoter p21 (waf1/cip1) expression demonstrating that epigenomic events can influence subsequent gene expression patterns which culminates in cells being blocked in the G2 phase of the cell cycle. A variety of constituents in foods have also been reported to influence epigenetics and vice versa.

It is not currently known whether additive or antagonistic responses occur among dietary components that have the same molecular target. For example, both garlic organosulfur compounds and sulforaphane, which is present in broccoli, induce expression of detoxifying enzymes via the binding of the transcription factor Nrf2 to the antioxidant response element (ARE), which is located in the promoter region of related genes. Can these findings be interpreted to mean that if an individual consumes sufficient organosulfur compounds, then sulforaphane will no longer have any anticancer effects? Or, might other molecular targets be important?

12.5.2 Selenium

International evidence points to an inverse association between selenium status and mortality from cancer of the colon, rectum, prostate, breast, ovary, lung, and leukemia [30]. Data from most case–control and cohort studies show a possible protective relationship of the mineral for lung and prostate cancer, but have not been overly convincing for other cancer sites, including breast and colon/rectum [5]. A meta-analysis suggests that selenium may afford some protection against lung cancer in populations where average selenium levels are low [31]. The evidence for these findings is greatest in studies using toenail selenium as an indicator of status [31].

Cohort studies also have identified low baseline serum or toenail selenium concentrations as a risk factor for prostate cancer [32, 33]. Interestingly, an intervention study supports the protective effects of selenium against cancer. The Nutritional Prevention of Cancer (NPC) Trial was a double-blind, placebo-controlled trial in which 1,312 patients with a history of skin cancer were randomized to receive either 200 μg selenium as selenized yeast or a placebo for an average of 4.5 years [34]. Secondary end-point analyses show that the mineral resulted in a significant reduction in total cancer mortality (RR=0.5), total cancer incidence (RR=0.63), and incidence of lung (RR=0.54), colorectal (RR=0.42), and prostate (RR=0.37) cancer [34]. Selenium supplementation led to a significant 49% reduction in incidence among those in the lowest tertile of baseline plasma selenium (<105 ng/mL) and to a nonsignificant 30% reduction in incidence among those in the second tertile. However, for those with baseline selenium in the highest tertile (<122 ng/mL), selenium supplementation was associated with a nonsignificant 20% increase in cancer incidence. These results suggest that there may be a narrow window for the most beneficial dose of dietary selenium. Waters et al. [35] have demonstrated this phenomenon in terms of a nonlinear U-shaped relationship between toenail selenium concentration as a biomarker for exposure and DNA damage in the prostate of dogs. Importantly, this U-shaped relationship between intake or exposure and biological function has also been seen with other dietary components including vitamin D [36].

The Selenium and Vitamin E Cancer Prevention Trial (SELECT) randomized 35,533 healthy men to either 200 μg/day selenomethionine, 400 IU vitamin E, both supplements, or placebos in order to try and replicate the results of the NPC trial. However, in SELECT, neither prostate cancer nor other cancer end points differed significantly among the four treatments [37]. Differences in baseline selenium status or formulation of the selenium supplement may have contributed to the differences observed between the two studies. In fact, in the SELECT trial, median baseline serum levels were 135 ng/mL compared to 114 ng/mL observed in the NPC trial and 78% of subjects would have been classified as being in the upper tertile of baseline selenium in the NPC trial (the group that had a nonsignificant increased risk with selenium supplementation). Another important difference between the two studies is the form of selenium provided. The NPC trial utilized selenium-enriched yeast whereas the SELECT trial used selenomethionine. Although selenium-enriched yeast is predominantly selenomethionine, minor selenium compounds in the yeast, such as Se-methyl-selenocysteine, but not selenomethionine, can be metabolized to alpha-keto acids which function as a histone deacetylase inhibitor and leads to de-repression of silenced tumor suppressor proteins [38]. These data suggest that different metabolites may have different biological effects.

12.5.3   *Folate*

Folates are a group of water-soluble B vitamins found in high concentrations in green leafy vegetables and whole grain cereals. Folic acid, the synthetic form, is used to fortify manufactured cereal products (such as flour), grains, and spreads. Folate-rich foods may protect against pancreatic cancer, and at least some evidence suggests that these foods also protect against esophageal and colorectal cancers [5].

The relationship between dietary folate and cancer serves as an important example for the importance of the timing of exposure and diet–gene interactions. The mechanisms by which dietary folate can modulate cancer development are related to the sole biochemical function of folate—mediating the transfer of one-carbon moieties. In this role, folate is an important factor in DNA synthesis, stability, integrity, and repair. If dietary folate is limited, the balance of purine and pyrimidine DNA precursors is altered, and normal DNA repair is inhibited. Moreover, uracil, which is not normally present in DNA, is misincorporated into the DNA molecule in place of thymidine, resulting in DNA strand breakage, chromosomal damage, and malignant transformation. Furthermore, cytosine methylation is altered, leading to global DNA hypomethylation and/or changes in gene-specific methylation and inappropriate protooncogene activation. A growing body of evidence from cell culture, animal, and human studies indicates that folate deficiency is associated with DNA strand breaks, impaired DNA repair, and increased mutations, and that folic acid supplementation can correct some of these defects. However, a common polymorphism in methylenetetrahydrofolate reductase (MTHFR), a key protein which controls whether folate is partitioned towards DNA precursor synthesis or DNA methylation, can potentially modify the relationship between folate status and cancer. The most common variant in the *MTHFR* gene, *C677T*, causes a valine for alanine substitution in the protein and reduced enzyme activity in the heterozygotes (CT; 35%) and homozygotes (TT; 70%) [39]. Studies suggest that there is no clear relationship between plasma folate and colorectal adenomas among those with the CC or CT genotype for MTHFR; thus, only a subset of the population (i.e., those with the TT genotype) may benefit from an increased folate intake [40]. These results demonstrate that not all respond identically to bioactive food components. Furthermore, mutations in another folate-metabolizing enzyme, thymidylate synthase, appear to modulate folate intake and colon cancer risk [41]. Possibly 50–100 genes, either directly or indirectly, are involved with folate metabolism. The affected sites include receptors, binding proteins, enzymes, tissue-specific gene products, and downstream factors that rely upon folate-derived metabolites. As a result many factors may be involved in determining if this vitamin is an important dietary variable. The variability within the human genome means that there are thousands of polymorphisms which may determine the biological response to folate.

Folate also serves as an excellent example that dietary components may have different biological effects when given as a normal dietary constituent (folate) or as a supplement (folic acid) and that the effect may be different in normal compared to transformed cells. Animal studies and clinical observations suggest that folate possesses dual modulatory effects on carcinogenesis depending on the timing and dose of folate intervention [42]. Folate deficiency in normal epithelial tissues appears to predispose them to neoplastic transformation, and modest levels of folate supplementation suppress the development of tumors in normal tissues [42]. In contrast, data from animal models, human intervention studies, and analyses of cancer incidence data suggest that supplementation with synthetic folic acid may promote the growth of initiated cells [43, 44]. Rodent studies report a reduction in early markers of colon cancer, such as aberrant crypt foci, when folic acid is given prior to initiation of lesions [45], but cancer development is accelerated if folic acid is given after the emergence of lesions; this is presumably through the provision of DNA precursors for cancer cell growth [46]. Recent findings from several large-scale human observational or placebo-controlled trials indicate that supplemental folic acid increases risk of cancer at several sites, including the breast [47], lung [48], and prostate [43]. Combined high-dose folic acid and vitamin B_{12} supplementation (5 and 1.25 mg/day daily, respectively, for 6 months) had detrimental effects on biomarkers of genomic stability, including increased uracil misincorporation and tumor suppressor gene promoter methylation, in rectal biopsies from colorectal adenoma patients [49]. Overall, these types of observations suggest that the optimal timing, dose, and form of nutrient intervention need to be established for safe and effective cancer prevention in humans.

12.5.4 Carotenoids

Folate is not the only nutrient where high-dose supplementation may potentially have adverse effects on cancer risk. Common green, yellow/red, and yellow/orange vegetables and fruits contain a host of carotenoids. These include lutein, cryptoxanthin, lycopene, β-carotene, α-carotene, and zeaxanthin. Epidemiological studies have reported that foods with higher amounts of carotenoids may reduce the risk of mouth, pharynx, and larynx and lung cancer [5]. Foods containing lycopene possibly protect against prostate cancer [5]. High intakes of fruits and vegetables rich in β-carotene or high plasma concentrations of the nutrient usually have a significant inverse association with lung cancer risk [5]. Adding to this epidemiological evidence, findings from studies on animals demonstrated that β-carotene inhibits cancer-related events, such as the induction of stimulation of intercellular communication via gap junctions, which can have a role in the regulation of cell growth, differentiation, and apoptosis. This evidence provided strong support for testing the effect of β-carotene supplements on lung cancer in randomized intervention trials, as was done in the Alpha-Tocopherol Beta-Carotene Study (ATBC) [50], the Physician's Health Study [51], and the Beta Carotene and Retinol Efficacy Trial (CARET) [52]. Unexpectedly, results from the ATBC and CARET studies showed adverse treatment effects in terms of increased lung cancer incidence in high-risk subjects.

The different results obtained in supplementation trials compared to cohort studies may reflect that fruits and vegetables contain, in addition to β-carotene, numerous other compounds that may be protective against cancer. In fact, β-carotene may simply be a marker for the actual protective substances in fruits and vegetables. Alternately, β-carotene may have different effects when consumed as a supplement rather than in the food supply. It is possible that a protective association present in dietary intake amounts of carotenoids is lost or reversed by the pharmacological levels present in supplementation trials. In one animal study, low-dose β-carotene was protective against smoking-induced changes in p53, while high doses promoted these changes [53]. Moreover, it may be that a subgroup of humans is particularly vulnerable to excess β-carotene. The ATBC, CARET, and PHS studies illustrate that definitive evidence of both safety and efficacy is required for individual constituents of fruits and vegetables before dietary guidelines can be proposed. Thus, consumption of supplements for cancer prevention might have unexpected adverse effects and that consumption of the relevant nutrients through the diet is preferred.

12.6   Dietary Fiber

The term dietary fiber encompasses a complex mix of mostly nondigestible plant cell compounds with variable effects on gut physiology. Dietary fiber from different sources varies in composition, and it is unlikely that all will be equally protective against cancer. Fiber exerts several biological effects in the gastrointestinal tract, including dilution of the fecal content, decreased transit time, and increased stool bulk. Fermentation products, especially short-chain fatty acids, such as butyrate, are produced by the gut microflora from a wide range of dietary fibers. Butyrate can induce apoptosis, cell cycle arrest, and differentiation of tumor cells. But with non-transformed colon cells, in contrast, butyrate is a growth factor and nutrient. An important mechanism by which butyrate causes biological effects in colon cancer cells is through inhibition of histone deacetylase activity. This leads to hyperacety-lation of histones resulting in transcriptional dysregulation and silencing of genes that are involved in the control of cell cycle progression, differentiation, apoptosis, and cancer development.

Sixteen cohort and 91 case–control studies have investigated the relationship between dietary fiber and cancer [5]. Most of these studies showed decreased risk with increased fiber intake; a meta-analysis of cohort data showed a 10% decreased risk of colorectal cancer per 10 g/day increased intake of fiber [5]. A pooled analysis of 8,100 colorectal cancer cases among 730,000 participants, followed up for 6–20 years, showed a nonsignificant decreased risk for the groups that consumed the most dietary fiber [54]. In contrast, the protective effect of dietary fiber against colon cancer has not been observed in randomized intervention trials. Adopting a diet that is low in fat and high in fiber, fruits, and vegetables did not affect the risk of recurrence of colorectal adenomas [55] and did not alter rectal mucosal cell pro-liferation rates [56]. A Cochrane meta-analysis of five randomized controlled trials of increased dietary fiber found no difference between intervention and control groups for the development of adenomas [57]. The different results obtained between the observational and intervention studies may reflect the fact that the intervention studies utilized high-risk individuals, did not use a sufficiently large intake range to detect a response, or that the duration of exposure was insufficient to detect a differ-ence. Taking the evidence as a whole, therefore, the link between dietary fiber and colon cancer remains inconclusive.

The effect of dietary fiber on mammary and prostate cancer risk has also been inconsistent. Whereas a high intake of fiber was protective against breast cancer in the Malmo Diet and Cancer Cohort [58], dietary fiber and fiber fractions did not affect breast cancer risk in the Nurses' Health Study [59]. A large case–control study conducted in Italy found a moderate but significant inverse association between selected types of dietary fiber and prostate cancer risk [60]. The association was strongest for cellulose and for soluble and vegetable fibers [60].

The assessment of a cancer-protective effect for dietary fiber can be complicated by correlations among dietary fiber, dietary fat, and caloric intakes (i.e., high-fiber diets may be relatively low in fat and calories). Another confounding factor is the possible effect on risk caused by micronutrients and phytochemicals in high-fiber foods.

12.7   Meat Intake

Meat, including all animal flesh apart from fish and seafood, can be further classified as either red (beef, pork, lamb, and goat) or poultry. The latter usually has more white than red muscle fibers. The term processed meat refers to meat preserved by smoking, curing, or salting, or addition of chemical preservatives [5]. The WCRF/AICR report suggests that there is convincing evidence that red meat and processed meat are causally related to about a 20% increase in colorectal cancer [5]. However, it is important to realize that increased meat intake likely will not have the same effect on all individuals. Evidence exists that a combination of multiple SNPs in four cytochrome P-450 enzymes, which is present in almost 5% of the population, may pose a risk; it was associated with a more than 40-fold increased risk of colorectal cancer in individuals who consumed a large amount of red meat (>5 times/week) compared to those who did not [61].

There is evidence that an increased cancer risk may not be a function of meat per se but may reflect a high intake of animal fat and/or carcinogens generated through various meat cooking or processing methods. The high energy density of meat can increase the likelihood of obesity, itself a major risk factor for cancer. In fact, total meat consumption is positively associated with weight gain in a cohort of almost 275,000 men and women [62]. After adjustment for estimated energy intake, an increase in meat intake (250 g/day) leads to a 2-kg higher weight gain after 5 years [62]. Cooking methods may foster the formation of carcinogens, including polycyclic aromatic hydrocarbons (PAH) and heterocyclic amines (HCA) [5]. Carcinogenic nitroso-compounds may occur in some processed meats.

Over 100 distinct PAH are formed when organic substances like meat are burnt. These compounds are formed from the pyrolysis of fats that occurs when fat drips from meat onto hot coal, forming smoke that is redeposited on the meat surface. Eleven PAH compounds have been classified as carcinogenic to laboratory animals and as suspect carcinogens in humans [63]. The second class of compounds found in cooked meats is the HCA. These are formed during high-temperature cooking by pyrolysis of proteins, amino acids, or creatinine. The amount in the diet can be substantial and is influenced by cooking habits such that prolonged high-temperature cooking of meats results in the greatest content. Epidemiologic studies have linked HCA with cancers of the colorectum, breast, prostate, lung, and pancreas [64]. Polymorphisms in specific genes associated with metabolism or detoxification of HCA (e.g., *CYP1A1*, *CYP1A2*, *GSTM1*, and *NAT2*) may explain variations in genetic susceptibility among individuals [65]. In view of the possible role of HCA in human carcinogenesis, minimizing exposure seems prudent, i.e., avoiding overheating and overcooking.

Nitrites and nitrates are often used as preservatives in meats and other "cured" products. These additives are not carcinogenic in experimental animals; however, nitrate can interact with dietary substances, such as amines or amides, to produce *N*-nitroso compounds (nitrosamines and nitrosoamides) which are potent carcinogens in animals and probably humans [66]. Epidemiologic studies have demonstrated a direct relationship between nitrosamine exposure and cancer of the stomach, esophagus, nasopharynx, urinary bladder, liver, and brain [66]. When 14 volunteers consumed a diet high in red meat (325 g/day), compared to an isocaloric high fish diet (375 g/day), there was a significantly higher amount of nitroso compounds excreted in the feces (9 μmol/day vs. 1.7 μmol/day, respectively) [67]. Several naturally occurring foods and their constituents, including tea, garlic, and cruciferous vegetables, may inhibit the formation of endogenous nitrosamines [68]. This reduction in carcinogen formation may contribute to the generally protective effect of fruit and vegetables on cancer risk since vitamin C may reduce their formation while other compounds, such as allyl sulfur, may reduce their bioactivation to agents which bind to DNA and thereby lead to the initiation phase of cancer.

Heme iron from animal sources is better absorbed than iron from plant sources, and thus animal food is important in minimizing this nutritional deficiency. However, excess heme iron in the colon may irritate the mucosa and alter the normal rates of proliferation/exfoliation, circumstances that increase the risk for the development of colon cancer [69]. There is also evidence that heme iron in meat may foster the generation of free radicals through the Fenton reaction [70]. Multiple factors may influence the amount of free iron and thus free radicals, including the formation of iron-binding and transport proteins [71].

In a large prospective study of meat consumption and colorectal cancer risk, meat-derived increases in heme iron, nitrate/nitrite, and HCA were all associated with increased risk of colon cancer [72]. Nevertheless, it is important to remember that meat can be a valuable source of many nutrients, including protein, iron, zinc, selenium, and vitamins B_6 and B_{12}. Iron deficiency is the most common and widespread nutritional deficiency in the world. Therefore, it is important to limit red meat consumption rather than to avoid it. Also, it is important to consider the entire diet and thus interactions among different food groups; for example, fruit and vegetables decreasing the formation of nitrosamines.

12.8   Alcohol

Alcohol (ethanol) has been classified by the International Agency for Cancer Research (IARC) as a human carcinogen. Alcohol can be both a source of dietary energy and function as a drug; thus influencing both physical and mental performance. The WCRF/AICR panel judged that there is convincing evidence that alcoholic drinks increase mouth, pharynx and larynx, esophagus, colorectum (men), and breast cancer [5]. Alcoholic drinks are probably also a cause of liver cancer, and of colorectal cancer in women [5]. In a recent cohort study [73] men within the highest quintile of alcohol consumption (>115 g/week) had a 42% increased risk of cancer compared to those within the lowest quintile. The type of beverage consumed does not appear to influence risk and thus alcohol itself appears to be the primary agent leading to the transformation of cells to neoplastic lesions.

In humans, ethanol in alcoholic drinks is mainly oxidized in the liver by alcohol dehydrogenase (ADH) to acetaldehyde. Acetaldehyde is the most toxic metabolite of alcohol and is particularly damaging to cells. In experimental animals, acetaldehyde is a mutagen and carcinogen that causes DNA damage [74]. Acetaldehyde is subsequently metabolized to acetate, mainly by the enzyme aldehyde dehydrogenase 2 (ALDH2). Functional variants in genes involved in alcohol metabolism result in differences between individuals in exposure to carcinogenic acetaldehyde, suggesting possible interactions of genetic susceptibility and alcohol exposure in cancer. For example, there are two main variants of ALDH2, resulting from the replacement of glutamate at position 487 with lysine. The glutamate allele (also designated *ALDH2*1*) encodes a protein with normal catalytic activity, whereas the lysine allele (*ALDH2*2*) encodes an inactive protein [75]. The *ALDH2*2* allele is prevalent in Asians, with a frequency of up to 40%, whereas it does not exceed 5% in European and African populations [76]. As a result, individuals that are homozygous for the lysine allele have no detectable ALDH and experience facial flushing, tachycardia, nausea, and hypotension after alcohol drinking due to acetaldehyde accumulation [75]. Prospective studies in cancer-free alcoholics have shown that the hazard ratio for future aerodigestive tract cancer in individuals with the inactive protein is approximately 12 times higher than in individuals with the active protein [77]. A true understanding of the effect of dietary alcohol may be clouded because of the compounds found in alcoholic beverages, including flavonoids, which can potentially suppress tumorigenesis.

12.9   Summary and Conclusions

Mounting evidence continues to surface that food can have a profound effect on cancer risk and tumor behavior. The research highlighted in this chapter is only a small part of a large body of evidence linking diet and cancer. While a wealth of epidemiological and preclinical (animal and cell culture) investigations exist, more randomized prevention trials are needed. Overall, multiple cancer processes appear to be influenced by many dietary components, although the response is often variable. The overall response is likely dependent on literally thousands of bioactive components that occur in food and their interactions with other environmental factors and the consumer's genetics.

Expanding knowledge about the physiological consequences of nutrigenomics—which includes nutrigenetic (genetic profiles that modulate the response to food components), nutritional transcriptomics (influence of food components on gene expression profiles), and nutritional epigenomics (influence of food components on DNA methylation and other epigenetic events and vice versa)—will help identify those who will and will not respond to particular dietary interventions. New reports are constantly surfacing that population studies are underestimating the significance of diet in overall cancer prevention and therapy, and that subpopulations may be particularly sensitive to subtle changes in eating behaviors. To identify those who will benefit most from dietary change, more attention needs to be given to the identification of three types of biomarkers: (1) those reflecting exposures needed to bring about a desired response, (2) those which indicate a change in a physiologically relevant biological process which is linked to cancer, and (3) those which can be used to predict a personalized susceptibility based on interactions between food components (such as nutrients, phytochemicals, and non-nutrients) and interactions between genes and food components.

The responses to food and/or food components, which may be inhibitory or stimulatory, depending on the specific bioactive food component, are mediated through one and likely multiple, biological mechanisms. The identification and elucidation of the specific molecular sites of action for food components is critical for identifying those who will benefit maximally or be placed at risk from a particular dietary change. Until this information is available it remains prudent to eat a variety of foods and to maintain a healthy weight through appropriate caloric intake and exercise. As the science of nutrition unfolds, a clearer understanding will surely emerge about how food components modulate cancer, and how the food supply might be modified through agronomic approaches and/or biotechnology. While the challenges to unraveling the multiple relationships between diet and cancer prevention will be challenging, the societal and health benefits that will occur are enormous.

References

1. Jermal A, Center MM, DeSantis C, Ward EM. Global patterns of cancer incidence and mortality rates and trends. Cancer Epidemiol Biomarkers Prev. 2010;19:OF1–15.
2. World Health Organization. Ten Statistical highlights in global public health. World Health Statistics. Geneva: WHO; 2007.
3. Hanahan D, Weinberg RA. The hallmarks of cancer. Cell. 2000;100:57–70.
4. Espey DK, Wu XC, Swan J, et al. Annual report to the nation on the status of cancer, 1975–2004, featuring cancer in American Indians and Alaska Natives. Cancer. 2007;110:2119–52.

5. World Cancer Research Fund/American Institute for Cancer Research. Food, nutrition, physical activity, and the prevention of cancer: a global perspective. Washington, DC: AICR; 2007.
6. Doll R, Peto R. The causes of cancer: quantitative estimates of avoidable risk of cancer in the United States today. J Natl Cancer Inst. 1981;66:1191–308.
7. Wong HL, Seow A, Arakawa K, Lee HP, Yu MC, Ingles SA. Vitamin D receptor start codon polymorphism and colorectal cancer risk: effect modification by dietary calcium and fat in Singapore Chinese. Carcinogenesis. 2003;24:1091–5.
8. Siezen CL, van Leeuwen AI, Kram NR, Luken ME, van Kranen HJ, Kampman E. Colorectal adenoma risk is modified by the interplay between polymorphisms in arachidonic acid pathways genes and fish consumption. Carcinogenesis. 2005;26:449–57.
9. Calle EE, Rodriquez C, Walker-Thurmond K, Thun MJ. Overweight, obesity and mortality from cancer is a prospectively studied cohort of U.S. adults. N Engl J Med. 2003;348: 1625–38.
10. Teucher B, Rohrmann S, Kaaks R. Obesity: focus on all-cause mortality and cancer. Maturitas. 2010;65:112–6.
11. Powolny AA, Wang S, Carlton PS, Hoot DR, Clinton SK. Interrelationships between dietary restriction, the IGF-1 axis, and expression of vascular endothelial growth factor by prostate adenocarcinoma in rats. Mol Carcinog. 2008;47:458–65.
12. Renehan AG, Zwahlen M, Minder C, O'Dwyer ST, Shalet SM, Egger M. Insulin-like growth factor (IGF-1), IGF binding protein-3, and cancer risk: systematic review and meta-regression analysis. Lancet. 2004;363:1346–53.
13. Bray GA. Medical consequences of obesity. J Clin Endocrinol Metab. 2004;89:2583–9.
14. Hursting SD, Smith SM, Lashinger LM, Harvey AE, Perkins SN. Calories and carcinogenesis: lessons learned from 30 years of calorie restriction research. Carcinogenesis. 2010;31:83–9.
15. Colman RJ, Anderson RM, Johnson SC, Kastman EK, et al. Caloric restriction delays disease onset and mortality in rhesus monkeys. Science. 2009;325:201–4.
16. Howell A, Chapman M, Harvie M. Energy restriction for breast cancer prevention. Recent Results Cancer Res. 2009;181:97–111.
17. Longo VD, Fontana L. Calorie restriction and cancer prevention: metabolic and molecular mechanisms. Trends Pharmacol Sci. 2009;31:89–98.
18. Centers for Disease Control and Prevention (CDC). Prevalence of regular physical activity among adults—United States, 2001 and 2005. MMWR Morb Mortal Wkly Rep. 2007;56: 1209–12.
19. Rogers CJ, Berrigan D, Zaharoff DA, Hance KW, et al. Energy restriction and exercise differentially enhance components of systemic and mucosal immunity in mice. J Nutr. 2008; 138:115–22.
20. IARC Working Group. IARC Working Group on the evaluation of cancer-preventive strategies. In: Vanio H, Bianchini F, editors. IARC handbooks of cancer prevention, weight control and physical activity, vol. 6. Lyon: IARC Press; 2002. p. 1–315.
21. U.S. Department of Health and Human Services. Objectives 22-2 and 22-3. In: Healthy people 2010. Washington, DC: Department of Health and Human Services; 2000.
22. U.S. Department of Health and Human Services. 2008. Physical activity guidelines for Americans. Hyattsville, MD: Department of Health and Human Services. http://www.health. gov/paguidelines. Accessed 9 Aug 2011.
23. Hill JO, Wyatt HR, Reed GW, Peters JC. Obesity and the environment: where do we go from here? Science. 2003;299:853–5.
24. Davis CD. Mechanisms for the cancer-protective effects of bioactive dietary components in fruits and vegetables. In: Berdanier CD, Dwyer J, Feldman EB, editors. Handbook of nutrition and food. 2nd ed. Boca Raton, FL: CRC Press; 2007. p. 1187–210.
25. Wildman REC, Wildman REC, editors. Handbook of nutraceuticals and functional foods. Boca Raton, FL: CRC Press LLC; 2004.
26. Powolny AA, Singh SV. Multitargeted prevention and therapy of cancer by diallyl trisulfide and related *Allium* vegetable-derived organosulfur compounds. Cancer Lett. 2008;269:305–14.

27. Shukla Y, Kaira N. Cancer chemoprevention with garlic and its constituents. Cancer Lett. 2007;247:167–81.
28. Nagini S. Cancer chemoprevention by garlic and its organosulfur compounds-panacea or promise? Anticancer Agents Med Chem. 2008;8:313–21.
29. Druesne-Pecollo N, Chaumonetet C, Pagniez A, Vaugelade P, Bruneau A, Thomas M, Cherbuy C, Duee PH, Martel P. In vivo treatment by diallyl disulfide increases histone acetylation in rat colonocytes. Biochem Biophys Res Commun. 2007;354:140–7.
30. Schrauzer GN, White DA, Schneider CI. Cancer mortality correlations studies III: statistical associations with dietary selenium intakes. Bioinorg Chem. 1997;7:23–34.
31. Zhuo H, Smith AH, Steinmaus C. Selenium and lung cancer: a quantitative analysis of heterogeneity in the current epidemiological literature. Cancer Epidemiol Biomarkers Prev. 2004;13:771–8.
32. Klein EA. Selenium: epidemiology and basic science. J Urol. 2004;171:S50–3.
33. Li H, Stampfer MJ, Giovannucci EL, et al. A prospective study of plasma selenium levels and prostate cancer risk. J Natl Cancer Inst. 2004;96:696–703.
34. Clark LC, Combs GF, Turnbull BW, et al. Effects of selenium supplementation for cancer prevention in patients with carcinoma of the skin. A randomized controlled trial. JAMA. 1996;276:1957–63.
35. Waters DJ, Shen S, Cooley DM, Bostwick DG, et al. Effects of dietary selenium supplementation on DNA damage and apoptosis in canine prostate. J Natl Cancer Inst. 2003;95:237–41.
36. Toner C, Milner JA, Davis CD. The vitamin D and cancer conundrum: aiming at a moving target. J Am Diet Assoc. 2010;110:1492–500.
37. Lippman SM, Klein EA, Goodman PJ, Lucia MS, et al. Effect of selenium and vitamin E on risk of prostate cancer and other cancers: the Selenium and Vitamin E Cancer Prevention Trial (SELECT). JAMA. 2009;30:39–51.
38. Pinto JT, Lee JI, Sinha R, Macewan ME, Cooper AJ. Chemopreventive mechanisms of alpha-keto acid metabolites of naturally occurring organoselenium compounds. Amino Acids. 2011;41:29–41.
39. Frosst P, Blom HJ, Milos R, et al. Candidate genetic risk factor for vascular disease: a common mutation in methylenetetrahydrofolate reductase. Nat Genet. 1995;10:111–3.
40. Marugame T, Tsuji E, Kiyohara C, et al. Relation of plasma folate and methyltetrahydrofolate reductase C677T polymorphism to colorectal adenomas. Int J Epidemiol. 2003;32:64–6.
41. Ulrich CM, Curtin K, Potter JD, Bigler J, Caan B, Slattery ML. Polymorphisms in the reduced folate carrier, thymidylate synthase, or methionine synthase and risk of colon cancer. Cancer Epidemiol Biomarkers Prev. 2005;14:2509–16.
42. Kim YI. Role of folate in colon cancer development and progression. J Nutr. 2003;133:3731s–9.
43. Mason JB, Dickstein A, Jacques PF, et al. A temporal association between folic acid fortification and an increase in colorectal cancer rates may be illuminating important biological principles: a hypothesis. Cancer Epidemiol Biomarkers Prev. 2007;16:1325–9.
44. Figueiredo JC, Grau MV, Haile RW, et al. Folic acid and risk of prostate cancer: results from a randomized clinical trial. J Natl Cancer Inst. 2009;101:432–5.
45. Kim YI. Folate and colorectal cancer: an evidence-based critical review. Mol Nutr Food Res. 2007;51:267–92.
46. Song J, Sohn K-J, Medline A, Ash C, Gallinger S, Kim Y-I. Chemopreventive effects of dietary folate on intestinal polyps in Apc+/MSH-/-mice. Cancer Res. 2000;6:3191–9.
47. Stolzenberf-Solomon RZ, Chang SC, Leitzmann MF, et al. Folate intake, alcohol use and postmenopausal breast cancer risk in the prostate lung colorectal and ovarian screening trial. Am J Clin Nutr. 2006;83:895–904.
48. Ebbing M, Bonaa KH, Nygard O, et al. Cancer incidence and mortality after treatment with folic acid and vitamin B12. JAMA. 2009;302:2119–26.
49. van den Donk M, Pellis L, Crott JW, et al. Folic acid and vitamin B12 supplementation does not favorably influence uracil misincorporation and promoter methylation in rectal mucosa DNA of subjects with previous colorectal adenomas. J Nutr. 2007;137:2114–20.

50. Heinonen OP, Huttunen IK, Albanes D, et al., for the Alpha-Tocopherol Beta-Carotene Cancer Prevention Study Group. The effect of vitamin E and beta carotene on the incidence of lung cancer and other cancers in male smokers. N Engl J Med. 1994;330:1029–35.

51. Hennekens CH, Buring IE, Manson IE, et al. Lack of effect of long-term supplementation with beta carotene on the incidence of malignant neoplasms and cardiovascular disease. N Engl J Med. 1996;334:1145–9.

52. Omenn OS, Goodman GE, Thomquist MD, et al. Effects of a combination of beta carotene and vitamin A on lung cancer and cardiovascular disease. N Engl J Med. 1996;334:1150–5.

53. Liu C, Russell RM, Wang XD. Low dose beta-carotene supplementation of ferrets attenuates smoke-induced lung phosphorylation of JNK, p38 MAPK and p53 proteins. J Nutr. 2004;134:2705–10.

54. Haas P, Machado MJ, Anton AA, Silva ASS, DeFranciso A. Effectiveness of whole grain consumption in the prevention of colorectal cancer: meta-analysis of cohort studies. Int J Food Sci Nutr. 2009;60:1–13.

55. Schatzkin A, Lanza E, Corle D, et al. Lack of effect of a low-fat, high-fiber diet on the recurrence of colorectal adenomas. N Engl J Med. 2000;342:1149–55.

56. Pfeiffer R, McShane L, Wargovich M, et al. The effect of a low-fat, high fiber, fruit and vegetable intervention on rectal mucosal proliferation. Cancer. 2003;98:1161–8.

57. Asano TK, McLeod RS. Dietary fibre for the prevention of colorectal adenomas and carcinomas (Cochrane review). London: Wiley; 2004.

58. Mattison I, Wirfalt E, Johansson U, Bullberg B, Olsson H, Berglund G. Intakes of plant foods, fibre and fat and risk of breast cancer—a prospective study in the Malmo diet and cancer cohort. Br J Cancer. 2004;90:122–7.

59. Holmes MD, Liu S, Hankinson SE, Coldizt GA, Hunter DJ, Willett WC. Dietary carbohydrates, fiber and breast cancer risk. Am J Epidemiol. 2004;159:732–9.

60. Pelucchi C, Talamini R, Galeone C, et al. Fibre intake and prostate cancer risk. Int J Cancer. 2004;109:278–80.

61. Kury S, Buecher B, Robiou-du-Pont S, et al. Combinations of cytochrome P450 gene polymorphisms enhancing the risk for sporadic colorectal cancer related to red meat consumption. Cancer Epidemiol Biomarkers Prev. 2007;16:1460–7.

62. Vergnaud A-C, Norat T, Romaguera D, et al. Meat consumption and prospective weight change in participants of the EPIC-PANACEA study. Am J Clin Nutr. 2010;92:398–407.

63. Goldamn R, Shields PG. Food mutagens. J Nutr. 2003;133:965S–73.

64. Snyderwine EG, Sinha R, Felton JS, Ferguson LR. Highlights of the eighth international conference on carcinogenic/mutagenic N-substituted aryl compounds. Mutat Res. 2002; 506–507:1–8.

65. Murtaugh MA, Ma K, Sweeney C, Caan BJ, Slattery ML. Meat consumption patterns and preparation, genetic variants of metabolic enzymes, and their association with rectal cancer in men and women. J Nutr. 2004;134:776–84.

66. Ferguson LR. Natural and human-made mutagens and carcinogens in the human diet. Toxicology. 2002;181–182:79–82.

67. Joosen AM, Lecommandeur E, Kuhnle GG, Aspinall SM, Kap L, Rodwell SA. Effect of dietary meat and fish on endogenous nitrosation, inflammation and genotoxicity of faecal water. Mutagenesis. 2010;25:243–7.

68. Sutandyo N. Nutritional carcinogenesis. Acta Med Indones. 2010;42:36–42.

69. Sesnick AL, Termont DS, Kleibeuker JH, Van der Meer R. Red meat and colon cancer: the cytotoxic and hyperproliferative effects of dietary heme. Cancer Res. 1999;59:5704–9.

70. Tappel A. Heme of consumed ret meat can act as a catalyst of oxidative damage and could initiate colon, breast and prostate cancers, heart disease and other diseases. Med Hypotheses. 2007;68:562–4.

71. Mole DR. Iron homeostasis and its interaction with prolyl hydroxylases. Antioxid Redox Signal. 2010;12:445–58.

72. Cross AJ, Ferrucci LM, Risch A, et al. A large prospective study of meat consumption and colorectal cancer risk: an investigation of potential mechanisms underlying this association. Cancer Res. 2010;70:2406–14.

73. Tariola AT, Kurl S, Dyba T, Laukkanen JA, Kauhanen J. The impact of alcohol consumption on the risk of cancer among men: a 20-year follow-up study from Finland. Eur J Cancer. 2010;46:1488–92.
74. Seitz HK, Becker P. Alcohol metabolism and cancer risk. Alcohol Res Health. 2007;30:44–7.
75. Brooks PJ, Enoch M-E, Goldman D, Li T-K, Yokoyama A. The alcohol flushing response: an unrecognized risk factor for esophageal cancer from alcohol consumption. PLoS Med. 2009;3:258–63.
76. Brennan P, Lewis S, Hashibe M, et al. Pooled analysis of alcohol dehydrogenase genotypes and head and neck cancer: a HuGE review. Am J Epidemiol. 2004;159:1–16.
77. Yokoyama A, Omori T, Yokoyama T, et al. Risk of squamous cell carcinoma of the upper aerodigestive tract in cancer-free alcoholic Japanese men: an endoscopic follow-up study. Cancer Epidemiol Biomarkers Prev. 2006;15:2209–15.

Chapter 13
Health Benefits of Phytochemicals
in Whole Foods

Rui Hai Liu

Keywords Phytochemicals • Antioxidants • Cardiovascular disease • Cancer
• Fruit • Vegetables • Whole grains • Dietary supplements

Key Points

- Regular consumption of fruit and vegetables, as well as whole grains, is strongly associated with reduced risk of developing chronic diseases such as cardiovascular disease (CVD), cancer, type 2 diabetes, cataracts, and age-related functional decline.
- Phytochemicals are defined as bioactive nonnutrient plant compounds in fruit, vegetables, grains, and other plant foods, which have been linked to reducing the risk of major chronic diseases. Phytochemicals are classified into carotenoids, phenolics, alkaloids, nitrogen-containing compounds, and organosulfur compounds.
- Oxidative stress can cause oxidative damage to large biomolecules such as proteins, DNA, and lipids, resulting in an increased risk for cancer and CVD. To prevent or slow down the oxidative stress induced by free radicals, sufficient amounts of antioxidants need to be consumed.
- The additive and synergistic effects of phytochemicals in fruit, vegetables, and whole grains are responsible for their potent antioxidant and anticancer activities. The benefit of a diet rich in fruit, vegetables, and whole grains is attributed to the complex mixture of phytochemicals present in these and other whole foods.
- Dietary modification by increasing the consumption of a wide variety of fruit, vegetables, and whole grains daily is a practical strategy for consumers to

R.H. Liu, MD, PhD (✉)
Department of Food Science, Cornell University, 119A Stocking Hall, Ithaca,
NY 14853-7201, USA
e-mail: RL23@cornell.edu

N.J. Temple et al. (eds.), *Nutritional Health: Strategies for Disease Prevention*,
Nutrition and Health, DOI 10.1007/978-1-61779-894-8_13,
© Springer Science+Business Media, LLC 2012

optimize their health and reduce the risk of chronic diseases. Antioxidants are best acquired through the consumption of whole foods, not from expensive dietary supplements.

13.1 Introduction

Cardiovascular disease (CVD) and cancer remain as the top two leading causes of death in the United States and industrialized countries. Epidemiological studies consistently show that regular consumption of fruit and vegetables, as well as whole grains, is strongly associated with reduced risk of developing chronic diseases such as CVD, cancer, type 2 diabetes, cataracts, and age-related functional decline [1–3]. It is estimated that one-third of all cancer deaths in the United States could be avoided through appropriate dietary modification [3–5]. This suggests that change in dietary behavior, such as increasing consumption of fruit, vegetables, nuts, and whole grains, is a practical strategy to significantly reducing the incidence of chronic diseases. In addition, primary prevention of chronic diseases through dietary modification may be as effective, and less costly, than the secondary treatments commonly employed.

The value of adding citrus fruit, carotene-rich fruit and vegetables, and cruciferous vegetables to the diet for reducing the risk of cancer was specifically highlighted by the Nation al Academy of Sciences in 1982 [6]. The National Academy of Sciences on diet and health recommended consuming five or more servings of fruit and vegetables daily for reducing the risk of both cancer and heart disease in 1987 [7]. From this, the Five-a-Day program was developed as a tool to increase public awareness of the health benefits of fruit and vegetable consumption and promote adequate intakes of known vitamins. The 2010 Dietary Guidelines for Americans [8] recommends that most people should eat at least nine servings (4.5 cups) of fruits and vegetables a day based on a 2,000 kcal diet. This consists of four servings (2 cups) of fruits and five servings (2.5 cups) of vegetables. Plant-based foods, such as fruit, vegetables, and whole grains, which contain significant amounts of bioactive phytochemicals, may provide desirable health benefits beyond basic nutrition to reduce the risk of chronic diseases [9, 10]. While references to the importance of increased dietary fruit, vegetables, and nuts as part of reducing chronic disease are legion, implementation has not followed. This chapter will discuss the mechanisms by which phytochemicals appear to prevent the development of these chronic diseases.

13.2 Phytochemicals

The "phyto" of the word phytochemicals is derived from the Greek word "phyto," which means plant. Therefore, phytochemicals are plant chemicals. Phytochemicals are defined as bioactive nonnutrient plant compounds in fruit, vegetables, grains, and other plant foods. They have been linked to reducing the risk of major chronic diseases [10]. It is estimated that more than 5,000 individual phytochemicals have been identified in fruits, vegetables, and whole grains, but a large percentage are still unknown and need to be identified before we can fully understand the health benefits of phytochemicals in whole foods [9]. Phytochemicals may provide this protective effect by a variety of mechanisms including limiting oxidative stress induced by free radicals that are involved in the etiology of a wide range of chronic diseases [11, 12]. Phytochemicals differ widely in their compositions and ratios in fruits, vegetables, nuts, and grains. Their mechanisms are often complimentary to one another and probably work in synergy to each other, therefore it is suggested that in order to receive the greatest health benefits, people should consume a wide variety of plant-based foods daily [9, 10].

Phytochemicals can be classified into broad categories as phenolics, alkaloids, nitrogen-containing compounds, organosulfur compounds, phytosterols, and carotenoids (Fig. 13.1; [10]). Of these phytochemical groups, the phenolics and the carotenoids are the best understood.

13.2.1   Phenolics

Phenolics are compounds possessing one or more aromatic ring with one or more hydroxyl groups. They are generally categorized as phenolic acids, flavonoids, stilbenes, lignans, coumarins, and tannins (Fig. 13.1; [10]). In plants phenolics provide essential functions in the production and growth, defense against pathogens, parasites, and predators, as well as contributing to plant color. In addition to their roles in plants, phenolic compounds in our diet may provide health benefits associated with reduced risk of chronic diseases. Among the 11 common fruits consumed in the United States, cranberry has the highest total phenolic content, followed by apple, red grape, strawberry, pineapple, banana, peach, lemon, orange, pear, and grapefruit [13]. Among the ten common vegetables consumed in the United States, broccoli possesses the highest total phenolic content, followed by spinach, yellow onion, red pepper, carrot, cabbage, potato, lettuce, celery, and cucumber [14]. It is estimated that flavonoids account for approximately two-thirds of the phenolics in our diet, and the remaining one-third are from phenolic acids.

13.2.1.1   Flavonoids

Flavonoids are a group of phenolic compounds with antioxidant activity that have been identified in fruit, vegetables, and other plant foods, and have been linked to reducing the risk of major chronic diseases [10]. More than 4,000 distinct flavonoids have been identified. They commonly have a generic structure consisting of two aromatic rings (A and B rings) linked by three carbons that are usually in an oxygenated heterocycle ring or C ring (Fig. 13.2). Differences in the generic structure

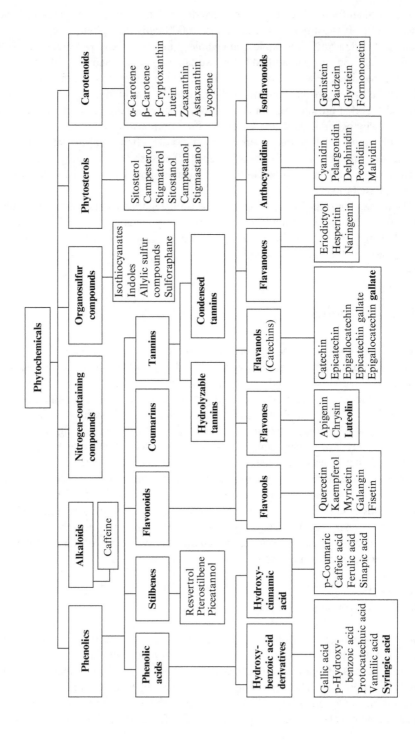

Fig. 13.1 Classification of dietary phytochemicals

Fig. 13.2 The generic structure
of flavonoids

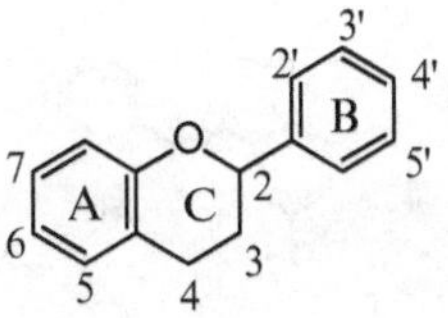

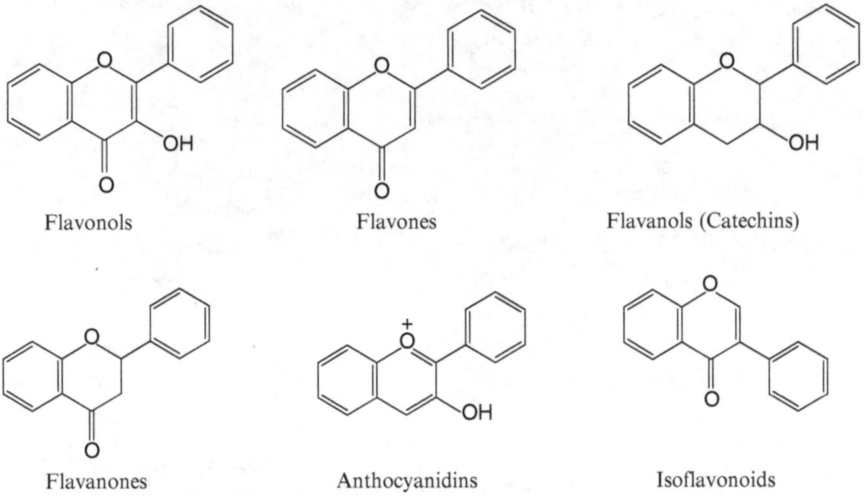

Flavonols   Flavones   Flavanols (Catechins)

Flavanones   Anthocyanidins   Isoflavonoids

Fig. 13.3 Structures of main classes of dietary flavonoids

of the heterocycle C ring classify them as flavonols, flavones, flavanols (catechins), flavanones, anthocyanidins, and isoflavonoids (Fig. 13.3). Common flavonoids in the diet include flavonols (quercetin, kaempferol, and myricetin), flavones (luteolin and apigenin), flavanols (catechin, epicatechin, epigallocatechin [EGC], epicatechin gallate [ECG], and epigallocatechin gallate [EGCG]), flavanones (naringenin), anthocyanidins (cyanidin and malvidin), and isoflavonoids (genistein and daidzein) (Fig. 13.4). Flavonoids are most frequently found in nature as conjugates in glyco-sylated or esterified forms, but can occur as aglycones, especially as a result of the effects of food processing or digestion in the gut. Many different glycosides can be found in nature as more than 80 different sugars have been discovered bound to flavonoids [15]. Anthocyanidins give the red and blue colors associated with some fruit, vegetables, and whole grains.

Human intake of all flavonoids is estimated at a few 100 mg [16] to 650 mg per day [17]. The total average intake of flavonols (quercetin, myricetin, kaempferol) and flavones (luteolin, apigenin) was estimated as 23 mg/day, of which quercetin contributed about 70%, kaempferol 17%, myricetin 6%, luteolin 4%, and apigenin 3% [18].

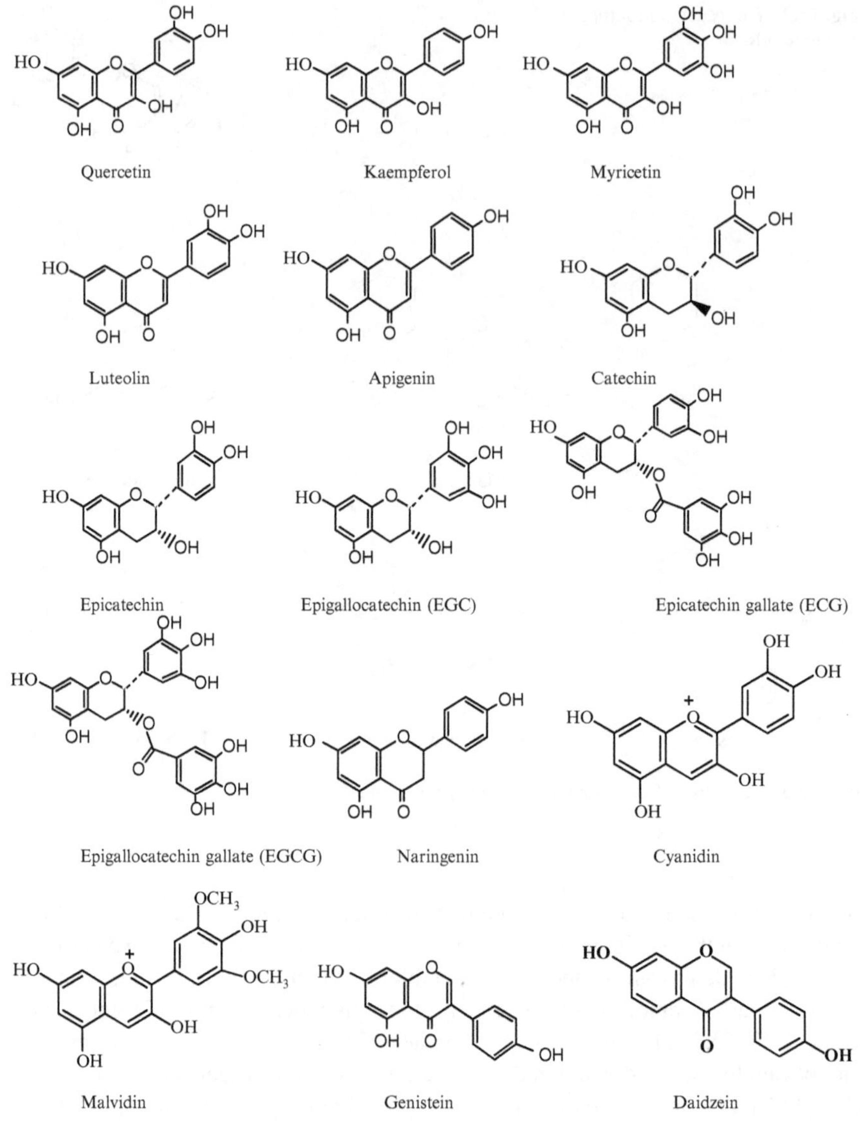

Fig. 13.4 Chemical structures of common dietary flavonoids

13.2.1.2 Phenolic Acids

Phenolic acids are one of the major sources of dietary phenolics. They can be subdivided into two major groups, hydroxybenzoic acids and hydroxycinnamic acids (Fig. 13.5; [10]). Hydroxybenzoic acid derivatives include *p*-hydroxybenzoic, protocatechuic, vannilic, syringic, and gallic acids. They are commonly present in the

a Benzoic acid derivatives

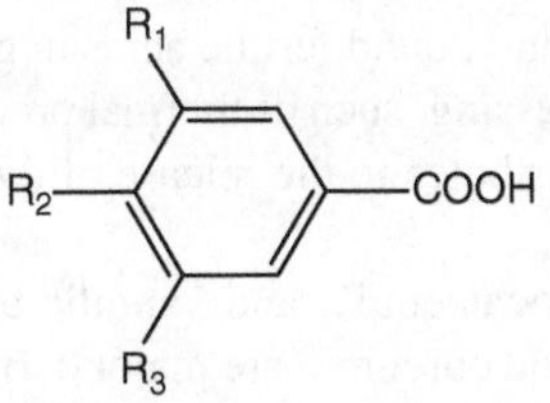

Benzoic acid	Substitutions		
derivatives	R_1	R_2	R_3
Benzoic acid	H	H	H
p-Hydroxybenzoic acid	H	OH	H
Protocatechuic acid	H	OH	OH
Vannilic acid	CH_3O	OH	H
Syringic acid	CH_3O	OH	CH_3O
Gallic acid	OH	OH	OH

b Cinnamic acid derivatives

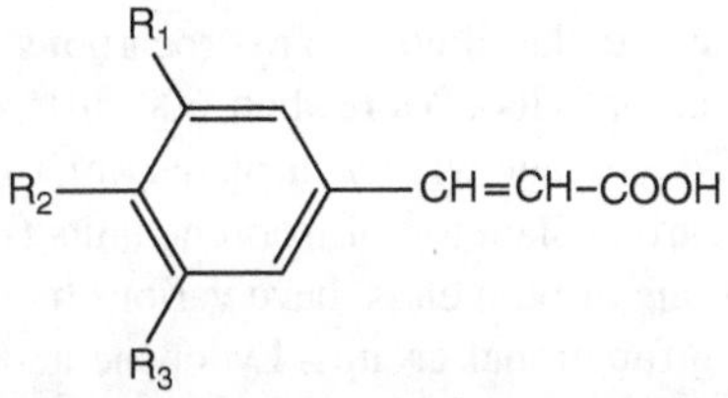

Cinnamic acid	Substitutions		
derivatives	R_1	R_2	R_3
Cinnamic acid	H	H	H
p-Coumaric acid	H	OH	H
Caffeic acid	OH	OH	H
Ferulic acid	CH_3O	OH	H
Sinapic acid	CH_3O	OH	CH_3O

Fig. 13.5 Structures of common phenolic acids: (**a**) benzoic acid and derivatives; (**b**) cinnamic acid and derivatives

bound form in foods and are typically components of a complex structure like lignins and hydrolyzable tannins [10]. They can also be found in the form of sugar derivatives and organic acids in plant foods.

Hydroxycinnamic acid derivatives include *p*-coumaric, caffeic, ferulic, and sinapic acids (Fig. 13.5). They are mainly present in the bound form, linked to cell wall structural components such as cellulose, lignin, and proteins through ester bonds [10]. Ferulic acids occur primarily in the seeds and leaves of plants, mainly covalently conjugated to mono- and disaccharides, plant cell wall polysaccharides, glycoproteins, polyamines, lignin, and insoluble carbohydrate biopolymers. Wheat bran is a good source of ferulic acids, which are esterified to hemicellulose of the cell walls. Free, soluble-conjugated, and bound ferulic acids in grains are present in the ratio of 0.1:1:100 [19]. Food processing, such as thermal processing, pasteurization, fermentation, and freezing, contributes to the release of these bound phenolic acids [20].

Caffeic, ferulic, p-coumaric, protocatechuic, and vannilic acids are present in almost all plants. Chlorogenic acids and curcumin are major derivatives of hydroxycinnamic acids present in plants. Chlorogenic acids are the ester of caffeic acid and are the substrate for enzymatic oxidation leading to browning, particularly in apples and potatoes. Curcumin is made of two ferulic acids linked by a methylene in a diketone structure and is the major yellow pigment of spice turmeric and mustard.

13.2.2   Carotenoids

Carotenoids are nature's most widespread pigments with yellow, orange, and red colors, and have also received substantial attention because of both their provitamin and antioxidant roles. Carotenoids are classified into hydrocarbons (carotenes) and their oxygenated derivatives (xanthophylls). More than 600 different carotenoids have been identified in nature. They occur widely in plants, microorganisms, and animals. Carotenoids have a 40-carbon skeleton of isoprene units (Fig. 13.6; [10]). The structure may be cyclized at one or both ends, have various hydrogenation levels, or possess oxygen-containing functional groups. Lycopene and β-carotene are examples of acyclized and cyclized carotenoids, respectively. Carotenoid compounds most commonly occur in nature in the all-trans form. The most characteristic feature of carotenoids is the long series of conjugated double bonds forming the central part of the molecule. This gives them their shape, chemical reactivity, and light-absorbing properties. β-Carotene, α-carotene, and β-cryptoxanthin have provitamin A activity and can be converted to vitamin A (retinol) after being metabolized. Zeaxanthin and lutein are the major carotenoids in the macular region (yellow spot) of the retina in humans. A diet rich in zeaxanthin and lutein has been associated with reduced risk of developing cataract and macular degeneration.

Orange and yellow vegetables and fruits, including carrots, sweet potatoes, pumpkin, winter squash, papaya, mango, and cantaloupe, are rich sources of the carotenoid β-carotene. Dark green leaf vegetables, including spinach, kale, turnip greens, and collards, are rich sources of lutein and zeaxanthin. Tomatoes, watermelons, pink grapefruits, apricots, and pink guavas are the most common sources of lycopene. It has been estimated that 85% of American lycopene intake comes from processed tomato products such as ketchup, tomato paste, and tomato soup. Carotenoid pigments play important functions in photosynthesis and photoprotection in plants. The photoprotection role of carotenoids originates from their ability to quench and inactivate reactive oxygen species such as singlet oxygen formed

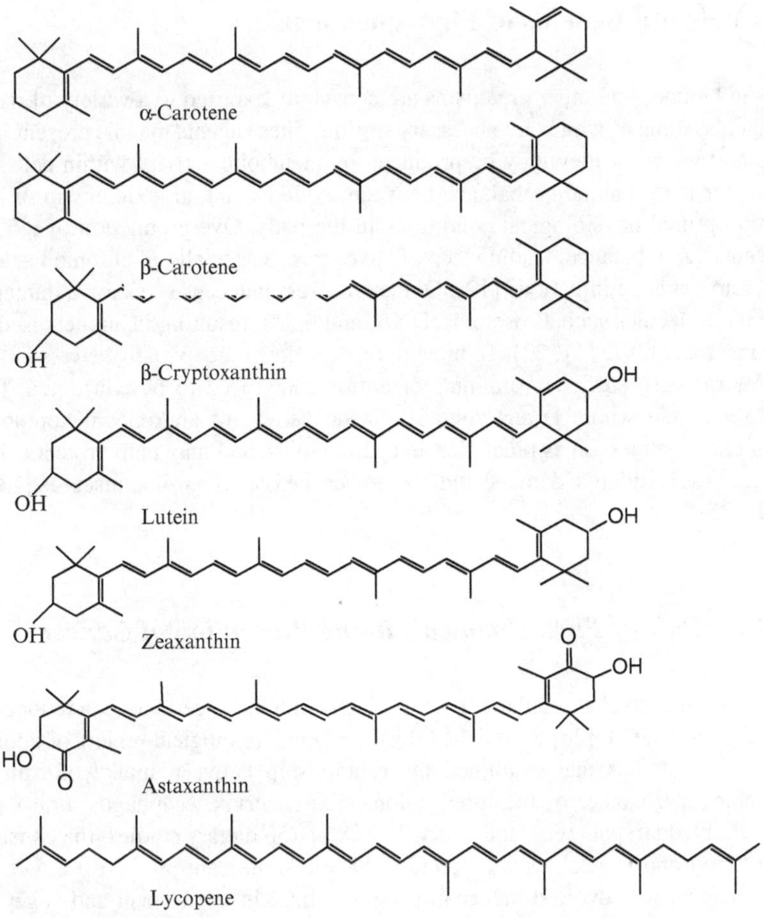

Fig. 13.6 Chemical structures of common dietary carotenoids

from exposure to light and air. This photoprotection role is also associated with its antioxidant activity in human health. Carotenoids can react with free radicals and become radicals themselves. Their reactivity depends on the length of the chain of conjugated double bonds and the characteristics of the end groups. Carotenoid radicals are stabilized by delocalization of unpaired electron over the conjugated polyene chain of the molecules. This also allows additional reactions to occur at many sites on the radical [21]. Carotenoids are especially powerful against singlet oxygen generated from lipid peroxidation or radiation. Astaxanthin, zeaxanthin, and lutein are excellent lipid-soluble antioxidants that scavenge free radicals, especially in a lipid-soluble environment. Carotenoids at sufficient concentrations can prevent lipid oxidation and related oxidative stress.

13.3 Health Benefits of Phytochemicals

Cells in humans and other organisms are constantly exposed to a variety of oxidizing agents, some of which are necessary for life. These agents may be present in air,
food, and water, or they may be produced by metabolic activity within cells. The
key factor is to maintain a balance between oxidants and antioxidants in order to
sustain optimal physiological conditions in the body. Overproduction of oxidants
can cause an imbalance, leading to oxidative stress, especially in chronic bacterial,
viral, and parasitic infections [12]. Oxidative stress can cause oxidative damage to
large biomolecules such as proteins, DNA, and lipids, resulting in an increased risk
for cancer and CVD [11, 22]. To prevent or slow down the oxidative stress induced
by free radicals, sufficient amounts of antioxidants need to be consumed. Fruit,
vegetables, and whole grains contain a wide variety of antioxidant compounds
(phytochemicals), such as phenolics and carotenoids, and may help protect cellular
systems from oxidative damage and also lower the risk of chronic diseases [13, 14,
19, 23–25].

13.3.1 Role of Phytochemicals in the Prevention of Cancer

Evidence suggests that regular consumption of fruit and vegetables can reduce cancer risk. Block et al. [26] established this in an epidemiological review of approximately 200 studies that examined the relationship between intake of fruit and
vegetables and cancer of the lung, colon, breast, cervix, esophagus, oral cavity,
stomach, bladder, pancreas, and ovary. In 128 of 156 dietary studies, the consumption of fruit and vegetables was found to have a significant protective effect. The
risk of cancer was twofold higher in persons whose intake of fruit and vegetables
was low compared to those with a high intake. Significant protection was found in
24 of 25 studies for lung cancer. Fruit was significantly protective in cancer of the
esophagus, oral cavity, and larynx. In 26 of 30 studies, there was a protective effect
of fruit and vegetable intake for cancer of the pancreas and stomach, and in 23 of 38
studies for colorectal and bladder cancer. A prospective study involving 9,959 men
and women in Finland showed an inverse association between the intake of
flavonoids and incidence of cancer at all sites combined [27]. After a 24-year follow-up, the risk of lung cancer was reduced by 50% in the highest quartile of
flavonol intake. Consumption of quercetin from onions and apples was found to be
inversely associated with lung cancer risk [28]. The effect of onions was particularly
strong against squamous cell carcinoma.

Boyle et al. [29] showed that increased plasma levels of quercetin following a
meal of onions was accompanied by increased resistance to strand breakage by
lymphocyte DNA and decreased levels of some oxidative metabolites in the urine.

Carcinogenesis is a multistep process, and oxidative damage is linked to the
formation of tumors through several mechanisms [12, 22]. Oxidative stress induced
by free radicals causes DNA damage, which, when left unrepaired, can lead to base

Table 13.1 Proposed mechanisms by which dietary phytochemicals may prevent cancer

Antioxidant activity
  Scavenge free radicals and reduce oxidative stress
  Inhibit nitrosation and nitration
  Prevent DNA binding
Inhibition of cell proliferation
Induction of cell differentiation
Inhibition of oncogene expression and induction of tumor suppress gene expression
Induction of cell cycle arrest and apoptosis
Inhibition of signal transduction pathways
Enzyme induction and enhancing detoxification
  Phase II enzyme
  Glutathione peroxidase (GPX)
  Catalase
  Superoxide dismutase (SOD)
Enzyme inhibition
  Cyclooxygenase-2 (COX-2)
  Inducible nitric oxide synthase (iNOS)
  Xanthine oxide
  Phase I enzyme (block activation of carcinogens)
Enhancement of immune functions and surveillance
Anti-angiogenesis
Inhibition of cell adhesion and invasion
Regulation of steroid hormone metabolism
Regulation of estrogen metabolism
Antibacterial and antiviral effects

mutation, single and double strand breaks, DNA cross-linking, and chromosomal breakage and rearrangement [22]. This potentially cancer-inducing oxidative damage might be prevented or limited by dietary antioxidants found in fruit and vegetables. Studies to date have demonstrated that phytochemicals in common fruit and vegetables can have complementary and overlapping mechanisms of action (Table 13.1), including scavenging free radicals, regulation of gene expression in cell proliferation and apoptosis, modulation of detoxification enzymes, stimulation of the immune system, regulation of hormone metabolism, and antibacterial and antiviral effects [30, 31].

13.3.2   Role of Phytochemicals in the Prevention of Cardiovascular Disease

Several epidemiological studies have examined the role of phytochemicals on CVD prevention. Dietary flavonoid intake was significantly inversely associated with mortality from coronary heart disease (CHD), and had an inverse relation (weaker but significant) with incidence of myocardial infarction (MI) [32]. Dietary flavonoid intake was also inversely associated with CHD mortality [33]. Intake of apples and onions, both high in quercetin, was inversely correlated with total and CHD mortality [34]. In a recent Japanese study, the total intake of flavonoids (quercetin, myricetin, kaempferol, luteolin, and ficetin) was inversely correlated with the plasma total cholesterol and low-density lipoprotein (LDL) cholesterol concentrations [35]. As a single phytochemical, quercetin intake was inversely related to total cholesterol and LDL plasma levels. Joshipura et al. [36] reported that total intake of both fruit and vegetables was each associated with decreased risk for CHD. The inverse associations between total consumption of fruit and vegetables and CHD were observed for above intake of more than 4 servings/day. The Women's Health Study subjects had a relative risk of 0.68 for CVD when comparing the highest vs. the lowest quintiles of fruit and vegetable intake, and the relative risk for MI was only 0.47. It was estimated that there was a 20–30% reduction in risk of CVD associated with high fruit and vegetable intake [37]. Most recently, in a study involving subjects from the National Health and Nutrition Examination Survey Epidemiologic Follow-up Study, there was a 27% lower CVD mortality with consumption of fruit and vegetables at least three times/day compared to only once/day. Fruit and vegetable intake was inversely associated with incidence of stroke, stroke mortality, CHD mortality, CVD mortality, and all-cause mortality [38].

Mechanisms for the prevention of atherosclerosis by dietary antioxidants in fruit and vegetables have been proposed. In the LDL oxidation hypothesis, oxidized LDL cholesterol has been suggested as the atherogenic factor that contributes to CVD [39, 40]. When circulating LDLs are present at high levels, they infiltrate the artery wall and increase intimal LDL, which can then be oxidized by free radicals. This oxidized LDL in the intima is more atherogenic than native LDL and serves as a chemotactic factor in the recruitment of circulating monocytes and macrophages. Oxidized LDL is typically taken up by macrophage scavenger receptors, thus inducing the formation of inflammatory cytokines and promoting cell proliferation, cholesterol ester accumulation, and foam cell formation. Gruel-like, lipid-laden foam cell accumulation in the blood vessel, forming fatty streak, would cause further endothelial injury and lead to atherosclerotic disease. Since oxidized LDL plays a key role in the initiation and progression of atherosclerosis, giving dietary supplements of antioxidants capable of preventing LDL oxidation has been an important therapeutic approach. Dietary antioxidants that are incorporated into LDL are themselves oxidized when the LDL is exposed to pro-oxidative conditions; this occurs before any extensive oxidation of the sterol or polyunsaturated fatty acids can occur [41]. Dietary antioxidants might therefore retard the progression of atherosclerotic lesions. In addition, phytochemicals have been shown to have roles in the reduction of platelet aggregation, modulation of cholesterol synthesis and absorption, and reduction of blood pressure. Recently, C-reactive protein, a marker of systemic inflammation, has been reported to be a stronger predictor of CVD than is LDL cholesterol [42], suggesting that inflammation is a critical factor in CVD. Inflammation not only promotes initiation and progression of atherosclerosis, but also causes acute thrombotic complications of atherosclerosis [43]. In a 2005 Italian study, the relationship between high-sensitivity C-reactive protein (hs-CRP) and the total antioxidant capacity (TAC) of a diet was studied. Even within groups controlled for dietary factors, TAC was found to be significantly higher among those with a low level of plasma hs-CRP when compared to subjects with high levels of plasma hs-CRP [44]. This indicates that the TAC of a specific diet is independently and inversely correlated with hs-CRP, and that this could be one of the mechanisms behind the protective effects of fruits and vegetables against CVD. Dietary phytochemicals can lower C-reactive protein dramatically. Both fruits and vegetables were inversely associated with plasma CRP concentrations [45]. Therefore, the anti-inflammatory activity of phytochemicals may play an important role in prevention of CVD.

13.4 Health Benefits of Phytochemicals in Whole Foods: Food Synergy

The hypothesis that dietary antioxidants lower the risk of chronic disease has been developed from epidemiological studies. These have consistently shown that consumption of whole foods, such as fruit, vegetables, and whole grains, is strongly associated with reduced risk of developing chronic diseases. Therefore, it is reasonable for scientists to isolate and identify the bioactive compounds responsible and hope to find the "magic bullet" to prevent those chronic diseases. The key question here is whether a purified phytochemical has the same health benefit as the compound does when its source is a food or a mixture of foods. It is now widely believed that the actions of the dietary supplements alone do not explain the observed health benefits of diets rich in fruit and vegetables; this is because, taken alone, the individual antioxidants studied in clinical trials do not appear to have consistent preventive effects [46–48]. The isolated pure compound either loses its bioactivity or may not behave the same way as the compound in whole foods. For example, numerous investigations have shown that the risk of cancer is inversely related to the consumption of green and yellow vegetables and fruit. Since β-carotene is present in abundance in these vegetables and fruit, it has been extensively investigated as a possible cancer-preventive agent. However, the role of carotenoids as anticancer supplements has recently been questioned as a result of several clinical studies [46, 49–51]. In one study, the incidence of nonmelanoma skin cancer was unchanged in patients receiving a β-carotene supplement [49]. In other studies, smokers gained no benefit from supplemental β-carotene with respect to lung cancer incidence and may even have suffered a significant increase in lung cancer and total mortality [46, 51]. In The Heart Outcomes Prevention Evaluation (HOPE) study, patients at high risk for cardiovascular events were given 400 IU/day vitamin E or a placebo for 4.5 year. No difference was found in deaths from cardiovascular causes, MIs, or deaths from CHD or strokes between the two groups [52]. In the Cambridge Heart Antioxidant Study (CHAOS), patients with CHD were given 400 or 800 IU α-tocopherol or a placebo for a median of 510 day. α-Tocopherol intake was associated with a significantly reduced risk of MI; however, it insignificantly increased risk of cardiovascular death [47]. Vitamin E supplementation had no effect on the endpoints of death, MI, or stroke for the patients who had recently suffered a MI [53]. Vitamin C supplements also failed to lower the incidence of cancer or

CHD [54, 55]. In a recent clinical trial, the potential of selenium and vitamin E for preventing prostate cancer was studied. Both these compounds were found to be beneficial when patterns of increased intake in diet were studied in many previous studies. However, in this clinical trial an increased incidence of prostate cancer was found in the group taking vitamin E, and of type 2 diabetes in the group taking selenium [56]. In both cases the increased incidence was close to significant ($P=0.06$ and $P=0.16$). The trial was therefore stopped due to ethical and safety concerns.

Phytochemical extracts from fruit and vegetables have recently been shown to have potent antioxidant and antiproliferative effects, and the combination of phytochemicals from fruit and vegetables has been proposed to be responsible for the potent antioxidant and anticancer activity of these foods [13, 14, 57]. The antioxidant value of 100 g apples is equivalent to 1,500 mg of vitamin C [57]; this is far higher than the total antioxidant activity of the 5.7 mg of vitamin C in 100 g of apples. Thus, the vast majority of the antioxidant activity comes from phytochemicals, not vitamin C. The natural combination of phytochemicals in fruit and vegetables is responsible for its potent antioxidant activity. Apple extracts also contain bioactive compounds that inhibit tumor cell growth in vitro. Phytochemicals in apples with peel (50 mg/mL on a wet basis) inhibit colon cancer cell proliferation by 43%. However, this was reduced to 29% when apple without peel was tested [57]. Whole apple extracts prevented mammary cancer in a rat model in a dose-dependent manner at the doses comparable to human consumption of 1, 3, and 6 apples a day [58]. This study demonstrated that whole apple extracts effectively inhibited mammary cancer growth in the rat model, thus consumption of apples may be an effective strategy for cancer protection. Recently, a study examining the possible additive, synergistic, or antagonistic interactions among phytochemicals yielded results suggesting that the apple phytochemical extracts and quercetin 3-beta-D-glucoside (Q3G) in combination possesses a synergistic effect against MCF-7 human breast cancer cell proliferation [59].

Different species and varieties of fruit, vegetables, and whole grains have different phytochemical profiles [13, 14, 19, 60, 61]. Therefore, consumers should obtain their phytochemicals from a wide variety of these foods for optimal health benefits. Temple and Gladwin reviewed more than 200 cohort and case–control studies that provided risk ratios (RR) concerning intake of fruit and vegetables and risk of cancer [62]. They concluded that cancer prevention is best achieved by the consumption of a wide variety of fruit and vegetables, although one group of fruit and vegetables may dominate a particular cancer. To improve their nutrition and health, consumers should be obtaining antioxidants from their diet and not from expensive nutritional supplements, which do not contain the balanced combination of phytochemicals found in fruit, vegetables, whole grains, and other whole foods. The health benefits from the consumption of fruit, vegetables, and whole grains extend beyond lowering the risk of developing cancers and CVD: it also has preventive effects on other chronic diseases such as cataracts, age-related macular degeneration, central neurodegenerative diseases, and type 2 diabetes [10].

The additive and synergistic effects of phytochemicals in fruit and vegetables have been proposed to be responsible for their potent antioxidant and anticancer activities [9, 10]. The benefit of a diet rich in fruit, vegetables, and whole grains is attributed to the complex mixture of phytochemicals present in these and other whole foods [13, 14, 57, 61]. This partially explains why no single antioxidant can replace the combination of natural phytochemicals in these foods in achieving the observed health benefits. There are thousands of phytochemicals present in whole foods. These compounds differ in molecular size, polarity, and solubility, which may affect the bioavailability and distribution of each phytochemical in different macromolecules, subcellular organelles, cells, organs, and tissues. This balanced natural combination of phytochemicals present in fruit, vegetables, and whole grains cannot simply be mimicked by pills or tablets.

Research progress in antioxidant and bioactive compounds has boosted the dietary supplement and nutraceutical industries (see Chap. 20 by Temple). The use of dietary supplements is growing, especially among baby-boomer consumers. However, many of these supplements have been developed based on the research results derived from biochemical and chemical analyses and studies, in vitro cell culture studies, and in vivo animal experiments, without human intervention studies. The health benefits of natural phytochemicals at the low levels present in fruit, vegetables, and whole grains does not mean that these compounds are more effective or safe when they are consumed at a higher dose, even in a pure dietary supplement form. Generally speaking, higher doses increase the risk of toxicity. The basic principle of toxicology is that any compound can be toxic if the dose is high enough; dietary supplements are no exception. Therefore, a thorough understanding of the efficacy and long-term safety of many dietary supplements needs further investigation.

It is also important to differentiate pharmacological doses from physiological (or nutritional) doses. Pharmacological doses are used clinically to treat specific diseases in certain situations; physiological (or nutritional) doses, such as the doses naturally present in whole foods, are much lower and are used to improve or maintain optimal health. Pharmacological doses can sometimes be toxic for long-term use. In the case of antioxidant nutrients and common dietary supplements, the proper physiological (or nutritional) doses should follow the Dietary Reference Intakes (DRI). We do not have DRI for phytochemicals or of many dietary supplements. It is therefore unwise to take mega-doses of purified phytochemicals as supplements to improve nutritional status or maintain health before the appearance of strong supporting evidence.

13.5 Summary

Dietary modification by increasing the consumption of a wide variety of fruit, vegetables, and whole grains daily is a practical strategy for consumers to optimize their health and reduce the risk of chronic diseases. Use of dietary supplements, functional foods, and nutraceuticals is increasing as industry is responding to consumers' demands, while the benefits of whole foods are quite apparent, the efficacy of nutraceuticals remains to be validated. However, there is a need for more information about the health benefits and possible risks of dietary supplements so as to ensure their efficacy and safety. Phytochemical extracts from fruit and vegetables have strong antioxidant and antiproliferative activities, and the major part of total antioxidant activity is from the combination of phytochemicals. The additive and synergistic effects of phytochemicals in fruit and vegetables are responsible for their potent antioxidant and anticancer activities. The benefit of a diet rich in fruit, vegetables, and whole grains is attributed to the complex mixture of phytochemicals present in these and other whole foods. This explains why no single antioxidant can replace the combination of natural phytochemicals in fruit and vegetables and achieve their health benefits. Therefore, the evidence suggests that antioxidants are best acquired through whole food consumption, not from expensive dietary supplements. Further research on the health benefits of phytochemicals in whole foods is warranted.

References

1. Temple NJ. Antioxidants and disease: more questions than answers. Nutr Res. 2000;20: 449–59.
2. Willet WC. Diet and health: what should we eat. Science. 1994;254:532–7.
3. Willett WC. Balancing life-style and genomics research for disease prevention. Science. 2002;296:695–8.
4. Doll R, Peto R. Avoidable risks of cancer in the United States. J Natl Cancer Inst. 1981;66: 1197–265.
5. Willet WC. Diet, nutrition, and avoidable cancer. Environ Health Perspect. 1995;103:165–70.
6. National Academy of Sciences. Committee on diet, nutrition, and cancer, assembly of life sciences, national research council. Diet, nutrition, and cancer. Washington: National Academy Press; 1982.
7. National Academy of Sciences. Committee on diet and health, national research council. Diet and health: implications for reducing chronic disease risk. Washington: National Academy Press; 1989.
8. USDA. Dietary Guidelines for Americans 2010. Available at http://www.cnpp.usda.gov/dietaryguidelines.html (2011). Last Accessed 11 Feb 2011.
9. Liu RH. Health benefits of fruits and vegetables are from additive and synergistic combination of phytochemicals. Am J Clin Nutr. 2003;78:517S–20.
10. Liu RH. Potential synergy of phytochemicals in cancer prevention: mechanism of action. J Nutr. 2004;134:3479S–85.
11. Ames BN, Gold LS. Endogenous mutagens and the causes of aging and cancer. Mutat Res. 1991;250:3–16.
12. Liu RH, Hotchkiss JH. Potential genotoxicity of chronically elevated nitric oxide: a review. Mutat Res. 1995;339:73–89.
13. Sun J, Chu Y-F, Wu X, Liu RH. Antioxidant and antiproliferative activities of fruits. J Agric Food Chem. 2002;50:7449–54.
14. Chu Y-F, Sun J, Wu X, Liu RH. Antioxidant and antiproliferative activities of vegetables. J Agric Food Chem. 2002;50:6910–6.
15. Hollman PCH, Arts ICW. Flavonols, flavones and flavanols—nature, occurrence and dietary burden. J Sci Food Agric. 2000;80:1081–93.
16. Hollman PCH, Katan MB. Dietary flavonoids: intake, health effects and bioavailability. Food Chem Toxicol. 1999;37:937–42.

17. Kuhnau J. The flavonoids. A class of semi-essential food components: their role in human nutrition. World Rev Nutr Diet. 1976;24:117–91.
18. Hertog MGL, Hollman PCH, Katan MB, Kromhout D. Intake of potentially anticarcinogenic flavonoids and their determinants in adults in The Netherlands. Nutr Cancer. 1993; 20:21–9.
19. Adom KK, Liu RH. Antioxidant activity of grains. J Agric Food Chem. 2002;50:6182–7.
20. Dewanto V, Wu X, Liu RH. Processed sweet corn has higher antioxidant activity. J Agric Food Chem. 2002;50:4959–64.
21. Britton G. Structure and properties of carotenoids in relation to function. FASEB J. 1995;9: 1551–8.
22. Ames BN, Shigenaga MK, Gold LS. DNA lesions, inducible DNA repair, and cell division: the three key factors in mutagenesis and carcinogenesis. Environ Health Perspect. 1993;101 Suppl 5:35–44.
23. Wang H, Cao GH, Prior RL. Total antioxidant capacity of fruits. J Agric Food Chem. 1996;44:701–5.
24. Vinson JA, Hao Y, Su X, Zubik L, Bose P. Phenol antioxidant quantity and quality in foods: fruits. J Agric Food Chem. 2001;49:5315–21.
25. Adom KK, Sorrells ME, Liu RH. Phytochemicals and antioxidant activity of wheat varieties. J Agric Food Chem. 2003;51:7825–34.
26. Block G, Patterson B, Subar A. Fruit, vegetables, and cancer prevention: a review of the epidemiological evidence. Nutr Cancer. 1992;18:1–29.
27. Knekt P, Jarvinen R, Seppanen R, et al. Dietary flavonoids and the risk of lung cancer and other malignant neoplasms. Am J Epidemiol. 1997;146:223–30.
28. Le Marchand L, Murphy SP, Hankin JH, Wilkens LR, Kolonel LN. Intake of flavonoids and lung cancer. J Natl Cancer Inst. 2000;92:154–60.
29. Boyle SP, Dobson VL, Duthie SJ, Kyle JAM, Collins AR. Absorption and DNA protective effects of flavonoid glycosides from an onion meal. Eur J Nutr. 2000;39:213–23.
30. Dragsted LO, Strube M, Larsen JC. Cancer-protective factors in fruits and vegetables: biochemical and biological background. Pharmacol Toxicol. 1993;72:116–35.
31. Waladkhani AR, Clemens MR. Effect of dietary phytochemicals on cancer development. Int J Mol Med. 1998;1:747–53.
32. Hertog MGL, Feskens EJM, Hollman PCH, Katan MB, Kromhout D. Dietary antioxidant flavonoids and risk of coronary heart disease: The Zutphen Elderly Study. Lancet. 1993;342: 1007–11.
33. Hertog MGL, Kromhout D, Aravanis C, et al. Flavonoid intake and long-term risk of coronary heart disease and cancer in the Seven Countries Study. Arch Intern Med. 1995;155:381–6.
34. Knekt P, Jarvinen R, Reunanen A, Maatela J. Flavonoid intake and coronary mortality in Finland: a cohort study. Br Med J. 1996;312:478–81.
35. Arai Y, Watanabe S, Kimira M, Shimoi K, Mochizuki R, Kinae N. Dietary intakes of flavonols, flavones and isoflavones by Japanese women and the inverse correlation between quercetin intake and plasma LDL cholesterol concentration. J Nutr. 2000;131:2243–50.
36. Joshipura KJ, Hu FB, Manson JE, et al. The effect of fruit and vegetable intake on risk for coronary heart disease. Ann Intern Med. 2001;134:1106–14.
37. Liu S, Manson JE, Lee I-M, et al. Fruit and vegetable intake and risk of cardiovascular disease: the Women's Health Study. Am J Clin Nutr. 2000;72:922–8.
38. Bazzano LA, He J, Ogden LG, et al. Fruit and vegetable intake and risk of cardiovascular disease in US adults: the first National Health and Nutrition Examination Survey Epidemiologic Follow-up Study. Am J Clin Nutr. 2002;76:93–9.
39. Berliner J, Leitinger N, Watson A, Huber J, Fogelman A, Navab M. Oxidized lipids in atherogenesisi: formation, destruction and action. Thromb Haemost. 1997;78:195–9.
40. Witztum JL, Berliner JA. Oxidized phospholipids and isoprostanes in atherosclerosis. Curr Opin Lipidol. 1998;9:441–8.
41. Sanchez-Moreno C, Jimenez-Escrig A, Saura-Calixto F. Study of low-density lipoprotein oxidizability indexes to measure the antioxidant activity of dietary polyphenols. Nutr Res. 2000;20:941–53.

42. Ridker PM, Rifai N, Rose L, Buring JE, Cook NR. Comparison of C-reactive protein and low-density lipoprotein cholesterol levels in the prediction of first cardiovascular events. N Engl J Med. 2002;347:1557–65.

43. Libby P, Ridker PM, Maseri A. Inflammation and atherosclerosis. Circulation. 2002;105:1135–43.

44. Brighenti F, Valtuena S, Pellegrini N, et al. Total antioxidant capacity of the diet is inversely and independently related to plasma concentration of high-sensitivity C-reactive protein in adult Italian subjects. Br J Nutr. 2005;93:619–25.

45. Esmaillzadeh A, Kimiagar M, Mehrabi Y, Azadbakht L, Hu FB, Willett WC. Fruit and vegetable intakes, C-reactive protein, and the metabolic syndrome. Am J Clin Nutr. 2006;84:1489–97.

46. Ommen GS, Goodman GE, Thomquist MD, Barnes J, Cullen MR. Effects of a combination of β-carotene and vitamin A on lung cancer and cardiovascular disease. N Engl J Med. 1996;334:1150–5.

47. Stephens NG, Parsons A, Schofield PM, Kelly F, Cheeseman K, Mitchinson MJ. Randomized controlled trial of vitamin E in patients with coronary disease: Cambridge Heart Antioxidant Study (CHAOS). Lancet. 1996;347:154–60.

48. Yusuf S, Dagenais G, Pogue J, Bosch J, Sleight P. Vitamin E supplementation and cardiovascular events in high-risk patients. The Heart Outcomes Prevention Evaluation Study Investigators. N Engl J Med. 2000;342:154–60.

49. Hennekens CH, Buring JE, Manson JE, Stampfer M, Rosner B. Lack of effect of long- term supplementation with β-carotene on the incidence of malignant neoplasms and cardiovascular disease. N Engl J Med. 1996;334:1145–9.

50. Greenberg ER, Baron JA, Stuckel TA, Stevens MM, Mandel JS. A clinical trial of β-carotene to prevent basal cell and squamous cell cancers of the skin. N Engl J Med. 1990;323:789–95.

51. The α-Tocopherol, β-Carotene Cancer Prevention Study Group. The effect of vitamin E and β-carotene on the incidence of lung cancer and other cancers in male smokers. N Engl J Med. 1994;330:1020–35.

52. The HOPE Investigators. Vitamin E supplementation and cardiovascular events in high-risk patients. N Engl J Med. 2000;342:154–60.

53. GISSI-Prevenzione Investigators. Dietary supplementation with n-3 polyunsaturated fatty acids and vitamin E after myocardial infarction: results of the GISSI-Prevenzione Trial. Lancet. 1999;354:447–55.

54. Blot WJ, Li JY, Taylor PR, et al. Nutrition intervention trials in Linxian, China: supplementation with specific vitamin/mineral combinations, cancer incidence, and disease-specific mortality in the general population. J Natl Cancer Inst. 1993;85:1483–92.

55. Salonen JT, Nyyssonen K, Salonen R, et al. Antioxidant supplementation in artherosclerosis prevention (ASAP) study: a randomized trial of the effect of vitamins E and C on 3-year progression of carotid atherosclerosis. J Intern Med. 2000;248:377–86.

56. Lippman SM, Klein EA, Goodman PJ, et al. Effects of selenium and vitamin E on risk of prostate cancer and other cancers: the Selenium and Vitamin E Cancer Prevention Trial (SELECT). JAMA. 2009;301:39–51.

57. Eberhardt MV, Lee CY, Liu RH. Antioxidant activity of fresh apples. Nature. 2000;405:903–4.

58. Liu RH, Liu J, Chen B. Apples prevent mammary tumors in rats. J Agric Food Chem. 2005;53:2341–3.

59. Yang J, Liu RH. Synergistic effect of apple extracts and quercetin 3-β-D-glucoside combination on antiproliferative activity in MCF-7 human breast cancer cells in vitro. J Agric Food Chem. 2009;57:8581–6.

60. He XJ, Liu RH. Triterpenoids isolated from apple peels have potent antiproliferative activity and may be partially responsible for apple's anticancer activity. J Agric Food Chem. 2007;55:4366–70.

61. He XJ, Liu RH. Phytochemicals of apple peels: isolation, structure elucidation, and their antiproliferative and antioxidant activities. J Agric Food Chem. 2008;56:9905–10.

62. Temple NJ, Gladwin KK. Fruits, vegetables, and the prevention of cancer: research challenges. Nutrition. 2003;19:467–70.

Chapter 14
Food Synergy: A Paradigm Shift in Nutrition Science

David R. Jacobs, Jr. and Norman J. Temple

Keywords Food synergy • Dietary patterns • Epidemiology • Mechanistic research • Reductionism

Key Points

- A major focus of nutrition science has centered on identifying the various macronutrients, vitamins, and minerals and then discovering their mode of action.
- Many associations between diet and risk of disease are best understood by looking at food as a whole and not merely as a collection of individual nutrients and other bioactive substances.
- Because of the limitations of epidemiology, it is extremely difficult in many cases to identify the substances in food that account for protection against disease. This is especially the case with phytochemical-rich plant foods, including fruit, vegetables, nuts, legumes, and whole-grain cereal foods.
- Research into foods and dietary patterns is needed. Such findings can be directly translated into public recommendations. They also serve as a scientific anchor point to which studies of food components must conform.

This chapter critically evaluates approaches to explaining the nature of the relationship between diet and disease. Most research has focused on studying single substances: macronutrients, micronutrients, as well as the many other bioactive

D.R. Jacobs, Jr., PhD (✉)
Division of Epidemiology and Community Health, School of Public Health, University of Minnesota, 1300 South Second Street, Suite 300, Minneapolis, MN 55454, USA
e-mail: Jacobs@epi.umn.edu

N.J. Temple, PhD
Centre for Science, Athabasca University, Athabasca, AB, T9S 3A3, Canada
e-mail: normant@athabascau.ca

N.J. Temple et al. (eds.), *Nutritional Health: Strategies for Disease Prevention*,
Nutrition and Health, DOI 10.1007/978-1-61779-894-8_14,
© Springer Science+Business Media, LLC 2012

substances present in food, either beneficial or harmful. Here we argue the case for turning our attention to food as a whole and to dietary patterns. This concept, known as food synergy, defined as additive or more than additive influences of foods and food constituents on health, is a powerful tool to help explain many nutrition-related diseases and how best to prevent and treat them. This chapter is based on previous publications from one of us (DJ) [1–5].

14.1   The Concept of Nutrient Deficiency Diseases

Up until the 1950s a major focus of nutrition science centered on identifying the various vitamins and minerals and then discovering their mode of action. The underlying philosophy can be summarized as follows:

1. There is a simple cause-and-effect relationship between deficiency of a nutrient and the associated specific disease.
2. Each nutrient deficiency disease can be explained in terms of the role played by that nutrient, especially in the areas of biochemistry and physiology.
3. The nutrient deficiency can be prevented (and often reversed) by giving that nutrient in an isolated (pure) form.

Classic examples are vitamin C in relation to scurvy and iron in relation to anemia. This concept is still very much alive and is seen in the vitamin and mineral chapters of every textbook used in college nutrition courses. The concept can be applied not only to vitamins and minerals but also to protein and essential fats.

In recent years the concept has shown its continuing value in several areas. Research has firmly established that supplementary folic acid (a form of folate) is protective against neural tube defects (NTD), a group of congenital disorders that include spina bifida [6]. While NTD cannot be characterized as a vitamin-deficiency disease in the mother, it seems clear that a low maternal intake of folate creates a deficiency condition in the fetus that hinders normal development of the nervous system, thereby allowing the condition to emerge [7]. As a result of this discovery addition of folic acid to refined grain products is now mandatory in the USA and Canada. Research on vitamin D is another illustrative example. There was little progress in this area for several decades after the relationship between vitamin D and rickets was discovered. But in recent years strong evidence has emerged that the vitamin has a protective action against the risk of cancer and several other diseases (Chap. 17).

The relationship between single substances and disease risk extends well beyond disease protection by micronutrients; there are many other bioactive substances in food, either beneficial or harmful. Sodium (as salt) is an especially clear example of this. A high intake of the mineral, which is the case for the great majority of the population, plays an important role in the causation of hypertension ([8, 9]; see Chap. 11) and cardiovascular disease (CVD) [9, 10]. Strong evidence also suggests that salt significantly increases the risk of stomach cancer [11]. *Trans* fatty acids are another example. This type of dietary fat adversely affects multiple cardiovascular risk factors and increases the risk of coronary heart disease (CHD) [12, 13]. Another bioactive substance in food is alcohol. High intakes cause multiple harmful effects. But as explained in the next chapter evidence has emerged over the last 2 decades that alcohol lowers the risk of CHD and possibly several other conditions. The active ingredient is alcohol itself, not the other substances commonly found in alcoholic beverages.

In many respects this reductionist approach to nutrition, called nutritionism by Scrinis [14], has not provided satisfactory answers to nutritional questions. Of special note, as discussed below, are studies in which nutrients were derived from food intake, found to be "protective" observationally (e.g., [15]), then used in higher doses as supplements in clinical trials. The purified nutrients in relatively high dose did not work as expected according to reductionist logic.

14.2   Food Synergy: An Alternative Paradigm

14.2.1   The Emergence of the Concept of Food Synergy

A strong argument has been made that to properly explain the many associations between diet and risk of disease we should view food as a whole, and not merely as a collection of individual nutrients and other bioactive substances. Over the past couple of decades this concept has emerged as an alternative concept to the one discussed above that focuses on individual nutrients and bioactive substances in food. In a nutshell—figuratively (and literally in the case of nuts)—food synergy has provided a better explanation for many nutrition-related diseases and how best to prevent and treat them. This represents a paradigm shift in our understanding of nutrition science.

Compelling support for food synergy comes from examining complex dietary patterns in relation to disease risk. The following examples illustrate this.

14.2.2   The Mediterranean Diet

One such dietary pattern that has been much studied in relation to CHD is the Mediterranean diet. While this diet varies from country to country around the Mediterranean, major features include a relatively high intake of vegetables, fruit, legumes, nuts, fish, whole-grain cereals, and olive oil. Conversely, the diet is typically low in meat (especially red meat).

Mente and colleagues carried out a systematic review on the relationship between diet and risk of CHD [16]. They examined some 26 nutrients, foods, and dietary patterns for the strength of the associations seen in cohort studies. They reported that of all the associations with risk of CHD, either positive or negative, the strongest one was for the Mediterranean diet. The Mediterranean diet has the following attributes: first, it manifests a strong protection against the risk of CHD; and second, it has a complex nutritional composition, rich in phytochemicals but low in saturated fat, heme iron, and many other substances found in meat. It appears highly likely that multiple dietary components and multiple pathways are responsible. Based on this it can be reasonably argued that this diet–disease association provides strong support for the food synergy concept.

14.2.3 The Alternate Healthy Eating Index

Another healthy eating dietary pattern is the Alternate Healthy Eating Index [17]. Findings from the Nurses' Health Study showed that middle-aged women adhering to this dietary pattern have a much reduced risk of death from CVDs, from cancer, and from all causes combined [18].

14.2.4 The A Priori Healthy Diet Score

A novel index based only on foods is the a priori Healthy Diet Score [19]. It was formulated as the sum of ranks of food groups that had been judged to be favorable or unfavorable for health by knowledgeable persons experienced in nutrition and nutritional epidemiology. It was related to reduced odds for myocardial infarction [19], to changes in intermediate risk markers [20, 21], to common carotid intima media thickness, mediated by waist circumference [20], and to incident diabetes, mediated by waist circumference [22].

14.2.5 Western Dietary Pattern

Just as every movie with a hero also needs a villain, so a healthy dietary pattern needs an unhealthy one. That role is played by the "Western" pattern. Such a diet is high in red meat, processed meat, refined cereals, French fries, and desserts. A publication from the Nurses' Health Study linked this dietary pattern to an elevated risk of death from CVDs, from cancer, and from all causes combined [23].

14.2.6 Meat and Health

The above dietary patterns are wide ranging. The meat content of these dietary patterns is just one factor among many. However, there has been much interest for many decades regarding the relationship between meat consumption and health. Vegetarian diets have been advocated by many people over the years as a healthier alternative to a meat-based diet. The most compelling evidence supporting the strong health benefits of a reduced intake of meat comes from a cohort study of half a million middle-aged and elderly Americans [24]. The findings clearly show that consumption of red meat (i.e., beef and pork) and processed meat are associated with higher risk of death from CVDs, from cancer, and from all causes combined.

There are several possible food constituent-based reasons that might explain these results, including the dietary intake of saturated fat, iron, and of various amino acids, and low intake of phytochemicals. The researchers in this study did adjust their hazard ratios for many confounding variables. This is important as meat consumption in this population was associated with a generally poor diet and unhealthy lifestyle. Many nutritional factors are probably involved in the connection between meat and risk of death from diverse causes. Accordingly, meat consumption should best be viewed from the perspective of food synergy.

14.2.7   DASH Diet

This diet was developed as a treatment for hypertension [25]. Key features are a generous intake of fruit, vegetables, and low-fat dairy products, combined with a reduced intake of meat, and therefore saturated fat. As described in Chap. 11 this food synergy approach has proven effective as a treatment for hypertension [25].

14.2.8   Food Synergy and Disease

In each of the above cases we see strong evidence that the relationship between diet and disease risk is best explained by focusing on foods rather than substances present in food. This applies both to prevention and to treatment. It also applies both to dietary patterns and to single food groups, such as meat. Our best explanation lies in the great complexity of food and dietary patterns, the thousands of different substances present, and the many pathways that connect food with the etiology of disease.

14.3   Food Synergy: A Research Pespective

14.3.1   The Limitations of Epidemiology

The evidence considered above supports the case for food synergy as an explanation for many diet–disease associations. However, there is a separate line of argument that also supports the food synergy concept. Nutrition research methods have limited power to identify which nutrients or other bioactive substances in a complex food are likely to be responsible for particular health benefits. The reasons for this were explained in Chap. 1 by the authors of this chapter. The major problem is that nutrients and other bioactive substances are not distributed randomly in foods. Instead, they are mostly associated with each other. In other words, focusing on one substance causes confounding due to the presence of many other substances. This problem is especially acute with the multitude of substances found in fruit and vegetables and other phytochemical-rich plant foods. These include folate, vitamin C, potassium, fiber, and, of course, a great many other phytochemicals, many of which have been little studied or not even identified. Because of this, it is unlikely that epidemiological studies will ever be able to identify, for example, whether lycopene prevents prostate cancer or if α-carotene prevents colon cancer.

Cereal fiber and whole-grain cereal foods pose a similar challenge and an interesting counterpoint. This was illustrated in a study by Jacobs et al. [26]. They observed that dietary fiber from whole-grain cereals has a stronger protective association with disease than does the same amount of fiber from refined cereals. The proposed explanation is because of the phytochemicals present in whole grains. This indicates that epidemiological studies cannot even state with confidence whether dietary fiber really has an independent protective association with disease risk, beyond its direct effect in the large intestine. However, the suggestion is that there is something of health value in the whole grain, which is a conclusion about a food, rather than a nutrient.

We see, therefore, that the problem of confounding makes it extremely difficult to identify which nutrients or other bioactive substances deserve the credit for the health benefits of fruit, vegetables, and whole-grain cereals.

One obvious way to circumvent this problem is to carry out randomized clinical trials (RCTs) on single substances. However, these are extremely costly and usually take several years. They are only appropriate, therefore, when dealing with dietary components where there is already strong supporting evidence. Besides, as argued in depth in Chap. 1 on research design, the many differences between drugs and foods heavily influence research design. One must be very careful when undertaking an RCT on a single substance derived from food to be clear about whether it is even possible to answer the question being asked about the health implications of a single nutritional substance. As well, one must consider whether that RCT answer would help elucidate what food people should eat. We will now illustrate these points with some examples.

The saga of antioxidants provides perhaps the best illustration of the limits of epidemiology. It was discovered that β-carotene (derived as a weighted average of β-carotene-containing foods) was inversely related to lung cancer [15]. Subsequent epidemiological studies led to the widely held view that β-carotene may be effective as a chemopreventive agent against a range of cancers [27]. At around the same time epidemiological evidence linked two other antioxidant vitamins—vitamins C and E—with protection against disease. Vitamin C was reported to have a negative association with risk of cancer [28] while several cohort studies observed that intake of vitamin E has a modest protective association with risk of CHD [16, 29]. Following these findings all three antioxidant vitamins were studied as disease preventatives; this involved administering these substances in a purified form at doses typically several times higher than the RDA. The results of long-term RCTs on β-carotene appeared in the mid-1990s and these have consistently shown that supplements are ineffective for the prevention of cancer [30]. Likewise, supplements of vitamin E have little or no preventive action against CHD [16]. Findings for all three antioxidants were actually adverse: an increase in total mortality of about 5–6% [31, 32]. The likely explanation for these findings is that the negative associations seen in epidemiological studies were entirely due to confounding by or interaction with phytochemicals and other substances that are found in the same foods as the antioxidant vitamins.

It is very likely that there is a complex food synergy at work in all the examples discussed above. In other words, the true reason that fruit, vegetables, and whole-grain cereals are protective against cancer and CHD is because of a complex interaction induced by a wide variety of nutrients and other bioactive substances. But let us suppose for one moment that the active ingredients are limited to a mere three or four phytochemicals. Because of the limitations of epidemiology, as explained above, it is extremely difficult to identify these substances with any confidence. For that reason, it hardly matters if the actual number of anticarcinogenic phytochemicals is 3 or 300. In contrast, a finding that a certain food or dietary pattern influences health is feasible, informative for future thinking, and of great practical value.

The limitations of epidemiology are also shown by studies investigating the relationship between homocysteine, folate, and CHD. Several epidemiological studies had revealed that blood homocysteine levels are correlated with risk of CHD and other CVDs [33, 34]. As supplements of folic acid (the form of folate used in supplements) are effective at lowering the blood homocysteine level [35, 36], it was hypothesized that this intervention will therefore be protective against CHD. Separate from this, epidemiological studies had indicated that dietary intake of folate has a strong inverse association with risk of CHD [16]. Indeed, in the systematic review carried out by Mente et al. [16], of all the nutrients, foods, and dietary patterns examined, folate had one of the strongest associations with risk of CHD, based on the findings from cohort studies. However, when the results of RCTs appeared, the optimistically expected results showed no indication that folic acid supplements prevent CHD [16, 35], and some RCTs have even suggested adverse effects on cancer and other diseases ([36, 37]; see Chap. 12 by Davis and Milner). The lesson here is that the studies that attempted to explain the association between blood homocysteine levels and risk of CHD did not confirm a simple, critical, causal role for homocysteine. The protective association between folate and CHD is most likely explained as one more case of confounding. These supplement studies do not necessarily imply no value for folate when obtained from foods, because nutrients obtained from (whole) foods are in natural balance, certified by evolution.

14.3.2   The Limitations of Mechanistic Research in Explaining the Effects of Food on Health

Some might argue that what cannot be achieved by epidemiology can be accomplished by laboratory-based research with the goal of explaining disease in terms of its causative mechanisms. For example, studies of the biochemical action of diverse phytochemicals on the processes of carcinogenesis will (supposedly) help identify which ones are potentially chemopreventive and should therefore be tested in RCTs. Similarly, while epidemiological research has provided indications that particular vitamins prevent CHD (and therefore, by implication, also help prevent atherosclerosis), this can be firmly established by studies of the processes of atherosclerosis at a cellular level.

Chapter 1 explained the serious limitations of mechanistic research on nutrients, often referred to as reductionism. We will illustrate this by returning to two of the antioxidant vitamins discussed above. Many studies were carried out during the 1980s that investigated the effects of β-carotene on body systems possibly related to cancer. This included studies of antioxidant action [38] and immune function [39, 40]. However, the dubious relevance of these studies to the relationship between diet and cancer became obvious when RCTs demonstrated that supplemental β-carotene does not prevent human cancer likewise with vitamin E. Here, the focus of research was on the ability of vitamin E to retard the oxidation of LDL [41]. But as vitamin E has shown little or no effectiveness in preventing CHD, it is hard to discern the practical value of the mechanistic research.

14.4   Future Research Directions

While there is nothing intrinsically wrong with an approach that searches for simple answers and roles of specific molecules, there is much to be said for accepting a finding about a food or food pattern as important information that answers a nutritional question. It is not necessary to reduce foods to constituents in order to understand that diet does affect health and to make policy for better eating. We should take a food synergy perspective, think foods first [4], working on the assumption that as we have little idea which substances are involved, the only practical approach is to assume that all nutrients and other bioactive substances in phytochemical-rich plant foods play a role in giving protection against cancer, CHD, and other chronic diseases. Even if there are simple reductionist answers to nutritional questions in generally well-nourished people, due to practical circumstances of solving the immensely complex problem of interacting food constituents, we are unlikely to make major progress in the reductionist mode in the near future.

What is the best way for researchers to design investigations to achieve a better understanding of how to maintain health? Based on the arguments presented here the answer lies in a two-stage strategy. First, epidemiological studies are required to identify which dietary patterns or foods have an apparent cause-and-effect relationship with disease. Such studies are quite reliable for that purpose. In the second stage RCTs need to be carried out in order to test either whole diets or individual foods.

Carrying out such RCTs presents serious challenges. These would need to be long term. Moreover, blinding is all but impossible as it is fairly obvious what is being eaten. Compliance to fixed diets may be an even larger problem. Imagine requiring that for several years coffee drinkers abstain from coffee or those who dislike coffee consume it regularly; or that meat lovers become vegetarian or vegetarians omnivores. However, such studies are feasible. For example, a 4-month trial using the DASH diet as a treatment for elevated blood pressure achieved excellent compliance [42]. The Women's Health Initiative set a much more ambitious target with the aim of persuading healthy women to make major changes in their diets and maintain them for 6 year [43]. For example, they aimed to reduce total fat intake to 20% of calories. However, the actual change was only about half of this.

14.5   Conclusions

Food which emanates from a living organism is a mixture of constituents, but not a random mixture. Rather, the particular mixture has proven adequate through evolution for the life of the organism eaten. To the extent that the organism has been eaten for a long time, evolution has also tested the mixture of constituents as a food for the eater.

There is still much to be gained from research that investigates individual nutrients and bioactive substances in food and then attempts to determine their role in health and disease. This is especially valuable in cases where problems of confounding are relatively small and it is therefore possible to investigate nutrients or other substances as single variables. An interesting example is vitamin D, where sun exposure is important in addition to food.

In several cases research on single nutrients has lead to important measures that have improved public health. We see this with folate in relation to spina bifida, although there are reservations about whether the good findings for NTDs carry over to the whole population for protection from chronic diseases of adulthood. Other notable cases of single-nutrient solutions to health problems are iron supplementation for iron deficiency anemia and vitamin B_{12} supplementation for the elderly. However, these instances usually occur in situations of relative deficiency.

We must stay open to the possibility of more such cases appearing. It is entirely possible that a small number of phytochemicals will be proven to have valuable health-enhancing actions. In such cases they could have potential as drugs. Indeed, some evidence of this has already been documented: lutein is showing promise for improving eye health [44, 45] and soy isoflavones for improving bone health [46]. Whether these nutrients would be useful as supplements in the general population is an unanswered question; or, perhaps they should be thought of as drugs.

But, increasingly, we are seeing the limitations of this approach. There is a strong case for placing much more emphasis on food synergy and regarding findings for foods or dietary patterns as final answers to questions. This can include dietary patterns, such as the Mediterranean diet, or single foods, such as red meat. It is not clear whether food synergy reflects a true mathematical synergistic relationship (i.e., the whole risk or benefit is greater than the sum of the parts) or else is simply an additive effect. At minimum, though, even in the absence of mathematical synergy, foods are complex mixtures, tested by evolution, which we would not come to by constituting them de novo from individual constituents.

We are sympathetic to attempts to properly understand the detailed causes of chronic diseases. However, we believe that the complexity of metabolism and pathology is such that nutrition research is a long way from being able to achieve this goal. Epidemiologic studies of nutrients are often misleading because they miss the context of the whole food and diet pattern. In other words, both epidemiologic study of nutrients and mechanistic research are inferior strategies for achieving valuable breakthroughs that lead to improvements in public health through improved diet. Despite that, mechanistic research attracts far more resources than food-based research. We conclude, therefore, that improved infrastructure for food-oriented research would be most valuable.

References

1. Messina M, Lampe JW, Birt DF, et al. Reductionism and the narrowing nutrition perspective: time for reevaluation and emphasis on food synergy. J Am Diet Assoc. 2001;101:1416–9.
2. Jacobs DR, Murtaugh MA. It's more than an apple a day: an appropriately processed plant-centered dietary pattern may be good for your health. Am J Clin Nutr. 2000;72:899–900.
3. Jacobs DR, Steffen LM. Nutrients, foods, and dietary patterns as exposures in research: a framework for food synergy. Am J Clin Nutr. 2003;78(Suppl):508S–13S.
4. Jacobs Jr DR, Tapsell LC. Food, not nutrients, is the fundamental unit in nutrition. Nutr Rev. 2007;65:439–50.
5. Jacobs Jr DR, Gross MD, Tapsell LC. Food synergy: an operational concept for understanding nutrition. Am J Clin Nutr. 2009;89:1543S–8S.
6. Wolff T, Witkop CT, Miller T, Syed SB; U.S. Preventive Services Task Force. Folic acid supplementation for the prevention of neural tube defects: an update of the evidence for the U.S. Preventive Services Task Force. Ann Intern Med. 2009;150:632–9.
7. Jacobs Jr DR, Mursu J, Meyer KA. The importance of food. Arch Pediatr Adolesc Med. 2012;166(2):187–8.
8. Sacks FM, Svetkey LP, Vollmer WM, et al.; The DASH-Sodium Collaborative Research Group. Effects on blood pressure of reduced dietary sodium and the dietary approaches to stop hypertension (DASH) diet. DASH-Sodium Collaborative Research Group. N Engl J Med. 2001;344:3–10.
9. He FJ, MacGregor GA. A comprehensive review on salt and health and current experience of worldwide salt reduction programmes. J Hum Hypertens. 2009;23:363–84.
10. Cook NR, Cutler JA, Obarzanek E, et al. Long term effects of dietary sodium reduction on cardiovascular disease outcomes: observational follow-up of the trials of hypertension prevention (TOHP). BMJ. 2007;334:885–8.

11. World Cancer Research Fund/American Institute for Cancer Research. Food, nutrition, physical activity, and the prevention of cancer: a global perspective. Washington: AICR; 2007.

12. Mozaffarian D, Aro A, Willett WC. Health effects of trans-fatty acids: experimental and observational evidence. Eur J Clin Nutr. 2009;63 Suppl 2:S5–21.

13. Mozaffarian D, Katan MB, Ascherio A, Stampfer MJ, Willett WC. Trans fatty acids and cardiovascular disease. N Engl J Med. 2006;354:1601–13.

14. Scrinis G. On the ideology of nutritionism. Gastronomica J Food Culture. 2008;8(1):39–48.

15. Shekelle RB, Lepper M, Liu S, et al. Dietary vitamin A and risk of cancer in the Western Electric study. Lancet. 1981;2:1185–90.

16. Mente A, de Koning L, Shannon HS, Anand SS. A systematic review of the evidence supporting a causal link between dietary factors and coronary heart disease. Arch Intern Med. 2009;169:659–69.

17. McCullough ML, Feskanich D, Stampfer MJ, et al. Diet quality and major chronic disease risk in men and women: moving toward improved dietary guidance. Am J Clin Nutr. 2002;76:1261–71.

18. van Dam RM, Li T, Spiegelman D, Franco OH, Hu FB. Combined impact of lifestyle factors on mortality: prospective cohort study in US women. BMJ. 2008;337:a1440.

19. Lockheart MS, Steffen LM, Rebnord HM, et al. Dietary patterns, food groups and myocardial infarction: a case-control study. Br J Nutr. 2007;98:380–7.

20. Nettleton JA, Schulze MB, Jiang R, Jenny NS, Burke GL, Jacobs DR. A priori-defined dietary patterns and markers of cardiovascular disease risk in the multi-ethnic study of atherosclerosis (MESA). Am J Clin Nutr. 2008;88:185–94.

21. Jacobs DR, Sluik D, Rokling-Andersen MH, Anderssen SA, Drevon CA. Association of 1-y changes in diet pattern with cardiovascular disease risk factors and adipokines: results from the 1-y randomized Oslo Diet and Exercise Study. Am J Clin Nutr. 2009;89:509–17.

22. Nettleton JA, Steffen LM, Ni H, Liu K, Jacobs Jr DR. Dietary patterns and risk of incident type 2 diabetes in the Multi-Ethnic Study of Atherosclerosis (MESA). Diabetes Care. 2008;31:1777–82.

23. Heidemann C, Schulze MB, Franco OH, van Dam RM, Mantzoros CS, Hu FB. Dietary patterns and risk of mortality from cardiovascular disease, cancer, and all causes in a prospective cohort of women. Circulation. 2008;118:230–7.

24. Sinha R, Cross AJ, Graubard BI, Leitzmann MF, Schatzkin A. Meat intake and mortality: a prospective study of over half a million people. Arch Intern Med. 2009;169:562–71.

25. Svetkey LP, Simons-Morton D, Vollmer WM, et al. Effects of dietary patterns on blood pressure: subgroup analysis of the Dietary Approaches to Stop Hypertension (DASH) randomized clinical trial. Arch Intern Med. 1999;159:285–93.

26. Jacobs DR, Pereira MA, Meyer KA, Kushi LH. Fiber from whole grains, but not refined grains, is inversely associated with all-cause mortality in older women: the Iowa women's health study. J Am Coll Nutr. 2000;19(3 Suppl):326S–30S.

27. Temple NJ, Basu TK. Role of beta-carotene in the prevention of cancer—a review. Nutr Res. 1988;8:685–701.

28. Block G. Vitamin C status and cancer. Epidemiologic evidence of reduced risk. Ann N Y Acad Sci. 1992;669:280–90.

29. Knekt P, Ritz J, Pereira MA, O'Reilly EJ, et al. Antioxidant vitamins and coronary heart disease risk: a pooled analysis of 9 cohorts. Am J Clin Nutr. 2004;80:1508–20.

30. Druesne-Pecollo N, Latino-Martel P, Norat T, et al. Beta-carotene supplementation and cancer risk: a systematic review and metaanalysis of randomized controlled trials. Int J Cancer. 2010;127:172–84.

31. Bjelakovic G, Nikolova D, Gluud LL, Simonetti RG, Gluud C. Mortality in randomized trials of antioxidant supplements for primary and secondary prevention: systematic review and meta-analysis. JAMA. 2007;297:842–57.

32. Bjelakovic G, Nikolova D, Gluud LL, Simonetti RG, Gluud C. Antioxidant supplements for prevention of mortality in healthy participants and patients with various diseases. Cochrane Database Syst Rev. 2008;(2):CD007176.

33. Seshadri N, Robinson K. Homocysteine, B vitamins, and coronary artery disease. Med Clin North Am. 2000;84:215–37.
34. Stubbs PJ, Al-Obaidi MK, Conroy RM, et al. Effect of plasma homocysteine concentration on early and late events in patients with acute coronary syndromes. Circulation. 2000;102: 605–10.
35. Clarke R, Halsey J, Lewington S, et al.; B-Vitamin Treatment Trialists' Collaboration. Effects of lowering homocysteine levels with B vitamins on cardiovascular disease, cancer, and cause-specific mortality: meta-analysis of 8 randomized trials involving 37 485 individuals. Arch Intern Med. 2010;170:1622–31.
36. Bazzano LA, Reynolds K, Holder KN, He J. Effect of folic acid supplementation on risk of cardiovascular diseases: a meta-analysis of randomized controlled trials. JAMA. 2006;296: 2720–6.
37. Ebbing M, Bønaa KH, Nygård O, et al. Cancer incidence and mortality after treatment with folic acid and vitamin B12. JAMA. 2009;302:2119–26.
38. Burton GW, Ingold KU. Beta-carotene: an unusual type of lipid antioxidant. Science. 1984;224: 569–73.
39. Schwartz J, Suda D, Light G. Beta carotene is associated with the regression of hamster buccal pouch carcinoma and the induction of tumor necrosis factor in macrophages. Biochem Biophys Res Commun. 1986;136:1130–5.
40. Rhodes J. Human interferon action: reciprocal regulation by retinoic acid and beta-carotene. J Natl Cancer Inst. 1983;70:833–7.
41. Asmis R, Llorente VC, Gey KF. Prevention of cholesteryl ester accumulation in P388D1 macrophage-like cells by increased cellular vitamin E depends on species of extracellular cholesterol. Conventional heterologous non-human cell cultures are poor models of human atherosclerotic foam cell formation. Eur J Biochem. 1995;233:171–8.
42. Blumenthal JA, Babyak MA, Hinderliter A, et al. Effects of the DASH diet alone and in combination with exercise and weight loss on blood pressure and cardiovascular biomarkers in men and women with high blood pressure: the ENCORE study. Arch Intern Med. 2010;170: 126–35.
43. Howard BV, Van Horn L, Hsia J, et al. Low-fat dietary pattern and risk of cardiovascular disease: the Women's Health Initiative Randomized Controlled Dietary Modification Trial. JAMA. 2006;295:655–66.
44. Ma L, Lin XM, Zou ZY, Xu XR, Li Y, Xu R. A 12-week lutein supplementation improves visual function in Chinese people with long-term computer display light exposure. Br J Nutr. 2009;102:186–90.
45. Ma L, Lin XM. Effects of lutein and zeaxanthin on aspects of eye health. J Sci Food Agric. 2010;90:2–12.
46. Ma DF, Qin LQ, Wang PY, Katoh R. Soy isoflavone intake increases bone mineral density in the spine of menopausal women: meta-analysis of randomized controlled trials. Clin Nutr. 2008;27:57–64.

Chapter 15
What Are the Health Implications of Alcohol Consumption?

Norman J. Temple

Keywords Alcohol • Fetal alcohol syndrome • Cancer • Obesity • Coronary heart disease • Blood pressure • Stroke • Pattern of drinking

Key Points

- Consumption of alcohol is associated with many effects harmful to health, including accidents, violence, suicide, fetal alcohol syndrome, and cancer.
- There is little evidence linking alcohol with weight gain.
- Strong evidence indicates that moderate consumption of alcohol is protective against coronary heart disease. There is also evidence suggesting that alcohol in moderation may be protective against several other conditions including elevated blood pressure, gallstones, loss of bone mineral density, hearing loss, dementia, benign prostatic hyperplasia, type 2 diabetes, and lung disease.
- For many conditions the relationship is J-shaped: the lowest risk is associated with consumption of alcohol in moderation whereas risk climbs with higher levels of alcohol consumption.
- For people aged 50–80 year the lowest risk of death is seen at an alcohol intake of around 1.0–1.5 drinks/day in men and 0.5–1.0 drinks/day in women. However, for people below age 40 year alcohol consumption does not reduce mortality.
- Alcohol is most protective when consumed in small regular amounts rather than binge drinking.

The harmful effects of alcohol are far better known than the beneficial effects. This is scarcely surprising: it requires no training in epidemiology to recognize the devastating harm that often comes with both drunkenness and chronic alcohol abuse.

N.J. Temple, PhD (✉)
Centre for Science, Athabasca University, Athabasca, AB, T9S 3A3, Canada
e-mail: normant@athabascau.ca

N.J. Temple et al. (eds.), *Nutritional Health: Strategies for Disease Prevention*,
Nutrition and Health, DOI 10.1007/978-1-61779-894-8_15,
© Springer Science+Business Media, LLC 2012

However, findings that have emerged in recent years have uncovered several surprising associations between moderate intake of alcohol and enhanced health and well-being.

In this chapter we use the American definition of a drink, namely 12.5–13.0 g of alcohol. This quantity of alcohol is approximately the amount contained in 12 oz (356 g) of regular beer, 4–5 oz (118–148 g) of wine, or 1.5 oz (42 g) of spirits. We also use the USDA dietary guidelines' definition of moderate alcohol consumption as up to two drinks a day for men and one drink a day for women.

15.1 Harmful Effects of Alcohol

15.1.1 Accidents, Violence, and Suicide

It is well established that abuse of alcohol is associated with accidents, violence, and suicide. The most dramatic evidence of this has come from Russia. Following the collapse of the Soviet Union in 1989 Russia experienced serious economic decline and much political turmoil. During the early 1990s life expectancy fell by 4 years in men and by 2 years in women. A major factor in this was apparently widespread alcohol abuse, particularly binge drinking, which led to large increases in deaths from accidents, homicide, and suicide, as well as cardiovascular disease [1, 2]. This severe health crisis has not abated. In 2006, in comparison with Western Europe, the mortality rate for Russians aged 15–54 year was five times higher for men and three times higher for women. Alcohol is apparently the dominant factor that explains this [3].

In the United States in 2006 alcohol was a factor in about 17,900 fatal crashes, or about 40% of all fatal crashes [4]. This figure includes alcohol intake by persons other than the driver, such as a pedestrian. This was the highest level since 1992, suggesting that the trend towards lower levels of alcohol-related car crashes has gone into reverse.

15.1.2 Chronic Alcohol Abuse

For many persons, years of alcohol abuse eventually leads to chronic health and nutritional problems. Alcohol is rich in calories and typically devoid in nutrients, especially alcohol- and sugar-rich hard liquors. The body often compensates for the high caloric intake by decreasing the stimulus to eat regular nutrient-rich foods. As a result, there is a high probability of malnutrition, especially of folate and thiamin. The thiamin deficiency associated with alcohol abuse is known as Wernicke–Korsakoff syndrome. Liver disease is also a likely result with a downward spiral from fatty liver, to alcoholic hepatitis, and, eventually, to cirrhosis.

15.1.3　Fetal Alcohol Syndrome

Pregnancy is another situation where alcohol misuse can have tragic consequences. This induces fetal alcohol syndrome (FAS). FAS encompasses a variety of symptoms including prenatal and postnatal growth retardation, abnormal facial features, and an increased frequency of major birth defects. Children born with FAS never recover.

A subclinical form of FAS is known as fetal alcohol effects (FAE). Children with FAE may be short or have only minor facial abnormalities, or develop learning disabilities, behavioral problems, or motor impairments.

FAS occurs at a level of alcohol intake that in a nonpregnant woman would not be considered alcohol abuse. Four drinks per day poses a real threat of FAS, although one or two drinks per day may still retard growth; the epidemiological data are weaker and somewhat inconsistent at these lower levels of consumption. Although women who have an occasional drink during pregnancy should not fear they are doing irreparable harm to their fetus, it is now generally accepted that any woman who is or may become pregnant should abstain from alcohol.

15.1.4　Cancer

Alcohol much increases the risk of several types of cancer. An alcohol intake at the high end of moderation (two drinks per day in women, four in men) is associated with RRs for different types of cancer as follows: 1.16 for colorectal cancer 1.6 for breast, 1.8 for mouth and pharynx, 2.4 for esophagus, and 3.0 for liver [5]. Our best evidence is that lower intakes of alcohol produce proportionately smaller RRs. For all cancer combined a significant risk is seen starting at an alcohol intake of two drinks per day, with a RR of 1.22 at 4 drinks per day [6].

Many studies have investigated the mechanisms by which alcohol may enhance risk of cancer. It is likely that metabolites of alcohol, such as acetaldehyde, are carcinogenic. Alcohol can act as a solvent enhancing the penetration of carcinogens into cells. For breast cancer, it is less likely that ethanol is directly toxic since the increase has been seen at relatively low levels. It is more likely that alcohol influences circulating estrogen levels which may impact on disease occurrence [7, 8].

Emerging evidence also indicates that alcohol, even in moderation, may suppress circulating folate levels, which could impact on DNA synthesis and gene expression. These findings, combined with various other epidemiological and experimental evidence, led to speculation that a low dietary intake of folate may increase risk of cancer while supplements of folic acid are anticarcinogenic. However, recent evidence from several cohort and randomized trials indicates that the reality is much more complex. Indeed, supplemental folic acid apparently increases the risk of cancer at several sites, including the breast, lung, and prostate. The topics of alcohol and folate in relation to cancer are further discussed in Chap. 12 by Davis and Milner.

15.1.5   Obesity

Alcohol, of course, is a source of calories (7 kcal/g). It is important to remember that alcoholic beverages also contain carbohydrates that add additional calories. A half liter of wine contains about 350 kcal while three cans of beer supply about 250–450 kcal, clearly enough to tip the energy balance well into positive territory. These numbers explain the popularity of low-calorie "light beers". It is predictable, therefore, that alcohol consumption should be associated with excess weight gain. But, is this actually the case in the real world?

Intervention studies are inconclusive. Alcohol consumption causes an increase in energy intake [9]. However, Cordain et al. [10] reported that when men consumed 35 g/day of alcohol (a little less than three glasses of wine) for a period of 6 week, this did not affect body weight. Similarly, a follow-up study confirmed this when overweight women consumed 25 g/day of alcohol, 5 day/week, for 10 week [11].

Several long-term cohort studies have been carried out. In a cohort study of 16,600 men aged 40–75, change in alcohol intake was not associated with change in waist circumference over 9 year of follow-up [12]. In a cohort study of 19,200 women of normal BMI at baseline, alcohol intake displayed a clear negative association with risk of becoming overweight or obese over the following 13 year [13]. In sharp contrast, other cohort studies have reported a positive association between alcohol consumption and weight gain [14]. At present, therefore, it is far from clear whether alcohol intake is a risk factor for weight gain.

15.2   Protective Effects of Alcohol

15.2.1   Coronary Heart Disease

A convincing body of evidence suggests that the risk of coronary heart disease (CHD) is reduced by 20–40% in persons who consume alcohol in moderation [5].

In some populations this association can be skewed if individuals at higher risk for CHD reduce or eliminate alcohol consumption due to a diagnosis of a related chronic disease (e.g., hypertension or diabetes). This is frequently described as the "sick quitter" syndrome and can create a spurious artificial inverse association between alcohol and CHD [15]. Because conditions such as hypertension and diabetes increase the risk of CHD by two to threefold, a study which does not take these conditions into account may find that moderate drinkers have as much as 50–70% less heart disease. However, even in large cohort studies where "sick quitters" are removed or moderate drinkers are compared to lifelong abstainers, alcohol has been found to have strong cardiovascular benefit [16, 17].

There has been much speculation that wine may be more potent than beer or spirits in preventing CHD. This is largely based on findings from ecological studies (i.e., countries such as France with a high intake of wine tend to have relatively low rates of CHD) [18]. It has been repeatedly shown that such associations can easily be spurious. This is indicated by the findings from case–control and cohort studies: these show no clear trend for one type of alcohol to be more consistently associated with protection from CHD [16, 18]. Where one type of alcohol does manifest a stronger association than other types, this is likely due to confounding by such factors as smoking and drinking pattern or to differences in other lifestyle factors such as eating patterns or physical activity.

A major cohort study in women indicates that HDL-C and HbA$_{1c}$ explain most of the association between alcohol and CHD, while fibrinogen played a much lesser role [16]. However, in men HDL-C was the dominant factor; it accounts for about half of the association. The other significant factors were HbA$_{1c}$, fibrinogen, triglycerides, and adiponectin.

15.2.2 Blood Pressure and Stroke

A relatively high alcohol intake (>4 drinks/day) has been shown to be associated with elevated blood pressure [19, 20] and an increased risk of stroke [5]. However, evidence from cohort studies suggests that the association between alcohol and hypertension may be J-shaped such that light and moderate drinkers have a modestly reduced risk of developing the condition [21]. Although the results of large epidemiological studies have not been consistent, the data as a whole indicate that there is also a J-shaped relationship between alcohol intake and risk of stroke [5]. The protective effect of moderate consumption of alcohol is seen only for ischemic stroke (the more common type), not for hemorrhagic stroke [5].

More work is needed to determine if drinking patterns influence risk of stroke (i.e., frequent consumption of small amounts of alcohol vs. binge drinking).

15.2.3 Impotence

The relationship between excessive alcohol intake and poor erectile function is well known. As Shakespeare put it: "It provokes the desire, but takes away from the performance" (Macbeth). But, as in the case of alcohol and blood pressure, recent findings have revealed an apparently beneficial effect, or at least no ill effects, of moderate alcohol consumption [22].

Although erectile dysfunction was originally thought to be purely psychogenic in nature, 80–90% is likely due to biological factors that may share a similar profile to atherosclerosis.

15.2.4 Gallstones

Most studies that have examined this question have reported a protective association between alcohol and risk of gallstones. For instance, Leitzmann et al. [23] observed that men who consume alcohol frequently (5–7 day/week) have a reduced risk of gallstones but not those who consume alcohol less frequently (1–2 day/week). These findings indicate that frequency of alcohol consumption rather than quantity is the critical factor.

15.2.5 Bone Health

While findings are not consistent, several studies have reported that compared to nondrinking, moderate alcohol intake is associated with higher bone mineral density, especially in postmenopausal women [24–26]. This suggests that alcohol may help prevent osteoporosis. However, as osteoporosis is so dependent on lifetime diet, physical activity, obesity, and other factors, it is probable that alcohol does not play an important role. In contrast to the situation with osteoporosis, high levels of drinking cause loss of balance and falls leading to an increased risk of hip or wrist fracture.

15.2.6 Hearing Loss

A cross-sectional study of subjects aged 50–91 year reported that moderate alcohol intake was associated with better hearing [27]. Again, like bone health, many other environmental and genetic effects play a more important role in the etiology of hearing loss.

15.2.7 Cognitive Function and Dementia

It is well known that heavy drinking has a damaging effect on brain function. Nevertheless, alcohol manifests a J-shaped relationship with the decline in brain function with aging. In several studies, mostly carried out on older adults, moderate consumption of alcohol was associated with enhanced cognitive ability or a slower rate of decline with aging. This effect is generally more pronounced in women [28–32].

While results have not been entirely consistent, several cohort studies have reported a protective association between moderate alcohol consumption and the development of dementia (mainly Alzheimer's disease) [32–35].

15.2.8   Benign Prostatic Hyperplasia

A cohort study reported that moderate alcohol intake (2.5–4 drinks/day) was associated with a reduced risk of benign prostatic hyperplasia (RR of 0.59) [36]. The mechanisms for this action are more speculative, but may include the effects of alcohol on steroid hormone levels.

15.2.9   Diabetes

Cohort studies have suggested that alcohol may be protective against type 2 diabetes. Moderate consumers have a 33–56% reduced risk of developing the condition [37]. Interestingly, several studies have suggested that moderate consumption of alcohol among men and women with type 2 diabetes is also associated with a much reduced risk of subsequent CHD [38–41], the number one killer of diabetics.

15.2.10   Lung Disease

Alcohol may also be protective against chronic obstructive pulmonary disease (COPD). A cohort study of middle-aged men in Finland, Netherlands, and Italy revealed a protective association between alcohol intake and risk of death from COPD [42]. The lowest risk was seen at an intake of up to about three drinks per day. Alcohol intake has also been observed to manifest a protective association with emphysema in smokers [43].

15.3   Effect of Alcohol on Total Mortality

The effects of alcohol consumption represent a complex mixture of harm and benefit. Overall, alcohol certainly causes more deaths than it prevents. In an estimate by Danaei et al. [5] alcohol causes 90,000 deaths per year in the USA. These are due to a variety of causes, the main ones being traffic accidents and other injuries, violence, chronic liver disease, cancer, alcohol use disorders, and hemorrhagic stroke. At the same time alcohol prevents 26,000 deaths from CHD, ischemic stroke, and diabetes. Alcohol therefore causes 64,000 more deaths than it prevents.

When intake is moderate and binge drinking is avoided, the beneficial health effects of alcohol on the cardiovascular system outweigh most detrimental effects. As a result, the net effect of alcohol on total mortality is a J-shaped curve with minimum mortality associated with a moderate intake of alcohol but with a rising curve as consumption increases, especially when there is binge drinking. As an example, a major study by the American Cancer Society on subjects with a mean age of 56 year reported that in each sex, persons consuming one drink daily had a risk of death from all causes about 20% below those of nondrinkers [44]. To put this in perspective, among American men and women aged 35–69 year, a moderate consumption of alcohol prevents approximately one death for every six deaths caused by smoking [44].

The benefits of alcohol are most apparent in the middle-aged and elderly. This is because alcohol reduces risk of CHD and stroke, the first and third leading cause of death, respectively, in that age group. By contrast the leading cause of death in Americans under age 40 year is accidents, with homicide and suicide also being other major causes, especially in males. These are all associated with alcohol. This age effect is illustrated by a report from the Nurses' Health Study. A moderate intake of alcohol has a protective relationship with total mortality in women aged over 50 year (RR is 0.80–0.88) but is associated with a doubling of the risk of death in those aged 34–39 year [45]. Similar findings were reported from England and Wales: a net favorable mortality outcome was seen only in men over age 55 and women over 65 year [46].

The alcohol intake corresponding to the nadir for mortality is still unclear but in people aged 50–80 year is around 1.0–1.5 drinks/day in men and 0.5–1.0 drinks/day in women [47, 48]. However, as this is based on self-reported intake, which represents a substantial underestimation, the true nadir is almost certainly higher [47].

15.4   Drinking Patterns

More recently, research has focused on the importance of pattern of drinking on risk of health outcomes. Not surprisingly, alcohol is most protective when consumed in small regular amounts rather than binge or episodic drinking. This was demonstrated in cohort studies in the USA [49] and Finland [50]. People who engaged in occasional heavy drinking had a higher risk of death than persons with the same alcohol intake but who did not engage in binge drinking. Similar observations were made on cardiovascular disease in Canada. The data from that study revealed that while alcohol consumption has a protective association with both CHD and hypertension, binge drinking increases the risk of both, especially in men [51]. In a study of US male health professionals, frequency of consumption (day/week) was more important than quantity consumed. Men who consumed alcohol at least 5 day/week had the lowest risk of both type 2 diabetes [52] and myocardial infarction [16], regardless of the total amount consumed. These findings are hardly surprising: many dietary components cause no harm in small, frequent doses but are toxic when a large dose is taken.

15.5   Conclusions

Clearly, alcohol can do much good but also much harm. It is important to bear in mind that the harmful effects of alcohol frequently occur at a much younger age than the benefits. Consequently, if the effects of alcohol are measured in terms of quality years of life (lost or gained), then the harm done to one (usually younger) person by alcohol may be far greater than the benefit gained by another (usually older) person.

The large majority of the harmful effects of alcohol can be avoided by sensible drinking, by drinking in moderation, and by the avoidance of alcohol when driving. For the person who can drink sensibly and can avoid alcohol's negative side, alcohol can be of considerable benefit. Like so much else in life, it is a matter of balance. While alcohol should perhaps not be prescribed [53], neither should it be proscribed.

Researchers in Australia estimated that for people aged over 60 year, the cost per life year gained by moderate consumption of alcohol was A$5,700 in men and A$19,000 in women [54]. This translates to about $7,000 in men and $23,000 in women in 2011 US dollars. On this basis alcohol can be considered a cost-effective medication. For instance, it is many times more cost-effective than is medication with statins for treatment of hypercholesterolemia [55].

The findings discussed in this chapter have implications for public health policy. But what are those implications? One possible policy is the following: all adults aged over 40 year should be encouraged to consume moderate amounts of alcohol daily, unless there is a specific reason to the contrary, such as religion, medication use, or a history of alcohol abuse. The problem with such a policy is the risk of causing a rise in the prevalence of alcohol abuse. Typically, about 5–10% of people in any society where alcohol is available become abusers of the beverage. The actual proportion is related to the mean alcohol intake: the higher the mean alcohol intake, the higher is the proportion of alcohol abusers [56]. Thus, a policy that encourages greater use of alcohol will likely also lead to more problems associated with abuse.

Arguably, the most prudent policy is one that explains that alcohol in moderation will likely have several health benefits for people in middle age and above, while also stressing the hazards of abuse.

Acknowledgment This chapter is a revision of a previous version co-authored with Eric Rimm and published in the second edition of this book. The author thanks Dr Rimm for his substantial contributions.

References

1. Leon DA, Chenet L, Shkolnikov VM, et al. Huge variation in Russian mortality rates 1984–94: artefact, alcohol, or what? Lancet. 1997;350:383–8.
2. Walberg P, McKee M, Shkolnikov V, Chenet L, Leon DA. Economic change, crime, and mortality crisis in Russia: regional analysis. BMJ. 1998;317:312–8.
3. Zaridze D, Brennan P, Boreham J, et al. Alcohol and cause-specific mortality in Russia: a retrospective case-control study of 48,557 adult deaths. Lancet. 2009;373:2201–14.
4. Reuters. U.S. drunken driving deaths up, total fatalities down. Available at http://www.reuters.com/article/idUSN2540572120070525 (2010). Last Accessed 26 July 2010.
5. Danaei G, Ding EL, Mozaffarian D, et al. The preventable causes of death in the United States: comparative risk assessment of dietary, lifestyle, and metabolic risk factors. PLoS Med. 2009;6:e1000058.

6. Bagnardi V, Blangiardo M, Vecchia CL, Corrao G. A meta-analysis of alcohol drinking and cancer risk. Br J Cancer. 2001;85:1700–5.

7. Hankinson SE, Willett WC, Manson JE, et al. Alcohol, height, and adiposity in relation to estrogen and prolactin levels in postmenopausal women. J Natl Cancer Inst. 1995;87: 1297–302.

8. Dorgan JF, Baer DJ, Albert PS, et al. Serum hormones and the alcohol-breast cancer association in postmenopausal women. J Natl Cancer Inst. 2001;93:710–5.

9. Yeomans MR. Alcohol, appetite and energy balance: is alcohol intake a risk factor for obesity? Physiol Behav. 2010;100:82–9.

10. Cordain L, Bryan ED, Melby CL, Smith MJ. Influence of moderate daily wine consumption on body weight regulation and metabolism in healthy free-living males. J Am Coll Nutr. 1997;16:134–9.

11. Cordain L, Melby CL, Hamamoto AE, et al. Influence of moderate chronic wine consumption on insulin sensitivity and other correlates of syndrome X in moderately obese women. Metabolism. 2000;49:1473–8.

12. Koh-Banerjee P, Chu NF, Spiegelman D, Rosner B, Colditz G, Willett W, et al. Prospective study of the association of changes in dietary intake, physical activity, alcohol consumption, and smoking with 9-y gain in waist circumference among 16 587 US men. Am J Clin Nutr. 2003;78:719–27.

13. Wang L, Lee IM, Manson JE, Buring JE, Sesso HD. Alcohol consumption, weight gain, and risk of becoming overweight in middle-aged and older women. Arch Intern Med. 2010;170:453–61.

14. Mozaffarian D, Hao T, Rimm EB, Willett WC, Hu FB. Changes in diet and lifestyle and long-term weight gain in women and men. N Engl J Med. 2011;364:2392–404.

15. Shaper AG, Wannamethee G, Walker M. Alcohol and mortality in British men: explaining the U-shaped curve. Lancet. 1988;2:1267–73.

16. Mukamal KJ, Jensen MK, Grønbaek M, Stampfer MJ, Manson JE, Pischon T, et al. Drinking frequency, mediating biomarkers, and risk of myocardial infarction in women and men. Circulation. 2005;112:1406–13.

17. Rimm E. Alcohol and cardiovascular disease. Curr Atheroscler Rep. 2000;2:529–35.

18. Rimm EB, Klatsky A, Grobbee D, Stampfer MJ. Review of moderate alcohol consumption and reduced risk of coronary heart disease: is the effect due to beer, wine, or spirits? BMJ. 1996;312:731–6.

19. Corrao G, Bagnardi V, Zambon A, La Vecchia C. A meta-analysis of alcohol consumption and the risk of 15 diseases. Prev Med. 2004;38:613–9.

20. Puddey IB, Beilin LJ, Vandongen R, Rouse IL, Rogers P. Evidence for a direct effect of alcohol consumption on blood pressure in normotensive men. A randomized controlled trial. Hypertension. 1985;7:707–13.

21. Thadhani R, Camargo Jr CA, Stampfer MJ, Curhan GC, Willett WC, Rimm EB. Prospective study of moderate alcohol consumption and risk of hypertension in young women. Arch Intern Med. 2002;162:569–74.

22. Bacon CG, Mittleman MA, Kawachi I, Giovannucci E, Glasser DB, Rimm EB. Sexual function in men older than 50 years of age: results from the health professionals follow-up study. Ann Intern Med. 2003;139:161–8.

23. Leitzmann MF, Giovannucci EL, Stampfer MJ, et al. Prospective study of alcohol consumption patterns in relation to symptomatic gallstone disease in men. Alcohol Clin Exp Res. 1999;23:835–41.

24. Rapuri PB, Gallagher JC, Balhorn KE, Ryschon KL. Alcohol intake and bone metabolism in elderly women. Am J Clin Nutr. 2000;72:1206–13.

25. Feskanich D, Korrick SA, Greenspan SL, Rosen HN, Colditz GA. Moderate alcohol consumption and bone density among postmenopausal women. J Womens Health. 1999;8:65–73.

26. Macdonald HM, New SA, Golden MH, Campbell MK, Reid DM. Nutritional associations with bone loss during the menopausal transition: evidence of a beneficial effect of calcium,

alcohol, and fruit and vegetable nutrients and of a detrimental effect of fatty acids. Am J Clin Nutr. 2004;79:155–65.

27. Popelka MM, Cruikshanks KJ, Wiley TL, et al. Moderate alcohol consumption and hearing loss: a protective effect. J Am Geriatr Soc. 2000;48:1273–8.

28. Britton A, Singh-Manoux A, Marmot M. Alcohol consumption and cognitive function in the Whitehall II Study. Am J Epidemiol. 2004;160:240–7.

29. Leroi I, Sheppard JM, Lyketsos CG. Cognitive function after 11.5 years of alcohol use: relation to alcohol use. Am J Epidemiol. 2002;156:747–52.

30. Kalmijn S, van Boxtel MP, Verschuren MW, Jolles J, Launer LJ. Cigarette smoking and alcohol consumption in relation to cognitive performance in middle age. Am J Epidemiol. 2002;156:936–44.

31. Stampfer MJ, Kang JH, Chen J, Cherry R, Grodstein F. Effects of moderate alcohol consumption on cognitive function in women. N Engl J Med. 2005;352:245–53.

32. Espeland MA, Gu L, Masaki KH, et al. Association between reported alcohol intake and cognition: results from the Women's Health Initiative Memory Study. Am J Epidemiol. 2005;161:228–38.

33. Ruitenberg A, van Swieten JC, Witteman JC, et al. Alcohol consumption and risk of dementia: the Rotterdam Study. Lancet. 2002;359:281–6.

34. Huang W, Qiu C, Winblad B, Fratiglioni L. Alcohol consumption and incidence of dementia in a community sample aged 75 years and older. J Clin Epidemiol. 2002;55:959–64.

35. Mukamal KJ, Kuller LH, Fitzpatrick AL, Longstreth Jr WT, Mittleman MA, Siscovick DS. Prospective study of alcohol consumption and risk of dementia in older adults. JAMA. 2003;289:1405–13.

36. Platz EA, Rimm EB, Kawachi I, et al. Alcohol consumption, cigarette smoking, and risk of benign prostatic hyperplasia. Am J Epidemiol. 1999;149:106–15.

37. Howard AA, Arnsten JH, Gourevitch MN. Effect of alcohol consumption on diabetes mellitus: a systematic review. Ann Intern Med. 2004;140:211–9.

38. Tanasescu M, Hu FB, Willett WC, Stampfer MJ, Rimm EB. Alcohol consumption and risk of coronary heart disease among men with type 2 diabetes mellitus. J Am Coll Cardiol. 2001;38:1836–42.

39. Ajani UA, Gaziano JM, Lotufo PA, et al. Alcohol consumption and risk of coronary heart disease by diabetes status. Circulation. 2000;102:500–5.

40. Solomon CG, Hu FB, Stampfer MJ, et al. Moderate alcohol consumption and risk of coronary heart disease among women with type 2 diabetes mellitus. Circulation. 2000;102:494–9.

41. Valmadrid CT, Klein R, Moss SE, Klein BE, Cruickshanks KJ. Alcohol intake and the risk of coronary heart disease mortality in persons with older-onset diabetes mellitus. JAMA. 1999;282:239–46.

42. Tabak C, Smit HA, Rasanen L, et al. Alcohol consumption in relation to 20-year COPD mortality and pulmonary function in middle-aged men from three European countries. Epidemiology. 2001;12:239–45.

43. Pratt PC, Vollmer RT. The beneficial effect of alcohol consumption on the prevalence and extent of centrilobular emphysema. Chest. 1984;85:372–7.

44. Thun MJ, Peto R, Lopez AD, Monaco JH, Henley SJ, Heath Jr CW, et al. Alcohol consumption and mortality among middle-aged and elderly U.S. adults. N Engl J Med. 1997;337:1705–14.

45. Fuchs CS, Stampfer MJ, Colditz GA, et al. Alcohol consumption and mortality among women. N Engl J Med. 1995;332:1245–50.

46. Britton A, McPherson K. Mortality in England and Wales attributable to current alcohol consumption. J Epidemiol Community Health. 2001;55:383–8.

47. White IR. The level of alcohol consumption at which all-cause mortality is least. J Clin Epidemiol. 1999;52:967–75.

48. Lee SJ, Sudore RL, Williams BA, Lindquist K, Chen HL, Covinsky KE. Functional limitations, socioeconomic status, and all-cause mortality in moderate alcohol drinkers. J Am Geriatr Soc. 2009;57:955–62.

49. Rehm J, Greenfield TK, Rogers JD. Average volume of alcohol consumption, patterns of drinking, and all-cause mortality: results from the US National Alcohol Survey. Am J Epidemiol. 2001;153:64–71.

50. Laatikainen T, Manninen L, Poikolainen K, Vartiainen E. Increased mortality related to heavy alcohol intake pattern. J Epidemiol Community Health. 2003;57:379–84.

51. Murray RP, Connett JE, Tyas SL, et al. Alcohol volume, drinking pattern, and cardiovascular disease morbidity and mortality: is there a U-shaped function? Am J Epidemiol. 2002;155:242–8.

52. Conigrave KM, Hu BF, Camargo Jr. CA, Stampfer MJ, Willett WC, Rimm EB. A prospective study of drinking patterns in relation to risk of type 2 diabetes among men. Diabetes. 2001;50: 2390–5.

53. Wannamethee SG, Shaper AG. Taking up regular drinking in middle age: effect on major coronary heart disease events and mortality. Heart. 2002;87:32–6.

54. Simons LA, McCallum J, Friedlander Y, Ortiz M, Simons J. Moderate alcohol intake is associated with survival in the elderly: the Dubbo Study. Med J Aust. 2000;173:121–4.

55. Thompson A, Temple NJ. The case for statins: has it really been made? J R Soc Med. 2004; 97:461–4.

56. Colhoun H, Ben-Shlomo Y, Dong W, Bost L, Marmot M. Ecological analysis of collectivity of alcohol consumption in England: importance of average drinker. BMJ. 1997;314:1164–8.

Chapter 16
Health Claims and Dietary Recommendations for Nonalcoholic Beverages

Ted Wilson

Keywords Beverages • Coffee • Tea • Milk • Fruit juice • Soft drinks • Energy drinks

Key Points

- Water intake whether from a tap or bottle is important, but the value of bottled water is clouded by cost, environmental impact, and the dubious nutritional value of supplemental ingredients.
- Green tea appears associated with reductions in cancer and cardiovascular disease risk and intake should be promoted at 3–4 cups/day. Black tea also may be beneficial.
- Coffee consumption does not hurt health and may actually improve health. Intake should be limited to 2–3 cups/day.
- Milk and fruit juices should be promoted for improved health.
- Sports drinks may provide important benefits for hydration status during extended exercise.
- Soft drinks should be avoided because they are associated with reduced milk consumption, increased energy intake, and increased weight gain.
- There is some suggestion that artificial sweetened drinks may not serve their desired purpose in that they may decrease satiety and increase risk of metabolic abnormalities.
- Consumption of weight-loss beverages may be beneficial, but energy drinks should be avoided.

T. Wilson, PhD (✉)
Department of Biology, Winona State University, Winona, MN 54603, USA
e-mail: twilson@winona.edu

N.J. Temple et al. (eds.), *Nutritional Health: Strategies for Disease Prevention*,
Nutrition and Health, DOI 10.1007/978-1-61779-894-8_16,
© Springer Science+Business Media, LLC 2012

16.1   Introduction

Beverages play a major role in determining nutritional health. Indeed, water represents as much as 60% of the body weight in a lean person but only 45% in the obese. Water from a tap or bottle remains the most popular beverage on the planet. Determining the nutritional consequences of a beverage is complicated by the myriad of additional contents mixed in with our water. Nonalcoholic beverages include coffee, tea, milk, juices, soft drinks, energy drinks, sports drinks, drinks for weight management, and of course water. Beverages provide about one-fifth of our daily energy intake, with the greatest intake occurring in 19- to 39-year olds [1]. These beverages may contain sugars, fats, minerals, and vitamins from natural or supplemental origin; these substances can alter the taste and nutritional consequences of beverages. Alcohol is another potential ingredient in beverages which can have dramatic health consequences. Alcoholic beverages were covered in the previous chapter.

The nutritional impact of beverages covers a broad area and has been described in a previous book from this book series [2]. This chapter briefly reviews the subject.

16.2   Water

The dietary recommended intake (DRI) of water for nonexercising persons is 3.7 L/day for men and 2.7 L/day for women [3]. This includes water obtained from both beverages and food. Water needs during exercise increase by about 1 mL water for each kcal burned. Ironically, while water is one of the most common compounds on the earth, access to quality potable water that is not tainted by pathogens such as *Cholerae* or toxins such as mercury is a major concern in many regions of the world.

A variety of health promotion water products also have recently appeared on the market. While they may be useful as a display of wealth or social status, their nutritional value appears to be inversely proportional to their cost and they should be avoided. These products typically contain water, of course, but also supplemental vitamins, minerals, herbs, and taste essences. Some of these "water" products actually contain a significant quantity of sugar making them not too dissimilar to the soda they are intended to "replace." This led to a successful lawsuit by the Center for Science in the Public Interest which claimed a fraudulent health claim [4]. Examples of currently supplemented water products from the Coca-Cola are the "VitaminWater" products and of the PHENOM® a series from Pepsi Co./General Nutrition Centers Inc. Calorie-free flavored water products are also popular, such as the calorie-free Aquafina series from Pepsi Co.

While the "taste" may be appealing, these are not products that should be promoted with the consumer, if health is a goal. The topic of the marketing of supplements of dubious nutritional value is a topic discussed at greater length by Temple in Chap. 20. It is ironic that bottled mineral waters are often processed from municipal water supplies and can be exorbitantly priced. For example, whereas the cost of a gallon of gasoline for one's car in the USA is about $4, the cost of various water products (calculated as cost per gallon) is $7.32 for Fiji Water (*pure water imported from Fiji*), $6.32 for Glaceau Smartwater (*condensed from water vapor*), and $3.64 for naturally sweetened Vitaminwater (*50 kcal/serving*).

The expanding use of bottled waters also creates environmental concerns. Vast numbers of plastic bottles are used for this purpose, most of which are nonreusable and end up in landfills. Environmental problems are made much worse when the bottled water is imported from distant places such as Europe or Fiji, leaving a major carbon footprint. Often these are made of glass, which adds weight. If people want to clean their tap water, a far cheaper and environmentally better approach is to simply filter tap water (e.g., using a Brita®).

16.3   Tea

After water, tea is the most popular beverage in the world. Countries where tea drinking is especially popular include Britain, China, Japan, and several other countries in Asia. Its popularity in the United States is increasing, in part as a consequence of the favorable health benefits attributed to it. Leaves from the tea plant *Camellia sinensis* are the source of the primary tea types (green, black, and oolong). For green tea, the plant leaves are steamed and parched after picking in order to prevent oxidation of the catechins present in the leaf. To produce black tea, the plant leaves are picked and then allowed to wither indoors, ferment and oxidize. Oolong tea is produced by "semi-fermenting" the green leaves, resulting in a tea that is chemically a mixture of green and black teas.

Even though each of these nonherbal teas is derived from *C. sinensis*, qualitative and quantitative chemical differences result from the different processing techniques. Freshly brewed green tea contains (-)-epigallocatechin-3-gallate (EGCG) and other phenolics and black tea contains lower levels of these polyphenolic compounds [5]. EGCG makes up more than 40% of the total polyphenolic mixture and appears to be the polyphenol most responsible for green tea's beneficial effects. Maximal plasma concentrations are achieved for EGCG 1.3–2.4 h after consumption. EGCG is classified by the FDA as "generally recognized as safe" (GRAS) and is a popular food additive and nutraceutical supplement. Tea also contains caffeine, though considerably less than in coffee. It also contains theophylline, a substance similar to caffeine in both its chemistry and pharmacological effects.

Green tea shows much potential as an anticancer agent; this is because many epidemiological studies have reported a protective association with the risk of cancer [5] and cardiovascular disease [6]. Black tea has also been demonstrated to be protective against cardiovascular disease by helping improve endothelial cell function and vasodilation [7]. The recent green tea meta-analysis by Wang et al. suggested that the protective relationship is strongest for green tea and slightly protective for black tea [8]. However, much of the generally optimistic epidemiological evidence concerning green tea is based heavily on studies in Japan and China where many people drink 8 or more cups/day. As a general guideline, consumption of up to 3–4 cups/day should be recommended as safe.

16.4 Coffee

It is reasonable to believe that low-to-moderate coffee consumption (≤3 cups/day) should be safe for the typical consumer. Overall, epidemiological evidence shows the absence of any appreciable association between coffee intake and most common neoplasms; indeed, an inverse relationship between coffee consumption and colorectal cancer risk may exist [9]. Coffee consumption has also been associated with neutral or moderately beneficial effects on cardiovascular disease risk and overall mortality [10], although consumption may cause a slight increase in systolic blood pressure for moderate levels of consumption (1–3 cups/day) [11]. In addition, recent evidence suggests that coffee consumption may be beneficial for preventing type 2 diabetes [12, 13] and prevention of cognitive decline [14], although these effects may be due simply to the caffeine contained in coffee.

The content of caffeine and phenolics is dependent on the type of coffee (American coffee, espresso, or mocha), the amount of coffee used, the duration of brewing, and the type of coffee bean used. Caffeic and ferulic acids have antioxidant and other biological activities and have been suggested to reach a plasma concentration of 114 and 96 nM, respectively, about an hour after consumption [15]. The amount of caffeine present in a cup varies from 30 mg in a weakly prepared decaffeinated instant coffee to 350 mg for a brew steeped from dark robusta beans, with 80–125 mg being more typical. It is noteworthy that decaffeination does not remove all of the caffeine; there are large variations in residual caffeine in "decaffeineated" coffee. The decaffeination process is associated with a reduced phenolic content and the introduction of compounds, such as nitric acids and formaldehyde, which may have deleterious health effects, with the water method perhaps being the mildest but also resulting in the greatest loss of phenolic content.

When estimating coffee consumption, it is important to consider the size of the container, the habit of refilling the cup, the variability of coffee drinking between different days (weekdays/weekends), and seasonal differences in intake. A general guideline for coffee is at most 2–3 cups/day.

16.5 Milk

Milk has long been recognized as a way to improve calcium intake and bone health, especially when it is fortified with vitamin D. The high calcium content in milk and its lipid content, which improves absorption of lipid-soluble vitamins, make it ideal as a medium for vitamin D fortification. Milk is also an excellent source of potassium, magnesium, and protein. The popularity of various recommendations for milk has had its ups and downs and ups over the last 20 years.

Milk consumption by persons who lack lactase in their intestine results in lactose intolerance. While persons of south-east Asian decent are most troubled by this condition, intolerance can also be seen in many people of northern European decent.

The problem can be overcome by use of lactose-free dairy products. Many such products are available.

The DASH study suggested that low-fat dairy consumption may reduce blood pressure [16]. Milk does not appear to affect overall mortality and may reduce the risk of cardiovascular disease, an effect that is probably stronger for low-fat milk [17]. The effect of milk on the lipoprotein profile is in part dependent on the fat content, with most studies showing slight increases in both LDL and HDL, making its effects somewhat neutral [18].

Milk consumption is relatively neutral with regard to weight change, despite the fat and calories it contains [19]. In overweight persons, consumption of skim milk has been suggested to decrease both appetite and energy intake during a subsequent meal [20]. Milk has also been inversely correlated with the risk of developing insulin resistance [21] and type 2 diabetes [22]. Milk consumption is arguably most important in younger persons who are developing bone mass. In the last few decades milk consumption among this age group has gradually declined at the same time that soft drink consumption, and obesity, has steadily increased. Milk consumption should be promoted in persons who are not lactose intolerant. A general guideline for milk consumption should be 2–4 cups/day with a recommendation of low-fat milk (preferably 1% or fat-free) as a way to promote decreased intake of both energy and saturated fat.

16.6   Fruit and Vegetable Juices

The benefits of eating fruit and vegetables for blood pressure reduction were demonstrated by the DASH study [16]. Persons in the lowest quartile of fruit and vegetable consumption are in the highest quartile for CVD and cancer. These findings support the Five-A-Day program to improve consumption in the USA. Fruits and vegetables are an excellent source of many vitamins and minerals, as well as phytochemical compounds, such as quercetin-3-galactoside from cranberries. This topic is examined in detail in Chap. 13 by Rui Hai Liu. For reasons of ease and convenience, converting a whole fruit or vegetable to a juice can be a useful way to improve consumption, although this may occur at the cost of losing some of the fiber in the original whole fruit or vegetable.

16.6.1   Fruit Juices

Juices are a popular way to increase fruit consumption because of their widespread accessibility, ease of storage, modest cost, and, of course, their enjoyable taste and eating pleasure. The new USDA recommendations at MyPlate suggest using 100% fruit juice as equivalent to a fruit serving for the purpose of the plate [23]. The energy content of apple juice and orange juice (OJ) (110 kcal/240 mL serving) is about 6% higher than that of cola drinks. The energy content of white grape juice is even higher, around 150 kcal/240 mL. As mentioned previously, caution is therefore warranted if a consumer wishes to consume more than two cups of juice per day.

16.6.2 Health Benefits of Citrus Juice

OJ is the most commonly consumed citrus juice in the USA. It is also the most nutrient-dense of the popular fruit juices. An 240 mL provides 72 mg of vitamin C (120% of Daily Value). OJ is also a good source of potassium (450 mg or 13% of DV), folate (60 μg or 15% of DV), and thiamin (0.15 mg or 10% of DV). Grapefruit juice is another citrus juice consumed by a great many people. It contains a similar amount of vitamin C to OJ, but a lower concentration of potassium and the B-vitamins folate, thiamin, and niacin. It also has a different profile of phenolic acids which may be responsible for an alteration of drug metabolism.

Epidemiological data, clinical investigations, and animal studies provide strong evidence that citrus juice consumption is beneficial with respect to coronary heart disease (CHD), cancer, and overall mortality. Potassium functions to maintain intracellular fluid balance and, as such, a high intake is associated with lower blood pressure and a reduced risk of stroke. Consumption of OJ may help lower the LDL/HDL ratio and components of citrus juice may also decrease LDL oxidation, thereby reducing the risk of CHD. However, because the vitamin C in OJ readily oxidizes following exposure to air, it should be consumed within 1 week of opening [24].

16.6.3 Health Effects of Other Fruit Juices

Cranberry juice has been used in folk medicine for millennia. Recent clinical studies have confirmed its usefulness for the prevention of urinary tract infections. The active antibacterial agents are proanthocyanidins specific to cranberries which prevent bacterial adhesion to the urinary tract. These antiadhesive effects may also be associated with oral and gastric health benefits. Consumption of cranberry products is associated with beneficial antioxidant [25], vasodilatory [26], and antiplatelet aggregation properties that may make these products a viable substitute to red wine and Concord grape juice for protection from CHD. However, consumer and researcher understanding of how cranberries affect human health is limited; reasons for this include the large range of product sweeteners used and the differences in the amount of cranberry juice actually present in these beverages, which can range from 3 to 27% v/v.

Concord and purple grape juices contain an array of polyphenolic compounds that are similar to but not identical to those in red wine. The biological effects of grape juice have been demonstrated to include a small improvement in plasma lipid profile, vasodilation, and antiplatelet aggregatory activities. Pomegranate has been associated with improved vasodilation and a hypocholesterolemic effect.

16.6.4   Vegetable Juices

Tomato juice has been popular for decades. It has been linked to health benefits such as protection against CHD and prostate cancer, but while the preliminary data look promising causality remains to be proven in long-term human trials [27]. Unfortunately, its health benefits are reduced by the large content of added salt (as high as 560–660 mg sodium/cup). However, in recent years numerous brands of vegetable juices have appeared on the market that are low in salt. This is usually prominently stated on the label. V8 is one well-known brand. These juices can have a relatively low energy content (50 kcal/cup as compared to 110 kcal/cup in OJ and apple juice). It is also important to recognize that many vegetable drinks include pear, white grape, or other juices as a source of sweeteners. Sugar calories are indeed sugar calories, regardless of whether they come from high-fructose corn syrup (HFCS) or pear juice.

16.6.5   Recommendations for Consumption of Fruit and Vegetable Juices

Fruit and vegetable juices clearly offer many health benefits. However, consumers should be reminded that whole fruits are better nutritionally, especially with respect to the control of energy intake. This is because they often do not contain the fiber component of the original whole fruit. It is far easier to consume 200 kcal from OJ or apple juice than from whole oranges or apples. In this regard low-salt vegetable juices are preferable as they have only half the energy content of fruit juices. Total juice consumption recommendations are difficult to make. However, as with all things, moderation is best. Intake should probably not exceed one to two 240 mL servings/day; persons seeking to lose weight should be especially cautious.

16.7   Sports Beverages

A variety of beverages are consumed by athletes for a variety of reason, a topic also discussed in Chap. 19 on sports nutrition by Kreider, Schwarz, and Leutholtz.

Compromises in physical and mental performance can occur with the loss of body water representing as little as 2% of body weight. Athletes are prone to dehydration and sports beverages can and do improve hydration status. The key component to this, surprisingly, is taste: if people like the taste, then they are more likely to drink more of the beverage, thereby leading to greater improvements in their hydration status.

Many beverages include sodium and glucose, the presence of which permits the intestine to co-transport the two substances into the blood. In addition, as sodium and glucose maintain spheres of hydration, this also enhances the rapid absorption of water from the intestine. However, beverages that contain an excessive amount of sugar, such as soft drinks, can actually promote a reduction in gastric emptying and an osmotic effect in the intestine that can lead to solvent drag of water into the intestine and dehydration.

Sports beverages can improve electrolyte status, although the effects are most significant for exercise of long duration (e.g., a half marathon or mowing a lawn on a hot summer day for 2 h). However, if a nonathletic person is adequately hydrated and has a proper electrolyte balance prior to beginning an exercise session, consuming electrolytes from a sports beverage is unlikely to significantly improve physical performance during exercise of long duration.

It is generally recommended that everyone should carry out 30 min/day of moderate exercise, such as walking at a brisk pace. Consumers should pay close attention to the energy content of sports drinks. For example, Gatorade provides 125 kcal per 500 mL bottle. Exercising people should therefore be reminded that for the purpose of weight management, such drinks can negate most of the caloric benefits of the exercise. Unless it is an especially hot day and they exercise for a prolonged time period, then dehydration is unlikely.

16.8   Soft Drinks and Sugar-Sweetened Beverages

If we increase our consumption of sugar-sweetened beverages (SSBs; soft drinks, fruit punch, etc.), several things predictably occur. We tend to have more dental caries. We tend to decrease our consumption of more nutrient-dense cow's milk. We tend to see a decrease in bone density. We also tend to increase our overall energy intake because soft drinks are less satiating than solid foods. SSBs can contribute to a positive energy balance and in adolescents this can represent 17% of average energy intake [27]. In this regard the NHANES data set reports that in 2005–2006 adolescents aged 13–19 years consumed an average of 242 kcal/day of soft and fruit drinks [28]. Encouragingly, SSB consumption by youths appears to have begun to drop slightly, though this change has not been observed in adults.

Adult and adolescent obesity is increasing in the USA, but the exact reason for this is probably multifactorial. Soft drink consumption is probably one contributing factor. In adults consumption of SSBs is positively and tightly correlated with weight gain, the risk for developing metabolic syndrome and type 2 diabetes [29]. SSBs have been reported to contribute to increased body mass index in children and adolescents [30]. Blaming SSBs for all childhood obesity represents an apparently low-hanging fruit. In contrast, data from the 5-year-long data collection in Project EAT suggest that there was no direct association between soft drink consumption and adolescent weight gain, although weight gain was associated with those who consumed low-calorie soft drinks or who consumed little or no whole milk [31].

Cardiovascular disease risk may also be associated with the consumption of SSBs. Data from the Nurses' Health Study suggest a positive association between

SSB intake and risk of CHD [32]. After 24 years of follow-up, women who consumed two or more servings of SSBs per day had a 35% higher risk of developing CHD compared with those who consumed less than one serving/month.

HFCS is a common sweetener used in the beverage industry. While the health effects of HFCS remains controversial for many, it appears that it does not contribute to direct negative health effects [33]. Furthermore, it appears that consumption of fructose in the human diet (≤50 g/day) does not create deleterious effects [34]. Caution may be warranted as epidemiological studies move to more definitive controlled clinical trials.

In a perfect world, we would have zero consumption of SSBs. Instead, water would have many splendid tastes making caloric beverages irrelevant. Water consumption would also be supported by billion dollar advertizing campaigns. In the light of reality it is perhaps sensible to suggest limiting SSBs to no more than 1 or 2 cups/day, though none is clearly preferable. Given that the consumption of diet soft drinks is not typically associated with reductions in body weight, their consumption should be limited in a similar fashion. Furthermore, as mentioned previously with respect to the VitaminWater product, even a beverage that is marketed as "water" can contain sugar and calories.

A "solution" to the SSB problem would seem to be to find a safe, acaloric sweetener that could be added to water to provide a soft drink-like beverage that was really just water, but satiated in the sense of taste. This idea is the origin of diet soft drinks and other products that contain reduced calories because they include an artificial sweetener. Artificial sweeteners currently on the market include aspartame, sucralose, and saccharine. As noted above, these products do not appear to solve the problem; use of artificially sweetened beverages was associated with weight gain in the Project EAT study [31]. In addition, three prospective epidemiologic studies have found positive associations between use of diet soft drinks and incident metabolic abnormalities including diabetes [35–37]. It should be noted that epidemiological studies of general use of artificial sweeteners have not been able to identify which sweetener each participant used.

16.9 Weight Loss and Weight-Management Beverages

For several reasons, meal-replacement beverages may have a place in the regular nutrition of many persons. A variety of meal-replacement beverages (e.g., Slim-Fast, Met-Rx, and Atkins Nutritionals) provide consumers with a convenient way to consume a relatively balanced nutritional intake that supplies about 200 kcal along with a typically large protein intake, as well as minerals and vitamins. Liquid meal replacements are readily available, do not require refrigeration, and are often quite palatable. However, to improve palatability some beverages actually contain generous amounts of fat or sugar.

Older persons tend to consume more food during the meal following consumption of a liquid meal replacement, making these beverages useful for elderly persons attempting to gain weight, but detrimental for overweight persons attempting to lose weight [38]. While liquid meal replacements may be useful for some persons, the best nutritional advice for most people is to consume a balanced diet that emphasizes a variety of foods consumed in their solid form.

16.10   Health Effects of Energy Drinks

The increasing popularity of what are popularly termed "energy drinks" is an American and global phenomenon. They are consumed for perceived enhancements in mental acuity, wakefulness, and physical performance. Regardless of whether their consumption provides a measureable change in physiological/mental status, their popularity is very great with net sales of several billion dollars per year.

Energy drinks constitute a class of beverages whose ingredients are not uniform. While caffeine is a primary ingredient in most brands, its content is quite variable ranging from 50 to 500 mg per container [39]. Energy drinks generally contain a variety of other compounds with potential for altering physiological/mental activity. These may include taurine (neurotransmitter function) and various B vitamins. In addition, most contain sugar, although some are nearly calorie-free; the energy content ranges from 10 to 150 kcal/8 oz serving.

While energy drinks are commonly believed to have significant physiological effects, there is relatively scant documentation of this. Echocardiographic evidence suggests that Red Bull® consumption may improve stroke volume in persons with heart failure [40]. Surprisingly, in a study of 70 college-aged subjects who consumed a 240 mL serving of Red Bull or placebo, the author (TW) did not observe any statistically significant changes in heart rate, ECG QRT segments/intervals, or blood pressure during the 2 h following consumption [41]. However, given the consistency of anecdotal reports that link cardiac pathologies associated with energy drinks when consumed with alcohol, caution seems warranted [42].

What is tangible is that inclusion of large amounts of caffeine in combination with alcohol in an energy drink-like format can lead to an inability of the consumer to perceive sobriety. This can lead to motor vehicle accidents, poor decision-making skills, and acts of sexual violence which lead the US FDA to send four letters of caution to makers of these beverages [43]; these cautionary letters are expected to reduce the availability of alcoholic energy drinks that also contain caffeine.

16.11   The Future

It is estimated that over 25,000 new beverage products entered the world market in 2010 [44]. A more complete description of beverage and food products that are becoming popular can be found in Chap. 21. Few clinical trials for the impact of established beverage products on health have been published; clinical trials for the newer products are nearly nonexistent. The ever-changing formulations that manufacturers put into their beverages make clinical trials difficult. What is tested today may have a formula or manufacturing method that differs next year when the results of a trial are published. Given this reality, it remains important for clinicians to be able to provide the "best estimates" of the health impact of the beverages described in this chapter.

16.12 Conclusions

Nonalcoholic beverages constitute a substantial proportion of our diet and can have a major impact on our health. No single beverage can replace water, that ubiquitous beverage, but even drinking water has its faults when we consider the presence of potential pollutants and environmental impacts. Tea appears to have beneficial effects and coffee may have some beneficial effects as well. Consumption of low-fat milk has been determined to promote better health. Fruit and vegetable juices provide health benefits and have the potential to get consumers closer to the Five-A-Day goal, although the consumer may be better eating the raw fruit if possible. Sugar-sweetened soft drinks appear to promote weight gain and diminish the opportunity for a varied and high nutrient intake by providing energy almost entirely from sugar. Artificially-sweetened soft drinks should in principle solve the problem of sugar-sweetened soft drinks, but doubts have been raised about their safety. Sports drinks may be helpful for improving hydration and electrolyte status, but only during sustained exercise and may also increase caloric intake. With regards to energy drinks, conclusive evidence to support health concerns regarding their consumption generally does not exist; caution should remain the operative word. Water enters our bodies in a variety of forms and the materials dissolved in the water are important to consider with respect to their impact on our health. The vitamins, minerals, phytochemicals, and other materials dissolved in the beverages we consume require our attention as we seek to make the best dietary recommendations possible.

References

1. Nielson SJ, Popkin BM. Changes in beverage intake between 1977 and 2001. Am J Prev Med. 2004;27:205–10.
2. Wilson T, Temple NJ, editors. Beverages in health and nutrition. Totowa: Humana Press; 2004.
3. Dietary Reference Intakes: Electrolytes and Water. National Academy of Sciences. Institute of Medicine. Food and Nutrition Board. Available at http://www.iom.edu/Global/News%20. Announcements/~/media/442A08B899F44DF9AAD083D86164C75B.ashx (2011). Last Accessed 8 July 2011.
4. Court Finds Coke in Violation of Various FDA Regs and Denies its Motion to Dismiss the Lawsuit. Available at http://www.cspinet.org/new/201007231.html (2011). Last Accessed 8 July 2011.

5. Carlson JR, Bauer BA, Vincent A, Limburg PJ, Wilson T. Reading the tea leaves: anticarcino-genic properties of (-)-epigallocatechin-3-gallate. Mayo Clin Proc. 2007;82:725–32.
6. Basu A, Lucas EA. Mechanisms and effects of green tea on cardiovascular health. Nutr Rev. 2007;65:361–75.
7. Jochmann N, Lorenz M, Krosigk A, et al. The efficacy of black tea in ameliorating endothelial function is equivalent to that of green tea. Br J Nutr. 2008;99:863–8.
8. Wang ZM, Zhou B, Wang YS, et al. Black and green tea consumption and the risk of coronary artery disease: a meta-analysis. Am J Clin Nutr. 2011;93:506–15.
9. Higdon JV, Frei B. Coffee and health: a review of recent human research. Crit Rev Food Sci Nutr. 2006;46:101–23.
10. Lopez-Garcia E, van Dam RM, Li TY, Rodriguez-Artalejo F, Hu FB. The relationship of cof-fee consumption with mortality. Ann Intern Med. 2008;148:904–14.
11. Zhang Z, Hu G, Caballero B, Appel L, Chen L. Habitual coffee consumption and risk of hyper-tension: a systematic review and meta-analysis of prospective observational studies. Am J Clin Nutr. 2011;93:1212–9.
12. Campos H, Baylin A. Coffee consumption and risk of type 2 diabetes and heart disease. Nutr Rev. 2007;65:173–9.
13. Huxley R, Lee CM, Barzi F, et al. Coffee, decaffeinated coffee, and tea consumption in relation to incident type 2 diabetes mellitus: a systematic review with meta-analysis. Arch Intern Med. 2009;169:2053–63.
14. Laitala VS, Kaprio J, Koskenvuo M, Räihä I, Rinne JO, Silventoinen K. Coffee drinking in middle age is not associated with cognitive performance in old age. Am J Clin Nutr. 2009;90: 640–6.
15. Williamson G, Dionisi F, Renouf M. Flavanols from green tea and phenolic acids from coffee: critical quantitative evaluation of the pharmacokinetic data in humans after consumption of single doses of beverages. Mol Nutr Food Res. 2011;55:864–73.
16. Appel LJ, Moore TJ, Obarzanek E, et al. A clinical trial of the effects of dietary patterns on blood pressure. N Engl J Med. 1997;336:1117–24.
17. Soedamah-Muthu SS, Ding EL, Al-Delaimy WK, Hu FB, Engberink MF, Willett WC, et al. Milk and dairy consumption and incidence of cardiovascular diseases and all-cause mortality: dose–response meta-analysis of prospective cohort studies. Am J Clin Nutr. 2011;93:158–71.
18. Ohlsson L. Dairy products and plasma cholesterol levels. Food Nutr Res. 2010;54. doi:10.3402/fnr.v54i0.5124.
19. Mozaffarian D, Hao T, Rimm EB, Willett WC, Hu FB. Changes in diet and lifestyle and long-term weight gain in women and men. N Engl J Med. 2011;364:2392–404.
20. Dove ER, Hodgson JM, Puddey IB, Beilin LJ, Lee YP, Mori TA. Skim milk compared with a fruit drink acutely reduces appetite and energy intake in overweight men and women. Am J Clin Nutr. 2009;90:70–5.
21. Tremblay A, Gilbert JA. Milk products, insulin resistance syndrome and type 2 diabetes. J Am Coll Nutr. 2009;28:91S–102.
22. Tong X, Dong JY, Wu ZW, Li W, Qin LQ. Dairy consumption and risk of type 2 diabetes mel-litus: a meta-analysis of cohort studies. Eur J Clin Nutr. 2011;65:1027–31.
23. What foods are in the Fruit Group? Available at http://www.choosemyplate.gov/foodgroups/fruits.html (2011). Last Accessed 16 June 2011.
24. Johnston CS, Bowling DL. Stability of ascorbic acid in commercially available orange juices. J Am Diet Assoc. 2002;102:525–9.
25. Maher MA, Mataczynski H, Stephaniak HM, Wilson T. Cranberry juice induces nitric oxide dependent vasodilation and transiently reduces blood pressure in conscious and anaesthetized rats. J Med Food. 2000;3:141–7.
26. Wilson T, Porcari JP, Maher MA. Cranberry juice inhibits metal- and non-metal initiated oxi-dation of low density lipoprotein. J Nutraceut Function Med Foods. 1999;2:5–14.
27. Tan HL, Thomas-Ahner JM, Grainger EM, et al. Tomato-based food products for prostate cancer prevention: what have we learned? Cancer Metastasis Rev. 2010;29:553–68.

28. Popkin BM. Sugar and artificial sweeteners: seeking the sweet truth. In: Wilson T, Temple NJ, Bray GA, Stuble-Boyle M, editors. Nutrition guide for physicians. New York: Humana/Springer Press Inc; 2010. p. 25–38.
29. Popkin BM. Patterns of beverage use across the lifecycle. Physiol Behav. 2010;100:4–9.
30. Malik VS, Popkin BM, Bray GA, Després JP, Willett WC, Hu FB. Sugar-sweetened beverages and risk of metabolic syndrome and type 2 diabetes: a meta-analysis. Diabetes Care. 2010;33:2477–83.
31. Vanselow MS, Pereira MA, Neumark-Sztainer D, Raatz SK. Adolescent beverage habits and changes in weight over time: findings from Project EAT. Am J Clin Nutr. 2009;90:1489–95.
32. Fung TT, Malik V, Rexrode KM, Manson JE, Willett WC, Hu FB. Sweetened beverage consumption and risk of coronary heart disease in women. Am J Clin Nutr. 2009;89:1037–42.
33. White JS. Misconceptions about high-fructose corn syrup: is it uniquely responsible for obesity, reactive dicarbonyl compounds, and advanced glycation endproducts? J Nutr. 2009;139:1219S–27.
34. Rizkalla SW. Health implications of fructose consumption: a review of recent data. Nutr Metab (Lond). 2010;7:82.
35. Lutsey PL, Steffen LM, Stevens J. Dietary intake and the development of the metabolic syndrome: the Atherosclerosis Risk in Communities study. Circulation. 2008;117:754–61.
36. Nettleton JA, Lutsey PL, Wang Y, Lima JA, Michos ED, Jacobs Jr DR. Diet soda intake and risk of incident metabolic syndrome and type 2 diabetes in the Multi-Ethnic Study of Atherosclerosis (MESA). Diabetes Care. 2009;32:688–94.
37. Dhingra R, Sullivan L, Jacques PF, et al. Soft drink consumption and risk of developing cardio-metabolic risk factors and the metabolic syndrome in middle- aged adults in the community. Circulation. 2007;116:480–8.
38. Stull AJ, Apolzan JW, Thalacker-Mercer AE, Iglay HB, Campbell WW. Liquid and solid meal replacement products differentially affect postprandial appetite and food intake in older adults. J Am Diet Assoc. 2008;108:1226–30.
39. Reissig CJ, Strain EC, Griffiths RR. Caffeinated energy drinks—a growing problem. Drug Alcohol Depend. 2009;99:1–10.
40. Baum M, Weiss M. The influence of a taurine containing drink on cardiac parameters before and after exercise measured by echocardiography. Amino Acids. 2001;20:75–82.
41. Ragsdale F, Gronli TD, Batool SN, et al. Effect of red bull energy drink on cardiovascular and renal function. Amino Acids. 2010;38:1193–2000.
42. Di Rocco JR, During A, Morelli PJ, Heyden M, Biancaniello TA. Atrial fibrillation in healthy adolescents after highly caffeinated beverage consumption: two case reports. J Med Case Reports. 2011;5:18.
43. Arria AM, O'Brien MC. The "high" risk of energy drinks. JAMA. 2011;305:600–1.
44. Sloan AE. What's next in beverages. Food Technol. 2011;3:19.

Chapter 17
Trends in Dietary Recommendations: Nutrient Intakes, Dietary Guidelines, Food Guides, Food Labels, and Dietary Supplements

Geraldine Cuskelly, Jayne V. Woodside, and Norman J. Temple

Keywords Recommendations for nutrient intake • Dietary guidelines • Food guides • Food-based dietary guidelines • Food labels • Dietary supplements

Key Points

- One type of dietary recommendation concerns nutrient intake. An example is the Dietary Reference Intakes used in the USA and Canada. These are most suited for health professionals and play an important role in food labels.
- Dietary guidelines are published in different countries. The goals of these are to prevent nutrient deficiencies and to reduce the risk of chronic diseases. Several sets of dietary guidelines include quantitative recommendations that focus on nutrients. They are usually intended for health professionals who in turn translate them into practical advice for the public. Other sets of dietary guidelines are based entirely on foods rather than nutrients; these are known as food-based dietary guidelines.

G. Cuskelly, PhD
Institute for Agri-Food and Land Use, School of Biological Sciences,
Queen's University Belfast, David Keir Bldg, Stranmillis Road, Belfast BT9 5AG, UK
e-mail: g.cuskelly@qub.ac.uk

J.V. Woodside, PhD
Nutrition and Metabolism Research Group, School of Medicine, Dentistry and Biomedical Sciences, Centre for Public Health, Institute of Clinical Science, Queen's University Belfast, Block B, Grosvenor Road, Belfast BT12 6BJ, UK
e-mail: j.woodside@qub.ac.uk

N.J. Temple, PhD (✉)
Centre for Science, Athabasca University, Athabasca, AB, T9S 3A3, Canada
e-mail: normant@athabascau.ca

N.J. Temple et al. (eds.), *Nutritional Health: Strategies for Disease Prevention*,
Nutrition and Health, DOI 10.1007/978-1-61779-894-8_17,
© Springer Science+Business Media, LLC 2012

- Many countries have published food guides that provide advice on the overall diet. These are particularly suited for the general public. One example is MyPlate.
- There is evidence suggesting that certain dietary supplements may be beneficial, especially, calcium, vitamin D, and fish oil.

17.1 Introduction

The ultimate goal of nutritional research is the provision of diets that will give the best chance of long-term health. This chapter explores the various types of nutrition recommendations. They all share a goal of improving public health.

We start by looking at the most basic type of recommendations: those that concern the quantities of essential nutrients that people should consume.

Other types of nutrition recommendations focus on the key aspects of the diet (i.e., foods and their major components). These dietary guidelines have emerged since the late 1960s. They strive to translate research findings into guidelines that are evidence-based and authoritative. These guidelines are of two general types. One is based on particular nutrients or food components, especially fat, sugar, dietary fiber, salt, and alcohol ("no more than 30% of energy from fat") and is directed mainly at health professionals. The general message filters through to the public. The other type of dietary guidelines is advice on how to make food choices ("eat five servings a day of fruit and vegetables"). The aim of such recommendations is that the majority of the general population can understand and accept them.

This chapter examines the various attempts that have been made in a number of countries to formulate such recommendations. Unfortunately, conflicts can arise between the recommendations of, on the one hand, nutritional scientists and government agencies who formulate dietary guidelines and, on the other hand, the wishes of parts of the food industry who may see their profits threatened if the proposed guidelines are published and widely followed. This can result in dietary guidelines being modified under pressure from industry lobby groups. Though it is desirable for all relevant stakeholders to be involved either in development of, or consultation on, food guidelines, no single stakeholder group should exert undue influence on the process.

After discussing dietary guidelines, attention is turned to various types of food guides. As we shall see, there are a wide variety in use around the world.

That section is followed by an examination of food labels. This area has been the focus of much research in recent years with the generation of many new ideas, both for the front of packs and the reverse.

In the final section of the chapter, we discuss the merits of using supplements.

17.2   Recommendations for Nutrient Intake

Numerous countries produce their own sets of recommendations concerning the intake of nutrients. They are most useful for health professionals who then translate the recommendations into practical advice for the general public. The information they provide reaches the public in different ways. In particular, they are a key part of dietary recommendations. This means that dietary guidelines, such as the national food guide, must be designed so that a person who follows them is assured of obtaining enough of every nutrient. Nutrient intake recommendations also play an important role in food labels.

17.2.1   USA and Canada

The Recommended Dietary Allowances (RDA) are estimates of the daily amounts of nutrients considered necessary to meet the needs of the great majority of healthy people. The RDA have been in use in the USA for several decades, longer in fact than the other types of dietary guidelines described in this chapter. In the late 1990s, a new set of recommendations were formulated: the Dietary Reference Intakes (DRI). The DRI were developed by the Institute of Medicine (IOM) of the National Academy of Sciences together with Health Canada and are used in both the USA and Canada. The DRI include several sets of tables of nutrient recommendations, one of which is the RDA. For those nutrients where there is insufficient information to establish an RDA, estimates are made, known as Adequate Intakes (AI). Values of RDA and AI are given for 14 vitamins and 15 minerals. Values are also given for energy, carbohydrates, essential (n-3 and n-6) fatty acids, protein, dietary fiber, and water. Tables are broken down by age and sex. Specific recommendations are given for women who are pregnant or lactating.

The DRI include two other sets of tables. The Estimated Average Requirements (EAR) are the mean requirements of a group for each particular nutrient. The EAR for a nutrient will meet the nutrient need of half the population within that age and gender group. The other table within the DRI gives Tolerable Upper Intake Levels (UL). This is the maximum amount of potentially toxic nutrients that appears safe for most healthy people to consume on a regular basis. Whereas RDA and AI are targets to aim for, UL provides a warning against overconsumption of nutrients. UL values are (presumably) commonly used by food manufacturers when they add nutrients to foods and by the manufacturers of dietary supplements. In each case selecting quantities of nutrients based, in part, on UL values is intended to help minimize the risk of delivering a toxic excess of nutrients. For a number of nutrients, a UL has not been determined due to inadequate data on adverse effects of overconsumption (vitamin K, chromium, cyanocobalamin, pantothenic acid, potassium, riboflavin, and thiamin). For these nutrients, the IOM recommends that consumption should be from food only.

As noted above, Canada collaborated with the USA in the development of DRI. An important reason behind this move was to help harmonize food labels across the world's largest trading border.

17.2.2   United Kingdom

The Scientific Advisory Committee on Nutrition (SACN) is tasked with advising the UK Health Department and the Food Standards Agency on nutrition. However, as part of government spending cutbacks, this work has been brought under the direct control of the Department of Health.

The Dietary Reference Values (DRV) were set in a 1991 report [1]. This report took a new approach by defining a range of values for each nutrient to reflect variability in requirements within the population. Nutrient recommendations are stated as Lower Reference Nutrient Intake (LRNI), EAR, and Reference Nutrient Intake (RNI); these represent the lower limit, mean, and upper limits, respectively, of nutrient requirements in a population. We now examine these terms in more detail:

- The DRV are based on the assessment of the distribution of nutrient requirements for a population. They aid interpretation of dietary information on both groups and individuals. However, strictly speaking, the DRV are intended primarily to provide a comparator for nutrient intakes of groups only.
- The EAR is the mean requirement of a group for a particular nutrient or for energy. About half the population will usually need more than the EAR and half less.
- The RNI is the amount of a nutrient (calculated as mean EAR + 2SD) which is sufficient for almost all individuals in a group (about 97%). It therefore exceeds the requirement of most people; habitual intakes above the RNI are almost certain to be adequate.
- The LRNI is the amount of a nutrient or energy (calculated as mean EAR–2SD), which is sufficient for only a few individuals in a group who have low needs. Habitual intakes below the LRNI by an individual will almost certainly be inadequate.

Although the criteria for determining the above values have not always been backed by strong evidence, it was judged to be the best possible and set a framework for amendments as new information from research becomes available.

The DRV has obvious similarities to the American DRI: EAR is essentially the same in both sets of guides while the RNI (of the DRV) is equivalent to the RDA and AI (of the DRI).

Of the population's average percentage of total energy (including alcohol), total fat should contribute no greater than 33%, of which saturated fat should be not more than 10%, polyunsaturated fat 6% (not more than 10%), and *trans* fatty acids not more than 2%. Nonstarch polysaccharides (the term used in the UK for dietary fiber) should be 18–24 g/day (corresponding to around 30–35 g/day dietary fiber) and nonmilk extrinsic sugars should not exceed 60 g/day or 10% of total dietary energy.

These guidelines can be used for several diverse purposes: yardsticks for surveys (RNI), in guidance of dietary composition (RNI), for food labels (RNI), and to provide a general guide in assessing the adequacy of an individual's diet (LRNI/RNI).

17.2.3   Comment

Other countries use variations of the above systems. Often only one set of values are given for nutrient intake recommendations, the equivalent of RDA and RNI, although different terms are used by different agencies.

In affluent countries and communities, intakes that meet the RDA (or RNI) can be assumed for most nutrients for most people. In this situation, the RDA serves as a warning in cases of undernutrition. In underdeveloped countries and communities, by contrast, malnutrition is often widespread. In that context, nutritional policy is generally focused on attempts to reach the RDA (or an equivalent set of recommendations concerning essential nutrients) for as many people and nutrients as possible.

17.3   Dietary Guidelines

Around the world, different agencies publish sets of dietary guidelines that are aimed at the improvement of population diets. In highly developed countries, the primary goal is to reduce the impact of chronic disease.

Dietary guidelines fall into two distinct groups:

- Some include several quantitative recommendations that focus on nutrients, such as setting a target for sodium of <1,500 mg/day. Recommendations of this type are most useful for health professionals. However, the aim is that the advice filters down to the general population.
- Other sets of dietary guidelines are based entirely on foods rather than nutrients; they are therefore referred to as food-based dietary guidelines (FBDG). They are written in clear simple language, are nonquantitative, and can be understood by the general public.

17.3.1   Dietary Guidelines that Include Quantitative Recommendations

First, we examine some examples of dietary guidelines that include quantitative recommendations and that are therefore intended mainly for health professionals.

17.3.1.1 USA

American Heart Association Recommendations

Several professional bodies in the USA publish dietary recommendations. Amongst the most well known are those from the American Heart Association (AHA) which, not surprisingly, have a focus on the prevention of cardiovascular disease (CVD). Their 2006 revision [2] includes the following recommendations:

- Consume a diet rich in vegetables and fruits.
- Choose whole-grain, high-fiber foods.
- Consume fish, especially oily fish, at least twice a week.
- Limit your intake of saturated fat to <7% of energy, *trans* fat to <1% of energy, and cholesterol to <300 mg/day.
- Minimize your intake of beverages and foods with added sugars.
- Choose and prepare foods with little or no salt.
- If you consume alcohol, do so in moderation.

Previous versions of these guidelines stressed that the total intake of fat should be <30% of energy. That recommendation has now disappeared; this reflects a clear trend seen with various expert organizations that now focus on the type of dietary fat rather than quantity. The recommendation for sodium intake in the above report was <2,300 mg/day, but in a 2010 revision the AHA lowered this to <1,500 mg/day [3].

Dietary Guidelines for Americans

These are published by the Departments of Agriculture (USDA) [4]. The most significant differences from the AHA recommendations are as follows: saturated fat is less restricted (<10% of energy); sodium (<2,300 mg/day for half of the population but <1,500 mg/day for children, adults aged >51, and persons at risk); and alcohol (up to one drink a day for women and up to two per day for men).

Researchers at the Department of Nutrition of the Harvard School of Public Health have disputed whether the Dietary Guidelines for Americans are the best ones for the prevention of chronic disease. They base this verdict on the findings from large cohort studies. They compared two sets of composite food scores: the Health Eating Index (based on the Dietary Guidelines for Americans) and the Alternative Healthy Index (based on the findings from large cohort studies). The Alternative Healthy Index manifested a stronger negative association with risk of death from cancer and CVD, and with all-cause mortality [5, 6]. The actual food content of the Alternative Healthy Index is similar to the Mediterranean diet and is described below when we discuss the Healthy Eating Pyramid (HEP).

17.3.1.2 World Health Organization

The 2003 dietary guidelines from the World Health Organization (WHO) [7, 8] are broadly similar to the American ones. The one notable exception is that the recommendation for total fat is 15–30% of energy. The recommendation for carbohydrate is 55–75% of energy. Setting an upper limit of 75% for carbohydrate, which is a high figure, reflects the fact that the target populations for WHO, and also of the Food and Agriculture Organization (FAO), include the populations of developing countries, who consume high-carbohydrate staple foods, such as rice, as a major component of their diets.

Other recommendations are as follows: protein 10–15% of energy, sugars (including those present in fruit juice) <10%, n-6 polyunsaturated fats 5–8%, n-3 fats 1–2%, saturated fat <10%, and *trans* fatty acids <1%. The maximum recommended salt intake is 5 g/day. There is no specific recommendation for dietary fiber but rather a general recommendation for fruit, vegetables, and whole grain foods.

17.3.2   Food-Based Dietary Guidelines

In contrast to the dietary guidelines described above, FBDG are based entirely on foods rather than nutrients, are nonquantitative, and can be readily understood by the general public. FBDG were recommended for health promotion by a FAO/WHO international consultation published in 1998 [9]. In support of this approach, Mozaffarian and Ludwig [10] recently argued that dietary guidelines should be based on foods rather than nutrients.

FBDG are culturally sensitive and take the customary dietary pattern of consumers into account. They vary from country to country depending on local needs. FBDG used in developing countries often contain statements along the following lines: "Drink lots of clean, safe water," "Foods consumed should be safe and clean," and "Use only iodized salt."

A recent report described the use of FBDG in Chile, South Africa, Germany, and New Zealand [11]. The FAO has compiled a website that gives the FBDG used in many countries around the world [12].

17.3.2.1   Canada

Whereas the USA has several agencies that publish dietary recommendations, Canada is the very opposite. Until a few years ago, the official guidelines were Nutrition Recommendations for Canadians. They were broadly similar to the guidelines from the AHA. A process was then initiated for a major revision but for reasons that are entirely unclear this work was abandoned. Currently, the only dietary guidelines are Canada's Guidelines for Healthy Eating [13], which are as follows:

- Enjoy a variety of foods.
- Emphasize cereals, breads, other grain products, vegetables, and fruits.
- Choose low-fat dairy products, lean meats, and foods prepared with little or no fat.
- Achieve and maintain a healthy body weight by enjoying regular physical activity and healthy eating.
- Limit salt, alcohol, and caffeine.

17.3.2.2   UK

In the UK, the *Eight tips for healthy eating* [14] are as follows:

1. Base your meals on starchy foods
2. Eat lots of fruit and veg
3. Eat more fish
4. Cut down on saturated fat and sugar
5. Eat less salt
6. Get active and be a healthy weight
7. Don't get thirsty
8. Don't skip breakfast

When comparing the dietary guidelines produced by different agencies, both quantitative recommendations as well as FBDG, we see notable differences in the inclusion of such dietary components as cholesterol, sugar, and fat. These differences point to the debates that have taken place among nutrition experts and other stakeholders but that have not yet been settled.

17.4   Food Guides

Numerous countries have published food guides that provide advice on the overall diet for the general public. They tend to center on a colored diagram or poster. They are, in essence, FBDG in a graphical format. Several of these were reviewed by Kaufer-Horwitz et al. [15]. Several examples are described below.

17.4.1   USA

Until 2005, the guide disseminated to the general public in the USA was the Food Guide Pyramid. It was a simple matter to look at this one-page document and determine how many servings should be eaten from each food group, much as one reads a TV guide. But this all changed with the launch of MyPyramid, which was in use from 2005 until 2011. Unlike all other food guides around the world this one required the use of the internet. The user entered his or her profile (age, sex, and physical activity) and then received a personalized set of diet recommendations. The obvious challenge with this food guide was the matter of user accessibility. It seems certain that there are millions of people who are willing to read a simple, printed food guide, but either cannot or would not use a website for this purpose.

In most countries, such as Canada for example, the government health department is responsible for writing the food guide but in the USA they work for the Department of Agriculture. That department has a serious conflict of interest: it must help make farming and food production profitable (which often means boosting the sale of less than healthy foods) while at the same time advising people how to eat for health. As a result, there is strong evidence that both the old and new pyramids are compromises between these two opposing forces. The nutrition experts from the Harvard School of Public Health have been highly critical of MyPyramid and the food industry lobbying that played a role in its development [16]. Marion Nestle of New York University described MyPyramid as "a disaster" [17].

It seems that the USDA listened to the criticism. In 2011 they abandoned MyPyramid. In its place they went over to a simple pictorial design called MyPlate [18]. The food guide is depicted as a plate with food sectors. Unlike MyPyramid, it places little emphasis on how many servings should be eaten from each food group. This new food guide is much easier to understand.

17.4.2   Harvard's Healthy Eating Pyramid and Healthy Eating Plate

In view of the criticisms leveled against MyPyramid (and its predecessor), it should come as little surprise that alternative food guides have been developed in the USA. A well-known one is the HEP, produced by the Department of Nutrition of the Harvard School of Public Health [16, 19]. Just as MyPyramid is based on Dietary Guidelines for Americans, so HEP is based on the Alternative Healthy Index which was discussed above. The visual design of HEP is similar to the Food Guide Pyramid but with one notable exception: it does not specify the number of servings from each food group. Instead, it tells users to: "Forget about numbers and focus on quality."

In 2011, after the USDA switched from MyPyramid to MyPlate, Harvard followed suite and switched from HEP to their own plate design, called Healthy Eating Plate [20].

17.4.3   Canada's Food Guide

A fully revised version of Canada's Food Guide was published in 2007 [21]. It is similar to the old Food Guide Pyramid with respect to the number of servings from each food group. It is based on a chart and is simple to use. There are several notable features. The recommended number of servings of fruit and vegetables (which are lumped together in one food group) has now overtaken grains. Supplements are specifically recommended for particular groups: 400 IU (10 µg)/day of vitamin D for people over age 50 (remember this is for people living in Canada) and a multi-vitamin containing folic acid for women who could become pregnant and those who are pregnant or breastfeeding. Anyone wishing to use this food guide should request a printed copy as this makes using it much easier than reading it via the internet.

17.4.4 United Kingdom

The Eatwell Plate is a pictorial food guide that was produced by the Food Standards Agency [22]. It shows the proportion and types of foods that are needed to make up a healthy balanced diet. The food guide is depicted as a plate with food sectors, similar to MyPlate in the USA. It resembles the Canadian guide in that fruit and vegetables are combined into one group. Fatty and sugary foods are included in the pictorial representation. The Eatwell Plate is based on the Eat Well, Be Well tips (see Sect. 17.3.2.2).

17.4.5 Other Countries

The Australian and Swedish food guides are plate-shaped and are similar to MyPlate and the UK guide. The guide used in China and Korea is a pagoda that resembles the Food Guide Pyramid that was formerly used in the USA.

The Dietary Guidelines for Japanese, released in 2000, provide the basics of a healthy diet for the people of Japan. In July 2005, the Ministry of Health, Labour and Welfare and the Ministry of Agriculture, Forestry and Fisheries of Japan jointly released a new pictorial guide, The Japanese Food Guide Spinning Top, to help people implement the Dietary Guidelines for Japanese [23].

17.4.6 Comment

These various food guides all convey essentially the same message. If followed, then the diet will almost certainly provide all nutrients in amounts that meet RDA recommendations. One notable exception to this rule appears to be vitamin D, a topic discussed later in the chapter. The food guides encourage consumption of at least five servings of fruit and vegetables daily. The newer revisions of the American and Canadian food guides have put much more emphasis on whole grains but have stopped short of stressing that only whole grains (not refined grains) should be eaten.

Several differences between the food guides are noteworthy. An important difference concerns the use of quantitative recommendations. Canada's Food Guide emphasizes how many servings each person should consume as did MyPyramid. However, several food guides used in different countries, including the new MyPlate (USA), Healthy Eating Plate (Harvard), and the Eatwell Plate (UK), give very little attention to numbers of servings. Some might argue that specifying the recommended numbers of servings might assist consumers in selecting a healthier diet. However, to what extent individuals pay attention to the recommended amounts is not known. As with several other aspects of the construction of these food guides, this is an area that deserves investigation.

Fruit and vegetables are grouped together in some guides (Canada, UK, China, Korea, Portugal, and Mexico) but are in separate groups in others (Germany, USA, Sweden, and Australia). In the USA and Canada, potatoes go with other vegetables, but in Sweden potatoes and root vegetables are given their own group distinct from other vegetables. Several countries—Korea, UK, Portugal, Germany, and Mexico—include potatoes with grains. Keeping potatoes separate from other vegetables is probably sensible as potatoes are poor in phytochemicals, have a high glycemic index, and are often consumed as fries (and may therefore have a high content of oxidized fat).

Another inconsistency is in the placement of legumes. In Canada, beans are included with meat, hence the name of that food group: "meat and alternatives." MyPlate, by contrast, allows beans to be counted with either "protein foods" or with vegetables. This schism in the placement of legumes reflects the fact that legumes are a low-fat, protein-rich alternative to meat but are also a good source of fiber and various nutrients such as folate.

One problem with food guides is that manufactured products often contain mixtures of diverse foods, or their derivatives, along with several additives. Accordingly, placing manufactured foods in a food guide is often difficult. How does a shopper perceive a label on, for example, "Breakfast Apple Crisp?" Is this food a valuable means to obtain fruit or is it a significant source of fat and sugar? Similarly, many functional foods and fortified foods, far more commonly available now, would be difficult to assign to a single sector of most of the food guides.

Some food guides are limited by not including foods representative of diverse cultures, such as food eaten by people from Africa or southeast Asia. To address this in the USA, MyPlate lists a wide range of foods in each food group. This has not been possible with the UK Eatwell Plate.

17.4.7 Traffic Lights Food Guide

Food guides typically categorize foods into two broad classes: those that are recommended and those that should be eaten only in limited quantities. But nutrition science informs us that many foods belong somewhere in between. It therefore makes more sense to divide food into three distinct groups. One of us (NT) has designed a food guide based on this approach (see Table 17.1).

The design of the traffic lights food guide is radically different from the food guides described above. One outstanding feature is its simplicity. Within each food group, foods have been categorized as follows: green (eat freely, based on recommended amounts), amber (eat in limited amounts), or red (these are treats; eat little or none). This food guide design is a logical development of traffic food labels (see next section), which are becoming increasingly used in the UK [24] and other countries. By using a similar system for both food labels and a food guide, the general public should find food selection to be much simplified.

Table 17.1 Traffic lights food guide

Food group	Green (eat freely based on recommended amounts)	Amber (eat in limited amounts)	Red (eat little or none)
Fruit/vegetables	Nearly all fruits and vegetables	Potatoes, fruit juice	French fries
Grain products	Whole grains, such as whole wheat bread, oats, dark rye bread, and popcorn	Refined cereals, such as white rice, white bread, and corn flakes	Cookies, cakes, popcorn with salt/butter
Milk products	Skim and 1% milk, fortified soy milk	2% milk, low-fat cheese	Whole milk, regular cheese, cream cheese, ice cream
Meat, fish, beans, nuts	Fish, beans, lentils, nuts	Lean beef, chicken	Bacon, red meat (all beef and pork products unless labeled as lean), eggs
Oils, fats	Most vegetable oils, soft margarine (preferably from canola oil or soy oil)		Hard margarine, butter

Eat a mixture of foods from the different food groups while carefully following the rules given below:

Key rules for a healthy diet

1. Eat only enough to satisfy your appetite. If you are gaining excess weight or you wish to lose weight, then eat less (and exercise more)
2. Eat five to ten servings a day of grain products. Of this, at least three servings (preferably more) should be whole grains. One serving is a slice of bread, a cup of breakfast cereal, or half a bagel
3. Eat five to ten servings a day of whole fruit and vegetables. One serving is an apple, a banana, a cup of salad, or half a cup of other vegetables. In addition, up to one cup of juice (two servings) may be consumed. Aim for a mixture of different types of fruit and vegetables. Fresh or frozen is better than canned
4. Consume two or three servings a day of milk products (more for adolescents and women who are pregnant or breastfeeding). One serving is a cup of milk or yogurt, or an ounce and a half (45 g) of cheese
5. Consume one to three servings a day of meat, fish, beans, peas, lentils, and nuts. A serving is 3 oz (90 g) of fish or meat, or half a cup of cooked beans
6. Aim for about three teaspoons a day of margarine, oils, and salad dressing, or double that if you eat little or no other sources of polyunsaturated fats, such as nuts or fish
7. Minimize your consumption of sugar. This includes sugar in coffee and soft drinks, and drinks labeled as fruit beverage. Also, minimize your consumption of foods rich in both fat and sugar, such as cakes and donuts
8. Cut down on the amount of salt. Remember: most salt in the diet comes from processed foods, such as most types of bread, margarine, and canned foods
9. It is OK to consume alcohol provided that this is done responsibly. Never drink and drive, never drink if pregnant, and don't get drunk. An acceptable intake is 1 drink a day for women and 2 drinks a day for men. One drink is approximately 12 oz (360 g) of beer, 4–5 oz (120–150 g) of wine, or 1.5 oz (42 g) of spirits. (Note: these figures are based on the American definition of a "drink." The size of a "drink" varies between countries and may be 20% smaller than indicated by these figures.)

17.5   Food Labels

Developed countries have regulations that specify what information must be stated on food labels. Generally, the labeling regulations only apply to food that is sold in packages, such as cans or cardboard boxes, whereas many foods that are not packaged by the manufacturer, such as fresh meat and fish, do not require a label.

Food labels are an important topic as they play a vital role in determining the food choices made by millions of people. Alas, as we shall see, the great majority of labels fail, by a wide margin, to achieve their potential to positively influence the national diet.

17.5.1   Why Food Labels Cause Confusion

Labels are of two types:

- Front-of-pack (FOP) labels inform the buyer of the brand name and the type of food (such as Quaker Oats).
- Back-of-pack (BOP) labels provide details of the nutritional composition of the food (such as 185 mg sodium per 35 g serving). To help the consumer interpret the information, the amounts are also stated as percent of recommended daily intake (called Daily Values in the USA and Canada). This part of the label also lists the ingredients in the food, usually in order by amount (main ingredient first). In the UK, it is mandatory to include "the big four" on packaged foods (energy, protein, carbohydrate, and fat). Many food manufacturers also include four additional nutrients (sugars, saturated fat, salt, and one other).

Food labels in common use, both in Europe and North America, can be a source of confusion for consumers. Here are the major problems:

- FOP labels often give misleading names to foods. Soft drinks may be the most egregious example. In North America, only pure fruit juice can be called "juice." However, there are several imitation juice products that contain no more than 20% actual juice; many contain none at all. These pseudo-juice products are, in reality, sugar solutions with added colors and flavors. Despite being little different from cola drinks, they have names that suggest real fruit, such as fruit beverage, fruit nectar, and fruit cocktail. Adding to the confusion, the brand name may also be suggestive of real fruit (e.g., Sunny Delight).
- The BOP label lists the ingredients in the food in order by amount but seldom gives the actual quantity of each ingredient. What this can mean in practice is that a manufacturer sells a juice containing mainly apple juice (which is cheap) with some added berry juice (which costs much more). But as many customers prefer berry juice, the FOP label will likely say, in large letters, "made with real berries" and have large images of berries. The list of ingredients will merely

Table 17.2 Examples of information found on food labels

Brand	Type of food	Serving size	Calories	Sodium (mg)	Percentage of daily value
A	Crackers	4 crackers (20 g)	80		
B	Crackers	5 crackers (32 g)	130		
C	Sardines	58 g (half can)		210	9
D	Sardines	106 g (whole can)		420	17

indicate that there is more apple juice than berry juice. Therefore, even if the customer realizes that the label is deceptive and tries to determine how much berry juice is actually present, this will be impossible.

- In addition to the list of ingredients, BOP labels also give a table with the content of selected food components (energy, fiber, sugar, fat, and some nutrients). However, food components that should be consumed in limited amounts (such as sodium, sugar, and saturated fat) are interspersed with others that are often lacking in the diet (such as dietary fiber and n-3 fat). The effect of this is almost certainly to make food labels more confusing for consumers.

- There are guidelines for serving sizes on food labels, but these are often ignored. What we often find is that labels for similar products use different serving sizes, thereby making it difficult for consumers to compare them. The problem is illustrated in Table 17.2. The information in the table has been extracted from actual labels of foods sold in Canada. Let us suppose a shopper wishes to buy crackers low in energy and sardines low in sodium. Brand A crackers contain 80 cal per 4 crackers (20 g), while brand B contains 130 cal per 5 crackers (32 g). The shopper might easily conclude—wrongly—that brand A is lower in energy. In fact, measured as Calories (kcal) per 100 g the two products are almost identical in energy density. A similar problem occurs with the sodium content of sardines. Brand C contains 210 mg sodium per 58 g serving (which is half of the 115 g can), while brand D contains 420 mg sodium per 106 g serving (which is the whole can). This means that the two brands have an almost identical content of sodium (as mg per 100 g). However, the shopper may decide to compare the cans by looking at the sodium content in terms of percentages of Daily Values. But because brand C has a much smaller serving size than brand D, the label states that it has much less sodium per serving (9% vs. 17%). This leads to a completely false conclusion.

Let us suppose that a typical shopper is in a supermarket and wishes to buy breakfast cereals. The time spent evaluating each of the choices on offer is typically no more than a few seconds. For that reason, it is the FOP labels that are crucial for making a choice. But as we have seen, FOP labels may give misleading information. Even if the shopper is especially diligent and carefully reads the BOP labels in order to make a careful selection, he or she is likely to end up being confused by the information. Indeed, research studies in various countries have revealed that the majority of people have problems understanding food labels [25]. This is especially the case with older adults and those with less education.

17.5.2 Improved Designs for Food Labels

It is clear that the types of food labels used in many countries fail to give consumers the information they require, and in a user-friendly format. This is a serious barrier that prevents consumers from making informed choices as to which food items are healthiest. In response to these challenges, several new designs have been proposed.

One alternative system that shows much promise has been developed in Britain and is based on traffic lights [24]. With this system, colored circles are placed on the FOP and indicate if the food has a high (red), medium (orange), or low (green) content of fat, saturated fat, sugars, and salt. The label also indicates the actual quantity of these substances per serving. Research studies have been carried out in which consumers have been asked to compare traffic light labels with more complicated systems. Traffic light labels are generally well liked and are very effective for helping people to assess the health value of a food [26–29].

One possible improvement to this system is to add an extra traffic light to indicate the overall health value of the food. In order to accomplish this, a standardized methodology for comparison of diverse foods is required. Several such systems have been proposed [30–34].

Little research has been done to determine whether a traffic lights design will result in people actually eating a healthier diet. Until that research is done, it is a matter of speculation as to whether this design is superior to the type of food guides discussed earlier.

Other tests have been conducted using even simpler FOP labels where the overall health value of a food is summarized as stars (or a similar symbol); healthier products are given more stars. This format obviously provides less information than traffic light labels. However, tests conducted in several European countries revealed that this system scores even better in terms of allowing people to correctly identify which foods are healthiest [26].

One system based on stars in called Guiding Stars. Each food receives from zero to three stars based on an overall health score. The system was tested in a supermarket chain in the northeast of the USA to determine its impact on people's actual shopping habits [35]. The stars were affixed to the shelf adjacent to the food rather than on the actual food label. This strategy bypasses the problem that manufacturers of unhealthy foods are most unlikely to announce that facts on food labels. The supermarkets also provided educational material to shoppers. At the start of the study 24.5% of foods sold carried at least one star, but after 2 years this rose to 25.9%, a significant increase. While this improvement is modest, it does provide support for the view that easy-to-understand food labels will lead to an increase in sales of healthier food items. In 2011 Loblaws, a major supermarket chain in Canada, announced that the system will be implemented in their stores.

Another variation of the above systems is a health tick. This indicates if a particular food is judged to be healthy. A version of this system was recently adopted by ten food companies based in the USA [36]. While being very simple for consumers to comprehend, it has the disadvantage that all foods are placed in one of only two categories. Traffic lights come much closer to the reality of nutrition science by using three categories while a system based on stars typically uses three or four categories. This may explain why, when tested in several European countries, the health tick system did not fare as well as either traffic lights or stars for helping people to identify which foods are healthiest [26].

A well-designed FOP should be combined with more detailed nutritional information (such as Guideline Daily Amount [GDAs]). In the view of Feunekes et al. [26]: "This will allow consumers to make a quick decision, whilst also providing detailed information if consumers desire this."

17.6   The Case for Supplements

The conventional wisdom is that we can obtain adequate amounts of all nutrients from a normal diet. In other words, following MyPyramid will result in a diet that meets the RDA (and likewise for guides in other countries). Moreover, there is a dearth of hard evidence that supplements of most vitamins and minerals have a favorable impact on health. But, on the other hand, for several nutrients (and fish oil) a case can be made for the merits of supplements. The evidence is strongest for people whose diets are marginally deficient, and in groups such as pregnant women and the elderly. Here, we examine this evidence.

17.6.1   Folic Acid

Folic acid provides perhaps the most compelling example that supplements of a nutrient may be warranted. Randomized controlled trials (RCTs) demonstrated that supplements of folic acid (the form of folate used in supplements) can prevent a substantial proportion of neural tube defects (NTD), such as spina bifida [37]. In 1996, in response to this evidence, the Food and Drug Administration (FDA) mandated that all grain foods in the USA be supplemented with folic acid. This has lead to a sharp drop in the incidence of NTD [38]. The likely reason that folic acid supplements achieve this benefit is because a substantial proportion of females have a poor folate intake which is a probable risk factor for NTD. Supplementation therefore corrects this problem.

17.6.2   Calcium

The RDA for calcium is 1,200 mg for adults aged over 50. This is equivalent to a liter of milk. There is no doubt that large numbers of people consume inadequate amounts of the mineral. This problem is most likely to occur in non-Caucasians (who frequently have lactose intolerance) and women (because of their lower food intake). Indeed, a survey in Canada revealed that a quarter of women have a calcium intake of under 500 mg/day [39].

While the evidence lacks consistency, it appears that supplemental calcium may afford protection against fractures in older adults [40, 41]. The most probable explanation for this is that a long-term low intake of calcium plays an important role in the development of osteoporosis. However, recent evidence has suggested that calcium supplements may cause an increased risk of CVD [42]. Calcium supplements are therefore not recommended for people where the risk of bone problems is relatively low, such as most men and black people.

17.6.3 Vitamin D

In the USA, Canada, and some other countries milk is fortified with this nutrient. Fish is another good source. But if milk and fish consumption are low, then vitamin D status may be poor since there are few other significant dietary sources. This is especially the case for those living in northern latitudes, such as Canada and the northern states of the USA, due to lack of ultraviolet light from sunshine for much of the year. Vitamin D insufficiency is most common in the elderly, people with dark skin, and those who expose little skin to the sun.

Solid evidence has emerged in recent years linking a poor vitamin D status with various health problems. RCTs have reported that in older adults supplemental vitamin D is protective against both falls [43] and fractures [44]. In the case of fractures, the presumed mechanism is by working in tandem with calcium to help prevent osteoporosis. An especially important development in recent years has been findings indicating that vitamin D may have a strong protective association with the risk of cancer, especially colon cancer [45]. Other evidence suggests that vitamin D helps reduce the risk of CVD and total mortality [45]. Some studies have linked the vitamin with protection against both type 1 [46] and type 2 diabetes [47]. Low levels of vitamin D are correlated with faster cognitive decline in older adults [48].

The ideal dose for prevention is much above the RDA in the USA (currently 15 µg or 600 IU/day at ages up to 70 years, and 20 µg or 800 IU/day at age >70 years). Based on the totality of the evidence, a supplemental dose of 25 µg (or 1,000 IU) per day is indicated for such groups as the elderly, people with dark skin, and those who expose little skin to the sun [49, 50]. This policy has already been partially implemented in Canada where all adults over age 50 are advised to take a vitamin D supplement, though at a lower dose than that suggested here. It is important to appreciate that the potent form of vitamin D is D_3. Many supplements have D_2 but that is only one quarter as potent.

It must be stressed that this whole area is controversial with some arguing that an optimal dose of supplement should be even higher [51, 52] while others are much more negative [53].

17.6.4 Fish Oil

Many cohort studies have reported an inverse association between fish consumption (especially fatty fish) and risk of CHD [54, 55]. These studies indicate that eating fish between once and five times per week is associated with a reduced risk of CHD, especially fatal CHD. Fish oil displays the same protective associations, although the cardioprotective benefits of fish and fish oils may be greater in populations with a low habitual fish consumption and those at greater risk of developing CHD. These findings have been, to some extent, replicated in RCTs [54, 55]. There is also some evidence supporting the use of n-3 PUFA supplements for secondary prevention of coronary events, particularly after MI, although benefit may be reduced in post-MI patients receiving optimal drug treatments. Current data from RCTs using n-3 PUFA for primary prevention of CVD are limited.

Many studies have suggested that long-chain n-3 fatty acids, the active ingredient in fish oil, exert several cardioprotective actions, lowering triglyceride levels, improving endothelial function (the endothelium is the surface of the arteries that faces the blood), and reducing thrombosis, inflammation, and arrhythmias. This subject is reviewed in Chap. 10 by McEvoy, Young, and Woodside.

These findings demonstrate impressive evidence that fish and fish oil are likely to have a protective action against the risk of CHD. For a person who chooses not to eat fish regularly, fish oil supplements will provide most, probably all, of the benefit.

17.6.5 Antioxidant Vitamins

Epidemiological studies have repeatedly indicated that intake of vitamin E has an inverse association with CVD [56]. Similarly, β-carotene has a clear protective association with risk of different types of cancer, especially lung cancer (see Chap. 12 by Davis and Milner). The obvious implication of these findings is that supplements of antioxidant vitamins will protect against CVD and cancer.

Several large RCTs have been conducted in which the three major antioxidant vitamins—β-carotene or vitamins C or E—have been tested. The major goal of most trials has been the prevention of either heart disease or cancer. The dose has typically been several times higher than the RDA. Meta-analyses have revealed no benefit of either β-carotene or vitamin E for CVD endpoints [57]. Supplementation with β-carotene failed to prevent cancer and appears to increase the incidence of lung and stomach cancers [58]. Moreover, a major meta-analysis concluded that supplementing with β-carotene or vitamins C or E leads to an *increase* of about 5–6% in all-cause mortality [59, 60]. These findings strongly suggest that while foods naturally rich in antioxidants and phytochemicals, such as fruit and vegetables, are excellent for the health, this benefit does not extend to purified antioxidants. The reasons for this are complex and poorly understood.

The saga of antioxidant vitamins serves as a reminder that we must always move with caution when making claims for the benefits of dietary supplements. What may appear to be a strong case for the value of a particular dietary supplement can easily collapse once the results of RCTs become available and we find that the supplement fails to deliver the promised healthy benefits.

17.6.6   Multivitamins

It seems likely that in most cases where supplements appear to be beneficial, this is only because large numbers of people have an inadequate diet and may therefore be marginally deficient in various nutrients. For that reason, our first priority should clearly be to encourage people to improve the overall quality of their diets. But at the same time, we need to be realistic and recognize that a substantial section of the population will continue to eat a diet that is far from ideal. For that reason a case can be made for a general recommendation to take a daily multivitamin, at least for specific population groups. This view was well formulated by Willett and Stampfer [61]:

> Given the greater likelihood of benefit than harm, and considering the low cost, we conclude that a daily multivitamin that does not exceed the RDA of its component vitamins makes sense for most adults. Substantial data suggest that higher intakes of folic acid, vitamin B_6, vitamin B_{12}, and vitamin D will benefit many people, and a multivitamin will ensure an adequate intake of other vitamins for which the evidence of benefit is indirect. A multivitamin is especially important for women who might become pregnant; for persons who regularly consume one or two alcoholic drinks per day; for the elderly, who tend to absorb vitamin B_{12} poorly and are often deficient in vitamin D; for vegans, who require supplemental vitamin B_{12}; and for poor urban residents, who may be unable to afford adequate intakes of fruit and vegetables.

17.6.7   Comment

There is considerable skepticism over the use of supplements within the nutrition community. The reasons for this are not hard to identify. One problem is that the supporting evidence is often contradictory and may fail to survive trial by RCT. This is well illustrated by the history of β-carotene and vitamins C and E. Another problem is the widespread practice of dishonest marketing of dietary supplements, as documented in Chap. 20 by Temple. It is therefore essential that recommendations for the use of supplements be based on solid evidence stemming from the results of well-conducted studies.

Ideally, supplements should be safe, effective, and cheap. Recent discoveries in nutrition have brought such a supplement several steps closer to becoming a reality.

It may well be that for maximum effectiveness some of the nutrients discussed here need to be consumed in amounts greater than is obtainable from a normal diet.

This may be especially the case for vitamin D for people living in northern latitudes, such as the northern states of the USA and Canada.

17.7   Conclusion

The various sets of recommendations described in this chapter are a work in progress. The three altogether different types of recommendations serve entirely distinct purposes.

The development of the DRI in the USA and Canada is dramatic in that one set of recommendations concerning intake of nutrients (the RDA) was replaced by several tables. Clearly, the committee that developed the DRI felt that the obvious loss in terms of simplicity was a necessary price to pay for giving health professionals a more complete set of tools, and a similar argument can be made for the introduction of DRV in the UK.

There are several challenges in developing recommendations concerning nutrient intake. One is differentiating between two distinct objectives: achieving a diet that is adequate in each nutrient so that there is *no risk of deficiency*, and achieving an *optimal intake for health.* While not explicitly stated, the section on supplements focuses more on the latter than the former.

The dietary guidelines concerned with countering chronic disease are intended mainly for health professionals. Accordingly, there is far less of a challenge in grabbing the attention of the reader. The guidelines place their focus on the most important nutritional issues. But we see notable differences in the inclusion of such dietary components as cholesterol, fiber, and fat. These contrasts in the various guidelines point to the debates that have taken place but have not yet been settled.

Food guides, such as the MyPyramid, are also a work in progress. They are directed at the general population and for that reason are meant to be easily understood and highly flexible, something that is far easier to aim for than to achieve. But as we have seen, the various guidelines often fail to put enough emphasis on various key areas of the diet, such as ensuring a generous intake of whole grains and on reducing consumption of red meat. There are also numerous differences between the guides in the placement of different foods. More research is clearly needed into how to optimize the presentation of the guides so that users are encouraged to consume the healthiest diet. As national diets around the world continue their rapid evolution, the food guides will also need to evolve.

Clearly, guidelines must address the key nutritional issues of their target population. In developed countries, the focus is primarily on the prevention of chronic diseases. But in less developed countries, malnutrition is still a widespread problem. However, such chronic diseases as obesity are becoming rapidly more prevalent in many less developed countries and food guidelines must stay abreast of these changes in population health.

One conclusion is inescapable: this entire field will be the scene of considerable debate, controversy, and research for years to come.

References

1. Committee on Medical Aspects of Food Policy (COMA). Dietary reference values for food energy and nutrients for the United Kingdom. London: HMSO; 1991.
2. American Heart Association Nutrition Committee, Lichtenstein AH, Appel LJ, Brands M, et al. Diet and lifestyle recommendations revision 2006: a scientific statement from the American Heart Association Nutrition Committee. Circulation. 2006;114:82–96.
3. Lloyd-Jones DM. Cardiovascular risk prediction: basic concepts, current status, and future directions. Circulation. 2010;121:1768–77.
4. U.S. Department of Agriculture. Diet guidelines, 2010. 2011. Available at http://www.cnpp.usda.gov/DGAs2010-PolicyDocument.html. Last Accessed 6 Aug 2011.
5. McCullough ML, Feskanich D, Stampfer MJ, et al. Diet quality and major chronic disease risk in men and women: moving toward improved dietary guidance. Am J Clin Nutr. 2002; 76:1261–71.
6. van Dam RM, Li T, Spiegelman D, Franco OH, Hu FB. Combined impact of lifestyle factors on mortality: prospective cohort study in US women. BMJ. 2008;337:a1440.
7. Report of a Joint WHO/FAO Expert Consultation. Diet, nutrition and the prevention of chronic diseases. Tech Rep Ser 916. Geneva: WHO; 2003.
8. Nishida C, Uauy R, Kumanyika S, Shetty P. The joint WHO/FAO expert consultation on diet, nutrition and the prevention of chronic diseases: process, product and policy implications. Public Health Nutr. 2004;7:245–50.
9. WHO (World Health Organization). Preparation and use of food-based dietary guidelines. Report of a joint FAO/WHO Consultation. Technical Report Series No. 880. Geneva: WHO; 1998.
10. Mozaffarian D, Ludwig DS. Dietary guidelines in the 21st century—a time for food. JAMA. 2010;304:681–2.
11. Keller I, Lang T. Food-based dietary guidelines and implementation: lessons from four countries—Chile, Germany, New Zealand and South Africa. Public Health Nutr. 2008;11:867–74.
12. FAO. Food-based dietary guidelines. 2010. Available at http://www.fao.org/ag/agn/nutrition/education_guidelines_zaf_en.stm. Last Accessed 16 Oct 2010.
13. Health Canada. Canada's guidelines for healthy eating. 2010. Available at http://www.hc-sc.gc.ca. Last Accessed 23 Mar 2010.
14. National Health Service. Eight tips for healthy eating. 2010. Available at http://www.nhs.uk/Livewell/Goodfood/Pages/eight-tips-healthy-eating.aspx. Last Accessed 7 May 2012.
15. Kaufer-Horwitz M, Valdés-Ramos R, Willett WC, Anderson A, Solomons NW. A comparative analysis of the scientific basis and visual appeal of seven dietary guideline graphics. Nutr Rev. 2005;25:335–47.
16. The Department of Nutrition, Harvard School of Public Health. The nutrition source. 2011. Available at http://www.thenutritionsource.org. Last Accessed 7 Aug 2011. [This website is now replaced by Ref. 20 below].
17. Nestle M. Eating made simple. Sci Am. 2007;297:60–3.
18. U.S. Department of Agriculture (USDA). MyPlate. 2011. Available at http://www.choosemyplate.gov/index.html. Last Accessed 7 Aug 2011.
19. Reedy J, Krebs-Smith SM. A comparison of food-based recommendations and nutrient values of three food guides: USDA's MyPyramid, NHLBI's dietary approaches to stop hypertension eating plan, and Harvard's healthy eating pyramid. J Am Diet Assoc. 2008;108:522–8.
20. The Department of Nutrition, Harvard School of Public Health. The nutrition source. 2011. Available at http://www.thenutritionsource.org. Last Accessed 7 Oct 2011.
21. Health Canada. Eating well with Canada's food guide. 2011. Available at http://www.hc-sc.gc.ca. Last Accessed 7 Aug 2011.
22. Eatwell Plate. 2011. Available at http://www.nhs.uk/Livewell/Goodfood/Pages/eatwell-plate.aspx. Last Accessed 7 Aug 2011.

23. Yoshiike N, Hayashi F, Takemi Y, Mizoguchi K, Seino F. A new food guide in Japan: the Japanese food guide Spinning Top. Nutr Rev. 2007;65:149–54.
24. Food Standards Agency. Traffic Light Labeling. 2010. Available at http://www.eatwell.gov.uk/foodlabels/trafficlights. Last Accessed 22 Mar 2010.
25. Cowburn G, Stockley L. Consumer understanding and use of nutrition labelling: a systematic review. Public Health Nutr. 2005;8:21–8.
26. Feunekes GI, Gortemaker IA, Willems AA, Lion R, van den Kommer M. Front-of-pack nutrition labelling: testing effectiveness of different nutrition labelling formats front-of-pack in four European countries. Appetite. 2008;50:57–70.
27. Gorton D, Ni Mhurchu C, Chen MH, Dixon R. Nutrition labels: a survey of use, understanding and preferences among ethnically diverse shoppers in New Zealand. Public Health Nutr. 2009;12:1359–65.
28. Jones G, Richardson M. An objective examination of consumer perception of nutrition information based on healthiness ratings and eye movements. Public Health Nutr. 2007;10: 238–44.
29. Kelly B, Hughes C, Chapman K, et al. Consumer testing of the acceptability and effectiveness of front-of-pack food labelling systems for the Australian grocery market. Health Promot Int. 2009;24:120–9.
30. Drewnowski A, Fulgoni V. Nutrient profiling of foods: creating a nutrient-rich food index. Nutr Rev. 2008;66:23–39.
31. Scarborough P, Rayner M, Stockley L. Developing nutrient profile models: a systematic approach. Public Health Nutr. 2007;10:330–6.
32. Sacks G, Rayner M, Stockley L, Scarborough P, Snowdon W, Swinburn B. Applications of nutrient profiling: potential role in diet-related chronic disease prevention and the feasibility of a core nutrient-profiling system. Eur J Clin Nutr. 2011;65:298–306.
33. NuVal™ Nutritional Scoring System. 2011. Available at http://www.nuval.com. Last Accessed 22 Apr 2011.
34. Chiuve SE, Sampson L, Willett WC. The association between a nutritional quality index and risk of chronic disease. Am J Prev Med. 2011;40:505–13.
35. Sutherland LA, Kaley LA, Fischer L. Guiding stars: the effect of a nutrition navigation program on consumer purchases at the supermarket. Am J Clin Nutr. 2010;91:1090S–4S.
36. Lupton JR, Balentine DA, Black RM, et al. The smart choices front-of-package nutrition labeling program: rationale and development of the nutrition criteria. Am J Clin Nutr. 2010;91:1078S–89.
37. MRC Vitamin Study Research Group. Prevention of neural tube defects: results of the Medical Research Council Vitamin Study. Lancet. 1991;338:131–7.
38. Wolff T, Witkop CT, Miller T, Syed SB; U.S. Preventive Services Task Force. Folic acid supplementation for the prevention of neural tube defects: an update of the evidence for the U.S. Preventive Services Task Force. Ann Intern Med. 2009;150:632–9.
39. Gray-Donald K, Jacobs-Starkey L, Johnson-Down L. Food habits of Canadians: reduction in fat intake over a generation. Can J Public Health. 2000;91:381–5.
40. Key TJ, Appleby PN, Spencer EA, Roddam AW, Neale RE, Allen NE. Calcium, diet and fracture risk: a prospective study of 1898 incident fractures among 34 696 British women and men. Public Health Nutr. 2007;10:1314–20.
41. Bischoff-Ferrari HA, Rees JR, Grau MV, Barry E, Gui J, Baron JA. Effect of calcium supplementation on fracture risk: a double-blind randomized controlled trial. Am J Clin Nutr. 2008;87:1945–51.
42. Bolland MJ, Grey A, Avenell A, Gamble GD, Reid IR. Calcium supplements with or without vitamin D and risk of cardiovascular events: reanalysis of the Women's Health Initiative limited access dataset and meta-analysis. BMJ. 2011;342:d2040.
43. Bischoff-Ferrari HA, Dawson-Hughes B, Staehelin HB, et al. Fall prevention with supplemental and active forms of vitamin D: a meta-analysis of randomised controlled trials. BMJ. 2009;339:b3692.

44. Bischoff-Ferrari HA, Willett WC, Wong JB, et al. Prevention of nonvertebral fractures with oral vitamin D and dose dependency: a meta-analysis of randomized controlled trials. Arch Intern Med. 2009;169:551–61.

45. Scragg R. Vitamin D and public health: an overview of recent research on common diseases and mortality in adulthood. Public Health Nutr. 2011;14:1515–32.

46. Zipitis CS, Akobeng AK. Vitamin D supplementation in early childhood and risk of type 1 diabetes: a systematic review and meta-analysis. Arch Dis Child. 2008;93:512–7.

47. Pittas AG, Dawson-Hughes B, Li T, et al. Vitamin D and calcium intake in relation to type 2 diabetes in women. Diabetes Care. 2006;29:650–6.

48. Llewellyn DJ, Lang IA, Langa KM, et al. Vitamin D and risk of cognitive decline in elderly persons. Arch Intern Med. 2010;170:1135–41.

49. Giovannucci E, Liu Y, Rimm EB, Hollis BW, Fuchs CS, Stampfer MJ, et al. Prospective study of predictors of vitamin D status and cancer incidence and mortality in men. J Natl Cancer Inst. 2006;98:451–9.

50. Johnson MA, Kimlin MG. Vitamin D, aging, and the 2005 dietary guidelines for Americans. Nutr Rev. 2006;64:410–21.

51. Cashman KD, Hill TR, Lucey AJ, et al. Estimation of the dietary requirement for vitamin D in healthy adults. Am J Clin Nutr. 2008;88:1535–42.

52. Cashman KD, Wallace JM, Horigan G, et al. Estimation of the dietary requirement for vitamin D in free-living adults >=64 y of age. Am J Clin Nutr. 2009;89:1366–74.

53. Slmomski A. IOM endorses vitamin D, calcium only for bone health, dispels deficiency claims. JAMA. 2011;305:453–6.

54. Van Horn L, McCoin M, Kris-Etherton PM, et al. The evidence for dietary prevention and treatment of cardiovascular disease. J Am Diet Assoc. 2008;108:287–331.

55. Mente A, de Koning L, Shannon HS, Anand SS. A systematic review of the evidence supporting a causal link between dietary factors and coronary heart disease. Arch Intern Med. 2009;169:659–69.

56. Young IS, Woodside JV. Antioxidants in health and disease. J Clin Pathol. 2001;54:176–86.

57. Vivekananthan DP, Penn MS, Sapp SK, Hsu A, Topol EJ. Use of antioxidant vitamins for the prevention of cardiovascular disease: meta-analysis of randomised trials. Lancet. 2003;361:2017–23.

58. Druesne-Pecollo N, Latino-Martel P, Norat T, et al. Beta-carotene supplementation and cancer risk: a systematic review and metaanalysis of randomized controlled trials. Int J Cancer. 2010;127:172–84.

59. Bjelakovic G, Nikolova D, Gluud LL, Simonetti RG, Gluud C. Mortality in randomized trials of antioxidant supplements for primary and secondary prevention: systematic review and meta-analysis. JAMA. 2007;297:842–57.

60. Bjelakovic G, Nikolova D, Gluud LL, Simonetti RG, Gluud C. Antioxidant supplements for prevention of mortality in healthy participants and patients with various diseases. Cochrane Database Syst Rev. 2008;(2):CD007176.

61. Willett WC, Stampfer MJ. Clinical practice. What vitamins should I be taking, doctor? N Engl J Med. 2001;345:1819–24.

Chapter 18
Population Nutrition and Health Promotion

Norman J. Temple and Marion Nestle

Keywords Health promotion • Smoking • Exercise • Intake of fruits and vegetables
• Prevention of coronary heart disease • Government policy • Socioeconomic status

Key Points

- Health promotion campaigns of various types have been conducted: in communities, at worksites, and in physician offices. The most common targets have been smoking, exercise, and dietary, such as intake of fruits and vegetables. The aim has most often been to reduce excess weight, lower the blood cholesterol, blood pressure, and blood glucose, and prevent coronary heart disease (CHD).
- Results of these campaigns have been mixed. Some have achieved very little while others have met with moderate success. Typically, target outcomes have been improved by a few percentage points and this should reduce the risk of CHD by about 5–15%.
- In the light of this limited success of individually oriented health promotion programs, we argue in support of government policy initiatives in order to improve population health. This includes use of taxes and subsidies to adjust the price of various foods so as to shift consumption patterns to healthier foods. Other policy measures can include restrictions on advertising of unhealthy food, especially to children, and improved food labeling.

N.J. Temple, PhD (✉)
Centre for Science, Athabasca University, Athabasca, AB, T9S 3A3, Canada
e-mail: normant@athabascau.ca

M. Nestle, PhD, MPH
Department of Nutrition, Food Studies, and Public Health, New York University,
411 Lafayette St, 5th Floor, New York, NY 10003-7035, USA

Department of Nutritional Sciences, Cornell University, Ithaca, New York
e-mail: marion.nestle@nyu.edu

N.J. Temple et al. (eds.), *Nutritional Health: Strategies for Disease Prevention*,
Nutrition and Health, DOI 10.1007/978-1-61779-894-8_18,
© Springer Science+Business Media, LLC 2012

- Policy measures along these lines are likely to meet with resistance from the food industry.
- Low socioeconomic status (SES), such as low income and poor education, is a major risk factor for poor health. This may be mediated via unhealthy lifestyle choices, such as a poor diet, as well as by psychological factors. Therefore, attempts to improve the population health will require action focused on engaging or affecting these disadvantaged people.

18.1   Introduction

It is now generally accepted that lifestyle—diet, tobacco use, exercise—has a major impact on health, especially lifestyle-related chronic diseases. However, there is a world of difference between awareness of these facts and their translation into preventive action.

While the focus of this chapter is on nutrition in relation to health promotion, we also examine other areas, especially smoking and exercise. This is necessary because many health promotion campaigns take a broad lifestyle approach and simultaneously tackle nutrition, exercise, and smoking.

Trends towards a healthier lifestyle and better health over recent decades have been inconsistent. In the USA, deaths from coronary heart disease (CHD) have fallen by half since their peak in the late 1960s. Yet, at the same time, the USA has been struck by an epidemic of obesity. Between 1976–1980 and 1988–1994 obesity among adults jumped from 14.5 to 22.9% [1]. This then climbed to 30.5% in 1999–2000 [2]. Since then the curve seems to have flattened: in 2007–2008, the prevalence of adults who were overweight was 68%, of whom 32% were obese [3]. Much the same trend is seen with American children and adolescents: an enormous rise in the prevalence of overweight and obesity between the years 1980 and about 2000, followed since then by a flatter curve [4]. A fast rising prevalence of overweight and obesity has also been reported from all other Western countries [5].

Despite vast amounts of dietary information being disseminated, there has been an underwhelming rate of progress in improving the American diet. Between 1970 and 2007, Americans made little change in their consumption of fruits and vegetables. It is currently around 2–3 servings/day, about half the recommended intake [6]. The trend for intake of whole grains has moved in the wrong direction and is now about one-third of the recommended intake [6].

This poor rate of progress in the area of diet should be seen as part of a more general problem that large sections of the population give a low priority to a healthy lifestyle. There was an impressive fall by about half in smoking rates in men in many Western countries starting around 1970. But in the USA this progress seems to have slowed to a crawl since around 1990: the proportion of Americans who smoke is stuck at about one in four [7]. About half of adults state that they engage in regular physical activity, a figure that has changed little in recent years [7–9]. Things are no better in the UK: about 30% of British adults reported that they engage in at least moderate physical activity for at least 30 min on at least 5 days each week [10].

18.2   Health Promotion Campaigns

During the 1970s, the intimate connection between lifestyle and health became increasingly apparent. As a result many people assumed that the next step was to disseminate this information to the public and exhort lifestyle changes, action deemed sufficient to bring about the necessary changes. Here we look at various types of health promotion campaigns, most of them focused on risk factors for cardiovascular disease (CVD).

18.2.1   Campaigns in Communities

A number of community interventions have used the mass media combined with various other methods to reach the target population. Three major projects were carried out in the USA during the 1980s. Their aims were to lower elevated levels of blood cholesterol, blood pressure, and weight, to cut smoking rates, and to persuade more people to take exercise. Each program lasted 5–8 years and succeeded in implementing its intervention on a broad scale, involving large numbers of programs and participants. In the Stanford Five-City Project, conducted by Farquhar et al. [11] in California, two intervention cities received health education via TV, radio, newspapers, other mass-distributed print media, direct education, and schools. On average each adult was exposed to 26 h of education, achieved at the remarkably low per capita cost of $4/year (i.e., about 800 times less than total health-care costs). A similar project was the Minnesota Heart Health Program which included three intervention cities and three control cities in the Upper Midwest [12]. A third project was the Pawtucket Heart Health Program in which the population of Pawtucket, Rhode Island received intensive education at the grass roots level: schools, local government, community organizations, supermarkets, and so forth, but without involving the media [13].

An analysis combined the results of the three studies so as to increase the sample size to 12 cities [14]. Improvements in blood pressure, blood cholesterol, BMI, and smoking were of very low magnitude and were not statistically significant; the estimated risk of CHD mortality was unchanged. These results are mirrored by two other community projects: little success was seen in the Heart To Heart Project in Florence, South Carolina [15] and the Bootheel Heart Health Project in Missouri [16].

One factor contributing to the lack of effect may have been secular trends; the projects took place at a time when American lifestyles were becoming generally more healthy and CHD rates were falling. This suggests that when a population starts receiving health education, even if little more than reports in the mass media and government policy pronouncements, large numbers of people will decide to adopt a healthier lifestyle. A health promotion campaign superimposed on such secular trends may have little *additional* benefit.

Fortunately, we have some examples of reasonably successful community projects for heart disease prevention. One of the earliest and most successful such projects was conducted in North Karelia, a region of eastern Finland which had an exceptionally high rate of the disease [17]. Indeed, Finnish men had the distinction of having the highest mortality rate in the world for CHD. The intervention began in 1972 before much health information had reached the population. Nutrition education was an important component of the intervention. Over the next few years, CHD rates in North Karelia fell sharply. Between 1972 and 2007, CHD mortality in middle-aged men fell by an astonishing 80% [18]. This can be largely explained by changes in risk factors: serum cholesterol declined by 22% while sharp decreases were also seen for blood pressure and smoking. An intensive educational campaign spread to the rest of Finland leading to a national drop in CHD rates [19].

Two other European studies also achieved some success. Positive results were seen in the German Cardiovascular Prevention Study [20], which took place from about 1985 to 1992, when there was no particular favorable trend in risk factors for the population as a whole. It was carried out in six regions of the former West Germany using a wide-ranging approach similar to that used in the American community studies. The intervention caused a small decrease in blood pressure and serum cholesterol (about 2%) and a 7% fall in smoking, but had no effect on weight. Action Heart was a community-based health promotion campaign conducted in Rotherham, England [21]. After 4 years, 7% fewer people smoked and 9% more drank low-fat milk, but there was no change in exercise habits, obesity, or consumption of wholemeal bread.

Two community campaigns are of particular interest because each was narrowly focused on changing only one aspect of lifestyle and used paid advertising as a major intervention strategy. The 1% Or Less campaign aimed to persuade the population of two cities in West Virginia to switch from whole milk to low-fat milk (1% or less) [22]. Advertising in the media was a major component of the intervention (at a cost of slightly less than a dollar per person) together with supermarket campaigns (taste tests and display signs), education in schools, as well as other community education activities. Low-fat milk sales, as a proportion of total milk sales, increased from 18 to 41% within just a few weeks. The intervention campaign was repeated in another city in West Virginia; this time only paid advertising was used [23]. Low-fat milk sales increased from 29 to 46% of total milk sales. An Australian intervention campaign also used paid advertising as a major component [24]. The campaign ran in the State of Victoria from 1992 to 1995 and aimed to increase consumption of fruits and vegetables. Significant increases in consumption of these foods were reported (fruits by 11% and vegetables by 17%).

Another area where campaigns have been narrowly focused on trying to change just one aspect of lifestyle has been those attempting to increase levels of exercise. Numerous such interventions have been carried out and some encouraging results have appeared. Several dozen interventions have tried to persuade people to engage in more walking. A systematic review of these concluded that people can be encouraged to walk more by interventions tailored to their needs, targeted at the most sedentary or at those most motivated to change, and delivered either at the level of the individual or household or through group-based approaches [25]. By this means, interventions can potentially increase walking by up to 30–60 min/week on average.

Several dozen interventions have also attempted to increase exercise levels among children and adolescents [26]. Results have been very mixed. The authors of a systematic review concluded that: "For adolescents, multicomponent interventions and interventions that included both school and family or community involvement have the potential to make important differences to levels of physical activity and should be promoted." [26].

A recent study that is worthy of attention is Romp & Chomp [27]. This is a community-wide, multisetting, multistrategy intervention conducted on young children in Australia from 2004 to 2008. The goals of the intervention were to reduce the prevalence of obesity by improving diets and encouraging children to engage in more exercise. This intervention was carried out in the city of Geelong with a target group of 12,000 children. Despite the ambitious scope of the intervention it did achieve its goals.

The investigators summarize the current situation as follows:

> Early-childhood settings in the intervention areas are now places in which fruit, vegetables, and water are promoted and packaged snacks and sweet drinks are restricted or discouraged. Driving these changes has been the implementation and enforcement of effective policy, cultural changes within organizations, and capacity-building with early-childhood teachers and caregivers. The consistency and continued reinforcement of messages across the community was a key factor in the success of the intervention, in addition to the capacity building of a willing and influential group of gatekeepers (early-childhood workers). Utilizing capacity-building and policy-based strategies also increases the potential of the intervention to benefit future cohorts of children.

Taken together, the community intervention studies indicate that small changes in cardiovascular risk factors can be made by the methods used to date. The evidence is suggestive that interventions focused on a small number of changes and using paid advertising can achieve much success when well planned.

18.2.2 Worksite Health Promotion

As an alternative to health promotion using a community intervention approach, other interventions have focused on the worksite. A pioneering project of this type, which started in 1976, was carried out in Europe by the World Health Organization. The project was conducted over 6 years in 80 factories in Belgium, Italy, Poland, and the UK with the aim of preventing CHD [28, 29]. The trial achieved modest risk factor reductions (1.2% for plasma cholesterol, 9% for smoking, 2% for systolic blood pressure, and 0.4% for weight); these were associated with a 10% reduction in CHD.

At around the same time, Live for Life was carried out by the Johnson & Johnson company in the USA. This comprehensive intervention was started in 1979 and lasted 2 years. Employees exposed to the program showed significant improvements in smoking behavior, weight, aerobic capacity, incidence of hypertension, days of sickness, and health-care expenses [30].

Another worksite project took place in New England [31]. Employees were encouraged to increase their intake of fiber and to reduce their fat intake. Compared with the control sites, the program had no effect on fiber intake but fat intake fell by about 3%. A few years later the research team reported that they succeeded in increasing employees' intake of fruits and vegetables by 19% (0.5 serving/day) using an approach that targeted employees and their families [32]. A similar project in Minnesota offered employees weight control and smoking cessation programs [33]. No effect was seen on weight but the prevalence of smoking was reduced by 2% more than occurred in the control worksites.

The above reports represent a small selection of large numbers of such interventions that have taken place. The American Heart Association recently reviewed the subject [34]. One of their conclusions is that interventions at the worksite can be highly cost-effective, saving employers several dollars for each dollar invested.

18.2.3   Health Promotion in the Physician's Office

In 1994, two British studies reported the effects of health promotion activities carried out by nurses in the offices of family physicians. The aim was to improve cardiovascular risk factors. Each study was a randomized trial aimed at cardiovascular screening and lifestyle intervention. Both studies achieved only modest changes despite intensive intervention. The OXCHECK study reported no significant effect on smoking or excessive alcohol intake but did observe small significant improvements in exercise participation, weight, dietary intake of saturated fat, and serum cholesterol [35, 36]. The Family Heart Study achieved a 12% lowering of risk of CHD (based on a risk factor score) [37]. Similar findings came from an American study where patients were given mailed personalized dietary recommendations, educational booklets, a brief physician endorsement, and motivational counseling by phone. After 3 months the intervention group had increased its consumption of fruits and vegetables by 0.6 serving/day but there was no change in intake of red meat or dairy products [38].

Wilcox et al. [39] reviewed 32 intervention studies carried out in a medical setting. They concluded that:

Overall, these interventions tended to produce modest but statistically significant effects for physical activity or exercise, dietary fat, weight loss, blood pressure, and serum cholesterol…. Whereas small by conventional statistical definitions, these findings are likely to be meaningful when considered from a public health perspective.

A variation of the above trials is the targeting of patients at high risk of CHD, probably the most cost-effective form of intervention [40]. A study from Sweden exemplifies this approach. Subjects at relatively high risk of CHD received either simple advice from their physician or intensive advice (five 90-min sessions plus an all-day session) [41]. The intensive advice had a modest impact; it reduced the risk of CHD by approximately 6%. Two highly successful randomized controlled trials, one in the USA and one in Finland, were carried out on overweight subjects with impaired glucose tolerance, the goal being to prevent the progression to type 2 diabetes [42, 43]. The interventions consisted of physical activity and dietary change. In both studies, the estimated risk reduction was about 58%. These studies are more fully described in Chap. 7 by Temple and Steyn. In general, interventions focused on high-risk subjects have been more successful than other interventions [44].

The major deficiency of the high-risk approach, as Rose [45] has pointed out, is that it only affects a minority of future cases: the 15% of men at "high risk" of CHD account for only 32% of future cases. Therefore, to achieve a major effect on CHD it is necessary to target the entire population. This logic also applies to other diseases related to diet and lifestyle, such as stroke and cancer.

18.2.4 Computer-Based Health Promotion

In recent years, many health promotion programs have been developed that use computers for delivering information. This is a diverse field with programs targeting exercise, diet, obesity, and smoking. Promising findings have been reported from many interventions [46–49].

18.2.5 Health Promotion and the Individual

What the above projects teach us is that appealing to individuals to change their lifestyles will be effective in some instances but not in others and can therefore be frustratingly difficult. While some projects have achieved a moderate degree of success, typically progress has amounted to no more than a few percentage points. This might be expected to reduce the risk of CHD by about 5–15%. While this is certainly beneficial, it will not, however, affect the majority of people at risk. Thus exhortations to the individual, whether via the media, in the community, at the worksite, or in the physician's office, are most unlikely to turn the tide of the chronic diseases of lifestyle.

Pennant et al. [50] recently carried out a systematic review that assessed the effectiveness of community programs for the prevention of CVD. They included only those interventions that targeted the whole population living within a defined geographic area. Their conclusions are similar to the comments made above. Overall, systolic blood pressure was reduced by 2.9 mmHg, total cholesterol level by 0.01 mmol/L, and smoking prevalence by 1.7%. The estimated decrease in 10-year CVD risk was 9.1%. This is relative risk, meaning the proportion of cases prevented. The estimated decrease in absolute risk was 0.65%, indicating that one case of CVD would be prevented during 10 years for every 150 people in the target population. The authors of this review were unable to identify factors that made program success more likely.

A remarkable feature about the studies reviewed by Pennant et al. [50] is that almost every one of them was done before the year 2000 (35 were carried out between 1970 and 2000 but only one between 2000 and 2008). This suggests that in the field of health promotion the hare has been replaced by a tortoise! The one area where significant progress has been made in recent years is the development of computer-based interventions.

Myriad factors influence people's lifestyle behavior besides concerns about how to protect health. Social factors, such as housing, employment, and income also shape people's attitudes, as does education. Advertising directly affects what people want and prices determine whether they can afford it. We are also creatures of habit and custom; resistance may therefore be expected when lifestyle modification demands changes in longstanding behavior and goes against fashion or peer pressure. We must also bear in mind that individuals have little control over many aspects of their physical environment, such as pollution, food contamination, and where and what kinds of foods are sold. It is probably naïve, therefore, to expect dramatic results from interventions that merely exhort the individual to lead a healthier lifestyle. Indeed, this has sometimes been characterized as "victim blaming."

This is in no way to dismiss interventions aimed at encouraging people to improve their lifestyle. Quite the contrary: minor changes can make valuable contributions to public health that more than justify the expense and effort involved. For instance, Jeffery and associates [33] concluded that a smoking cessation program at a worksite costs about $100–$200 per smoker who quits, whereas the cost to the employer for each employee who smokes is far greater. Similarly, Action Heart estimated that the cost per year of life gained was a mere 31 (British) pounds [21].

Health promotion, therefore, can be a cost-effective way to educate and persuade large numbers of people to lead a healthier lifestyle and thereby improve their health [51, 52]. More research is required to determine why different health promotion projects have achieved such varying levels of success. Would campaigns be more successful if the focus was on one lifestyle change rather than many? Is paid advertising the best means to utilize scarce resources?

18.3 Government Policy

18.3.1 The Case for Public Health Policies

While health promotion is a valuable and cost-effective means to improve the health of the population, it clearly has major limitations. We now turn our attention to an alternative strategy.

Effective interventions may need to tackle the factors that determine how people make food choices. Such interventions require the implementation of policies, especially by governments. In the words of Davey Smith and Ebrahim [53]: "... even with the substantial resources given to changing people's diets the resulting reductions in cholesterol concentrations are disappointing. [Health promotion programs] are of limited effectiveness. Health protection—through legislative and fiscal means—is likely to be a better investment." These words were published in 1998 but there is no reason to think differently today.

Governments have a variety of powers at their disposal that can be put into service. One approach, which relies entirely on voluntary cooperation, is to issue statements of policy. However, these can easily amount to no more than hollow declarations. This is well illustrated by government policies on tobacco which, for many years and in many countries, meant very little. On the other hand, policy statements can serve as a clarion call to action. For instance, British and American government policy on diet, lifestyle, and disease, in conjunction with the media and medical science, helped change the climate of opinion so that it is now widely accepted that, for example, people should exercise more and diets should contain more fruits and vegetables.

But governments have other powerful tools that can bring about positive changes in lifestyle across much of the population. This whole subject is discussed in more detail in Chap. 23 by Temple. Here we summarize the key points.

For many products, there is a relationship between price and sales. This has been clearly shown for tobacco and alcohol and certainly applies to food. By means of taxes and subsidies, fruits, vegetables, and wholegrain cereals can be made more attractively priced in comparison with less healthy choices, such as beverages containing sugar. This would most likely induce many people to shift their diets in a healthier direction.

Food advertising is another area where policy interventions might positively affect food choices. The annual worldwide advertising budgets in 2010 for Burger King, Coca-Cola, PepsiCo, and McDonald's were $392 million, $758 million, $1.01 billion, and $1.3 billion, respectively [54]. In stark contrast, the money allocated by federal and state governments for promoting consumption of unrefined foods, such as fruits, vegetables, whole grains, and beans, is miniscule. The extent to which these vast imbalances in advertising budgets affect people's actual diets is not known but is almost certainly significant [55]. Common sense dictates that if advertising did not work, the advertisers would not be wasting their money.

A particular issue is food advertising on children's TV. It is overwhelmingly (80–90%) for unhealthy food choices or for fast-food restaurants and helps boost sales of the advertised foods. This advertising has been linked to the risk of obesity in children and adolescents.

Advertising is but one part of the wider production and marketing strategy of the food industry. Manufacturers sell foods with less fat but the missing fat often reappears in foods that are often little more than concoctions of fat, sugar, white flour, and salt. The food industry promotes these foods because they are so profitable. At the same time food labeling is a minefield of confusion for large sections of the population, as detailed in Chap. 17. The system is, in theory, based on "consumer choice" but, in reality, choices are largely uninformed.

The above information compels the view that government policies in such areas as the pricing, advertising, and labeling of food may be an effective means to induce desirable changes in eating patterns.

Here we offer some specific suggestions as to how existing government policies could be modified along the above lines so as to encourage healthier diets [56].

1. Subsidies paid to milk producers could be changed to favor low-fat milk. Likewise, by the use of such means as subsidies, grading regulations, and labeling, and perhaps even taxation, the sale of low-fat meat could be encouraged over high-fat varieties.
2. There is much scope for improved food labels so as to facilitate purchase of foods with a low content of saturated fat, sugar, and salt. In addition, labeling and nutrition information should be extended to areas presently outside the system, such as fresh meat.
3. By means of regulations and rewards, schools could be encouraged to sell meals of superior health value while restricting the sale of junk food. Similar policies could be applied to other institutions under government control, such as the military, prisons, and cafeterias in government offices.
4. TV advertising could be regulated so as to control the content, duration, and frequency of advertising for unhealthy food products, especially when the target audience is children.

The approach discussed earlier was well put by Blackburn in an article published in 1992 [57]:

> ... A shift of focus to reducing, by policy change, many widespread practices that are life-threatening, while enhancing life-supportive practices, should redirect the currently misplaced emphasis on achieving "responsible" behavior and its purported difficulty. For example, local communities may more appropriately be considered to have a "youth tobacco access problem," approachable in part by regulation, than a "youth smoking problem," approachable mainly by education. Policy interventions may also be designed to .... make preventive practice more economical, as well as to encourage the development of more healthy products by industry. They may be a partial answer to another major paradox: while unhealthy personal behavior is medically discouraged for individuals, the whole of society legalizes, tolerates, and even encourages the same practices in the population.

We must at this point inject a note of caution. While the policy proposals discussed here appear to make excellent sense, there is a lack of solid research evidence to demonstrate their effectiveness [58].

The problem of lead pollution is an excellent illustration of what can be achieved by governmental action. In the 1970s, regulations implemented by the American government forced major reductions or removal of lead from gasoline, paint, water, and consumer products. As a result by the early 1990s, the blood level of the average American child was less than one quarter of what it had been in the late 1970s [59, 60]. Another remarkable success story concerns folic acid. After it was discovered that giving supplements of the vitamin to women during early pregnancy prevents neural tube defects (NTD), it became mandatory, starting in 1998, to add it to cereals in both the USA and Canada. This has almost certainly been responsible for a major reduction in the incidence of NTD by ~20–78% [61–63].

18.3.2 Barriers Against Public Health Policies

While many might consider the policies discussed here to be worthy of implementation, it must be appreciated that there are barriers that need to be overcome. In particular, industry profits enormously from the sale of highly processed food and has often shown itself to be resistant to change. In this regard, industry often secures government support.

The history of attempts to enact legislative control over tobacco illustrate how effective an industry can be when it utilizes a large budget, much of it used for contributions to political parties, in attempts to delay, dilute, or stop laws. There is clear evidence that this is a major reason why the US Congress has for decades been so lethargic when it comes to antismoking legislation. The inherent conflict of interest also promotes government inaction; cigarette companies support the government through tax revenues.

If the tobacco industry can achieve so many successes, then it will likely be much easier for the food industry to thwart interventions that threaten its profits. This is because the relationship between diet and disease is far less clear than is the case with tobacco. Indeed, there is ample evidence that governments are sympathetic to the wishes of the agricultural and food industries. Typically, while the health arm of governments encourages people to eat a healthier diet, the departments responsible for the agricultural and food industries are mainly focused on maintaining high production and sales.

There is considerable evidence of how industry has successfully pressured governments to bow to their wishes on questions of nutrition policy. As discussed by Nestle [64], the meat industry has been particularly effective in rewriting dietary guidelines. In the late 1970s, the goal was "eat less meat." This then became "choose lean meat." By 1992 people were encouraged to consume at least two or three servings daily. More recent US dietary guidelines urge reduction of "solid fats and added sugars" (SoFAS), continuing a long tradition of using nutrients as euphemisms for foods, in this case meat and sodas, for example [65].

The pressure exerted by the food industry in protection of its financial welfare is further explored in Chap. 22 by Nestle and Wilson.

18.3.3 National Nutrition Policies: The Example of Norway

These barriers to effective nutrition policies help explain why national governments have rarely implemented national nutrition policies. One of the rare examples is the case of Norway. The Norwegian Nutrition and Food Policy is a pioneering project that was implemented in 1976 [66]. It recognized the need to integrate agricultural, economic, and health policy. The policy included consumer and price subsidies, marketing measures, consumer information, and nutrition education in schools. Unfortunately, the policy clashed with policies aiming to stimulate agriculture. As a result subsidies went to pork, butter, and margarine rather than to potatoes, vegetables, and fruits. Despite these setbacks, the policy has achieved some success in moving the national diet in the intended direction [67].

Most governments have de facto nutrition policies for such things as school meals and food assistance to the poor, but these are usually disconnected from other areas of food policy. The massive U.S. farm legislation, for example, governs such matters as agricultural subsidies (roughly 15% of authorized spending), food assistance programs (67%), and the rest for conservation, organic production, research, and other programs [68]. Food advocates are becoming increasingly active in attempts to bring agricultural and nutrition assistance policies in line with health promotion policies.

18.3.4   Are Nutrition Policies Acceptable to the Public?

An important question concerns the extent to which the public would accept the suggested policies. The issues of seat belt use, drunk driving, and bans on smoking in many indoor public places illustrate that when legislation is implemented and the public is educated as to its importance, there is a high degree of acceptance.

18.4   Campaigns Against Obesity

It was established during the 1950s and 1960s that smoking is a major cause of disease and death. Finally, in the 1980s, governments in many countries started to take the problem seriously. Even in countries where smoking had been long accepted as almost normal, such as France, Ireland, and China, the governments have implemented tough polices, such as a ban on smoking in many indoor public places. This story is now starting to be repeated with obesity.

Many of the policies advocated above have come together in campaigns that have focused on curbing the obesity epidemic. An excellent—but, alas, rare—example of the implementation of a broad strategy comes from an intervention carried out in France [69]. Children in schools in two towns were given nutrition education. This program was launched in 1992 and expanded somewhat after 1997 to the adult population of the towns. From 1999 there was even wider community activity in support of more physical activity and a healthier lifestyle. At the same time there was much media interest. The BMI of children aged 5–12 years was measured in 2004 and compared with two other towns that received no intervention. The findings revealed that the children in the intervention towns had a lower BMI (15.7 vs. 16.5) and a lower prevalence of overweight or obesity (7.4% vs. 19.4% in boys; 10.4% vs. 16.0% in girls). This is a remarkable degree of success.

18.5   Socioeconomic Status and Health

One area of importance is the relationship between socioeconomic status (SES) and health. Low SES is strongly and consistently associated with a raised mortality rate. This applies to total mortality as well as to death from CHD and cancer. The risk ratios are in the range 1.5–4, clearly making SES a major determinant of health. Various measures of SES have been examined—income, social status of job, being unemployed, area of residence, and education—and each seems to manifest a similar relationship with mortality [70–76].

Various studies have investigated why SES is associated with increased mortality. In general, lower SES is associated with higher rates of smoking and a diet of lower nutritional quality. Is SES merely a proxy measure of lifestyle? Or does SES affect health by a more direct mechanism? This question is of much more than mere theoretical importance and has a bearing on health strategies. If people of low SES are unhealthy because they lead an unhealthy lifestyle, then the solution lies in encouraging changes in their lifestyles. But, if a low SES is intrinsically unhealthy, then the solution lies elsewhere.

Our best evidence is that both possibilities are partially correct. After correcting for confounding variables, especially smoking, exercise, blood cholesterol, blood pressure, and weight, most studies have found that the strength of the association between SES and mortality is reduced by about a quarter or a half [70, 73, 77, 78]. A recent British cohort study that included four separate assessments of lifestyle during the follow-up period, in addition to the baseline check, found that 72% of the association between SES and risk of death was now explained by lifestyle, especially exercise, diet, and alcohol intake [79]. These reports indicate that people with lower SES tend to lead a less healthy lifestyle and this partly explains their poorer health.

But this still leaves much of the association between SES and mortality unexplained, somewhere between one quarter and three quarters. Many studies carried out in Europe and North America have demonstrated that people of low SES tend to eat a less nutritious diet [80]. Consistent with this, Drewnowski [81] showed in his cost analysis that energy-dense foods, such as sugar, oil, fried potatoes, and refined grains, provide energy at far lower cost than lean meat, fish, fresh vegetables, and fruit. This helps explain why such conditions as hypercholesterolemia, hypertension, and obesity are associated with low SES. Nevertheless, it appears that much of the association between SES and mortality cannot be explained by lifestyle and must therefore be a more direct consequence of low SES.

Psychological factors appear to play an important role in explaining the association between SES and mortality [74, 82]. The psychological factor most closely associated with risk of poor health is lack of control at work [82–84]. We can speculate that other psychological factors, such as resentment, frustration, and a feeling of disempowerment, all contribute to poor health among low-income groups. Whatever the precise mechanisms, there is little doubt that structural elements of inequality within Western societies—economic, educational, social status—lead to reduced health.

But what should be done about this? An effective strategy to deal with the challenge of low SES must include efforts to reduce socioeconomic inequalities [85]. But if people of lower SES could be persuaded to adopt the same lifestyle, including diet, as those of higher SES, perhaps as much as half of the problem would likely disappear. Dietary advice is still worth some effort although approaches that change the food environment to make it easier for low-income groups to have access to and to afford healthier foods stand a better chance of being effective.

18.6   Government Policy: Some Final Comments

Based on the close association between various measures of SES and health, an essential component of enhancing a population's health must be measures to improve health-oriented policies, including the SES of the more deprived sections of the population. This means serious measures to counter such widespread problems as poverty and poor education, and reducing income inequities. In countries where there is a strong tradition of social welfare, appropriate measures can be undertaken by the government. Where more individualistic and business-oriented ideologies are the norm, as in the USA, implementing such measures presents a much greater challenge. The private sector would need to act, for example, through charitable and other nongovernmental organizations and private schools. The goal of achieving both a healthy population and a healthy economy would seem more difficult to realize under such governmental systems; nevertheless, a healthy workforce and population is ultimately in the interest of business. Such societies must also find a way to public health.

This viewpoint applies to the relationship between nutrition and the diseases related to it. When governments are focused on economic issues, they lose sight of nutrition policies, and national health can easily become a low priority. In that case the failure of the government and business sectors to work together for the public health may cause the loss of great opportunities for the prevention of such diseases as cancer and CHD. In these circumstances governments must be pressured to implement policies for the improvement of the national health.

The philosophy discussed here need not stop at nutrition: what applies to nutrition certainly applies to other areas of lifestyle, especially to smoking. Exercise also lends itself to policy initiatives. What is the point in telling people to exercise if there is a lack of appropriate facilities? What is the point in telling people to cycle if the roads are too dangerous for bikes? What is needed is a comprehensive view of human health that takes all such factors into consideration.

As the century unfolds people may look back with incredulity on today's world where narrow commercial interests and government *laissez-faire* predominate while the national health flounders. More optimistically, an innovative meshing of business interests, individualism, and recognition of community health needs will emerge.

References

1. Flegal KM, Carroll MD, Kuczmarski RJ, Johnson CL. Overweight and obesity in the United States: prevalence and trends, 1960–1994. Int J Obes. 1998;22:39–47.
2. Flegal KM, Carroll MD, Ogden CL, Johnson CL. Prevalence and trends in obesity among US adults, 1999–2000. JAMA. 2002;288:1723–7.
3. Flegal KM, Carroll MD, Ogden CL, Curtin LR. Prevalence and trends in obesity among US adults, 1999–2008. JAMA. 2010;303:235–41.
4. Ogden CL, Carroll MD, Flegal KM. High body mass index for age among US children and adolescents, 2003–2006. JAMA. 2008;299:2401–5.
5. Siedell JC. Obesity in Europe: scaling an epidemic. Int J Obes. 1995;19 Suppl 3:S1–4.
6. Krebs-Smith SM, Reedy J, Bosire C. Healthfulness of the U.S. food supply: little improvement despite decades of dietary guidance. Am J Prev Med. 2010;38:472–7.
7. King DE, Mainous 3rd AG, Carnemolla M, Everett CJ. Adherence to healthy lifestyle habits in US adults, 1988–2006. Am J Med. 2009;122:528–34.
8. Centers for Disease Control and Prevention (CDC). Physical activity trends—United States, 1990–1998. MMWR Morb Mortal Wkly Rep. 2001;50:166–9.
9. Centers for Disease Control and Prevention (CDC). Prevalence of regular physical activity among adults—United States, 2001 and 2005. MMWR Morb Mortal Wkly Rep. 2007;56: 1209–12.
10. Sproston K, Primatesta P. Health survey for England 2003. London: Stationery Office; 2004.
11. Farquhar JW, Fortmann SP, Flora JA, Taylor CB, Haskell WL, Williams PT, et al. Effects of communitywide education on cardiovascular disease risk factors. The Stanford Five-City Project. JAMA. 1990;264:359–65.
12. Luepker RV, Murray DM, Jacobs DR, et al. Community education for cardiovascular disease prevention: risk factor changes in the Minnesota Heart Health Program. Am J Public Health. 1994;84:1383–93.
13. Carleton RA, Lasater TM, Assaf AR, Feldman HA, McKinlay S, Pawtucket Heart Health Program Writing Group. The Pawtucket Heart Health Program: community changes in cardiovascular risk factors and projected disease risk. Am J Public Health. 1995;85:777–85.
14. Winkleby MA, Feldman HA, Murray DM. Joint analysis of three U.S. community intervention trials for reduction of cardiovascular risk. J Clin Epidemiol. 1997;50:645–58.
15. Goodman RM, Wheeler FC, Lee PR. Evaluation of the Heart To Heart Project: lessons from a community-based chronic disease prevention project. Am J Health Promot. 1995;9:443–55.
16. Brownson RC, Smith CA, Pratt M, et al. Preventing cardiovascular disease through community-based risk reduction: the Bootheel Heart Health Project. Am J Public Health. 1996;86: 206–13.
17. Puska P, Nissinen A, Tuomilehto J, et al. The community based strategy to prevent coronary heart disease: conclusions from the ten years of North Karelia project. Annu Rev Public Health. 1985;6:147–93.
18. Vartiainen E, Laatikainen T, Peltonen M, et al. Thirty-five-year trends in cardiovascular risk factors in Finland. Int J Epidemiol. 2010;39:504–18.
19. Valkonen T. Trends in regional and socio-economic mortality differentials in Finland. Int J Health Sci. 1992;3:157–66.
20. Hoffmeister H, Mensink GB, Stolzenberg H, et al. Reduction of coronary heart disease risk factors in the German Cardiovascular Prevention study. Prev Med. 1996;25:135–45.
21. Baxter T, Milner P, Wilson K, et al. A cost effective, community based heart health promotion project in England: prospective comparative study. BMJ. 1997;315:582–5.
22. Reger B, Wootan MG, Booth-Butterfield S, Smith H. 1% or less: a community-based nutrition campaign. Public Health Rep. 1998;113:410–9.
23. Reger B, Wootan MG, Booth-Butterfield S. Using mass media to promote healthy eating: a community-based demonstration project. Prev Med. 1999;29:414–21.

24. Dixon H, Boland R, Segan C, Stafford H, Sindall C. Public reaction to Victoria's "2 Fruit 'n' 5 Veg Day" campaign and reported consumption of fruit and vegetables. Prev Med. 1998;27: 572–82.

25. Ogilvie D, Foster CE, Rothnie H, et al. Interventions to promote walking: systematic review. BMJ. 2007;334:1204.

26. van Sluijs EM, McMinn AM, Griffin SJ. Effectiveness of interventions to promote physical activity in children and adolescents: systematic review of controlled trials. BMJ. 2007; 335:703.

27. de Silva-Sanigorski AM, Bell AC, Kremer P, et al. Reducing obesity in early childhood: results from Romp & Chomp, an Australian community-wide intervention program. Am J Clin Nutr. 2010;91:831–40.

28. World Health Organisation European Collaborative Group. European collaborative trial of multifactorial prevention of coronary heart disease: final report on the 6-year results. Lancet. 1986;1:869–72.

29. World Health Organisation European Collaborative Group. Multifactorial trial in the prevention of coronary heart disease. Eur Heart J. 1983;4:141–7.

30. Breslow L, Fielding J, Herrman AA, Wilbur CS. Worksite health promotion: its evolution and the Johnson & Johnson experience. Prev Med. 1990;1(9):13–21.

31. Sorensen G, Morris DM, Hunt MK, Hebert JR, Harris DR, Stoddard A, et al. Work-site nutrition intervention and employees' dietary habits: the Treatwell program. Am J Public Health. 1992;82:877–80.

32. Sorensen G, Stoddard A, Peterson K, et al. Increasing fruit and vegetable consumption through worksites and families in the Treatwell 5-a-Day Study. Am J Public Health. 1999;89:54–60.

33. Jeffery RW, Forster JL, French SA, et al. The Healthy Worker Project: a work-site intervention for weight control and smoking cessation. Am J Public Health. 1993;83:395–401.

34. Carnethon M, Whitsel LP, Franklin BA, et al. Worksite wellness programs for cardiovascular disease prevention: a policy statement from the American Heart Association. Circulation. 2009;120:1725–41.

35. Imperial Cancer Research Fund OXCHECK Study Group. Effectiveness of health checks conducted by nurses in primary care: results of the OXCHECK study after one year. BMJ. 1994; 308:308–12.

36. Imperial Cancer Research Fund OXCHECK Study Group. Effectiveness of health checks conducted by nurses in primary care: final results of the OXCHECK study. BMJ. 1995;310: 1099–104.

37. Family Heart Study Group. Randomised controlled trial evaluating cardiovascular screening and intervention in general practice: principal results of British Family Heart Study. BMJ. 1994;308:313–20.

38. Delichatsios HK, Hunt MK, Lobb R, Emmons K, Gillman MW. EatSmart: efficacy of a multifaceted preventive nutrition intervention in clinical practice. Prev Med. 2001;33(2 Pt 1):91–8.

39. Wilcox S, Parra-Medina D, Thompson-Robinson M, Will J. Nutrition and physical activity interventions to reduce cardiovascular disease risk in health care settings: a quantitative review with a focus on women. Nutr Rev. 2001;59:197–214.

40. Field K, Thorogood M, Silagy C, Normand C, O'Neill C, Muir J. Strategies for reducing coronary risk factors in primary care: which is most cost effective? BMJ. 1995;310:1109–12.

41. Lindholm LH, Ekbom T, Dash C, Eriksson M, Tibblin G, Schersten B. The impact of health care advice given in primary care on cardiovascular risk. BMJ. 1995;310:1105–9.

42. Tuomilehto J, Lindstrom J, Eriksson JG, et al. Prevention of type 2 diabetes mellitus by changes in lifestyle among subjects with impaired glucose tolerance. N Engl J Med. 2001;344:1343–50.

43. Knowler WC, Barrett-Connor E, Fowler SE, et al. Reduction in the incidence of type 2 diabetes with lifestyle intervention or metformin. N Engl J Med. 2002;346:393–403.

44. Ammerman AS, Lindquist CH, Lohr KN, Hersey J. The efficacy of behavioral interventions to modify dietary fat and fruit and vegetable intake: a review of the evidence. Prev Med. 2002;35: 25–41.

45. Rose G. The strategy of preventive medicine. Oxford: Oxford University Press; 1992.
46. Krebs P, Prochaska JO, Rossi JS. A meta-analysis of computer-tailored interventions for health behavior change. Prev Med. 2010;51:214–21.
47. Neville LM, O'Hara B, Milat AJ. Computer-tailored dietary behaviour change interventions: a systematic review. Health Educ Res. 2009;24:699–720.
48. Alexander GL, McClure JB, Calvi JH, et al. A randomized clinical trial evaluating online interventions to improve fruit and vegetable consumption. Am J Public Health. 2010;100: 319–26.
49. Enwald HP, Huotari ML. Preventing the obesity epidemic by second generation tailored health communication: an interdisciplinary review. J Med Internet Res. 2010;12:e24.
50. Pennant M, Davenport C, Bayliss S, Greenheld W, Marshall T, Hyde C. Community programs for the prevention of cardiovascular disease: a systematic review. Am J Epidemiol. 2010;172: 501–16.
51. Aldana SG. Financial impact of health promotion programs: a comprehensive review of the literature. Am J Health Promot. 2001;15:296–320.
52. Golaszewski T. Shining lights: studies that have most influenced the understanding of health promotion's financial impact. Am J Health Promot. 2001;15:332–40.
53. Davey Smith G, Ebrahim S. Dietary change, cholesterol reduction, and the public health— what does meta-analysis add? BMJ. 1998;316:1220.
54. Advertising Age. 100 Leading National Advertisers: 2011. Report, June 20, 2011. Available at http://adage.com/datacenter/marketertrees2011 (2011). Last Accessed 2 July 2011.
55. Nestle M, Wing R, Birch L, et al. Behavioral and social influence on food choice. Nutr Rev. 1998;56:S50–64.
56. Nestle M, Jacobson MF. Halting the obesity epidemic: a public health policy approach. Public Health Rep. 2000;115:12–24.
57. Blackburn H. Community programmes in coronary heart disease prevention health promotion: changing community behaviour. In: Marmot M, Elliott P, editors. Coronary heart disease. From aetiology to public health. Oxford: Oxford University Press; 1992. p. 495–514.
58. Finkelstein E, French S, Variyam JN, Haines PS. Pros and cons of proposed interventions to promote healthy eating. Am J Prev Med. 2004;27(3 Suppl):163–71.
59. Pirkle JL, Brody DJ, Gunter EW, et al. The decline in blood lead levels in the United States. JAMA. 1994;272:284–91.
60. Brody DJ, Pirkle JL, Kramer RA, et al. Blood lead levels in the US population. JAMA. 1994;272:277–83.
61. Honein MA, Paulozzi LJ, Mathews TJ, Erickson JD, Wong LY. Impact of folic acid fortification of the US food supply on the occurrence of neural tube defects. JAMA. 2001;285:2981–6.
62. Gucciardi E, Pietrusiak MA, Reynolds DL, Rouleau J. Incidence of neural tube defects in Ontario, 1986–1999. Can Med Ass J. 2002;167:237–40.
63. Liu S, West R, Randell E, et al. A comprehensive evaluation of food fortification with folic acid for the primary prevention of neural tube defects. BMC Pregnancy Childbirth. 2004;4:20.
64. Nestle M. Food politics. How the food industry influences nutrition and health. 2nd ed. Berkeley: University of California Press; 2007.
65. U.S. Department of Agriculture. Diet Guidelines, 2010. Available at http://www.cnpp.usda. gov/DGAs2010-PolicyDocument.html (2011). Last Accessed 6 Aug 2011.
66. Klepp K, Forster JL. The Norwegian Nutrition and Food Policy: an integrated approach to a public health problem. J Public Health Policy. 1985;6:447–63.
67. Norum KR, Johansson L, Botten G, Bjornboe G-EA, Oshaug A. Nutrition and food policy in Norway: effects on reduction of coronary heart disease. Nutr Rev. 1997;55:S32–9.
68. Johnson R, Monke J. What is the "Farm Bill?" Congressional Research Service Report RS 22131. January 3; 2011. Available at http://www.nationalaglawcenter.org/assets/crs/RS22131. pdf. Last Accessed 7 May 2012.
69. Romon M, Lommez A, Tafflet M, et al. Downward trends in the prevalence of childhood overweight in the setting of 12-year school- and community-based programmes. Public Health Nutr. 2009;12:1735–42.

70. Bucher HC, Ragland DR. Socioeconomic indicators and mortality from coronary heart disease and cancer: a 22-year follow-up of middle-aged men. Am J Public Health. 1995;85:1231–6.
71. Lin RJ, Shah CP, Svoboda TJ. The impact of unemployment on health: a review. Can Med Ass J. 1995;153:529–40.
72. Sorlie PD, Backlund E, Keller JB. US mortality by economic, demographic, and social characteristics: The National Longitudinal Mortality Study. Am J Public Health. 1995;85:949–56.
73. Davey Smith G, Neaton JD, Wentworth D, Stamler R, Stamler J. Socioeconomic differentials in mortality risk among men screened for the Multiple Risk Factor Intervention Trial: I. White men. Am J Public Health. 1996;86:486–96.
74. Lynch JW, Kaplan GA, Cohen RD, Tuomilehto J, Salonen JT. Do cardiovascular risk factors explain the relation between socioeconomic status, risk of all-cause mortality, cardiovascular mortality, and acute myocardial infarction? Am J Epidemiol. 1996;144:934–42.
75. Morris JN, Blane DB, White IR. Levels of mortality, education, and social conditions in the 107 local education authority areas of England. J Epidemiol Community Health. 1996;50: 15–7.
76. Mackenbach JP, Kunst AE, Cavelaars AEJM, Groenhof F, Geurts JJM. Socioeconomic inequalities in morbidity and mortality in Western Europe. Lancet. 1997;349:1655–9.
77. Morris JK, Cook DG, Shaper AG. Loss of employment and mortality. BMJ. 1994;308: 1135–9.
78. Pekkanen J, Tuomilehto J, Uutela A, Vartiainen E, Nissinen A. Social class, health behaviour, and mortality among men and women in eastern Finland. BMJ. 1995;311:589–93.
79. Stringhini S, Sabia S, Shipley M, Brunner E, Nabi H, Kivimaki M, et al. Association of socioeconomic position with health behaviors and mortality. JAMA. 2010;303:1159–66.
80. Darmon N, Drewnowski A. Does social class predict diet quality? Am J Clin Nutr. 2008; 87:1107–17.
81. Drewnowski A. Obesity and the food environment: dietary energy density and diet costs. Am J Prev Med. 2004;27(3 Suppl):154–62.
82. Marmot MG, Bosma H, Brunner E, Stansfield S. Contribution of job control and other risk factors to social variations in coronary heart disease incidence. Lancet. 1997;350:235–9.
83. North FM, Syme SL, Feeney A, Shipley M, Marmot M. Psychosocial work environment and sickness absence among British civil servants: The Whitehall II Study. Am J Public Health. 1996;86:332–40.
84. Johnson JV, Stewart W, Hall EM, Fredlund P, Theorell T. Long-term psychosocial work environment and cardiovascular mortality and among Swedish men. Am J Public Health. 1996;86: 324–31.
85. Pickett K, Wilkinson R. The spirit level: why greater equality makes societies stronger. New York: Bloomsbury Press; 2011.

Chapter 19
Optimizing Nutrition for Exercise and Sports

Richard B. Kreider, Neil A. Schwarz, and Brian Leutholtz

Keywords Athletic diet • Sports nutrition • Ergogenic aids • Nutrient timing

Key Points

- Most active individuals can meet energy, macronutrient, and micronutrient needs by consuming a balanced and nutrient-dense diet.
- Athletes engaged in intense training should maintain a diet with sufficient calories, carbohydrate, and protein to meet energy needs and optimize recovery.
- Athletes should consume a low-dose multi-vitamin during heavy training periods in order to ensure they ingest a sufficient amount of micronutrients to meet daily needs.
- Timing nutrient intake prior to, during, and following exercise can optimize exercise performance, recovery, and training adaptations.

R.B. Kreider, PhD, FACSM, FISSN (✉)
Exercise and Sport Nutrition Laboratory, Department of Health and Kinesiology,
Texas A&M University, 332A Blocker, College Station,
TX 77843-4243, USA
e-mail: rkreider@hlkn.tamu.edu

N.A. Schwarz, MS
Department of Health, Human Performance and Recreation, Baylor University,
1312 South 5th Street, Waco, TX 76798-7313, USA
e-mail: Neil_Schwarz@baylor.edu

B. Leutholtz, PhD, FACSM
Center for Exercise, Nutrition and Preventive Health Research, Department of Health,
Human Performance and Recreation, Baylor University,
1312 South 5th Street, Waco, TX 76798-7313, USA
e-mail: Brian_Leutholtz@baylor.edu

N.J. Temple et al. (eds.), *Nutritional Health: Strategies for Disease Prevention*,
Nutrition and Health, DOI 10.1007/978-1-61779-894-8_19,
© Springer Science+Business Media, LLC 2012

- Individuals engaged in intense exercise should consume enough water and/or glucose-electrolyte drinks to maintain hydration.
- Several nutrients have been reported to enhance exercise performance and/or training adaptations. Competitive athletes should only consider taking dietary supplements that have been shown to be safe and effective, are from reputable manufacturers that conduct tests for the presence of banned substances, and are allowed in their sport.

19.1 Introduction

The primary factors that affect exercise performance capacity include an individual's genetic endowment, the quality of training, and effective coaching (see Fig. 19.1). Beyond these factors, nutrition plays a critical role in optimizing performance capacity. In order for athletes to perform well, their training and diet must be optimal. If athletes do not train enough or have an inadequate diet, their performance may be decreased [1]. On the other hand, if athletes train too much, without a sufficient diet, they may be susceptible to becoming overtrained (see Fig. 19.2).

Because optimizing training and dietary practices are critical to peak performance, athletes have searched for various ways to improve exercise performance capacity through the use of *ergogenic aids*. An *ergogenic aid* is any training technique, mechanical device, nutritional practice, pharmacological method, or psychological technique that can improve exercise performance capacity and/or enhance training adaptations [2–4]. This includes aids that may help prepare an individual to

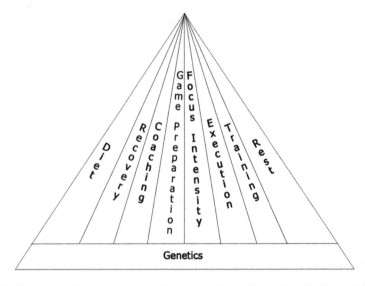

Fig. 19.1 Factors that affect performance. Reprinted with permission from Kreider et al. [4]

Fig. 19.2 Relationship of training volume/intensity to performance. Reprinted with permission from Kreider et al. [4]

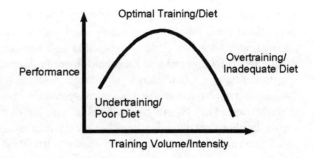

exercise, improve the efficiency of exercise, and/or enhance recovery from exercise. Ergogenic aids may also allow an individual to tolerate heavy training to a greater degree by helping them recover faster or help them stay healthy during intense training. This chapter presents an overview of the role that nutrition has on optimizing sport performance; describes nutritional guidelines that athletes should employ to optimize training adaptations; and evaluates the potential ergogenic value of various nutrients that have been proposed to improve exercise capacity and/or training adaptations. Guidelines presented have been adapted from several of our previous publications [2, 4, 5] and position stands that have been coauthored with the International Society of Sports Nutrition (ISSN) [2, 6–9].

19.2   Energy Demands for Active Individuals

The first component in optimizing training and performance through nutrition is to ensure that the athlete is consuming enough calories to offset energy expenditure [1, 2, 4, 10, 11]. People who participate in a general fitness program (e.g., exercising 30–40 min/day, three times per week) can generally meet nutritional needs following a normal diet (e.g., 1,800–2,400 kcal/day or about 25–35 kcal/kg/day for a 50–80 kg individual) because their caloric demands from exercise are not too great (e.g., 200–400 kcal/session) [2]. However, athletes involved in moderate levels of intense training (e.g., 2–3 h/day of intense exercise performed five to six times per week) or high volume intense training (e.g., 3–6 h/day of intense training in 1–2 workouts for 5–6 days/week) may expend 600–1,200 kcal/h or more during exercise. For this reason, their caloric needs may approach 45–60 kcal/kg/day for power athletes (2,700–7,200 kcal/day for a 50–100 kg athlete) and 50–80 kcal/kg/day for endurance athletes (2,500–6,400 kcal/day for a 50–80 kg athlete) [2]. For elite athletes, energy expenditure during heavy training or competition may be enormous [2, 4]. For example, energy expenditure for cyclists to compete in the Tour de France has been estimated as high as 12,000 kcal/day (150–200 kcal/kg/day for a 60–80 kg athlete) [12–14]. Additionally, caloric needs for large athletes (i.e., 100–150 kg) may range between 6,000 and 12,000 kcal/day, depending on the volume and intensity of different training phases [2, 4, 14].

Although some exercise physiologists and nutritionists argue that athletes can meet caloric needs simply by consuming a well-balanced diet, it is often very difficult for larger athletes and/or athletes engaged in high volume/intense training to be able to eat enough food in order to meet caloric needs. Maintaining an energy deficient diet during training often leads to significant weight loss (including muscle mass), illness, onset of physical and psychological symptoms of overtraining, and reductions in performance [15]. Nutritional analyses of athletes' diets have revealed that many are susceptible to maintaining negative energy intakes during training. Susceptible populations include runners, cyclists, swimmers, triathletes, gymnasts, skaters, dancers, wrestlers, boxers, and athletes attempting to lose weight too quickly [2, 10]. Additionally, female athletes have been reported to have a high incidence of eating disorders [10]. Consequently, it is important for the exercise physiologist working with athletes to ensure that athletes are well fed and consume enough calories to offset the increased energy demands of training and maintain body weight [2].

Although this sounds relatively simple, intense training often suppresses appetite and/or alters hunger patterns so that many athletes do not feel like eating [10]. Some athletes do not like to exercise within several hours after eating because of sensations of fullness and/or a predisposition to cause gastrointestinal distress. Further, travel and training schedules may limit food availability and/or the types of food that athletes are accustomed to eating. This means that care should be taken to plan meal times in concert with training as well as to make sure that athletes have sufficient availability of nutrient-dense foods throughout the day for snacking between meals (e.g., drinks, fruits, carbohydrate/protein bars) [7, 15]. For this reason, sport nutritionists often recommend that in order to meet energy needs athletes should consume 4–6 meals each day and snack between meals [7]. Use of nutrient-dense energy bars and high-calorie carbohydrate/protein supplements often provide a convenient way for athletes to supplement their diet in order to maintain energy intake during training.

19.3   General Macronutrient Guidelines for Athletes

The second component to optimizing training and performance through nutrition is to ensure that athletes consume the proper amounts of carbohydrate, protein, and fat in their diet [2]. Table 19.1 summarizes macronutrient guidelines for individuals initiating training through elite athletes. The following overviews macronutrient needs based on the level of training an individual is performing.

19.3.1   Carbohydrate

Carbohydrate serves as the primary fuel for high-intensity intermittent or prolonged exercise. Carbohydrate is stored in the muscle (about 15 g/kg) and liver

Table 19.1 General dietary guidelines for athletes

	Beginners (50–90 kg)	Intermediate (50–90 kg)	Advanced (50–100 kg)	Power athletes (60–120 kg)	Endurance athletes (50–80 kg)
Caloric intake					
kcal/kg/day	30–35	35–40	40–50	45–60	50–80
kcal	1,500–3,150	1,750–600	2,000–5,000	2,700–7,200	2,500–6,400
Carbohydrate					
% Energy	45–55	45–55	50–60	40–50	55–65
g/kg/day	3–5	4–6	5–8	5–8	7–13
Protein					
% Energy	13–17	11–15	10–16	10–15	10–12
g/kg/day	1–1.5	1–1.5	1.5–2.0	1.7–2.2	1.5–2.0
Fat					
% Energy	<30	25–30	22–30	<30	25–30
g/kg/day	1–1.2	1–1.35	1–1.5	1.5–2.0	1.5–2.2

Reprinted with permission from Kreider et al. [4]

(about 80–100 g). Intense exercise significantly depletes muscle and liver glycogen stores. The depleted stores are replenished from dietary carbohydrate. Unfortunately, when significant amounts of carbohydrate are depleted, it may be difficult to fully replenish carbohydrate levels within 1 day. Consequently, when athletes train once or twice per day over a period of days, carbohydrate levels may gradually decline, leading to fatigue, poor performance, and/or overtraining.

Athletes involved in moderate amounts of intermediate to advanced training programs typically need to consume a diet consisting of 45–60% carbohydrate (i.e., 4–8 g/kg/day) in order to maintain liver and muscle glycogen stores [2, 7, 11]. Research has also shown that athletes involved in high-volume intense training (e.g., 3–6 h/day of intense training in 1–2 workouts for 5–6 day/week) may need to consume 8–10 g/kg/day of carbohydrate (i.e., 400–1,500 g/day for 50–150 kg athletes) in order to maintain muscle glycogen levels. This would be equivalent to consuming 0.5–2.0 kg/day of spaghetti. Preferably, the majority of dietary carbohydrate should come from complex carbohydrates with a low to moderate glycemic index (e.g., grains, starches, fruits). However, since it is physically difficult to consume that much carbohydrate per day when an athlete is involved in intense training, many nutritionists and exercise physiologists recommend that athletes consume concentrated carbohydrate juices/drinks and/or consume high-carbohydrate supplements to meet carbohydrate needs. While this amount of carbohydrate is not necessary for the fitness-minded individual who only trains three to four times per week for 30–60 min, it is essential for competitive athletes engaged in intense moderate- to high-volume training. The general consensus in the scientific literature is the body can oxidize 1–1.1 g/min of carbohydrate or about 60 g/h [7]. The American College of Sports Medicine (ACSM) recommends ingesting 0.7 g/kg/h during exercise in a 6–8% solution (i.e., 6–8 g/100 mL of fluid). Harger-Domitrovich et al. [16] reported that 0.6 g/kg/h of maltodextrin optimized carbohydrate utilization [16]. This would be about 30–70 g of carbohydrate per hour for a 50–100 kg individual [17–19]. Studies also indicate that ingestion of additional amounts of carbohydrate does not further increase carbohydrate oxidation.

It should also be noted that exogenous carbohydrate oxidation rates have been shown to differ based on the type of carbohydrate consumed because they are taken up by different transporters [20–22]. For example, oxidation rates of disaccharides and polysaccharides like sucrose, maltose, and maltodextrins are high while fructose, galactose, trehalose, and isomaltulose are lower [23–25]. Ingesting combinations of glucose and sucrose or maltodextrin and fructose have been reported to promote greater exogenous carbohydrate oxidation than other forms of carbohydrate [20–26]. These studies generally indicate a ratio of 1–1.2 for maltodextrin to 0.8–1.0 fructose. For this reason, we recommend that care should be taken to consider the type of carbohydrate ingested prior to, during, and following intense exercise in order to optimize carbohydrate availability.

19.3.2   Protein

There has been considerable debate regarding protein needs of athletes [5, 8, 27–30]. Initially, it was recommended that athletes do not need to ingest more than the RDA for protein (i.e., 0.8–1.0 g/kg/day). However, research over the last decade has indicated that athletes engaged in intense training need to ingest about 1.5–2 times the RDA (i.e., 1.5–2.0 g/kg/day) in order to maintain protein balance. If an insufficient amount of protein is obtained from the diet, an athlete will maintain a negative nitrogen balance resulting in protein catabolism and slow recovery. Over time, this may lead to lean muscle wasting and training intolerance [2].

For people involved in a general fitness program, protein needs can generally be met by ingesting 0.8–1.0 g/kg/day of protein. It is generally recommended that athletes involved in moderate amounts of intense training consume 1–1.5 g/kg/day of protein (50–225 g/day for a 50–150 kg athlete), while athletes involved in high volume intense training consume 1.5–2.0 g/kg/day of protein (75–300 g/day for a 50–150 kg athlete) [2]. This protein need would be equivalent to ingesting 3–11 servings of chicken or fish per day for a 50–150 kg athlete. Although smaller athletes typically can ingest this amount of protein in their normal diet, larger athletes often have difficulty consuming this much dietary protein. Additionally, a number of athletic populations have been reported to be susceptible to protein malnutrition (e.g., runners, cyclists, swimmers, triathletes, gymnasts, dancers, skaters, wrestlers, boxers). Therefore, care should be taken to ensure that athletes consume a sufficient amount of quality protein in their diet in order to maintain nitrogen balance (e.g., 1.5–2 g/kg/day).

However, it should be noted that not all protein is the same. Proteins differ based on the source of the protein, its amino acid profile, and the methods of processing or isolating it [31]. These differences influence availability of amino acids that have been reported to possess biological activity (e.g., α-lactalbumin, β-lactoglobulin, glycomacropeptides, immunoglobulins, lactoperoxidases, lactoferrin) and the rate and metabolic activity of the protein [32]. For example, different types of proteins (e.g., casein and whey) are digested at different rates, and this directly affects catabolism and anabolism [31–34]. Therefore, care should be taken not only to make sure that the athlete consumes enough protein in their diet but also that the protein is high quality. The best dietary sources of low-fat, high-quality protein are light skinless chicken, fish, egg white, and skim milk (casein and whey) [31, 35]. The best sources of high-quality protein found in nutritional supplements are whey, colostrum, casein, milk proteins, and egg protein [31, 35]. Although some athletes may not need to supplement their diet with protein and some sports nutrition specialists may not think that protein supplements are necessary, it is common for a sports nutrition specialist to recommend that some athletes (e.g., power and strength) supplement their diet with protein in order to meet dietary protein needs and/or provide essential amino acids (EAA) following exercise in order to optimize protein synthesis.

The ISSN has recently adopted a position stand on protein that highlights the following points [8]:

1. Exercising individuals need approximately 1.4–2.0 g of protein per kg of body weight per day.
2. Concerns that protein intake within this range is unhealthy are unfounded in healthy, exercising individuals.
3. An attempt should be made to obtain protein requirements from whole foods, but supplemental protein is a safe and convenient method of ingesting high-quality dietary protein.
4. The timing of protein intake in the time period encompassing the exercise session has several benefits including improved recovery and greater gains in fat-free mass.
5. Protein residues such as branched-chain amino acids have been shown to be beneficial for the exercising individual, including increasing the rates of protein synthesis, decreasing the rate of protein degradation, and possibly aiding in recovery from exercise.
6. Exercising individuals need more dietary protein than their sedentary counterparts.

19.3.3 Fat

The dietary recommendations of fat intake for athletes are similar to or slightly greater than those recommended for nonathletes in order to promote health [2]. Maintenance of energy balance, replenishment of intramuscular triglycerides, and adequate consumption of essential fatty acids are of greater importance among athletes and allow for somewhat increased intake [36]. This depends on the athlete's training state and goals. For example, higher fat diets appear to maintain circulating testosterone concentrations better than low-fat diets [37, 38]. This may have relevance to the documented suppression of testosterone that may occur during volume-type overtraining [39]. Generally, it is recommended that athletes consume a moderate amount of fat (approximately 30% of their daily caloric intake), while increases up to 50% of energy can be safely ingested by athletes during regular high-volume training [36]. For athletes attempting to decrease body fat, however, it has been recommended that they consume 0.5–1 g/kg/day of fat. This is because some weight-loss studies indicate that people who are most successful in losing weight and maintaining the weight loss are those who ingest less than 40 g/day of fat in their diet, although this is not always the case [40]. Certainly, the type of dietary fat (e.g., *n*-6 vs. *n*-3; saturation state) is a factor in such research and could play an important role in any discrepancies [41, 42]. Strategies to help athletes manage dietary fat intake include teaching them which foods contain various types of fat so they can make better food choices and learn how to count fat grams.

19.3.4 Nutrient Timing

In addition to the general nutritional guidelines described earlier, research has also demonstrated that the timing and composition of meals consumed may play a role in optimizing performance, training adaptations, and preventing overtraining [2, 4]. In this regard, it takes about 3–4 h for carbohydrate to be digested and begin to be stored as muscle and liver glycogen. Consequently, preexercise meals should be consumed about 4–6 h prior to exercise [11]. This means that if an athlete trains in the afternoon, breakfast is the most important meal to top off muscle and liver glycogen levels. Research has also indicated that ingesting a light carbohydrate and protein snack 30–60 min prior to exercise (e.g., 50 g of carbohydrate and 5–10 g of protein) serves to increase carbohydrate availability toward the end of an intense exercise bout, increase availability of amino acids during exercise, maintain insulin levels, and decrease exercise-induced catabolism [27, 43, 44].

When exercise lasts more than 1 h, athletes should ingest glucose/electrolyte solution (GES) drinks in order to maintain blood glucose levels, help prevent dehydration, and reduce the immunosuppressive effects of intense exercise [45–48]. Within 30 min following intense exercise, athletes should consume carbohydrate and protein (e.g., 1 g/kg of carbohydrate and 0.5 g/kg of protein) as well as consume a high-carbohydrate meal within 2 h of the exercise [49, 50]. This nutritional strategy has been found to accelerate glycogen resynthesis as well as promote a more anabolic hormonal profile that may hasten recovery [49–51]. Finally, for 2–3 days prior to competition, athletes should taper training by 30–50% and consume 200–300 g/day of *extra* carbohydrate in their diet. This *carbohydrate loading* technique has been shown to supersaturate carbohydrate stores prior to competition and improve endurance exercise capacity [11]. Thus, the type of meal and timing of eating are important factors in maintaining carbohydrate availability during training and potentially decreasing the incidence of overtraining.

The ISSN has adopted a position stand on nutrient timing [7] that was summarized with the following points:

1. Prolonged exercise (>60–90 min) of moderate- to high-intensity exercise will deplete the internal stores of energy, and prudent timing of nutrient delivery can help offset these changes.
2. During intense exercise, regular consumption (10–15 fl oz) of a GES delivering 6–8% CHO (6–8 g CHO/100 mL fluid) should be consumed every 15–20 min to sustain blood glucose levels.
3. Glucose, fructose, sucrose, and other high-glycemic CHO sources are easily digested, but fructose consumption should be minimized as it is absorbed at a slower rate and increases the likelihood of gastrointestinal problems.
4. The addition of PRO (protein; 0.15–0.25 g/kg/day) to CHO at all time points, especially postexercise, is well tolerated and may promote greater restoration of muscle glycogen when carbohydrate intakes are suboptimal.
5. Ingestion of 6–20 g of EAA and 30–40 g of high-glycemic CHO within 3 h after an exercise bout and immediately before exercise has been shown to significantly stimulate muscle PRO synthesis.
6. Daily postexercise ingestion of a CHO+PRO supplement promotes greater increases in strength and improvements in lean tissue and body fat% during regular resistance training.
7. Milk PRO sources (e.g., whey and casein) exhibit different kinetic digestion patterns and may subsequently differ in their support of training adaptations.
8. Addition of creatine monohydrate to a CHO+PRO supplement in conjunction with regular resistance training facilitates greater improvements in strength and body composition as compared with when no creatine is consumed.
9. Dietary focus should center on adequate availability and delivery of CHO and PRO. However, including small amounts of fat does not appear to be harmful and may help to control glycemic responses during exercise.
10. Irrespective of timing, regular ingestion of snacks or meals providing both CHO and PRO (3:1 CHO:PRO ratio) helps to promote recovery and replenishment of muscle glycogen when lesser amounts of carbohydrate are consumed.

19.4 Proposed Nutritional Ergogenic Aids

Nutritional ergogenic aids include alterations in the composition of the diet, timing of eating, and/or supplementation of various macro- and micronutrients that may enhance performance. Nutritional strategies that improve the preparation for exercise, the efficiency of exercise, performance capacity, and/or enhance the recovery from exercise may be viewed as ergogenic. Consequently, the final nutritional strategy for enhancing training and/or performance capacity in athletes is the appropriate use of effective, safe, and legal nutritional ergogenic aids. Although research has demonstrated that some nutritional strategies and nutrients may affect exercise training and/or performance capacity, the majority of nutritional ergogenic aids marketed to athletes do not affect performance. The following reviews macro- and micronutrients that have been proposed to improve exercise capacity.

19.4.1 Carbohydrate and Carbohydrate By-products

As stated previously, dietary carbohydrate availability can significantly affect muscle and liver carbohydrate stores and performance capacity. For this reason, in addition to the dietary guidelines described earlier, a significant amount of research has been conducted on determining ways to optimize carbohydrate availability during exercise and/or spare muscle glycogen use during exercise. Generally, increasing availability of any form of carbohydrate has the potential to improve exercise capacity by serving as an exogenous fuel source. Table 19.2 describes the proposed ergogenic value and summary of research findings for several forms of carbohydrate and carbohydrate by-products that have been proposed to enhance exercise performance. Of the nutrients reviewed, glucose-electrolyte solution (GES) sport drinks possess the greatest potential to improve exercise capacity. Although some clinical and/or exercise benefits have been reported from corosolic acid, calcium D-glucarate, dihydroxyacetone phosphate (DHAP), fructose 1,6-diphosphate (FDP), polylactate, pyruvate, and ribose supplementation, it is our view that additional research is necessary to determine the efficacy of these nutrients before they are recommended for athletes.

19.4.2 Lipids and Lipid By-products

Since fat can serve as a primary fuel source during low- to moderate-intensity exercise and most people have a considerable amount of fat stored as potential energy, researchers have therefore investigated the effects of lipids and lipid by-products on exercise capacity and training. The basic rationale is that if fat oxidation can be increased during exercise, carbohydrate stores can be spared, exercise capacity can be improved, and/or a greater amount of fat can be burned during exercise. Table 19.3 presents research findings for selected lipids and lipid by-products that have been proposed to affect exercise performance. Of the nutrients presented, glycerol supplementation used as a means to hyper-hydrate athletes susceptible to dehydration appears to possess the most ergogenic potential. There is also accumulating evidence that conjugated linoleic acids (CLA) supplementation affects body composition during training and/or may possess health benefits [52] and that L-carnitine tartrate may help athletes tolerate training to a greater degree [53]. Based on current data, there appears to be limited ergogenic value of medium-chain triglyceride (MCT) supplementation. Finally, although there may be some health benefits from diets high in omega-3 fatty acids, there is no evidence that supplementation with them affects exercise performance.

19.4.3 Protein and Amino Acids

Amino acids are the foundation of protein in the body and are essential for the synthesis of tissue, specific proteins, hormones, enzymes, and neurotransmitters [27].

Table 19.2 Proposed nutritional ergogenic aids for athletes—carbohydrate and carbohydrate by-products

Nutrient	Theoretical ergogenic value	Summary of research findings/recommendations
Colosolic (corosolic) acid	Colosolic (corosolic) acid activates the "shuttle" that attracts glucose molecules, which are then transported into cells for energy. The result would be a stabilizing of blood glucose levels	It has been shown to be a glucose transport activator in vitro [63]. Additionally, it has been shown that it aids glucose uptake in vivo [64] and may be a beneficial addition to carbohydrate sports drinks
Dihydroxyacetone phosphate (DHAP)	Supplementation has been suggested to enhance glycolytic and oxidative metabolism. Additionally, DHAP and pyruvate supplementation have been suggested to promote fat loss and extend endurance exercise capacity by serving as a fuel during exercise	Few well-controlled studies have evaluated the ergogenic value of DHAP. Several studies have reported that DHAP supplementation (16–75 g/day) improves maximal VO_2 and promotes fat loss in obese individuals on hypocaloric diets [65]. However, more research is needed before definitive conclusions can be made
Fructose 1,6-diphosphate (FDP)	FDP serves as an intermediate step in glycolysis after the energy requiring steps of converting glucose to glucose-6-phosphate. Theorized to increase blood ATP and 2,3-diphosphoglycerate levels, enhance dissociation of oxygen from hemoglobin, and serve as an efficient source of carbohydrate to enhance exercise capacity	Studies indicate that FDP supplementation (0.25 g/kg) can serve as an effective fuel source during exercise [66]. Some studies indicate that exercise capacity may be improved in patients with peripheral vascular disease. However, recent studies in healthy subjects indicate that FDP supplementation has no advantage over other forms of carbohydrate
GES sport drinks	Ingesting sport drinks during prolonged exercise (e.g., 6–8 oz of 6–8% solution every 5–15 min) has been found to help maintain blood glucose availability to the muscle and improve time to exhaustion in moderately intense exercise bouts (e.g., 70% of VO_2 max) lasting 3–4 h	Numerous studies indicate that ingesting GES drinks during exercise maintains blood glucose levels, helps promote fluid retention, and decreases dehydration [67]. Recommended for exercise bouts lasting more than 60 min, particularly if exercising in hot/humid environments. Ingesting highly concentrated carbohydrate drinks (i.e., >15%) may slow gastric emptying and promote dehydration

(continued)

Table 19.2 (continued)

Nutrient	Theoretical ergogenic value	Summary of research findings/recommendations
Pinitol	D-Pinitol or (3-O-methyl-chiroinositol) is a plant extract that has been reported to stimulate glucose uptake into L6 cells [68]	A recent study found that co-ingestion of low doses of D-pinitol with creatine monohydrate augmented whole body creatine retention [69]. However, this may not lead to any ergogenic benefit as another study demonstrated that D-pinitol in addition to creatine showed no additional adaptations to resistance exercise vs. creatine alone [70]
Calcium D-glucarate, calcium glucarate glucaric acid	Supplementation with calcium D-glucarate, calcium glucarate, and glucaric acid have been purported to help enhance metabolism, recovery, and growth by helping the body eliminate toxins, waste products, and carcinogens. In addition, the theory suggests that since D-glucaric acid and CDG have been found to help eliminate excess estrogen, it may help bodybuilders improve androgen/estrogen balance [71]	D-Glucarate acid is an important nutrient that promotes the removal of toxins, carcinogens, and excess estrogen from the body. Although there may be some potential health benefits of supplementing the diet with CDG, there is no evidence that CDG supplementation at the recommended doses would affect training adaptations, body composition, or help athletes reduce excess estrogen levels [71]
Polylactate (PL)	PL is a semisoluble amino acid/lactate salt that has been theorized to be easily converted to pyruvate for entrance into the tricarboxylic acid (TCA) cycle and thereby enhance carbohydrate availability during endurance exercise	Few studies have evaluated the ergogenic value of PL supplementation. One study reported that in comparison to a placebo and maltodextrin trial, PL supplementation during exercise (7% GES solution) promoted higher pH and bicarbonate levels with no differences in performance [72]. Conversely, another study reported that addition of PL to a glucose polymer drink did not affect physiological responses to exercise or performance [73]

Pyruvate	Supplementation of pyruvate with DHAP has been suggested to enhance fat loss and extend endurance exercise capacity by serving as a fuel during exercise	Studies indicate that calcium pyruvate (6–25 g/day) with or without DHAP (16–75 g/day) supplementation promoted significantly more fat loss in obese individuals on hypocaloric diets [74, 75]. However, there are little data to support that the dosages currently marketed to promote fat loss (i.e., 0.5–2 g/day) affect body composition or exercise responses. In addition, the overall quantity of research examining calcium pyruvate is minimal at best thus it is not warranted to include calcium pyruvate as a weight-loss supplement
Ribose	Ribose is a naturally occurring five carbon sugar (pentose) that is primarily found in the body as a constituent of riboflavin (vitamin B_2), nucleic acids, nucleotides, and nucleosides. Supplementation has been theorized to increase ATP availability and recovery during intense exercise	Some medical studies indicate that ribose supplementation (10–60 g/day) can increase ATP availability in certain patient populations, blunt the ischemia threshold in heart patients, and enhance the predictive value of thallium exercise tests [76–78]. Most recent studies show no ergogenic value in athletes [79, 80]

Adapted from Kreider et al. [4]

Table 19.3 Proposed nutritional ergogenic aids—lipids and lipid by-products

Nutrient	Proposed ergogenic value	Summary of research findings
Conjugated linoleic acids (CLA)	CLA are essential fatty acids found primarily in fat from whole dairy products. Animal studies indicate that adding CLA to dietary feed decreases body fat, increases bone mass, has anticarcinogenic properties, enhances immunity, and inhibits atherosclerotic progression [81]. Consequently, CLA supplementation in humans has been suggested to help manage body composition, delay loss of bone, and provide health benefits	Animal studies are impressive [82, 83] and some human studies suggest benefits at some but not all dosages [84, 85]. Evidence is accumulating that the *trans*-10, *cis*-12 and *cis*-9, *trans*-11 isomers of CLA may enhance long-term weight loss and possess some health benefits [52]
Glycerol	Glycerol has been reported to promote fluid retention by decreasing urine formation. Glycerol added to water may be beneficial to hyperhydrate athletes prior to exercise in an effort to prevent dehydration during exercise	Although studies indicate that glycerol can significantly enhance body fluid, evidence is mixed on whether it can improve exercise capacity [86–88]. Little research has been done on glycerol in the last 5 years; however, a 2006 study agreed with previous findings that glycerol has little impact on performance [89]
L-Carnitine	Carnitine serves as a transporter of fatty acids from the cytosol into the mitochondria and helps modulate the metabolism of coenzyme-A (CoA). Studies indicate that fatty acid oxidation is regulated in part by the ability to shuttle fatty acids into the mitochondria for entrance into the TCA cycle. Consequently, L-carnitine supplementation has been theorized as a means of enhancing fat oxidation and sparing muscle glycogen during exercise as well as promoting fat loss	Although there are some data showing that L-carnitine supplementation may be beneficial for some patient populations, most well-controlled studies indicate that L-carnitine supplementation does not affect muscle carnitine content, fat metabolism, and/or weight loss in overweight or trained subjects [90, 91]. One study reported that L-carnitine supplementation during a period of intensified training helped athletes to better tolerate the training [53]

Medium-chain triglycerides (MCT)	MCT are shorter chain fatty acids that can easily enter the mitochondria of the cell and be converted to energy through fat metabolism [72]. Theoretically, MCT feedings should serve as an efficient fuel source for exercise possibly serving to enhance endurance capacity	Studies are mixed as to whether MCT can serve as an effective source of fat during exercise and/or improve exercise performance [92–95]. A 2001 study found that 60 g/day of MCT oil for 2 week was not sufficient at improving performance [96]. In fact Goedecke found that not only did MCT supplementation not improve performance, but, actually negatively affected sprint performance in trained cyclists [97]. These findings have been confirmed by others that MCT oils are not sufficient to induce positive training adaptations and may cause gastric distress [98]. One recent study found that MCT oil may positively affect RPE and lactate clearance [99]
Omega-3 fatty acids	These have been reported to serve as antioxidants, enhance immunity, and decrease risk of cardiovascular disease. Some have suggested that omega-3 FA supplementation in athletes would decrease muscle damage and help maintain immune function	Although there is evidence that omega-3 FA supplementation (1–3 g/day) may affect lipid peroxidation and immune responses, most studies indicate no ergogenic benefit on aerobic or anaerobic power [100]

Adapted from Kreider et al. [4]

Amino acids are also involved in the synthesis of energy through gluconeogenesis and regulation of numerous metabolic pathways. Consequently, it has been suggested that athletes may require additional protein in their diet in order to enhance muscle and tissue growth, the synthesis of hormones and enzymes necessary for energy metabolism, or serve as a potential energy substrate during exercise.

Table 19.4 describes the potential ergogenic value of amino acids which have been purported to affect exercise capacity and/or promote training adaptations.

Table 19.4 Proposed nutritional ergogenic aids—protein and amino acids

Nutrient	Proposed ergogenic value	Summary of research findings
α-Keto-glutarate (α-KG)	α-KG is an intermediate in the TCA cycle that is involved in aerobic energy metabolism. There is some clinical evidence that α-KG may serve as an anticatabolic nutrient after surgery [101, 102]	It is unclear whether α-KG supplementation during training affects training adaptations
α-Keto-iso-caproate (KIC)	KIC is a branched-chain keto acid that is a metabolite of leucine. In a similar manner as HMB, leucine and metabolites of leucine are believed to possess anticatabolic properties [103]	There is some clinical evidence that KIC may spare protein degradation in clinical populations [104, 105]. However, we are not aware of any studies that have evaluated the effects of KIC supplementation during training on body composition
Arginine, orni-thine, lysine	Clinical studies indicate that supplementation of these amino acids may stimulate growth hormone release serving to preserve muscle mass during bed rest. Additionally, some studies indicate that arginine supplementation improves immune status. Consequently, some have suggested that supplementation of these amino acids during training may increase muscle mass and strength gains	Recent studies indicate that supplementation with arginine, ornithine, and/or lysine (10–25 g/day) does not enhance the effect of exercise stimulation on either hGH or various measures of muscular strength or power in experienced weightlifters [106, 107]. Some recent studies suggest that arginine-based supplementation increases nitric oxide and may affect performance [108–110]
Aspartate, asparag-ine	These amino acids serve as precursors to oxaloacetate in the TCA cycle. Supplementation has been theorized to spare muscle glycogen use and enhance endurance performance capacity	Some well-controlled studies support the ergogenic value of aspartate and arginine supplementation on sparing muscle glycogen use and improving exercise capacity [111–113]. However, additional research is necessary

(continued)

Table 19.4 (continued)

Nutrient	Proposed ergogenic value	Summary of research findings
Branched-chain amino acids (BCAA)	Exercise-induced decreases in BCAA levels have been suggested to contribute to central fatigue as well as muscle catabolism. Supplementation of BCAA with sports drinks may increase BCAA availability and decrease the ratio of free tryptophan/BCAA. Theoretically, this may minimize serotonin production in the brain and delay central fatigue. Additionally, BCAA supplementation has been reported to decrease exercise-induced protein degradation and/or muscle enzyme release (an indicator of muscle damage) possibly by promoting an anticatabolic hormonal profile [44, 114]. BCAA has also been reported to stimulate protein synthesis after intense exercise [2, 7]	A number of studies have reported that BCAA supplementation (4–16 g) can affect physiological and psychological responses to exercise. However, it is unclear the degree to which these potentially beneficial effects may affect performance. In terms of training, there is some evidence to support contentions that BCAA supplementation may affect catabolism and body composition [115–117], particularly when training at altitude. In addition, BCAA and essential amino acid supplementation have been shown to stimulate protein synthesis after resistance exercise and lead to greater training adaptations over time [2, 7]
Creatine	The availability of phosphocreatine (PC) stores in the muscle significantly affects the amount of energy generated during brief periods of high-intensity exercise. Creatine supplementation has been shown to increase muscle creatine and PC content, help maintain ATP levels during exercise, and accelerate the rate of resynthesis of ATP during and following high-intensity, short-duration exercise leading to greater sprint performance and training adaptations	Numerous studies have indicated that creatine supplementation increases high-intensity exercise performance and muscle mass during training [118]. Compared to controls performance gains are typically 5–15% greater while muscle mass gains are 2–5 lb greater during 4–12 weeks of training [119, 120]. The gains in muscle mass appear to be a result of an improved ability to perform high-intensity exercise enabling an athlete to train harder and thereby promote greater training adaptations and muscle hypertrophy [121–123]. Although concerns have been raised about the safety and possible side effects of creatine supplementation [120, 124], recent long-term safety studies have reported no apparent side effects [125–127] and/or that creatine monohydrate may lessen the incidence of injury during training [128–131]. Consequently, supplementing the diet with creatine monohydrate and/or creatine-containing formulations seems to be a safe and effective method to increase muscle mass [9]

(continued)

Table 19.4 (continued)

Nutrient	Proposed ergogenic value	Summary of research findings
Essential amino acids (EAA)	Recent studies have indicated that ingesting 3–6 g of EAA prior to [132, 133] and/or following exercise stimulates protein synthesis [134–148]. Theoretically, this may enhance gains in muscle mass during training	Recent research indicates that ingesting EAA with carbohydrate immediately following resistance exercise in elderly subjects promoted significantly greater training adaptations as compared to waiting until 2 h after exercise to consume the supplement [149]. Although more data are needed, there is a strong theoretical rationale and some supportive evidence that EAA supplementation may enhance protein synthesis and training adaptations. Because EAAs include BCAAs it is probable that positive effects on protein synthesis from EAA ingestion are likely due to the BCAA content [139–148]
Glutamine	Glutamine has been reported to increase cell volume and stimulate synthesis of protein [150, 151] and glycogen [152]. Glutamine availability also directly affects lymphocytic function. Theoretically, glutamine supplementation prior to and/or following exercise (e.g., 6–10 g) may help to optimize cell hydration, protein synthesis, and maintain immune function leading to greater training adaptations	A study found that subjects who supplemented their diet with glutamine (5 g) and BCAA (3 g) enriched whey protein during training promoted about a 1 kg greater gain in muscle mass and greater gains in strength than ingesting whey protein alone [153]. While a 1 kg increase in lean body mass was observed, it is likely that these gains were due to the BCAAs that were added to the whey protein. In a well-designed 6-week investigation, Candow et al. [154] studied the effects of oral glutamine supplementation combined with resistance training in young adults. At the end of the 6-week intervention, the authors observed that glutamine supplementation during resistance training had no significant effect on muscle performance, body composition, or muscle protein degradation in young healthy adults. While there may be other beneficial uses for glutamine supplementation, there does not appear to be any scientific evidence that it supports increases in lean body mass or muscular performance
β-Alanine	Carnosine is a dipeptide comprised of the amino acids histidine and β-alanine naturally occurring in large amounts in skeletal muscles. Carnosine is believed to be one of the primary muscle-buffering substances available in skeletal muscle [54]	Studies have shown that β-alanine supplementation can increase the number of repetitions one can perform, increase lean body mass, increase knee extension torque, and training volume [54]

(continued)

Table 19.4   (continued)

Nutrient	Proposed ergogenic value	Summary of research findings
β-Hydroxy β-meth-ylbu-tyrate (HMB)	Leucine and metabolites of leucine such as α-ketoisocaproate (KIC) have been reported to inhibit protein degradation. The anticatabolic effects have been suggested to be regulated by the leucine metabolite HMB. Adding HMB to dietary feed improved carcass quality in sows and steers. It has been hypothesized that supplementing the diet with leucine and/or HMB may inhibit protein degradation during resistance training	Supplementing the diet with 1.5–3 g/day of calcium HMB has been reported to increase muscle mass and strength, particularly among untrained subjects initiating training [155–158] and the elderly [159]. Gains in muscle mass are typically 0.5–1 kg greater than controls during 3–6 weeks of training. There is also recent evidence that HMB may lessen the catabolic effects of prolonged exercise [160] and that there may be additive effects of co-ingesting HMB with creatine [161, 162]. However, the effects of HMB supplementation in athletes are less clear. Most studies conducted on trained subjects have reported nonsignificant gains in muscle mass possibly due to a greater variability in response of HMB supplementation among athletes [163–165]. Consequently, there is fairly good evidence showing that HMB may enhance training adaptations in individuals initiating training. However, additional research is necessary to determine whether HMB may enhance training adaptations in trained athletes
Ornithine-α-ketoglu-tarate (OKG)	OKG is believed to possess anabolic/catabolic effects. Animal and clinical studies suggest that patients administered OKG experience improved protein balance. Theoretically, it may provide some value for athletes engaged in intense training	A recent study reported that OKG supplementation (10 g/day) during 6 weeks of resistance training promoted greater gains in bench press [166]. However, no significant differences were observed in squat strength, training volume, or gains in muscle mass
Tryptophan	Tryptophan is an amino acid which increases during prolonged exercise as fatty acids are mobilized for fat oxidation. Increases in brain concentrations of tryptophan have been reported to contribute to fatigue as well as increase endogenous opioid production. Tryptophan supplementation has been theorized to help athletes tolerate pain and enhance endurance exercise capacity	Most studies indicate that increases in tryptophan in the blood and brain contribute to central fatigue. Although an initial study suggested that L-tryptophan supplementation improves endurance performance while exercising at 80% of maximal exercise capacity [167], another study indicated that L-tryptophan had no effect or promoted an ergolytic effect on performance [168]

Adapted from Kreider et al. [4]

Of the amino acids reviewed, without a doubt creatine has been consistently reported as one of most effective and safe nutritional supplements to enhance anaerobic exercise capacity, strength, and gains in muscle mass during training [9]. Studies have indicated that aspartate, branched-chain amino acids (leucine, isoleucine, and valine), EAA, glutamine, β-hydroxy β-methylbutyrate (HMB), and arginine may affect exercise capacity, enhance recovery, and/or promote greater training adaptations. However, not all studies report the ergogenic value of these supplements and additional research is needed. Finally, there is accumulating evidence that β-alanine supplementation can increase muscle carnosine levels and serve as an effective buffer during high-intensity exercise performance [54]. Although there may be some clinical applications, there appears to be little ergogenic value of ornithine, lysine, and tryptophan supplementation for athletes.

19.4.4   Vitamins

Vitamins are essential organic compounds which serve to regulate metabolic processes, energy synthesis, neurological processes, and prevent destruction of cells. There are two primary groups of vitamins: fat and water soluble. The fat-soluble ones include vitamins A, D, E, and K. The body stores these and therefore excessive intake may result in toxicity. Water-soluble vitamins are the B vitamins and vitamin C. Since these are water soluble, excessive intake is eliminated in urine, with few exceptions (e.g., vitamin B_6, which can cause peripheral nerve damage when consumed in excessive amounts).

Table 19.5 describes the RDA, proposed ergogenic benefit, and summary of research findings for fat- and water-soluble vitamins. Although research has demonstrated that specific vitamins may possess some health benefit (e.g., vitamins C and E, niacin, folic acid), few have been reported to directly provide ergogenic value for athletes. However, some vitamins may help athletes tolerate training to a greater degree by reducing oxidative damage (vitamins C and E) and/or help to maintain a healthy immune system during heavy training (vitamin C). Theoretically, this may help athletes tolerate heavy training leading to improved performance. The remaining vitamins reviewed appear to have little ergogenic value for athletes who consume a normal, nutrient-dense diet. Since dietary analyses of athletes have found deficiencies in caloric and vitamin intake, many sports nutritionists recommend that athletes consume a low-dose daily multivitamin and/or a vitamin enriched postworkout carbohydrate/protein supplement during periods of heavy training. A paper published in the American Medical Association also recommended that Americans consume a one-a-day, low-dose multivitamin in order to promote general health and well-being [55, 56]. We feel this is prudent advice particularly for active populations.

Table 19.5 Proposed nutritional ergogenic aids—vitamins

Nutrient	Recommended dietary allowances (RDA)	Proposed ergogenic value	Summary of research findings
Vitamin A	Males: 900 µg/day Females: 700 µg/day	Constituent of rhodopsin (visual pigment) and is involved in night vision. Some suggest that vitamin A supplementation may improve sport vision	No studies have shown that vitamin A supplementation improves exercise performance [169]
Vitamin D	5 µg/day (age <51)	Promotes bone growth and mineralization. Enhances calcium absorption. Supplementation with calcium may help prevent bone loss in osteoporotic populations	Co-supplementation with calcium may help prevent bone loss in athletes susceptible to osteoporosis [170]. However, vitamin D supplementation does not enhance exercise performance [169]
Vitamin E	15 mg/day	As an antioxidant, it has been shown to help prevent the formation of free radicals during intense exercise and prevent the destruction of red blood cells. This may improve or maintain oxygen delivery to the muscles during exercise. Some evidence suggests that it may reduce risk of heart disease or decrease incidence of recurring heart attacks	Numerous studies show that vitamin E supplementation can decrease exercise-induced oxidative stress [171–173]. However, most studies show no effects on performance at sea level. At high altitudes, vitamin E may improve exercise performance [174]. Additional research is necessary to determine whether long-term supplementation helps athletes to better tolerate training
Vitamin K	Males: 120 µg/day Females: 90 µg/day	Important in blood clotting. There is also some evidence that it may affect bone metabolism in postmenopausal women	Vitamin K supplementation (10 mg/day) in elite female athletes has been reported to increase calcium-binding capacity of osteocalcin and promote a 15–20% increase in bone formation markers and a 20–25% decrease in bone resorption markers, suggesting an improved balance between bone formation and resorption [175]
Thiamin (B$_1$)	Males: 1.2 mg/day Females: 1.1 mg/day	Coenzyme (thiamin pyrophosphate) in the removal of CO_2 from decarboxylic reactions from pyruvate to acetyl CoA and in TCA cycle. Supplementation is theorized to improve anaerobic threshold and CO_2 transport. Deficiencies may decrease efficiency of energy systems	Dietary availability of thiamin does not appear to affect exercise capacity when athletes have a normal intake [176]

(continued)

Table 19.5 (continued)

Nutrient	Recommended dietary allowances (RDA)	Proposed ergogenic value	Summary of research findings
Riboflavin (B₂)	Males: 1.3 mg/day Females: 1.7 mg/day	Constituent of flavin nucleotide coenzymes involved in energy metabolism. Theorized to enhance energy availability during oxidative metabolism	Dietary availability of riboflavin does not appear to affect exercise capacity when athletes have a normal intake [176]
Niacin (B₃)	Males: 16 mg/day Females: 14 mg/day	Constituent of coenzymes involved in energy metabolism. Theorized to blunt increases in fatty acids during exercise, reduce cholesterol, enhance thermoregulation, and improve energy availability during oxidative metabolism	Studies indicate that niacin supplementation (100–500 mg/day) can help decrease blood lipid levels and increase homocysteine levels in patients with elevated blood cholesterol [177, 178]. However, niacin supplementation (280 mg) during exercise has been reported to decrease exercise capacity by blunting the mobilization of fatty acids [179]
Pyridoxine (B₆)	1.3 mg/day (age <51)	Has been marketed as a supplement that will improve muscle mass, strength, and aerobic power in the lactic acid and oxygen systems. It also may have a calming effect that has been linked to an improved mental strength	In well-nourished athletes, pyridoxine failed to improve aerobic capacity or lactic acid accumulation [176]. However, when combined with vitamins B₁ and B₁₂, it may increase serotonin levels and improve fine motor skills that may be necessary in sports like pistol shooting and archery [180, 181]
Cyano-cobalamin (B₁₂)	2.4 µg/day	A coenzyme involved in the production of DNA and serotonin. DNA is important in protein and red blood cell synthesis. Theoretically, it would increase muscle mass, the oxygen-carrying capacity of blood, and decrease anxiety	In well-nourished athletes, no ergogenic effect has been reported. However, when combined with vitamins B₁ and B₆ cyanocobalamin has been shown to improve performance in pistol shooting [181]. This may be due to increased levels of serotonin, a neurotransmitter in the brain, which may reduce anxiety
Folic acid (folate)	400 µg/day	Functions as a coenzyme in the formation of DNA and red blood cells. An increase in red blood cells could improve oxygen delivery to the muscles during exercise. Believed to be important to help prevent birth defects and may help decrease homocysteine levels	Studies suggest that increasing dietary availability of folic acid during pregnancy can lower the incidence of birth defects [182]. Additionally, it may decrease homocysteine levels (a possible risk factor for heart disease) [183]. In well-nourished athletes without folate deficiency, folic acid did not improve exercise performance [184]

Pantothenic acid	5 mg/day	Acts as a coenzyme for acetyl coenzyme A (acetyl CoA). This may benefit aerobic or oxygen energy systems	Research has reported no improvements in aerobic performance with acetyl CoA supplementation. However, one study reported a decrease in lactic acid accumulation, without an improvement in performance [185]
β-Carotene	None	Serves as an antioxidant. Theorized to help minimize exercise-induced lipid peroxidation and muscle damage	Research indicates that ß-carotene supplementation with or without other antioxidants can help decrease exercise-induced peroxidation. Over time, this may help athletes tolerate training. However, it is unclear whether antioxidant supplementation affects exercise performance [173]
Vitamin C	Males: 90 mg/day Females: 75 mg/day	Used in a number of different metabolic processes in the body. It is involved in the synthesis of epinephrine, iron absorption, and is an antioxidant. Theoretically, it could benefit exercise performance by improving metabolism during exercise. There is also evidence that vitamin C may enhance immunity	In well-nourished athletes, vitamin C supplementation does not appear to improve physical performance [186, 187]. However, there is some evidence that supplementation (e.g., 500 mg/day) following intense exercise may decrease the incidence of upper respiratory tract infections [188–190]

Adapted from Kreider et al. [2]. RDA based on the 2002 Food & Nutrition Board, National Academy of Sciences-National Research Council recommendations

19.4.5   Minerals

Minerals are essential inorganic elements necessary for a host of metabolic processes. They serve as structure for tissue, important components of enzymes and hormones, and regulators of metabolic and neural control. Some minerals have been found to be deficient in athletes or become deficient in response to training and/or prolonged exercise. When mineral status is inadequate, exercise capacity may be reduced. Dietary supplementation of minerals in deficient athletes has generally been found to improve exercise capacity. Additionally, supplementation of specific minerals in nondeficient athletes has also been reported to affect exercise capacity.

Table 19.6 describes minerals that have been purported to affect exercise capacity in athletes. Of the minerals reviewed, several appear to possess health and/or ergogenic value for athletes under certain conditions. For example, calcium supplementation in athletes susceptible to premature osteoporosis may help maintain bone mass. There is also recent evidence that dietary calcium may play a role in fat loss and maintenance. Iron supplementation in athletes prone to iron deficiency and/or anemia has been reported to improve exercise capacity. Sodium phosphate loading has been reported to increase maximal oxygen uptake, anaerobic threshold, and improve endurance exercise capacity by 8–10% [57, 58]. Increasing dietary availability of salt (sodium chloride) during the initial days of exercise training in the heat has been reported to help maintain fluid balance and prevent dehydration. The ACSM recommendations for sodium levels (340 mg) represent the amount of sodium in less than 1/8 teaspoon of salt and meet recommended guidelines for sodium ingestion during exercise (300–600 mg/h or 1.7–2.9 g of salt during a prolonged exercise bout) [19, 59–61]. Finally, zinc supplementation during training has been reported to decrease exercise-induced changes in immune function. Consequently, somewhat in contrast to vitamins, there appear to be several minerals that may enhance exercise capacity and/or training adaptations for athletes under certain conditions. However, although ergogenic value has been purported for the remaining minerals, there is little evidence that boron, chromium, magnesium, or vanadium affect exercise capacity or training adaptations in healthy individuals eating a normal diet.

19.4.6   Water

The most important nutritional ergogenic aid for athletes is water [2]. Exercise performance can be significantly impaired when 2% or more of body weight is lost through sweat. For example, when a 70-kg athlete loses more than 1.4 kg of body weight during exercise (2% of body weight), performance capacity is often significantly decreased. Loss of more than 4% of body weight during exercise may lead to heat illness, heat exhaustion, heat stroke, and possibly death [62]. For this reason, it is critical that athletes consume a sufficient amount of water and/or GES sports drinks during exercise in order to maintain hydration status.

Table 19.6 Proposed nutritional ergogenic aids—minerals

Nutrient	RDA	Proposed ergogenic value	Summary of research findings
Boron	None	Boron has been marketed to athletes as a dietary supplement that may promote muscle growth during resistance training. The rationale was primarily based on an initial report that boron supplementation (3 mg/day) significantly increased β-estradiol and testosterone levels in postmenopausal women consuming a diet low in boron	Studies which have investigated the effects of 7 weeks of boron supplementation (2.5 mg/day) during resistance training on testosterone levels, body composition, and strength have reported no ergogenic value [191, 192]. There is no evidence at this time that boron supplementation during resistance training promotes muscle growth
Calcium	1,000 mg/day (age 19–50)	Involved in bone and tooth formation, blood clotting, and nerve transmission. Stimulates fat metabolism. Diet should contain sufficient amounts, especially in growing children/adolescents, female athletes, and postmenopausal women. Vitamin D needed to assist absorption	Calcium supplementation may be beneficial in populations susceptible to osteoporosis [193]. Additionally, calcium supplementation has been shown to promote fat metabolism and help manage body composition [194, 195]. Calcium supplementation provides no ergogenic effect on exercise performance
Chromium	Males: 35 μg/day Females: 25 μg/day (age 19–50)	Chromium, commonly sold as chromium picolinate, has been marketed with claims that the supplement will increase lean body mass and decrease body fat levels.	Animal research indicates that chromium supplementation increases lean body mass and reduces body fat. Early research on humans reported similar results [196]; however, more recent well-controlled studies reported that chromium supplementation (200–800 μg/day) does not improve lean body mass or reduce body fat [197, 198]
Iron	Males: 8 mg/day Females: 18 mg/day (age 19–50)	Iron supplements are used to increase aerobic performance in sports that use the oxygen system. Iron is a component of hemoglobin in the red blood cell, which is a carrier of oxygen	Most research shows that iron supplements do not appear to improve aerobic performance unless the athlete is iron depleted and/or has anemia [199]
Magnesium	Males: 420 mg/day Females: 320 mg/day	Activates enzymes involved in protein synthesis. Involved in ATP reactions. Serum levels decrease with exercise. Some suggest that magnesium supplementation may improve energy metabolism/ATP availability	Most well-controlled research indicates that magnesium supplementation (500 mg/day) does not affect exercise performance in athletes unless there is a deficiency [200, 201]

(continued)

Table 19.6 (continued)

Nutrient	RDA	Proposed ergogenic value	Summary of research findings
Phosphorus (phosphate salts)	700 mg/day	Phosphate has been studied for its ability to improve all three energy systems, primarily the oxygen system or aerobic capacity	Recent well-controlled research studies reported that sodium phosphate supplementation (4 g/day for 3 day) improved the oxygen energy system in endurance tasks [57, 58, 202]. There appears to be little ergogenic value of other forms of phosphate (i.e., calcium phosphate, potassium phosphate). More research is needed to determine the mechanism for improvement
Potassium	2,000 mg/day[a]	An electrolyte that helps regulate fluid balance, nerve transmission, and acid–base balance. Some suggest excessive increases or decreases in potassium may predispose athletes to cramping	Although potassium loss during intense exercise in the heat has been anecdotally associated with muscle cramping, the etiology of cramping is unknown [203, 204]. It is unclear whether potassium supplementation in athletes decreases the incidence of muscle cramping [205]. No ergogenic effects reported
Selenium	55 µg/day	Marketed as a supplement to increase aerobic exercise performance. Working closely with vitamin E and glutathione peroxidase (an antioxidant), selenium may destroy destructive free radical production of lipids during aerobic exercise	Although selenium may reduce lipid peroxidation during aerobic exercise, improvements in aerobic capacity have not been demonstrated [206, 207]

Sodium	500 mg/day[a]	An electrolyte that helps regulate fluid balance, nerve transmission, and acid–base balance. Excessive decreases in sodium may predispose athletes to cramping and hyponatremia	During the first several days of intense training in the heat, a greater amount of sodium is lost in sweat. Additionally, prolonged ultraendurance exercise may decrease sodium levels leading to hyponatremia. Increasing salt availability during heavy training in the heat has been shown to help maintain fluid balance and prevent hyponatremia [205, 208]
Vanadyl sulfate (vanadium)	None	Vanadium may be involved in reactions in the body that produce insulin-like effects on protein and glucose metabolism. Due to the anabolic nature of insulin, this has brought attention to vanadium as a supplement to increase muscle mass, enhance strength and power	Limited research has shown that type 2 diabetics may improve their glucose control; however, there is no proof that vanadyl sulfate has any effect on muscle mass, strength, or power [209, 210]
Zinc	Males: 11 mg/day Females: 8 mg/day	Constituent of enzymes involved in digestion. Associated with immunity. Theorized to reduce incidence of upper respiratory tract infections in athletes involved in heavy training	Studies indicate that zinc supplementation (25 mg/day) during training minimizes exercise-induced changes in immune function [47, 211–213]

Adapted from Kreider et al. [2]. RDA based on the 2002 Food & Nutrition Board, National Academy of Sciences-National Research Council recommendations

[a]Estimated minimum requirement

The normal sweat rate of individuals engaged in exercise ranges from 0.5 to 2.0 L/h depending on temperature, humidity, exercise intensity, and their sweat response to exercise [62]. This means that in order to maintain fluid balance and prevent dehydration, athletes need to ingest 0.5–2 L/h of fluid in order to offset weight loss. This requires frequent ingestion of 6–8 oz of cold water or a GES sports drink every 5–15 min during exercise [62]. Athletes should not depend on thirst to prompt them to drink because people do not typically get thirsty until they have lost a significant amount of fluid through sweat. Additionally, athletes should weigh themselves prior to and following exercise training to ensure that they maintain proper hydration [62]. The athlete should consume three cups of water for every pound lost during exercise in order to adequately rehydrate themselves after exercise [62]. Athletes should also train themselves to tolerate drinking greater amounts of water during training and make sure that they consume more fluid in hotter or humid environments. Preventing dehydration during exercise is one of the most effective ways to maintain exercise capacity. Finally, inappropriate and excessive weight-loss techniques (e.g., cutting weight in saunas, wearing rubber suits, severe dieting, vomiting, using diuretics) are extremely dangerous and should be prohibited. Sports nutrition specialists can play an important role in educating athletes and coaches about proper hydration methods and supervising fluid intake during training and competition.

19.4.7   Miscellaneous Substances

Table 19.7 describes a number of miscellaneous purported ergogenic substances used by athletes. Of the nutrients described, sodium bicarbonate, caffeine, and *Echinacea* appear to have the greatest potential to affect exercise performance and/or training adaptations. Sodium bicarbonate loading (0.3 g/kg of baking soda) prior to exercise has been consistently reported to enhance bouts of high-intensity exercise lasting from 1 to 3 min in duration (e.g., a 400–800 m run). Although some athletes may experience gastrointestinal distress, bicarbonate loading appears to be a highly effective ergogenic aid for athletes as long as they can tolerate the supplementation protocol.

Caffeine is a naturally occurring stimulant found in many foods consumed in the normal diet (e.g., coffee, tea, chocolate). Caffeine ingestion (3–9 mg/kg) prior to exercise has been reported to increase fat oxidation, spare muscle glycogen use, and enhance endurance exercise performance [6]. The ergogenic effects of caffeine appear to be more pronounced in nonhabitual caffeine users and in habitual users who abstain from consuming caffeine for about a week prior to competition. Although some athletic governing bodies have banned excessive intake of caffeine as an ergogenic aid, studies show that even when taken within the limits allowed by athletic governing bodies, caffeine may provide ergogenic benefit. Doses should be limited to 7 mg/kg or less to avoid a positive drug test for excess caffeine use.

Table 19.7 Proposed nutritional ergogenic aids—miscellaneous nutrients

Nutrient	Proposed ergogenic value	Summary of research findings
Alcohol	It has been studied for its use as a psychological stress reducer in precision sports such as riflery, archery, and dart throwing. It has also been investigated as an energy source	Some limited research supports an ergogenic effect in precision sports like riflery when about one drink of alcohol is consumed (a blood alcohol of 0.02%), 30–60 min prior to competition [214]. However, its use is illegal in these sports and not recommended
Alkaline salts (bicarbonate)	Sodium bicarbonate has been researched for its effect on improving power in sports that are of a short duration and use the lactic acid energy system	Bicarbonate loading (e.g., 0.3 g/kg taken 60–90 min prior to exercise or 5 g taken two times per day for 5 day) has been shown to be an effective way to buffer acidity during high-intensity exercise lasting 1–3 min [215, 216]. This can improve exercise capacity in events like the 400–800 m run or 100–200 m swim [217]. In elite male swimmers sodium bicarbonate supplementation significantly improved 200 m freestyle performance [218]. A 2009 study found similar improvements in performance in youth swimmers at distances of 50–200 m. Although bicarbonate loading can improve exercise, some people have difficulty with their stomach tolerating bicarbonate as it may cause gastrointestinal distress
Caffeine	Caffeine is a naturally derived stimulant found in many nutritional supplements, such as *Gaurana*, Bissey Nut, or Kola. Caffeine is also found in coffee, tea, soft drinks, energy drinks, and chocolate	Studies indicate that ingestion of caffeine (e.g., 3–9 mg/kg taken 30–90 min before exercise) can spare carbohydrate use during exercise and thereby improve endurance [219, 220]. Caffeine has also been shown to improve repeated sprint performance benefiting the anaerobic athlete [221, 222] and mental focus and performance [6]. People who drink caffeinated drinks regularly, however, appear to experience less ergogenic benefits from caffeine [223]. Additionally, some concern has been expressed that ingestion of caffeine prior to exercise may contribute to dehydration although recent studies have not supported this concern [221, 224, 225]. Caffeine doses above 9 mg/kg can result in urinary caffeine levels that surpass the doping threshold for many sport organizations

(continued)

Table 19.7 (continued)

Nutrient	Proposed ergogenic value	Summary of research findings
Choline (lecithin)	Choline is considered an essential nutrient that is needed for cell membrane integrity and to facilitate the movement of fats in and out of cells. It is also a component of the neurotransmitter acetylcholine and is needed for normal brain functioning, particularly in infants. For this reason, phosphatidyl choline (PC) has been purported as a potentially effective supplement to promote fat loss as well as improve neuromuscular function	There is some data from animal studies that support the potential value of PC as a weight-loss supplement [226]. However, it is currently unclear whether PC supplementation affects body composition in humans. Studies have reported no apparent effects of PC supplementation during endurance exercise on performance [227, 228]
Coenzyme Q10	It is found in the mitochondria and is involved in oxygen transport and ATP production. It is also an antioxidant that may help destroy free radicals during intense aerobic exercise	Coenzyme Q10 has been found to improve heart function, aerobic capacity, and exercise performance in patients with heart disease, but not in healthy athletes [229, 230]
Growth hormone releasing peptides (GHRP) and secretagogues	Research has indicated that GHRP and other nonpeptide compounds (secretagogues) appear to help regulate growth hormone (GH) release [231, 232]. These observations have served as the basis for development of nutritionally based GH stimulators (e.g., amino acids, pituitary peptides, "pituitary substances," *Mucuna pruriens*, broad bean, alpha GPC)	Although there is clinical evidence that pharmaceutical grade GHRPs and some nonpeptide secretagogues can increase GH and IGF-1 levels at rest and in response to exercise, it is currently unknown whether any of these nutritional alternatives would increase GH and/or affect training adaptations
Inosine	Inosine is a building block for DNA and RNA that is found in muscle. It has a number of potentially important roles that may enhance training and/or exercise performance [233]	Although there is some theoretical rationale, available studies indicate that inosine supplementation has no apparent effect on exercise performance capacity [234–236]

Isoflavones	These are naturally occurring nonsteroidal phytoestrogens that have a similar chemical structure as the ipriflavone (a synthetic flavonoid drug used in the treatment of osteoporosis) [237]. For this reason, soybeans (which is an excellent source of isoflavones) and isoflavone extracts have been investigated in the possible treatment of osteoporosis. More recently, the isoflavone extracts 7-isopropoxyisoflavone (ipriflavone) and 5-methyl-7-methoxy-isoflavone (methoxyisoflavone) have been marketed as "powerful anabolic" substances	Although there may be some beneficial role of isoflavones in maintenance of bone mass in women, there is currently no peer-reviewed data indicating that ipriflavone or methoxyisoflavone supplementation affects exercise, body composition, or training adaptations
Zinc-magnesium aspartate (ZMA)	Zinc and magnesium deficiency have been reported to reduce the production of testosterone and insulin-like growth factor (IGF-1). Athletes have been reported to have lower zinc and magnesium status. ZMA supplementation has been theorized to increase testosterone and IGF-1 leading to greater recovery, anabolism, and strength during training	One study reported that a zinc–magnesium formulation increased testosterone and IGF-1 (two anabolic hormones) leading to greater gains in strength in football players participating in spring training [238]. However, a recent study conducted in our lab was unable to reproduce these findings. More research is needed to evaluate the role of ZMA on body composition and strength during training

Adapted from Kreider et al. [2]

This is equivalent to ingesting 2–3 cups of premium coffee. Since caffeine serves as a mild diuretic, there has been some concern whether caffeine intake prior to exercise may hasten dehydration. However, recent studies indicate that caffeine does not appear to influence hydration status in active individuals who follow normal fluid intake guidelines [6].

Echinacea is an herb that has been reported to enhance immune function and decrease the severity, duration, and incidence of colds and upper respiratory tract infections. Since intense training may compromise immune function in athletes, some have suggested that *Echinacea* supplementation during heavy training may decrease the incidence of colds and infections, particularly when combined with vitamin C and/or zinc. Although there are data to support the immuno-enhancing effects of *Echinacea*, we are not aware of any studies that have determined whether use of *Echinacea* in athletes helps maintain immune function during intense training.

19.5 Summary

Dietary and nutritional practices of athletes can significantly affect exercise performance capacity. In order to optimize performance, athletes should (1) eat enough calories to offset energy expenditure (typically 50–80 kcal/kg/day); (2) consume the proper amount of carbohydrate (5–10 g/kg/day), protein (1.5–2.0 g/kg/day), and fat (0.5–1.5 g/kg/day); (3) ingest meals and snacks at appropriate time intervals prior to, during, and/or following exercise in order to provide energy for exercise as well as promote recovery following exercise; and (4) only consider using nutritional supplements that have been found to be an effective and safe means to improve performance capacity.

Table 19.8 presents the ISSN's recommendation regarding nutrients that have been shown to be effective; possibly effective; too early to tell; and apparently ineffective or potentially dangerous [2]. The foundation for good performance begins with a good diet and intelligent training. For strength/power athletes, research has indicated that water, carbohydrate, postexercise carbohydrate/protein intake, creatine monohydrate, sodium phosphate, and possibly sodium bicarbonate may have the greatest impact on optimizing performance and/or training adaptations. For endurance athletes, research suggests that carbohydrate loading, water/GES sports drinks, caffeine, sodium phosphate loading, and possibly use of glycerol in an attempt to hyper-hydrate prior to exercise may offer some ergogenic value. Sports drinks, carbohydrate, postexercise carbohydrate with protein or EAA, creatine, and HMB have been reported to help athletes tolerate exercise and training. In addition, postexercise carbohydrate, protein, EAA, and glutamine as well as vitamin C, zinc, and *Echinacea* may help athletes maintain a healthy immune system during training. Use of these strategies can help optimize performance and/or help athletes tolerate intense periods of training.

Table 19.8 Summary of categorization of dietary supplements based on available literature

Category	Muscle building supplements	Weight loss supplements	Performance enhancement
Apparently effective and generally safe	Weight-gain powders Creatine Protein EAA	Low-calorie foods, MRPs, and RTDs Ephedra, caffeine, and salicin-containing thermogenic supplements taken at recommended doses in appropriate populations (ephedra banned by FDA)	Water and sports drinks Carbohydrate Creatine Sodium phosphate Sodium bicarbonate Caffeine β-Alanine
Possibly effective	HMB (untrained individuals initiating training) BCAA	High-fiber diets Calcium Green tea extract CLA	Postexercise carbohydrate and protein EAA BCAA HMB Glycerol
Too early to tell	α-Ketoglutarate α-Ketoisocaproate Ecdysterones GHRP and secretagogues Ornithine Zinc/magnesium aspartate	Gymnema sylvestre, chitosan Phosphatidyl choline Betaine Coleus forskohlii DHEA Psychotropic nutrients/herbs	Medium chain triglycerides
Apparently not effective and/or dangerous	Glutamine Smilax Isoflavones Sulfo-polysaccharides (myostatin inhibitors) Boron Chromium CLA Gamma oryzanol Prohormones Tribulus terrestris Vanadyl sulfate (vanadium)	Calcium pyruvate Chromium (nondiabetics) HCA L-Carnitine Phosphates Herbal diuretics	Glutamine Ribose Inosine

Reprinted with permission from Kreider et al. [2]

Acknowledgments This chapter represents an update to a number of articles and book chapters researchers in the Exercise and Sport Nutrition Laboratory have developed over the years. In addition, the chapter incorporates guidelines developed as position stands by a number of colleagues for the International Society of Sports Nutrition. While we could not acknowledge everyone's contributions to this work, the authors would like to thank all of the students, colleagues, and research participants who have contributed to the research and exercise and sport nutrition guidelines presented in this chapter.

References

1. Kreider RB, Fry AC, O'Toole ML. Overtraining in sport. Champaign: Human Kinetics Publishers; 1998.
2. Kreider RB, Wilborn CD, Taylor L, et al. ISSN exercise & sport nutrition review: research & recommendations. J Int Soc Sports Nutr. 2010;7:7.
3. Williams M. Nutrition for health fitness and sport. 6th ed. Dubuque: McGraw-Hill; 2002.
4. Kreider R, Leutholtz B, Katch F, Katch V. Exercise and sport nutrition. Santa Barbara: Fitness Technologies Press; 2009. http://www.ExerciseAndSportNutrition.com.
5. Kreider RB, Campbell B. Protein for exercise and recovery. Phys Sportsmed. 2009;37:13–21.
6. Goldstein ER, Ziegenfuss T, Kalman D, et al. International society of sports nutrition position stand: caffeine and performance. J Int Soc Sports Nutr. 2010;7:5.
7. Kerksick C, Harvey T, Stout J, et al. International Society of Sports Nutrition position stand: nutrient timing. J Int Soc Sports Nutr. 2008;5:17.
8. Campbell B, Kreider RB, Ziegenfuss T, et al. International Society of Sports Nutrition position stand: protein and exercise. J Int Soc Sports Nutr. 2007;4:8.
9. Buford TW, Kreider RB, Stout JR, Greenwood M, Campbell B, Spano M, et al. International Society of Sports Nutrition position stand: creatine supplementation and exercise. J Int Soc Sports Nutr. 2007;4:6.
10. Berning JR. Energy intake, diet, and muscle wasting. In: Kreider RB, Fry AC, O'Toole ML, editors. Overtraining in sport. Champaign: Human Kinetics; 1998. p. 275–88.
11. Sherman WM, Jacobs KA, Leenders N. Carbohydrate metabolism during endurance exercise. In: Kreider RB, Fry AC, O'Toole ML, editors. Overtraining in sport. Champaign: Human Kinetics Publishers; 1998. p. 289–308.
12. Brouns F, Saris WH, Beckers E, et al. Metabolic changes induced by sustained exhaustive cycling and diet manipulation. Int J Sports Med. 1989;10 Suppl 1:S49–62.
13. Brouns F, Saris WH, Stroecken J, et al. Eating, drinking, and cycling. A controlled Tour de France simulation study, Part II. Effect of diet manipulation. Int J Sports Med. 1989;10 Suppl 1:S41–8.
14. Kreider RB. Physiological considerations of ultraendurance performance. Int J Sport Nutr. 1991;1:3–27.
15. Kreider RB. Nutritional considerations of overtraining. In: Stout JR, Antonio J, editors. Sport supplements: a complete guide to physique and athletic enhancement. Baltimore, MD: Lippincott, Williams & Wilkins; 2001. p. 199–208.
16. Harger-Domitrovich SG, McClaughry AE, Gaskill SE, Ruby BC. Exogenous carbohydrate spares muscle glycogen in men and women during 10 h of exercise. Med Sci Sports Exerc. 2007;39:2171–9.
17. Rodriguez NR, Di Marco NM, Langley S. American College of Sports Medicine position stand. Nutrition and athletic performance. Med Sci Sports Exerc. 2009;41:709–31.
18. Rodriguez NR, DiMarco NM, Langley S. Position of the American Dietetic Association, Dietitians of Canada, and the American College of Sports Medicine: nutrition and athletic performance. J Am Diet Assoc. 2009;109:509–27.
19. Sawka MN, Burke LM, Eichner ER, Maughan RJ, Montain SJ, Stachenfeld NS. American College of Sports Medicine position stand. Exercise and fluid replacement. Med Sci Sports Exerc. 2007;39:377–90.
20. Currell K, Jeukendrup AE. Superior endurance performance with ingestion of multiple transportable carbohydrates. Med Sci Sports Exerc. 2008;40:275–81.
21. Earnest CP, Lancaster SL, Rasmussen CJ, et al. Low vs. high glycemic index carbohydrate gel ingestion during simulated 64-km cycling time trial performance. J Strength Cond Res. 2004;18:466–72.
22. Jeukendrup AE, Moseley L. Multiple transportable carbohydrates enhance gastric emptying and fluid delivery. Scand J Med Sci Sports. 2010;20:112–21.

23. Jentjens RL, Jeukendrup AE. Effects of pre-exercise ingestion of trehalose, galactose and glucose on subsequent metabolism and cycling performance. Eur J Appl Physiol. 2003;88:459–65.
24. Jentjens RL, Venables MC, Jeukendrup AE. Oxidation of exogenous glucose, sucrose, and maltose during prolonged cycling exercise. J Appl Physiol. 2004;96:1285–91.
25. Venables MC, Brouns F, Jeukendrup AE. Oxidation of maltose and trehalose during prolonged moderate-intensity exercise. Med Sci Sports Exerc. 2008;40:1653–9.
26. Achten J, Jentjens RL, Brouns F, Jeukendrup AE. Exogenous oxidation of isomaltulose is lower than that of sucrose during exercise in men. J Nutr. 2007;137:1143–8.
27. Kreider RB. Dietary supplements and the promotion of muscle growth with resistance exercise. Sports Med. 1999;27:97–110.
28. Lemon PW, Tarnopolsky MA, MacDougall JD, Atkinson SA. Protein requirements and muscle mass/strength changes during intensive training in novice bodybuilders. J Appl Physiol. 1992;73:767–75.
29. Tarnopolsky MA, Atkinson SA, MacDougall JD, Chesley A, Phillips S, Schwarcz HP. Evaluation of protein requirements for trained strength athletes. J Appl Physiol. 1992;73:1986–95.
30. Tarnopolsky MA, MacDougall JD, Atkinson SA. Influence of protein intake and training status on nitrogen balance and lean body mass. J Appl Physiol. 1988;64:187–93.
31. Bucci L, Unlu L. Proteins and amino acid supplements in exercise and sport. In: Driskell J, Wolinsky I, editors. Energy-yielding macronutrients and energy metabolism in sports nutrition. Boca Raton: CRC Press; 2000. p. 191–212.
32. Boirie Y, Dangin M, Gachon P, Vasson MP, Maubois JL, Beaufrere B. Slow and fast dietary proteins differently modulate postprandial protein accretion. Proc Natl Acad Sci U S A. 1997;94:14930–5.
33. Boirie Y, Gachon P, Corny S, Fauquant J, Maubois JL, Beaufrere B. Acute postprandial changes in leucine metabolism as assessed with an intrinsically labeled milk protein. Am J Physiol. 1996;271(6 Pt 1):E1083–91.
34. Dangin M, Boirie Y, Garcia-Rodenas C, et al. The digestion rate of protein is an independent regulating factor of postprandial protein retention. Am J Physiol Endocrinol Metab. 2001;280:E340–8.
35. Kreider RB, Kleiner SM. Protein supplements for athletes: need vs. convenience. Your Patient Fitness. 2000;14:12–8.
36. Venkatraman JT, Leddy J, Pendergast D. Dietary fats and immune status in athletes: clinical implications. Med Sci Sports Exerc. 2000;32(7 Suppl):S389–95.
37. Dorgan JF, Judd JT, Longcope C, et al. Effects of dietary fat and fiber on plasma and urine androgens and estrogens in men: a controlled feeding study. Am J Clin Nutr. 1996;64:850–5.
38. Hamalainen EK, Adlercreutz H, Puska P, Pietinen P. Decrease of serum total and free testosterone during a low-fat high-fibre diet. J Steroid Biochem. 1983;18:369–70.
39. Fry AC, Kraemer WJ, Ramsey LT. Pituitary-adrenal-gonadal responses to high-intensity resistance exercise overtraining. J Appl Physiol. 1998;85:2352–9.
40. Pirozzo S, Summerbell C, Cameron C, Glasziou P. Should we recommend low-fat diets for obesity? Obes Rev. 2003;4:83–90.
41. Hu FB, Manson JE, Willett WC. Types of dietary fat and risk of coronary heart disease: a critical review. J Am Coll Nutr. 2001;20:5–19.
42. Vessby B. Dietary fat, fatty acid composition in plasma and the metabolic syndrome. Curr Opin Lipidol. 2003;14:15–9.
43. Cade JR, Reese RH, Privette RM, Hommen NM, Rogers JL, Fregly MJ. Dietary intervention and training in swimmers. Eur J Appl Physiol Occup Physiol. 1991;63:210–5.
44. Carli G, Bonifazi M, Lodi L, Lupo C, Martelli G, Viti A. Changes in the exercise-induced hormone response to branched chain amino acid administration. Eur J Appl Physiol Occup Physiol. 1992;64:272–7.
45. Burke LM. Nutrition for post-exercise recovery. Aust J Sci Med Sport. 1997;29:3–10.
46. Burke LM. Nutritional needs for exercise in the heat. Comp Biochem Physiol A Mol Integr Physiol. 2001;128:735–48.

47. Nieman DC. Nutrition, exercise, and immune system function. Clin Sports Med. 1999;18: 537–48.
48. Nieman DC, Fagoaga OR, Butterworth DE, et al. Carbohydrate supplementation affects blood granulocyte and monocyte trafficking but not function after 2.5 h or running. Am J Clin Nutr. 1997;66:153–9.
49. Tarnopolsky MA, Bosman M, Macdonald JR, Vandeputte D, Martin J, Roy BD. Postexercise protein-carbohydrate and carbohydrate supplements increase muscle glycogen in men and women. J Appl Physiol. 1997;83:1877–83.
50. Zawadzki KM, Yaspelkis III BB, Ivy JL. Carbohydrate-protein complex increases the rate of muscle glycogen storage after exercise. J Appl Physiol. 1992;72:1854–9.
51. Kraemer WJ, Volek JS, Bush JA, Putukian M, Sebastianelli WJ. Hormonal responses to consecutive days of heavy-resistance exercise with or without nutritional supplementation. J Appl Physiol. 1998;85:1544–55.
52. Campbell B, Kreider RB. Conjugated linoleic acids. Curr Sports Med Rep. 2008;7:237–41.
53. Volek JS, Kraemer WJ, Rubin MR, Gomez AL, Ratamess NA, Gaynor P. L-Carnitine L-tartrate supplementation favorably affects markers of recovery from exercise stress. Am J Physiol Endocrinol Metab. 2002;282:E474–82.
54. Culbertson J, Kreider R, Greenwood M, Cooke M. Effects of beta-alanine on muscle carnosine and exercise: a review of the current literature. Nutrients. 2010;2:75–98.
55. Fletcher RH, Fairfield KM. Vitamins for chronic disease prevention in adults: clinical applications. JAMA. 2002;287(23):3127–9.
56. Fairfield KM, Fletcher RH. Vitamins for chronic disease prevention in adults: scientific review. JAMA. 2002;287:3116–26.
57. Kreider RB, Miller GW, Schenck D, et al. Effects of phosphate loading on metabolic and myocardial responses to maximal and endurance exercise. Int J Sport Nutr. 1992;2:20–47.
58. Kreider RB, Miller GW, Williams MH, Somma CT, Nasser TA. Effects of phosphate loading on oxygen uptake, ventilatory anaerobic threshold, and run performance. Med Sci Sports Exerc. 1990;22:250–6.
59. Jeukendrup AE, Currell K, Clarke J, Cole J, Blannin AK. Effect of beverage glucose and sodium content on fluid delivery. Nutr Metab (Lond). 2009;6:9.
60. Rehrer NJ. Fluid and electrolyte balance in ultra-endurance sport. Sports Med. 2001;31: 701–15.
61. Sawka MN, Montain SJ. Fluid and electrolyte supplementation for exercise heat stress. Am J Clin Nutr. 2000;72(2 Suppl):564S–72.
62. Maughan RJ, Noakes TD. Fluid replacement and exercise stress. A brief review of studies on fluid replacement and some guidelines for the athlete. Sports Med. 1991;12:16–31.
63. Murakami C, Myoga K, Kasai R, et al. Screening of plant constituents for effect on glucose transport activity in Ehrlich ascites tumour cells. Chem Pharm Bull (Tokyo). 1993;41:2129–31.
64. Fukushima M, Matsuyama F, Ueda N, et al. Effect of corosolic acid on postchallenge plasma glucose levels. Diabetes Res Clin Pract. 2006;73:174–1747.
65. Stanko RT, Robertson RJ, Galbreath RW, Reilly Jr JJ, Greenawalt KD, Goss FL. Enhanced leg exercise endurance with a high-carbohydrate diet and dihydroxyacetone and pyruvate. J Appl Physiol. 1990;69:1651–6.
66. Myers J, Atwood JE, Forbes S, Evans B, Froelicher V. Effect of fructose 1,6-diphosphate on exercise capacity in patients with peripheral vascular disease. Int J Sports Med. 1990;11: 259–62.
67. Convertino VA, Armstrong LE, Coyle EF, et al. American College of Sports Medicine position stand. Exercise and fluid replacement. Med Sci Sports Exerc. 1996;28:i–vii.
68. Bates SH, Jones RB, Bailey CJ. Insulin-like effect of pinitol. Br J Pharmacol. 2000;130: 1944–8.
69. Greenwood M, Kreider RB, Rasmussen C, Almada AL, Earnest CP. D-pinitol augments whole body creatine retention in man. J Exerc Physiol Online. 2001;4(4):41–7. http://www.css.edu/users/tboone2/asep/GreenwoodNOVEMBER2001.pdf.

70. Kerksick CM, Wilborn CD, Campbell WI, et al. The effects of creatine monohydrate supplementation with and without D-pinitol on resistance training adaptations. J Strength Cond Res. 2009;23:2673–82.
71. Walaszek Z, Szemraj J, Narog M, et al. Metabolism, uptake, and excretion of a D-glucaric acid salt and its potential use in cancer prevention. Cancer Detect Prev. 1997;21:178–90.
72. Fahey T. The effects of ingesting polylactate or glucose polymer drinks during prolonged exercise. Int J Sport Nutr. 1991;1:49–56.
73. Swensen T, Crater G, Bassett Jr DR, Howley ET. Adding polylactate to a glucose polymer solution does not improve endurance. Int J Sports Med. 1994;15:430–4.
74. Kalman D, Colker CM, Wilets I, Roufs JB, Antonio J. The effects of pyruvate supplementation on body composition in overweight individuals. Nutrition. 1999;15:337–40.
75. Stanko RT, Arch JE. Inhibition of regain in body weight and fat with addition of 3-carbon compounds to the diet with hyperenergetic refeeding after weight reduction. Int J Obes Relat Metab Disord. 1996;20:925–30.
76. Gross M, Kormann B, Zollner N. Ribose administration during exercise: effects on substrates and products of energy metabolism in healthy subjects and a patient with myoadenylate deaminase deficiency. Klin Wochenschr. 1991;69:151–5.
77. Hegewald MG, Palac RT, Angello DA, Perlmutter NS, Wilson RA. Ribose infusion accelerates thallium redistribution with early imaging compared with late 24-hour imaging without ribose. J Am Coll Cardiol. 1991;18:1671–81.
78. Wagner DR, Gresser U, Zollner N. Effects of oral ribose on muscle metabolism during bicycle ergometer in AMPD-deficient patients. Ann Nutr Metab. 1991;35:297–302.
79. Kreider RB, Melton C, Greenwood M, et al. Effects of oral d-ribose supplementation on anaerobic capacity and selected metabolic markers in healthy males. Int J Sport Nutr Exerc Metab. 2003;13:87–96.
80. Op't Eijnde B, Van Leemputte M, Brouns F, et al. No effects of oral ribose supplementation on repeated maximal exercise and de novo ATP resynthesis. J Appl Physiol. 2001;91:2275–81.
81. Pariza MW, Park Y, Cook ME. The biologically active isomers of conjugated linoleic acid. Prog Lipid Res. 2001;40:283–98.
82. DeLany JP, West DB. Changes in body composition with conjugated linoleic acid. J Am Coll Nutr. 2000;19:487S–93.
83. Park Y, Albright KJ, Liu W, Storkson JM, Cook ME, Pariza MW. Effect of conjugated linoleic acid on body composition in mice. Lipids. 1997;32:853–8.
84. Blankson H, Stakkestad JA, Fagertun H, Thom E, Wadstein J, Gudmundsen O. Conjugated linoleic acid reduces body fat mass in overweight and obese humans. J Nutr. 2000;130:2943–8.
85. Gaullier JM, Berven G, Blankson H, Gudmundsen O. Clinical trial results support a preference for using CLA preparations enriched with two isomers rather than four isomers in human studies. Lipids. 2002;37:1019–25.
86. Inder WJ, Swanney MP, Donald RA, Prickett TC, Hellemans J. The effect of glycerol and desmopressin on exercise performance and hydration in triathletes. Med Sci Sports Exerc. 1998;30:1263–9.
87. Magal M, Webster MJ, Sistrunk LE, Whitehead MT, Evans RK, Boyd JC. Comparison of glycerol and water hydration regimens on tennis-related performance. Med Sci Sports Exerc. 2003;35:150–6.
88. Meyer LG, Horrigan Jr DJ, Lotz WG. Effects of three hydration beverages on exercise performance during 60 hours of heat exposure. Aviat Space Environ Med. 1995;66:1052–7.
89. Kavouras SA, Armstrong LE, Maresh CM, et al. Rehydration with glycerol: endocrine, cardiovascular, and thermoregulatory responses during exercise in the heat. J Appl Physiol. 2006;100:442–50.
90. Brass EP, Hiatt WR. Carnitine metabolism during exercise. Life Sci. 1994;54:1383–93.
91. Villani RG, Gannon J, Self M, Rich PA. L-carnitine supplementation combined with aerobic training does not promote weight loss in moderately obese women. Int J Sport Nutr Exerc Metab. 2000;10:199–207.

92. Angus DJ, Hargreaves M, Dancey J, Febbraio MA. Effect of carbohydrate or carbohydrate plus medium-chain triglyceride ingestion on cycling time trial performance. J Appl Physiol. 2000;88:113–9.
93. Calabrese C, Myer S, Munson S, Turet P, Birdsall TC. A cross-over study of the effect of a single oral feeding of medium chain triglyceride oil vs. canola oil on post-ingestion plasma triglyceride levels in healthy men. Altern Med Rev. 1999;4:23–8.
94. Goedecke JH, Elmer-English R, Dennis SC, Schloss I, Noakes TD, Lambert EV. Effects of medium-chain triaclyglycerol ingested with carbohydrate on metabolism and exercise performance. Int J Sport Nutr. 1999;9:35–47.
95. Jeukendrup AE, Thielen JJ, Wagenmakers AJ, Brouns F, Saris WH. Effect of medium-chain triacylglycerol and carbohydrate ingestion during exercise on substrate utilization and subsequent cycling performance. Am J Clin Nutr. 1998;67:397–404.
96. Misell LM, Lagomarcino ND, Schuster V, Kern M. Chronic medium-chain triacylglycerol consumption and endurance performance in trained runners. J Sports Med Phys Fitness. 2001;41:210–5.
97. Goedecke JH, Clark VR, Noakes TD, Lambert EV. The effects of medium-chain triacylglycerol and carbohydrate ingestion on ultra-endurance exercise performance. Int J Sport Nutr Exerc Metab. 2005;15:15–27.
98. Burke LM, Kiens B, Ivy JL. Carbohydrates and fat for training and recovery. J Sports Sci. 2004;22:15–30.
99. Nosaka N, Suzuki Y, Nagatoishi A, Kasai M, Wu J, Taguchi M. Effect of ingestion of medium-chain triacylglycerols on moderate- and high-intensity exercise in recreational athletes. J Nutr Sci Vitaminol (Tokyo). 2009;55:120–5.
100. Brilla LR, Landerholm TE. Effect of fish oil supplementation and exercise on serum lipids and aerobic fitness. J Sports Med Phys Fitness. 1990;30:173–80.
101. Hammarqvist F, Wernerman J, Ali R, Vinnars E. Effects of an amino acid solution enriched with either branched chain amino acids or ornithine-alpha-ketoglutarate on the postoperative intracellular amino acid concentration of skeletal muscle. Br J Surg. 1990;77:214–8.
102. Wernerman J, Hammarqvist F, Vinnars E. Alpha-ketoglutarate and postoperative muscle catabolism. Lancet. 1990;335:701–3.
103. Antonio J, Stout JR. Sport supplements. Philadelphia, PA: Lippincott, Williams and Wilkins; 2001.
104. Mitch WE, Walser M, Sapir DG. Nitrogen sparing induced by leucine compared with that induced by its keto analogue, alpha-ketoisocaproate, in fasting obese man. J Clin Invest. 1981;67:553–62.
105. Van Koevering M, Nissen S. Oxidation of leucine and alpha-ketoisocaproate to beta-hydroxy-beta-methylbutyrate in vivo. Am J Physiol. 1992;262(1 Pt 1):E27–31.
106. Procopio M, Maccario M, Savio P, et al. GH response to GHRH combined with pyridostigmine or arginine in different conditions of low somatotrope secretion in adulthood: obesity and Cushing's syndrome in comparison with hypopituitarism. Panminerva Med. 1998;40:13–7.
107. Wu G, Meininger CJ. Arginine nutrition and cardiovascular function. J Nutr. 2000;130:2626–9.
108. Bloomer RJ, Williams SA, Canale RE, Farney TM, Kabir MM. Acute effect of nitric oxide supplement on blood nitrate/nitrite and hemodynamic variables in resistance trained men. J Strength Cond Res. 2010;24:2587–92.
109. Bloomer RJ, Farney TM, Trepanowski JF, McCarthy CG, Canale RE, Schilling BK. Comparison of pre-workout nitric oxide stimulating dietary supplements on skeletal muscle oxygen saturation, blood nitrate/nitrite, lipid peroxidation, and upper body exercise performance in resistance trained men. J Int Soc Sports Nutr. 2010;7:16.
110. Bloomer RJ, Tschume LC, Smith WA. Glycine propionyl-L-carnitine modulates lipid peroxidation and nitric oxide in human subjects. Int J Vitam Nutr Res. 2009;79:131–41.
111. Colombani P, Wenk C, Kunz I, et al. Effects of L-carnitine supplementation on physical performance and energy metabolism of endurance-trained athletes: a double-blind crossover field study. Eur J Appl Physiol Occup Physiol. 1996;73:434–9.

112. Colombani PC, Bitzi R, Frey-Rindova P, et al. Chronic arginine aspartate supplementation in runners reduces total plasma amino acid level at rest and during a marathon run. Eur J Nutr. 1999;38:263–70.

113. Tuttle JL, Potteiger JA, Evans BW, Ozmun JC. Effect of acute potassium-magnesium aspartate supplementation on ammonia concentrations during and after resistance training. Int J Sport Nutr. 1995;5:102–9.

114. Coombes JS, McNaughton LR. Effects of branched-chain amino acid supplementation on serum creatine kinase and lactate dehydrogenase after prolonged exercise. J Sports Med Phys Fitness. 2000;40:240–6.

115. Bigard AX, Lavier P, Ullmann L, Legrand H, Douce P, Guezennec CY. Branched-chain amino acid supplementation during repeated prolonged skiing exercises at altitude. Int J Sport Nutr. 1996;6:295–306.

116. Candeloro N, Bertini I, Melchiorri G, De Lorenzo A. [Effects of prolonged administration of branched-chain amino acids on body composition and physical fitness]. Minerva Endocrinol. 1995;20:217–23.

117. Schena F, Guerrini F, Tregnaghi P, Kayser B. Branched-chain amino acid supplementation during trekking at high altitude. The effects on loss of body mass, body composition, and muscle power. Eur J Appl Physiol Occup Physiol. 1992;65:394–8.

118. Williams MH. Facts and fallacies of purported ergogenic amino acid supplements. Clin Sports Med. 1999;18:633–49.

119. Kreider RB. Effects of creatine supplementation on performance and training adaptations. Mol Cell Biochem. 2003;244:89–94.

120. Kreider RB, Melton C, Rasmussen CJ, et al. Long-term creatine supplementation does not significantly affect clinical markers of health in athletes. Mol Cell Biochem. 2003;244:95–104.

121. Volek JS, Duncan ND, Mazzetti SA, et al. Performance and muscle fiber adaptations to creatine supplementation and heavy resistance training. Med Sci Sports Exerc. 1999;31:1147–56.

122. Willoughby DS, Rosene J. Effects of oral creatine and resistance training on myosin heavy chain expression. Med Sci Sports Exerc. 2001;33:1674–81.

123. Willoughby DS, Rosene JM. Effects of oral creatine and resistance training on myogenic regulatory factor expression. Med Sci Sports Exerc. 2003;35:923–9.

124. Graham AS, Hatton RC. Creatine: a review of efficacy and safety. J Am Pharm Assoc (Wash). 1999;39:803–10.

125. Juhn MS, Tarnopolsky M. Potential side effects of oral creatine supplementation: a critical review. Clin J Sport Med. 1998;8:298–304.

126. Schilling BK, Stone MH, Utter A, Kearney JT, Johnson M, Coglianese R, et al. Creatine supplementation and health variables: a retrospective study. Med Sci Sports Exerc. 2001;33:183–8.

127. Taes YE, Delanghe JR, Wuyts B, Van De Voorde J, Lameire NH. Creatine supplementation does not affect kidney function in an animal model with pre-existing renal failure. Nephrol Dial Transplant. 2003;18:258–64.

128. Greenwood L, Greenwood M, Kreider R, et al. Effects of creatine supplementation on the incidence of cramping/injury during eighteen weeks of division I football training/competition. Med Sci Sports Exerc. 2002;34:S146.

129. Watsford ML, Murphy AJ, Spinks WL, Walshe AD. Creatine supplementation and its effect on musculotendinous stiffness and performance. J Strength Cond Res. 2003;17:26–33.

130. Greenwood M, Kreider RB, Greenwood L, Byars A. Cramping and injury incidence in collegiate football players are reduced by creatine supplementation. J Athl Train. 2003;38:216–9.

131. Greenwood M, Kreider RB, Melton C, et al. Creatine supplementation during college football training does not increase the incidence of cramping or injury. Mol Cell Biochem. 2003;244:83–8.

132. Tipton KD, Borsheim E, Wolf SE, Sanford AP, Wolfe RR. Acute response of net muscle protein balance reflects 24-h balance after exercise and amino acid ingestion. Am J Physiol Endocrinol Metab. 2003;284:E76–89.

133. Wolfe RR. Regulation of muscle protein by amino acids. J Nutr. 2002;132:3219S–24.
134. Biolo G, Williams BD, Fleming RY, Wolfe RR. Insulin action on muscle protein kinetics and amino acid transport during recovery after resistance exercise. Diabetes. 1999;48:949–57.
135. Miller SL, Tipton KD, Chinkes DL, Wolf SE, Wolfe RR. Independent and combined effects of amino acids and glucose after resistance exercise. Med Sci Sports Exerc. 2003;35:449–55.
136. Rasmussen BB, Tipton KD, Miller SL, Wolf SE, Wolfe RR. An oral essential amino acid-carbohydrate supplement enhances muscle protein anabolism after resistance exercise. J Appl Physiol. 2000;88:386–92.
137. Rasmussen BB, Wolfe RR, Volpi E. Oral and intravenously administered amino acids produce similar effects on muscle protein synthesis in the elderly. J Nutr Health Aging. 2002;6: 358–62.
138. Tipton KD, Rasmussen BB, Miller SL, et al. Timing of amino acid-carbohydrate ingestion alters anabolic response of muscle to resistance exercise. Am J Physiol Endocrinol Metab. 2001;281:E197–206.
139. Ferrando AA, Paddon-Jones D, Hays NP, et al. EAA supplementation to increase nitrogen intake improves muscle function during bed rest in the elderly. Clin Nutr. 2010;29:18–23.
140. Katsanos CS, Aarsland A, Cree MG, Wolfe RR. Muscle protein synthesis and balance responsiveness to essential amino acids ingestion in the presence of elevated plasma free fatty acid concentrations. J Clin Endocrinol Metab. 2009;94:2984–90.
141. Katsanos CS, Chinkes DL, Paddon-Jones D, Zhang XJ, Aarsland A, Wolfe RR. Whey protein ingestion in elderly persons results in greater muscle protein accrual than ingestion of its constituent essential amino acid content. Nutr Res. 2008;28:651–8.
142. Paddon-Jones D, Sheffield-Moore M, Katsanos CS, Zhang XJ, Wolfe RR. Differential stimulation of muscle protein synthesis in elderly humans following isocaloric ingestion of amino acids or whey protein. Exp Gerontol. 2006;41:215–9.
143. Katsanos CS, Kobayashi H, Sheffield-Moore M, Aarsland A, Wolfe RR. A high proportion of leucine is required for optimal stimulation of the rate of muscle protein synthesis by essential amino acids in the elderly. Am J Physiol Endocrinol Metab. 2006;291:E381–7.
144. Borsheim E, Kobayashi H, Traber DL, Wolfe RR. Compartmental distribution of amino acids during hemodialysis-induced hypoaminoacidemia. Am J Physiol Endocrinol Metab. 2006;290:E643–52.
145. Paddon-Jones D, Wolfe RR, Ferrando AA. Amino acid supplementation for reversing bed rest and steroid myopathies. J Nutr. 2005;135:1809S–12.
146. Katsanos CS, Kobayashi H, Sheffield-Moore M, Aarsland A, Wolfe RR. Aging is associated with diminished accretion of muscle proteins after the ingestion of a small bolus of essential amino acids. Am J Clin Nutr. 2005;82:1065–73.
147. Paddon-Jones D, Sheffield-Moore M, Zhang XJ, et al. Amino acid ingestion improves muscle protein synthesis in the young and elderly. Am J Physiol Endocrinol Metab. 2004;286: E321–8.
148. Volpi E, Kobayashi H, Sheffield-Moore M, Mittendorfer B, Wolfe RR. Essential amino acids are primarily responsible for the amino acid stimulation of muscle protein anabolism in healthy elderly adults. Am J Clin Nutr. 2003;78:250–8.
149. Esmarck B, Andersen JL, Olsen S, Richter EA, Mizuno M, Kjaer M. Timing of postexercise protein intake is important for muscle hypertrophy with resistance training in elderly humans. J Physiol. 2001;535(Pt 1):301–11.
150. Low SY, Taylor PM, Rennie MJ. Responses of glutamine transport in cultured rat skeletal muscle to osmotically induced changes in cell volume. J Physiol. 1996;492:877–85.
151. Rennie MJ, Ahmed A, Khogali SE, Low SY, Hundal HS, Taylor PM. Glutamine metabolism and transport in skeletal muscle and heart and their clinical relevance. J Nutr. 1996;126(Suppl): 1142S–9.
152. Varnier M, Leese GP, Thompson J, Rennie MJ. Stimulatory effect of glutamine on glycogen accumulation in human skeletal muscle. Am J Physiol. 1995;269(2 Pt 1):E309–15.
153. Colker CM. Effects of supplemental protein on body composition and muscular strength in healthy athletic male adults. Curr Ther Res. 2000;61:19–28.

154. Candow DG, Chilibeck PD, Burke DG, Davison KS, Smith-Palmer T. Effect of glutamine supplementation combined with resistance training in young adults. Eur J Appl Physiol. 2001;86:142–9.
155. Gallagher PM, Carrithers JA, Godard MP, Schulze KE, Trappe SW. Beta-hydroxy-beta-methylbutyrate ingestion, Part I: effects on strength and fat free mass. Med Sci Sports Exerc. 2000;32:2109–15.
156. Gallagher PM, Carrithers JA, Godard MP, Schulze KE, Trappe SW. Beta-hydroxy-beta-methylbutyrate ingestion, Part II: effects on hematology, hepatic and renal function. Med Sci Sports Exerc. 2000;32:2116–9.
157. Nissen S, Sharp R, Ray M, et al. Effect of leucine metabolite beta-hydroxy-beta-methylbutyrate on muscle metabolism during resistance-exercise training. J Appl Physiol. 1996;81:2095–104.
158. Panton LB, Rathmacher JA, Baier S, Nissen S. Nutritional supplementation of the leucine metabolite beta-hydroxy-beta- methylbutyrate (hmb) during resistance training. Nutrition. 2000;16:734–9.
159. Vukovich MD, Stubbs NB, Bohlken RM. Body composition in 70-year-old adults responds to dietary beta-hydroxy-beta-methylbutyrate similarly to that of young adults. J Nutr. 2001;131:2049–52.
160. Knitter AE, Panton L, Rathmacher JA, Petersen A, Sharp R. Effects of beta-hydroxy-beta-methylbutyrate on muscle damage after a prolonged run. J Appl Physiol. 2000;89:1340–4.
161. Jowko E, Ostaszewski P, Jank M, Sacharuk J, Zieniewicz A, Wilczak J, et al. Creatine and beta-hydroxy-beta-methylbutyrate (HMB) additively increase lean body mass and muscle strength during a weight-training program. Nutrition. 2001;17:558–66.
162. O'Connor DM, Crowe MJ. Effects of beta-hydroxy-beta-methylbutyrate and creatine monohydrate supplementation on the aerobic and anaerobic capacity of highly trained athletes. J Sports Med Phys Fitness. 2003;43:64–8.
163. Kreider RB, Ferreira M, Wilson M, Almada AL. Effects of calcium beta-hydroxy-beta-methylbutyrate (HMB) supplementation during resistance-training on markers of catabolism, body composition and strength. Int J Sports Med. 1999;20:503–9.
164. Ransone J, Neighbors K, Lefavi R, Chromiak J. The effect of beta-hydroxy beta-methylbutyrate on muscular strength and body composition in collegiate football players. J Strength Cond Res. 2003;17:34–9.
165. Slater G, Jenkins D, Logan P, et al. Beta-hydroxy-beta-methylbutyrate (HMB) supplementation does not affect changes in strength or body composition during resistance training in trained men. Int J Sport Nutr Exerc Metab. 2001;11:384–96.
166. Chetlin RD, Yeater RA, Ullrich IH, Hornsby WG, Malanga CJ, Byrner RW. The effect of ornithine alpha-ketoglutarate (OKG) on healthy, weight trained men. J Exerc Physiol Online. 2000;3(4). http://www.css.edu/users/tboone2/asep/ChetlinV.pdf.
167. Segura R, Ventura JL. Effect of L-tryptophan supplementation on exercise performance. Int J Sports Med. 1988;9:301–5.
168. Stensrud T, Ingjer F, Holm H, Stromme SB. L-tryptophan supplementation does not improve running performance. Int J Sports Med. 1992;13:481–5.
169. Williams MH. Vitamin supplementation and athletic performance. Int J Vitam Nutr Res Suppl. 1989;30:163–91.
170. Reid IR. Therapy of osteoporosis: calcium, vitamin D, and exercise. Am J Med Sci. 1996;312:278–86.
171. Appell HJ, Duarte JA, Soares JM. Supplementation of vitamin E may attenuate skeletal muscle immobilization atrophy. Int J Sports Med. 1997;18:157–60.
172. Goldfarb AH. Antioxidants: role of supplementation to prevent exercise-induced oxidative stress. Med Sci Sports Exerc. 1993;25:232–6.
173. Goldfarb AH. Nutritional antioxidants as therapeutic and preventive modalities in exercise-induced muscle damage. Can J Appl Physiol. 1999;24:249–66.
174. Tiidus PM, Houston ME. Vitamin E status and response to exercise training. Sports Med. 1995;20:12–23.

175. Craciun AM, Wolf J, Knapen MH, Brouns F, Vermeer C. Improved bone metabolism in female elite athletes after vitamin K supplementation. Int J Sports Med. 1998;19:479–84.

176. Fogelholm M, Ruokonen I, Laakso JT, Vuorimaa T, Himberg JJ. Lack of association between indices of vitamin B1, B2, and B6 status and exercise-induced blood lactate in young adults. Int J Sport Nutr. 1993;3:165–76.

177. Alaswad K, O'Keefe Jr JH, Moe RM. Combination drug therapy for dyslipidemia. Curr Atheroscler Rep. 1999;1:44–9.

178. Garg R, Malinow M, Pettinger M, Upson B, Hunninghake D. Niacin treatment increases plasma homocyst(e)ine levels. Am Heart J. 1999;138(6 Pt 1):1082–7.

179. Murray R, Bartoli WP, Eddy DE, Horn MK. Physiological and performance responses to nicotinic-acid ingestion during exercise. Med Sci Sports Exerc. 1995;27:1057–62.

180. Bonke D. Influence of vitamin B1, B6, and B12 on the control of fine motoric movements. Bibl Nutr Dieta. 1986;38:104–9.

181. Bonke D, Nickel B. Improvement of fine motoric movement control by elevated dosages of vitamin B1, B6, and B12 in target shooting. Int J Vitam Nutr Res Suppl. 1989;30:198–204.

182. Van Dyke DC, Stumbo PJ, Mary JB, Niebyl JR. Folic acid and prevention of birth defects. Dev Med Child Neurol. 2002;44:426–9.

183. Mattson MP, Kruman II, Duan W. Folic acid and homocysteine in age-related disease. Ageing Res Rev. 2002;1:95–111.

184. Weston PM, King RF, Goode AW, Williams NS. Diet-induced thermogenesis in patients with gastrointestinal cancer cachexia. Clin Sci (Lond). 1989;77:133–8.

185. Webster MJ. Physiological and performance responses to supplementation with thiamin and pantothenic acid derivatives. Eur J Appl Physiol Occup Physiol. 1998;77:486–91.

186. van der Beek EJ. Vitamin supplementation and physical exercise performance. J Sports Sci. 1991;9(Spec No):77–90.

187. van der Beek EJ, Lowik MR, Hulshof KF, Kistemaker C. Combinations of low thiamin, riboflavin, vitamin B6 and vitamin C intake among Dutch adults (Dutch Nutrition Surveillance System). J Am Coll Nutr. 1994;13:383–91.

188. Nieman DC. Exercise immunology: nutritional countermeasures. Can J Appl Physiol. 2001;26(Suppl):S45–55.

189. Pedersen BK, Bruunsgaard H, Jensen M, Krzywkowski K, Ostrowski K. Exercise and immune function: effect of ageing and nutrition. Proc Nutr Soc. 1999;58:733–42.

190. Petersen EW, Ostrowski K, Ibfelt T, et al. Effect of vitamin supplementation on cytokine response and on muscle damage after strenuous exercise. Am J Physiol Cell Physiol. 2001;280:C1570–5.

191. Ferrando AA, Green NR. The effect of boron supplementation on lean body mass, plasma testosterone levels, and strength in male bodybuilders. Int J Sport Nutr. 1993;3:140–9.

192. Green NR, Ferrando AA. Plasma boron and the effects of boron supplementation in males. Environ Health Perspect. 1994;102 Suppl 7:73–7.

193. Grados F, Brazier M, Kamel S, et al. Effects on bone mineral density of calcium and vitamin D supplementation in elderly women with vitamin D deficiency. Joint Bone Spine. 2003;70:203–8.

194. Zemel MB. Role of dietary calcium and dairy products in modulating adiposity. Lipids. 2003;38:139–246.

195. Zemel MB. Mechanisms of dairy modulation of adiposity. J Nutr. 2003;133:252S–6.

196. Hasten DL, Rome EP, Franks BD, Hegsted M. Effects of chromium picolinate on beginning weight training students. Int J Sport Nutr. 1992;2:343–50.

197. Campbell WW, Joseph LJ, Anderson RA, Davey SL, Hinton J, Evans WJ. Effects of resistive training and chromium picolinate on body composition and skeletal muscle size in older women. Int J Sport Nutr Exerc Metab. 2002;12:125–35.

198. Volpe SL, Huang HW, Larpadisorn K, Lesser II. Effect of chromium supplementation and exercise on body composition, resting metabolic rate and selected biochemical parameters in moderately obese women following an exercise program. J Am Coll Nutr. 2001;20:293–306.

199. Brutsaert TD, Hernandez-Cordero S, Rivera J, Viola T, Hughes G, Haas JD. Iron supplementation improves progressive fatigue resistance during dynamic knee extensor exercise in iron-depleted, nonanemic women. Am J Clin Nutr. 2003;77:441–8.
200. Bohl CH, Volpe SL. Magnesium and exercise. Crit Rev Food Sci Nutr. 2002;42:533–63.
201. Lukaski HC. Magnesium, zinc, and chromium nutrition and athletic performance. Can J Appl Physiol. 2001;26(Suppl):S13–22.
202. Cade R, Conte M, Zauner C, et al. Effects of phosphate loading on 2,3 diphosphoglycerate and maximal oxygen uptake. Med Sci Sports Exerc. 1984;16:263–8.
203. Morton DP, Callister R. Characteristics and etiology of exercise-related transient abdominal pain. Med Sci Sports Exerc. 2000;32:432–8.
204. Noakes TD. Fluid and electrolyte disturbances in heat illness. Int J Sports Med. 1998;19 Suppl 2:S146–9.
205. Shirreffs SM, Armstrong LE, Cheuvront SN. Fluid and electrolyte needs for preparation and recovery from training and competition. J Sports Sci. 2004;22:57–63.
206. Margaritis I, Tessier F, Prou E, Marconnet P, Marini JF. Effects of endurance training on skeletal muscle oxidative capacities with and without selenium supplementation. J Trace Elem Med Biol. 1997;11:37–43.
207. Tessier F, Margaritis I, Richard MJ, Moynot C, Marconnet P. Selenium and training effects on the glutathione system and aerobic performance. Med Sci Sports Exerc. 1995;27:390–6.
208. McCutcheon LJ, Geor RJ. Sweating. Fluid and ion losses and replacement. Vet Clin North Am Equine Pract. 1998;14:75–95.
209. Fawcett JP, Farquhar SJ, Thou T, Shand BI. Oral vanadyl sulphate does not affect blood cells, viscosity or biochemistry in humans. Pharmacol Toxicol. 1997;80:202–6.
210. Fawcett JP, Farquhar SJ, Walker RJ, Thou T, Lowe G, Goulding A. The effect of oral vanadyl sulfate on body composition and performance in weight-training athletes. Int J Sport Nutr. 1996;6:382–90.
211. Gibson RS, Heath AL, Ferguson EL. Risk of suboptimal iron and zinc nutriture among adolescent girls in Australia and New Zealand: causes, consequences, and solutions. Asia Pac J Clin Nutr. 2002;11 Suppl 3:S543–52.
212. Gleeson M, Bishop NC. Elite athlete immunology: importance of nutrition. Int J Sports Med. 2000;21 Suppl 1:S44–50.
213. Singh A, Failla ML, Deuster PA. Exercise-induced changes in immune function: effects of zinc supplementation. J Appl Physiol. 1994;76:2298–303.
214. Williams MH. Ergogenic and ergolytic substances. Med Sci Sports Exerc. 1992;24(9 Suppl):S344–8.
215. Kraemer WJ, Gordon SE, Lynch JM, Pop ME, Clark KL. Effects of multibuffer supplementation on acid-base balance and 2,3- diphosphoglycerate following repetitive anaerobic exercise. Int J Sport Nutr. 1995;5:300–14.
216. McNaughton L, Backx K, Palmer G, Strange N. Effects of chronic bicarbonate ingestion on the performance of high- intensity work. Eur J Appl Physiol Occup Physiol. 1999;80:333–6.
217. Matson LG, Tran ZV. Effects of sodium bicarbonate ingestion on anaerobic performance: a meta-analytic review. Int J Sport Nutr. 1993;3:2–28.
218. Lindh AM, Peyrebrune MC, Ingham SA, Bailey DM, Folland JP. Sodium bicarbonate improves swimming performance. Int J Sports Med. 2008;29:519–23.
219. Applegate E. Effective nutritional ergogenic aids. Int J Sport Nutr. 1999;9:229–39.
220. Graham TE. Caffeine and exercise: metabolism, endurance and performance. Sports Med. 2001;31:785–807.
221. Carr A, Dawson B, Schneiker K, Goodman C, Lay B. Effect of caffeine supplementation on repeated sprint running performance. J Sports Med Phys Fitness. 2008;48:472–8.
222. Glaister M, Howatson G, Abraham CS, Lockey RA, Goodwin JE, Foley P, et al. Caffeine supplementation and multiple sprint running performance. Med Sci Sports Exerc. 2008;40:1835–40.

223. Tarnopolsky MA, Atkinson SA, MacDougall JD, Sale DG, Sutton JR. Physiological responses to caffeine during endurance running in habitual caffeine users. Med Sci Sports Exerc. 1989;21:418–24.
224. Armstrong LE. Caffeine, body fluid-electrolyte balance, and exercise performance. Int J Sport Nutr Exerc Metab. 2002;12:189–206.
225. Falk B, Burstein R, Rosenblum J, Shapiro Y, Zylber-Katz E, Bashan N. Effects of caffeine ingestion on body fluid balance and thermoregulation during exercise. Can J Physiol Pharmacol. 1990;68:889–92.
226. Rama Rao SV, Sunder GS, Reddy MR, Praharaj NK, Raju MV, Panda AK. Effect of supplementary choline on the performance of broiler breeders fed on different energy sources. Br Poult Sci. 2001;42:362–7.
227. Buchman AL, Awal M, Jenden D, Roch M, Kang SH. The effect of lecithin supplementation on plasma choline concentrations during a marathon. J Am Coll Nutr. 2000;19:768–70.
228. Buchman AL, Jenden D, Roch M. Plasma free, phospholipid-bound and urinary free choline all decrease during a marathon run and may be associated with impaired performance. J Am Coll Nutr. 1999;18:598–601.
229. Kaikkonen J, Kosonen L, Nyyssonen K, et al. Effect of combined coenzyme Q10 and d-alpha-tocopheryl acetate supplementation on exercise-induced lipid peroxidation and muscular damage: a placebo-controlled double-blind study in marathon runners. Free Radic Res. 1998;29:85–92.
230. Svensson M, Malm C, Tonkonogi M, Ekblom B, Sjodin B, Sahlin K. Effect of Q10 supplementation on tissue Q10 levels and adenine nucleotide catabolism during high-intensity exercise. Int J Sport Nutr. 1999;9:166–80.
231. Bowers CY. Growth hormone-releasing peptide (GHRP). Cell Mol Life Sci. 1998;54:1316–29.
232. Camanni F, Ghigo E, Arvat E. Growth hormone-releasing peptides and their analogs. Front Neuroendocrinol. 1998;19:47–72.
233. Hargreaves M, McKenna MJ, Jenkins DG, et al. Muscle metabolites and performance during high-intensity, intermittent exercise. J Appl Physiol. 1998;84:1687–91.
234. McNaughton L, Dalton B, Tarr J. Inosine supplementation has no effect on aerobic or anaerobic cycling performance. Int J Sport Nutr. 1999;9:333–44.
235. Starling RD, Trappe TA, Short KR, et al. Effect of inosine supplementation on aerobic and anaerobic cycling performance. Med Sci Sports Exerc. 1996;28:1193–8.
236. Williams MH, Kreider RB, Hunter DW, et al. Effect of inosine supplementation on 3-mile treadmill run performance and VO2 peak. Med Sci Sports Exerc. 1990;22:517–22.
237. Messina M, Messina V. Soyfoods, soybean isoflavones, and bone health: a brief overview. J Ren Nutr. 2000;10:63–8.
238. Brilla LR, Conte V. Effects of a novel zinc-magnesium formulation on hormones and strength. J Exerc Physiol Online. 2000;3(4). http://www.css.edu/users/tboone2/asep/BrillaV.pdf.

Chapter 20
The Marketing of Dietary Supplements: Profit Before Health

Norman J. Temple

Hope springs eternal in the human breast

(A. Pope, Essay on man, Epistle i)

Keywords Dietary supplements • Health food stores • Multilevel marketing • Marketing methods • Herbal treatment • Nutrition Labeling and Education Act

Key Points

- Around half of people in North American use dietary supplements regularly.
- Supplements are marketed by a variety of different methods, including health food stores, multilevel marketing, TV advertising, bulk mail, spam e-mails, and internet websites.
- A large part of the marketing of supplements involves giving information that is unreliable or dishonest.
- There is very little regulation of the marketing of supplements in the USA and Canada.
- Advice is given for evaluating claims made for supplements and advice for the general public.

N.J. Temple, PhD (✉)
Centre for Science, Athabasca University, Athabasca, AB, T9S 3A3, Canada
e-mail: normant@athabascau.ca

N.J. Temple et al. (eds.), *Nutritional Health: Strategies for Disease Prevention*,
Nutrition and Health, DOI 10.1007/978-1-61779-894-8_20,
© Springer Science+Business Media, LLC 2012

20.1   Introduction

Dietary supplements refer to any substance taken in addition to regular food. Supplements include vitamins, minerals, amino acids, herbs, enzymes, and various substances extracted from plants and animals. They are sold as liquids, tablets, capsules, and powders. By definition these products are not conventional foods but are designed to supplement the diet.

Regulating the colossal and growing dietary supplement industry and protecting consumers from fraudulent and misleading products and health claims is a daunting enterprise. Clearly, dietary supplement manufacturers have succeeded in convincing consumers to buy their products, but an important question remains unanswered: What benefits do consumers receive from the billions of dollars they spend on supplements? This chapter examines the main methods of marketing supplements, provides examples of egregious marketing strategies and claims, reviews the current regulatory status of dietary supplements in Canada and the United States, and outlines ways in which health professionals can help consumers protect themselves from unscrupulous or even dishonest supplement marketers. What is especially remarkable is the meager amount of peer-reviewed research that has investigated this area and the effect that this lack of research has on our ability to make recommendations regarding the rational use of supplements.

20.2   Use of Dietary Supplements

Findings from National Health and Nutrition Examination Survey (2003–2006) revealed that 33% of adult Americans reported using vitamin/mineral supplements [1]. This survey also reported that herbal supplements were taken by about 20% of adult Americans. In total about half of all American take a dietary supplement regularly. Use of dietary supplements is also highly prevalent in Canada. Three in four Canadians take supplements and a third use them every day [2].

The people most likely to use dietary supplements are female, older, white, nonsmokers, regular exercisers, and better educated [3–8]. Supplement use tends to be higher among those with a higher income [7], although this is not a consistent finding [6]. Two studies reported that supplement use is highest in those who are healthiest [4], but one study indicated the reverse [7].

20.3   Common Supplements and Common Claims

A wide variety of supplements are sold, but space permits only a limited number to be discussed here. The quality of the evidence supporting their efficacy covers a wide spectrum.

20.3.1 Supplements with Strong Supporting Evidence

Supplements where the case for benefit is strong include fish oil, vitamin D, and calcium. Multivitamins—meaning pills containing a broad spectrum of vitamins and minerals—can be especially valuable for various groups of people, such as many elderly persons and many women during their reproductive years in order to achieve an adequate intake of iron and folic acid. These supplements are further discussed in Chap. 18 and other chapters.

Alas, most of the marketing of supplements is concentrated on products where the supporting evidence is far weaker but the profit potential far greater.

20.3.2 Herbs and Herbal Cocktails

St John's wort is one of the very few herbs where there is fairly solid supporting evidence of effectiveness. Evidence indicates that it may be effective in the treatment of mild to moderate depression. It has also been reported to have fewer side effects than drugs [9]. However, it poses a problem due to interaction with numerous medications [10].

However, the great majority of claims for the efficacy of herbal treatments fall flat when well-conducted studies are performed. Here are two prominent examples:

- Ginkgo biloba. This herb is used by many people in the belief that it helps maintain cognitive health in aging, especially for the memory. In 2009 the results were published of a randomized, double-blind trial in which 3,070 people across the USA, aged 72–96, were given either the herb or a placebo [11]. After 6 years there was no difference in the rate of cognitive decline between the two groups. No benefits were seen for memory function.
- Ginseng. This herb is widely used to improve energy and physical or cognitive performance. Despite claims dating back centuries and vast numbers of users who, presumably, are confident of its value, the supporting evidence is quite weak [9, 12].

Unlike conventional drugs, herbal supplements generally lack standardization of active ingredients. There can be much variation between different brands of what is supposedly the same herb due to such factors as the actual species of plant used, the part of the plant used, and the extraction method. For example, wide variations in the concentration of active components in different samples of ginseng and St. John's wort have been reported [13, 14]. It is also entirely possible that some manufacturers may simply be using cheap ingredients in place of expensive herbs. For these reasons an herb that was effective in a published study may be of no value when used by consumers.

Many supplements consist of a mixture of herbs. Often the label will give the ingredient list as a dozen or so herbs, each with a Latin name. As very little research has been conducted on mixtures of herbs, there is no good reason to be confident that such supplements will achieve any clinically valuable benefits. Moreover, such herbal cocktails pose a risk of inducing harmful side effects that will be very difficult to relate to any specific herb or herb combination. Polypharmacy is always hazardous, whether it is based on conventional drugs or herbal cocktails.

20.3.3 Antioxidants

Many supplements are sold with a claim of being "rich in antioxidants," the obvious implication being that such products will improve health or prevent disease. This can sound very impressive. Certainly, it is firmly established that foods naturally rich in antioxidants, such as fruit and vegetables, are excellent for the health. But the real reasons for this are complex and not universally understood. The weakness of the claim that more antioxidants equates to improved health or less risk of disease is disproven by the results of several large randomized clinical trials (RCTs) in which the three major antioxidant vitamins—β-carotene or vitamins C or E—have been given to patients. The dose has typically been several times higher than the RDA. A recent meta-analysis concluded that supplementing with these vitamins leads to an *increase* of about 5–6% in all-cause mortality [15, 16]. Therefore, when sellers of supplements state in their advertisements that a product is "rich in antioxidants," that is weak evidence that it will improve health or prevent disease.

20.3.4 Detoxification

For the sellers of supplements, detoxification is much like the word antioxidants: it provides a simple concept that most people can easily grasp and that can be used to provide an apparently scientific reason why a particular product will do wonders for the health. Detoxification is, of course, a well-established biochemical process. However, herbal treatments, in particular, are routinely sold with the promise that they will stimulate the liver—and perhaps some other organs as well—so that detoxification is accelerated and the body is cleansed. However, the claims being made that particular supplements will induce detoxification and lead to enhanced health are without any supporting evidence, at least none that this author has ever been able to find.

20.3.5 Boosting the Immune System

Many supplements come with the claim that they somehow stimulate the immune system. As with detoxification this is usually associated with herbs. For some herbs there is supporting evidence, Echinacea for example [17, 18]. But in most cases the claims come minus credibility.

The claims being made for supplements that supposedly boost the immune system resemble those for antioxidants and detoxification: in each case we see a marketing strategy based on a biomedical mechanism that is sufficiently simple that the average person can understand it.

20.3.6 Exotic Fruit Juices

In recent years several exotic fruit juices have appeared on the market. The main ones are acia, goji, mangosteen, and noni. They are sold by multilevel marketing and through health food stores (HFS). They invariably come with promises of wonderful benefits, though the supporting evidence appears to be very weak. Despite that, they are sold at an exorbitant price. For example, a health food store in Edmonton, Canada, where this author lives, charges about $50–60 per liter for these juices. By contrast, the local supermarket sells the common types of fruit and vegetable juices for less than $3 per liter.

20.3.7 Weight-Loss Products

With the huge obesity epidemic that has swept North America, it is scarcely surprising that supplement manufacturers have jumped on the bandwagon. New products appear with bewildering regularity. Typically, such products come with thin promises based on even thinner evidence. But what they do produce, very often, is a photo of a young woman with a BMI of about 20; the implicit message is obvious. Overall, the evidence for effectiveness of most of these supplements is weak [9].

20.3.8 A Repeating Story

What we see, time and time again, is weak evidence dressed up as solid science. The marketers of supplements like to use scientific evidence the way a drunk uses a lamp post: more for support than illumination. Sometimes the marketers go to the extreme and claim that their product cures almost anything and everything, even cancer.

The following are the types of evidence frequently referred to and that are used to promote the sale of supplements:

- A physiological or biochemical change in the body. Often this is based on mere speculation or even pure invention; for example, that health will be improved because a herbal mixture induces detoxification or because an exotic fruit juice is rich in antioxidants.
- Weak clinical evidence (e.g., a particular herb has been used for centuries as a treatment and therefore it must be effective; or evidence based on one or two small studies of dubious importance).

- Anecdotal evidence, often from an unqualified person with a serious conflict of interest. A slight variation of anecdotal evidence is the use of testimonials ("Jim from Miami says: 'Thanks to Speedy Fat Burn I have lost 25 lb in 1 month.'").

In order for claims to have real credibility we need to see hard scientific evidence. Ideally, there should be consistent evidence from well-conducted RCTs, with clinical endpoints, showing health benefits, and published in peer-reviewed journals. But such evidence is seldom available.

Dietary supplements are a multibillion dollar industry, but integral to its success has been the widespread use of blatantly misleading marketing. This strategy has been so successful because most of the population has a weak grasp of biomedical science [6].

20.3.9   Potential Hazards from Supplements

The supplement industry regularly boasts that its products are safe. Alas, this is often not the case. For example, many supplements contain undeclared active pharmaceutical agents. Cohen [19], in an editorial in the *New England Journal of Medicine*, published in 2009, wrote: "...more than 140 contaminated products have been identified, but these represent only a fraction of the contaminated supplements on the market." Similar findings came from a recent study in which a chemical analysis was conducted on traditional Ayurvedic medicines that were being sold in the USA via the internet. The findings revealed that 21 % of these herbal preparations exceeded one or more standards for acceptable daily intake of lead, mercury, or arsenic [20]. Now, it is certainly true that only a tiny fraction of users of supplements experience undesired side effects. However, they do occur; indeed, many deaths have been reported, especially from herbs [21]. Quite apart from hazardous contaminants, many herbs interact with various drugs. St. John's wort is especially problematic in that regard [10]. Compounding this problem patients often do not tell their physician about their use of supplements.

Another problem that is probably quite common, though hard numbers on this seem to be lacking, is that many people with a health problem that could be helped by a conventional medical treatment turn instead to ineffective supplements [20].

20.4   How Dietary Supplements Are Marketed

Dietary supplements are promoted using many types of marketing methods, including advertisement in newspapers (sometimes as multipage supplements), bulk mail ("junk mail"), spam e-mails, and internet websites, as well as by TV advertising. The actual selling of supplements is also carried out in diverse ways. They can be purchased in pharmacies, supermarkets, and HFS; directly from people engaged in multilevel marketing; and by mail order. This section describes some marketing strategies used by manufacturers and sellers.

20.4.1 Health Food Stores, Drugstores, and Supermarkets

HFS are a popular source of dietary supplements. They typically carry hundreds of different products. But here I describe my favorite: Cellfood. According to the leaflet that explains the wonders of the product, Cellfood "enhances the bioavailability of oxygen though its ability to 'dissociate' water molecules within the body—releasing nascent oxygen and hydrogen directly to the cells." The ingredient list includes almost the entire periodic table, even thallium, polonium, nitrogen, and the inert gases. The only elements missing are the ones well known to be toxic, such as lead and arsenic. Other ingredients include 16 amino acids and 32 enzymes. What is truly remarkable is that the recommended dose is a mere eight drops.

That some HFS staff act in dishonest, uninformed, or unethical ways has been demonstrated in several studies. One was conducted in Hawaii in 1998 [22]. A woman visited 40 HFS and in each one she stated that her mother had advanced breast cancer and requested advice. In 36 (90%) of the HFS, one or more products were recommended. The most common ones recommended were shark cartilage (recommended in 17 HFS), followed by essiac (an herbal treatment), maitake mushroom, CoQ10, and vitamin C. In all, 38 different products were recommended. No sales person mentioned potential adverse effects. The cost of the supplements of shark cartilage, essiac, and maitake ranged from US$300 to US$3,400. In a similar study conducted in Ontario, Canada, the investigator posed as a mother of a child with Crohn's disease [23]. Of the 32 HFS visited, 23 (72%) offered advice. The advice given was remarkably inconsistent, with 30 different herbs and nutritional supplements recommended. A third study was carried out in London, England [24]. Here, the researcher stated that she had been getting headaches lately. The researcher visited 29 HFS; of these, 25 (86%) recommended a specific product. Again, a variety of products were recommended. In addition, seven of the stores (24%) recommended diet or lifestyle change. One disturbing aspect of this study was that the researcher presented with what could be a serious disease, yet only seven of the HFS (24%) suggested a visit to a physician. By contrast, 18 of the HFS (62%) suggested a cause of the headaches.

In recent years it seems that every drugstore has given over a generous amount of shelf space to the display of supplements. The same is seen in the pharmacy section of supermarkets. However, these stores are quite different from HFS in one important respect: customers requesting advice are far less likely to be recommended to buy useless supplements. This is not surprising as pharmacists are trained health professionals and must abide by a code of ethics, This was recently confirmed in a study that I conducted in Canada [25]. Visits were made to 260 HFS and drugstores (or the pharmacy section of supermarkets). The results reveal that on 88% of times that questions were asked in HFS, the recommendations made were either unscientific (6%) or were poorly supported by the scientific literature (82%). By contrast, this occurred for only 27% of questions in drugstores/supermarkets. Conversely, on two-thirds of visits to drugstores/supermarkets, staff gave advice considered to be fairly accurate or accurate, but this seldom occurred in HFS (68% vs. 7%).

20.4.2   Multilevel Marketing

Many products—from Avon cosmetics to Tupperware—are sold by direct marketing, a strategy in which company salespeople recruit other salespeople. The foot soldiers and everyone up the chain get a commission for their sales. This marketing strategy is popular among manufacturers of dietary supplements. Its focus is profit, not consumer health. Although I am not aware of any actual surveys, my own observations have left me convinced that the great majority of the folks who are hired to sell supplements lack any real training in key areas of nutrition and health. This is also true of HFS.

Here is one example of this. A few years ago advertising leaflets were distributed here in Edmonton promoting a public lecture by Dr. Earl Mindell, the person behind the sale of goji juice by multilevel marketing. He was referred to as "Widely regarded as the world's #1 nutritionist" and the product as "The biggest discovery in nutrition in the last 40 years!"

Body Wise International provides our second example. This company rewards its "consultants" for selling products and also for recruiting other independent consultants. According to the company website, consultants can use this form of network marketing to generate "substantial incomes" [26]. The science behind some of the company's products is questionable, as evidenced in a public lecture given by Tim Tierney in 2002 in Edmonton. Tierney did not claim to be an expert but was giving the views of Dr. Jesse Stoff, who was described in the advertising leaflet for the lecture as "the world's leading viro-immunologist." A check of Medline, however, failed to produce a single publication by Stoff, though he has authored several populist nutrition books of dubious credibility.

Tierney made some rather strange statements in his lecture, including that milk causes osteoporosis. However, the real purpose of the lecture was the promotion of products sold by Body Wise. Tierney claimed that several people had been cured of cancer by the products. Indeed, the products seem to cure pretty much everything. For evidence of its efficacy, the audience was given, not references to papers in journals, but statements by half a dozen people describing fantastic cures experienced by themselves or others. The products come with money-back guarantees. The company website [26] refers to two unpublished research studies that examined immune function (number of natural killer cells) rather than clinical endpoints.

20.4.3   Advertising in the Media

Large amounts of advertising for supplements appears in newspapers and magazines. My local newspaper, the Edmonton Journal, regularly includes inserts that consist entirely of adverts for supplements. These direct readers either to internet websites or to HFS.

In the last few years there seems to be far more advertising of supplements during regular TV programs. As an example, ImmuGo is a product that is claimed to stimulate the immune system. During 2008 it was heavily advertised on American TV. According to the company website: "[It] was created by Vitamin Gene Arnold, the vitamin guru to Hollywood stars. For over 15 years, celebrities have been coming to Vitamin Gene for the latest breakthrough vitamins" [27].

Another form of distribution of many types of health-related products is by info-mercials. These TV programs are produced and paid for by commercial companies. They resemble regular TV programs but are, in reality, a form of advertising. They typically last for 30 min and air during the night.

20.4.4   Spam E-Mails

Spam e-mails are a cheap and easy way for manufacturers to promote their dietary supplements to tens of thousands, if not millions, of people. As a result large numbers of products are being touted, most of them of highly dubious value. In recent years vast numbers of spam e-mails have been sent out promoting sex-related herbal supplements. Spam e-mails typically work by directing the person to a website. There are large numbers of websites selling all types of supplements; they are, in effect, virtual HFS. They often flout U.S. law [28].

20.4.5   Internet Websites

According to the Federal Trade Commission (FTC) more than 90 million Americans use the Internet to find health information. The Internet is well suited to mass market-ing as it provides low-cost access to a global market. Unfortunately, its easy accessi-bility and openness make it an ideal forum for selling products of dubious value.

In an analysis of herbal products marketed on the Internet, it was discovered that among 338 websites that sold herbal remedies, 273 (81%) made one or more health claims [28]. Of these 273, over half (55%) broke US law by claiming that a product could treat, prevent, diagnose, or cure a specific disease, and half of them failed to show the standard federal disclaimer related to health claims.

20.4.6   The Object of the Exercise

The purpose of all this huge marketing enterprise is, of course, to maximize sales. As mentioned earlier, there are some supplements for which solid evidence exists justifying their use. Examples include vitamin D, calcium, and fish oil. Each of these costs around \$3 or \$4 per month. But go into a HFS, tell the salesperson that you do not have enough energy, you have an ache in your knee, and your mother died of cancer and you will likely be told to take a handful of supplements, each costing between \$20 and \$60 per month. This might easily add up to \$100–200 per month. And it is quite likely that the recommended supplements would have little or no beneficial effect on health.

20.5 Regulations on the Marketing of Supplements

The growth of the dietary supplement industry in recent years has created many regulatory challenges for federal agencies.

20.5.1 United States

The principal agency of the US government that regulates supplements is the Food and Drug Administration (FDA). However, product advertising is regulated by the FTC.

The Nutrition Labeling and Education Act (NLEA) of 1990 gave the FDA the authority to require the manufacturers of dietary supplements to furnish evidence that their products are safe [29]. The FDA was also empowered to approve the health claims for these products. However, enactment of the Dietary Supplement and Health Education Act (DSHEA) of 1994 freed dietary supplement manufacturers from many FDA regulations [29]. Whereas manufacturers were required by the NLEA regulations to prove that a dietary supplement is *safe*, the FDA must prove under DSHEA regulations that a supplement is *unsafe*. This shift in regulatory policy places burdens on a federal agency with important public health responsibilities but limited resources. Moreover, manufacturers are now free to make health-related claims (structure/function claims). However, manufacturers are not permitted to state explicitly that the product will cure or prevent a disease, and they must state that the FDA has not evaluated the agent. What this means in practice is as follows:

- A manufacturer may now claim that a supplement "boosts the immune system" (but cannot claim that it prevents infectious disease)
- That it "makes the body burn fat while you sleep" (but not that it cures obesity)
- That it "fights cholesterol" (but not that it prevents heart disease)

Needless to say, the distinction between the two types of claim is lost on most consumers. Put bluntly, the 1994 law, together with the lack of resources at the FDA to properly enforce regulations, has given supplement manufacturers *carte blanche* to employ deceptive marketing that often looks and smells like fraud.

Marion Nestle, of New York University, traced the history of how the supplement industry succeeded in pressuring Congress to pass DSHEA, thereby allowing the supplement industry almost unlimited freedom to market supplements [29]:

> Today, marketers of supplements are permitted to make practically any claim they want for the health benefits of their products. They are also permitted to vary the ingredient contents of their products with impunity. They do not have to remove potentially harmful products from the market unless taken to court by the FDA, and they do not have to prove that their products bestow the benefits claimed for them. This remarkable situation is the result of the industry's persistence and skill in generating public pressure on Congress to restrict the FDA's regulatory mission.

This lobbying effort has been hugely rewarding for the supplement industry. Use of dietary supplements increased sharply in the years following the passage of DSHEA [4, 5]; sales increased nearly 80% between 1994 and 2000 [6].

In 2003 the serious problems with the current regulatory climate were addressed by the editors of the *Journal of the American Medical Association* (*JAMA*) [30]:

> If dietary supplements have or promote .... biological activity, they should be considered to be active drugs. On the other hand, if dietary supplements are claimed to be safe because they lack or have minimal biological activity, then their ability to cause physiologic changes to support "structure/function" claims should be challenged, and their sale and distribution as products to improve health should be curtailed. Manufacturers of dietary supplements are trying to have it both ways. They claim their products are powerfully beneficial, on the one hand, but harmless on the other. To claim both makes no sense, and to claim either without trials demonstrating efficacy and safety is deceptive. The public should wonder why dietary supplements have effectively been given a free ride. New legislation is needed for defining and regulating dietary supplements.

A similar article was published in the *New England Journal of Medicine* in 2002 with a focus on herbal supplements [31], and a more general editorial in 2009 [19].

There has been one positive development. Around 2008–2010, new regulations were implemented that require manufacturers to follow certain "good manufacturing practices" [32]. Products must now contain the substances specified on the label and be free of contaminants. What effect this will have in practice awaits to be seen.

20.5.2   Canada

In 1999 Health Canada created a new organization, the Natural Health Products Directorate, to regulate dietary supplements. The directorate's mission is to ensure that Canadians have access to natural health products (NHP) that are safe, effective, and of high quality [33]. A key feature of the new regulations is that the requirements for safety and good manufacturing practices fall on the companies that manufacture, package, label, import, or distribute NHP. All manufacturers, importers, packagers, and labelers of NHP must now have site licenses, and any new NHP must have a product license. The regulations require a premarket review of products to assure Canadians that label information is truthful and health claims are supported by appropriate types of scientific evidence.

When these regulations were announced, the clear impression was given that the marketing of supplements was to be made much more honest. Hopefully, packages of supplements now contain what the label says they contain—nothing more and nothing less. But that is only one aspect of the problem. The greater problem is the widespread practice of dishonest marketing and here there is scant evidence of improvement. In other words, a bottle of vitamin E sold in Canada should now contain the vitamin in the amount stated on the label, but that is of little value if the people in HFS are still allowed to claim that it helps prevent disease, despite the lack of good supporting evidence. In that respect the situation in Canada is still every bit as bad as that in the USA.

20.6 Helping Consumers Make Informed Choices

Given the diverse and innovative forms of marketing of dietary supplements, consumers need help sorting through the health claims. This is especially the case as the great majority of people have a weak understanding of health science and how supplements might affect body functioning.

Untrustworthy claims for supplements often have the following features [34]:

1. A reliance on anecdotal evidence, especially testimonials.
2. A lack of supporting evidence from papers published in peer-reviewed journals.
3. The use of unscientific and unsupported theories to rationalize the use of supplements, such as claims that a supplement boosts immunity or detoxifies the body.
4. Money-back guarantees.
5. The use of claims that the product represents a "scientific breakthrough" or an "ancient remedy."
6. Touting the product as an effective treatment for a broad range of ailments. If things are too good to be true, they probably are.

The following simple rules are helpful in deciding whether to purchase a particular supplement:

- Whenever the above features are seen in connection with the marketing of a supplement, this should raise a red flag.
- In general, it is best to ignore advice given by persons who have a financial interest in selling supplements, especially when they appear to have no relevant qualifications. This includes staff in HFS and people engaged in multilevel marketing. Likewise, it is usually best to ignore all websites of supplement manufacturers and all advertising in the media, both print media and TV.
- Always use common sense. A healthy dose of skepticism is the consumer's best protection against fraudulent and misleading marketing.
- Before purchasing a supplement, independent advice should be obtained from a reliable source. There are a number of trustworthy websites that can be consulted; some are listed in Table 20.1. Legitimate health professionals can also be consulted, such as a physician, dietitian, or pharmacist.

Table 20.1 Internet resources on dietary supplements and health fraud

Dietary supplements

Mayo Clinic: http://www.mayoclinic.com

National Center for Complementary and Alternative Medicine (NCCAM): http://nccam.nih.gov/health/supplements/wiseuse.htm

Medline Plus: http://medlineplus.gov/

MEDLINE: http://www.ncbi.nlm.nih.gov/PubMed/

The National Cancer Institute's website gives reliable information about various supplements claimed to be effective in the prevention or treatment of cancer: http://www.cancer.gov/cancertopics/treatment/cam

Dietary Supplements. Available from Food and Nutrition Information Center, National Agricultural Library, June 2011: http://fnic.nal.usda.gov/

Dietary Supplements Alerts and Safety Information. Available from Food and Drug Administration (FDA), October 2009: www.cfsan.fda.gov/~dms/ds-warn.html

Healthfinder: www.healthfinder.gov

Supplement Watch: www.supplementwatch.com

Health fraud

National Council Against Health Fraud: www.ncahf.org

Quackwatch: www.quackwatch.org

(Note: the above two websites are closely linked.)

MEDLINEplus Guide to Healthy Web Surfing: www.nlm.nih.gov/medlineplus/healthywebsurfing.html

(A website that exposes fraud in the sale of various supplements, including herbs and weight-loss products, is www.dietfraud.com)

20.7 Conclusions

Consumers want to go to dietary heaven, but they do not want to die. Marketers of dietary supplements have rushed to fill this consumer demand, frequently using scientific evidence the way a drunk uses a lamppost: more for support than illumination. The marketing of dietary supplements is a situation where ethics, honesty, and a sincere desire to improve the health of consumers takes a backseat to maximizing sales and profit. Millions of consumers have embraced the use of dietary supplements with zeal, but they have turned a blind eye to issues of safety and efficacy.

It is entirely possible that on balance the supplement industry actually does more harm than good to people's health. This is because many people are persuaded to take worthless—and possibly harmful—supplements for the prevention and treatment of a health condition and forgo a beneficial intervention. Why eat extra portions of fruit and vegetables every day in order to lower your risk of cancer when you can much more easily take shark cartilage tablets? Why take flu shots when the same protection can be achieved with echinacea? Why cut down on your intake of hamburgers when cholesterol can be neutralized with garlic? The supplement industry has built an empire, but the emperor is almost naked.

But there is a positive side to dietary supplements. There is good supporting evidence for several supplements, such as vitamin D and fish oil. This was briefly reviewed earlier in the chapter. Occasionally, a supplement taken by large numbers of people based on weak evidence turns out to be effective. One such example is St. John's wort. It became popular for the treatment of mild to moderate depression, despite the meager quality of the supporting evidence. Nevertheless, it has stood up reasonably well to careful testing [35].

There is no good reason why we cannot have the best of both worlds: promote the use of supplements when there is good supporting evidence while proscribing against those without supporting evidence. A vital and urgent step in this direction is stricter regulation, including greater vigilance of the industry by government agencies. In the final analysis, an educated consumer is the best consumer when it comes to dietary supplements. And the need for education also applies to health professionals.

References

1. Bailey RL, Gahche JJ, Lentino CV, et al. Dietary supplement use in the United States, 2003–2006. J Nutr. 2011;141:261–6.
2. Ipsos Reid/Health Canada. Natural health product tracking survey—2010 final report. 2010. http://epe.lac-bac.gc.ca/100/200/301/pwgsc-tpsgc/por-ef/health/2011/135-09/report.pdf. Accessed 5 Aug 2011.
3. National Harris Interactive Survey. Widespread ignorance of regulation and labeling of vitamins, minerals and food supplements, according to a National Harris Interactive Survey. Released 23 Dec 2002. http://www.harrisinteractive.com/news/allnewsbydate.asp?NewsID=560. Accessed 5 Aug 2011.
4. Radimer K, Bindewald B, Hughes J, Ervin B, Swanson C, Picciano MF. Dietary supplement use by US adults: data from the National Health and Nutrition Examination Survey, 1999–2000. Am J Epidemiol. 2004;160:339–49.
5. Millen AE, Dodd KW, Subar AF. Use of vitamin, mineral, nonvitamin, and nonmineral supplements in the United States: the 1987, 1992, and 2000 National Health Interview Survey results. J Am Diet Assoc. 2004;104:942–50.
6. Blendon RJ, DesRoches CM, Benson JM, Brodie M, Altman DE. Americans' views on the use and of dietary supplements. Arch Intern Med. 2001;161:805–10.
7. Satia-Abouta J, Kristal AR, Patterson RE, Littman AJ, Stratton KL, White E. Dietary supplement use and medical conditions: the VITAL study. Am J Prev Med. 2003;24:43–51.
8. Gunther S, Patterson RE, Kristal AR, Stratton KL, White E. Demographic and health-related correlates of herbal and specialty supplement use. J Am Diet Assoc. 2004;104:27–34.
9. MayoClinic.com. http://www.mayoclinic.com. Accessed 6 April 2010.
10. Hammerness P, Basch E, Ulbricht C, et al.; Natural Standard Research Collaboration. St John's wort: a systematic review of adverse effects and drug interactions for the consultation psychiatrist. Psychosomatics. 2003;44:271–82.
11. Snitz BE, O'Meara ES, Carlson MC, et al.; Ginkgo Evaluation of Memory (GEM) Study Investigators. Ginkgo biloba for preventing cognitive decline in older adults: a randomized trial. JAMA. 2009;302:2663–70.
12. Vogler BK, Pittler MH, Ernst E. The efficacy of ginseng. A systematic review of randomised clinical trials. Eur J Clin Pharmacol. 1999;55:567–75.
13. Harkey MR, Henderson GL, Gershwin ME, Stern JS, Hackman RM. Variability in commercial ginseng products: an analysis of 25 preparations. Am J Clin Nutr. 2001;73:1101–6.
14. Draves AH, Walker SE. Analysis of the hypericin and pseudohypericin content of commercially available St John's Wort preparations. Can J Clin Pharmacol. 2003;10:114–8.

15. Bjelakovic G, Nikolova D, Gluud LL, Simonetti RG, Gluud C. Antioxidant supplements for prevention of mortality in healthy participants and patients with various diseases. Cochrane Database Syst Rev. 2008:CD007176.
16. Bjelakovic G, Nikolova D, Gluud LL, Simonetti RG, Gluud C. Mortality in randomized trials of antioxidant supplements for primary and secondary prevention: systematic review and meta-analysis. JAMA. 2007;297:842–57.
17. Zhai Z, Liu Y, Wu L, et al. Enhancement of innate and adaptive immune functions by multiple *Echinacea* species. J Med Food. 2007;10:423–34.
18. Melchart D, Linde K, Worku F, Sarkady L, Holzmann M, Jurcic K, et al. Results of five randomized studies on the immunomodulatory activity of preparations of *Echinacea*. J Altern Complement Med. 1995;1:145–60.
19. Cohen PA. American roulette—contaminated dietary supplements. N Engl J Med. 2009;361: 1523–5.
20. Saper RB, Phillips RS, Sehgal A, et al. Lead, mercury, and arsenic in US- and Indian-manufactured ayurvedic medicines sold via the Internet. JAMA. 2008;300:915–23.
21. Marty AT. Potentially fatal natural remedies. Am Fam Physician. 1999;59(1116):1119–20.
22. Gotay CC, Dumitriu D. Health food store recommendations for breast cancer patients. Arch Fam Med. 2000;9:692–9.
23. Calder J, Issenman R, Cawdron R. Health information provided by retail health food outlets. Can J Gastroenterol. 2000;14:767–71.
24. Vickers AJ, Rees RW, Robin A. Advice given by health food shops: is it clinically safe? J R Coll Physicians Lond. 1998;32:426–8.
25. Temple NJ, Eley D, Nowrouzi B. Advice on dietary supplements: a comparison of health food stores and pharmacies in Canada. J Am Coll Nutr. 2009;28:674–7.
26. Bodywise. www.bodywise.com. Accessed 9 March 2003. Formerly: www.healthywave.com.
27. ImmuGo. http://www.immugo.com. Accessed 19 Dec 2008.
28. Morris CA, Avorn J. Internet marketing of herbal products. JAMA. 2003;290:1505–9.
29. Nestle M. Food politics. How the food industry influences nutrition and health. 2nd ed. Berkeley, CA: University of California Press; 2007.
30. Fontanarosa PB, Rennie D, DeAngelis CD. The need for regulation of dietary supplements—lessons from ephedra. JAMA. 2003;289:1568–70.
31. Marcus DM, Grollman AP. Botanical medicines—the need for new regulations. N Engl J Med. 2002;347:2073–6.
32. Marra MV, Boyar AP. Position of the American Dietetic Association: nutrient supplementation. J Am Diet Assoc. 2009;109:2073–85.
33. Health Canada, Natural Health Products Directorate. Natural health products regulations. 2011. http://www.hc-sc.gc.ca/dhp-mps/prodnatur/about-apropos/index-eng.php. Accessed 5 Aug 2011.
34. Federal Trade Commission. 'Miracle' health claims: add a dose of skepticism. http://www.ftc.gov/bcp/edu/pubs/consumer/health/hea07.shtm. October 2001. Accessed 5 Aug 2011.
35. Memorial Sloan-Kettering Cancer Center. St. John's Wort. http://www.mskcc.org/mskcc/html/69385.cfm. Accessed 5 Aug 2011.

Chapter 21
Functional Food in the Marketplace: New Products, Availability, and Implications for the Consumer

Jill K. Rippe

Keywords Functional foods • Nutraceuticals • Marketing trends • New products

Key Points

- Functional foods emerged in the 1990s and have evolved over the last 20 years as a marketing category of health-focused food.
- Reducing or removing ingredients viewed as unhealthy in foods, such as salt (sodium) or certain fats and carbohydrates, creates a functionality that is defined by government regulations in categories termed "reduced," "lite," and "free."
- Examples of functionality by reduction or replacement are common for sodium, cholesterol, fat, *trans* fat, and digestible carbohydrates. Options for replacement include alternatives such as intense sweeteners and fibers.
- Adding bioactive ingredients viewed as healthy in foods (nutraceuticals and/or supplements) may help manufacturers make specified health claims. These vary by country.
- Examples of added functionality in foods are given for protein supplementation, cardiovascular health, bone and joint health, eye health, gut health, immunity, muscle fitness, weight management, mood, antiaging, and antioxidants.

J.K. Rippe, MS (✉)
Main Street Ingredients, 2340 Enterprise Avenue, La Crosse, WI 54602-1628, USA
e-mail: jill.r@mainstreetingredients.com

N.J. Temple et al. (eds.), *Nutritional Health: Strategies for Disease Prevention*,
Nutrition and Health, DOI 10.1007/978-1-61779-894-8_21,
© Springer Science+Business Media, LLC 2012

21.1 Introduction

The worldwide marketplace for new foods is a vast arena where many thousands of products are introduced each year and the majority fails within a few years. In this setting, functional foods have emerged and survived as a new marketing category of health-focused food. With the health impact of nutritional supplements critically reviewed in Chap. 20, this present chapter is focused on examples of actual food products and ingredients to give a sample of what is offered in this functional foods market. In Chap. 22 Nestle and Wilson discuss how political forces influence the choices available to the consumer.

Commercial interests create new foods to meet new opportunities in the marketplace. In the food industry, the word "new" itself is often used to improve demand or consumption. Public health alerts, popular diet trends, regulations, and governmental food policies can become opportunities. In 1990 mandatory food nutritional labeling was introduced in the United States [1] and similar regulations spread worldwide. A growing consumer awareness of process food and high levels of sodium, fat, cholesterol, and sugar coincided with regulatory approval to label products as "lite," "reduced," or "free," indicating reduction or removal of these ingredients perceived as unhealthy. This created functionality by reduction or elimination, often including replacement with alternative ingredients to provide saltiness, sweetness, or fat replacement.

A regulatory approval in the 1990s also led to making certain health claims on product labels, encouraging addition of functional ingredients such as soy protein, n-3 fatty acids (also known as omega-3 fatty acids), and plant stanol esters. Such functional ingredients with pharmaceutical-like activity became known as "nutraceuticals" [2] in some countries and "supplements" in others. Unlike whole foods that are intact and unrefined, the concepts of "reduction" and "addition" of ingredients encompass what is found in manufactured functional foods today. As there is no uniform agreement about what is healthy and acceptable in food, the criteria for ingredient and product approvals varies among countries and may therefore require manufacturers to gain approval country by country. Thus, many new functional foods or ingredients have limited availability across national boundaries; however, larger companies have the resources to broaden their markets worldwide.

What is new in food today will become obsolete tomorrow if it fails in any way. In other words, the quest for profit is the key driver of these developments. For that reason products mentioned as examples in this chapter may not even be available in any market by the time this book is published. Other products may have evolved into a "newer" form. For functional foods or ingredients, failure can happen for a number of reasons. They taste terrible. They cost too much. They suffer marketing mishaps. They do not provide the health benefits they proclaim. They are declared fraudulent, deceptive, or unsafe by individuals, public interest groups, or public institutions. What is new in food keeps evolving.

Presented here is a snapshot of commercially manufactured functional foods, along with brief commentary on definitions, technologies, and regulatory settings where applicable. Two broad categories of "subtracted/reduced" and "added" functionality are used to group these foods. A number of product and ingredient examples are given in each section. Company names as well as branded and trademarked products are printed in *italics*. Websites to locate more information are listed in Table 21.1.

Table 21.1 Product websites

Product company and brand name	Website[a]
Abbott Laboratories Ensure Muscle Health	http://ensure.com/
Abbott Laboratories Glucerna	http://glucerna.com/product/
Abbott Laboratories Zone Perfect	http://zoneperfect.com/
Agostoni Chocolate	http://www.agostonichocolate.com/
Alsosalt	http://www.alsosalt.com/
Arctic Zero	http://www.myarcticzero.com/
AstraREAL	http://www.astareal.com/
Attune Foods Munch	http://www.attunefoods.com/
Beauty Foods Chocolate Chai Renewal	http://www.beautyfoods.com/
Benecol	http://www.benecol.com/
Benelact	http://www.benelact.com/
Better Whey of Life	http://www.betterwheyoflife.com/
Bioactive Technologies Colostrum Sports Drink	http://www.bionz.com/
Bioferme Yosa	http://www.bioferme.fi/index.php?id=35
BioSophia Graindrops	http://www.organicmonitor.com/r2909.htm
Biovelop AB PromOat	http://www.biovelop.com/
Bolthouse Farms Protein Plus drink	http://bolthouse.com/
Bomba Energy SOL(e)UTION	http://www.bomba.at/languages.html
Borba Skin Balance Waters	http://www.borbaskinbalancewater.com/
BSN Lean Desert Protein	http://www.bsnonline.net/details/leandessertprotein.html
Burnbrae Farms Naturegg Omega Pro liquid eggs	http://www.burnbraefarms.com/consumer/our_products/egg_creations_can.htm
Ceba Foods Oatly Oat Drink	http://www.oatly.com/
Clif Luna & Builders Bars	http://www.clifbar.com/food/products_luna_bar/
ConAgra Foods Eggbeaters	http://www.eggbeaters.com/
Cyclamate	http://www.cyclamate.com/
Cytosport Muscle Milk	http://www.musclemilk.com/
Danone/Dannon DanActive, Activia & Danacol	http://www.danone.com/?lang=en
Diamond Foods Pringles	http://www.pringles.com/products/lightfatfree
Diamondcrystal SaltSense	http://www.diamondcrystalsalt.com/Culinary/Products/Salt-Sense/
Drank Extreme Relaxation Beverage	http://www.drankbeverage.com/
Dreamfields Low Carb Pasta	http://www.dreamfieldsfoods.com/find-dreamfields-pasta.html

(continued)

Table 21.1 (continued)

Product company and brand name	Website[a]
Earthling Organics Matcha Green Tea Bar	http://earthlingorganics.com/products
EAS Myoplex	http://eas.com/myoplex
Ecco Bella Health By Chocolate & Beauty Bliss	http://www.healthbychocolate.com/
Eggland's Best Eggs	http://www.egglandsbest.com/home.aspx
Emergen-C Joint Health	http://www.emergenc.com/index.php/products/specialty/joint-health
Equal	http://www.equal.com/
Farm Produce Marketing Orchard Maid drinkable yogurt	http://findarticles.com/p/articles/mi_m3301/is_9_103/ai_92137779/
First Foods Oat Supreme	http://www.first-foods.com/icecream.htm
Flash-5 Energy Bars	http://www.flash-5.com/
FloraGlo	http://www.floraglolutein.com/
Forward Foods Detour Bar	http://www.detourbar.com/
Frito-Lay Light Potato Chips & Taco Chips	http://www.fritolay.com/our-snacks/lays-light-original.html
Fruit-Sweetness	http://www.biovittoria.com/Live/biovittoria_1_1.php
Fuze Healthy Infusions Defensify	http://www.virtualvender.coca-cola.com/ft/index.jsp
GanedenBC	http://www.ganedenbc30.com/about_ganedenbc30.php
General Mills FiberOne	http://www.fiberone.com/
General Mills Total Cereal	http://www.totalcereal.com/
Genisoy	http://www.genisoy.com/
GFA Foods Smartbalance	http://www.smartbalance.com/
Glaceau Vitamin Water Focus	http://www.glaceau.com/
Glowelle Beauty Drink	http://www.glowelle.com/
Good Cacao Superfood Chocolate	http://www.goodcacao.com/home.html
Greens Plus Chia Greens Chia Energy Bar	http://greensplus.com/product_info.php/cPath/84_21_124_101/products_id/248
Groupe Biscuits Leclerc Praeventia	http://www.praeventia.ca/en/praeventia-cookies.php
GT's Enlightened Organic Raw Kombucha.	http://www.synergydrinks.com/kombucha_enlightened.aspx
Healthsmart Foods Chocolite	http://www.healthsmartfoods.com/
iChill	http://ichill.com/
Imagine Foods Rice Dream	http://www.tastethedream.com/products/rice_dream.php
Inneov Firmness	http://www.loreal-finance.com/eng/news/active-cosmetics-40.htm
Interstate Bakeries Wonderbread	http://www.wonderbread.com/white-bread.html
Isopure	http://www.theisopurecompany.com/
Jarrow Formula Brown Rice Protein	http://www.jarrow.com/product/264/Brown_Rice_Protein
Joint Juice	http://jointjuice.com/
Kashi U Black Currents & Walnuts	http://www.kashi.com/products/category/Cold%20Cereal
Kellogg's FiberPlus Antioxidants bars	http://www.kelloggsfiberplus.com/home.aspx

(continued)

Table 21.1 (continued)

Product company and brand name	Website[a]
Kemin Slendesta	http://www.kemin.com/nutraceuticals/slendesta
Kraft Lite & Reduced Fat Dairy Products	http://www.kraftrecipes.com/
La Tortilla Low Carb High Fiber Tortillas	http://www.latortillafactory.com/
Labradas Cookie Roll	http://www.labrada.com/
Lactaid	http://www.lactaid.com/
Lactel Jour Apres Jour	http://www.lactel.fr/chacun_son_lait/jaj. php?b=1&c=5&d=1
Lassonde Immunotec Research Prycena& Oasis Immuniforce	http://alassonde.com/en/products/oasis/health_break/ immuniforce.aspx
LifeTop culture straws	http://lifetop.eu/products/lifetop-straw
Lifeway Kefir, Basics Plus & BioKefir shots	http://www.lifeway.net/
Lipid Nutrition PinnoThin	http://www.lipidnutrition.com/ProductRange/ PinnoThin/
Live Great Foods Acclaim chocolate milk	http://www.dairyfoods.com/Articles/Dairy_News/ BNP_GUID_9-5-2006_A_10000000000000784975
Lomasalt	http://www.lohmann-chemikalien.de/
Marks & Spencer Super Juice	http://www.talkingretail.com/products/product-news/ marks-spencer-launches-new-cholesterol-lowering- super-juice
Martek DHA	http://www.martek.com/
Met Rx MRP	http://www.metrx.com/
Metagenics Ultrameal Rice Bar	http://www.metagenics.com/products/a-z-products-list/ UltraMeal-RICE-Bar
MiniChill	http://www.minichill.com/lab/
Minute Maid Heart Wise orange juice	http://www.minutemaid.com/
Modjo for Life	http://www.modjolife.com/start.php
MojoMilk	http://www.mojomilk.com/
Molico Beauty	http://www.nestle.com.br/site/marcas/Molico.aspx
MonaVie (M)mun	http://www.monavie.com/products/health-juices/mmun
Morton LiteSalt	http://www.mortonsalt.com/
Naka NutriHEART	http://www.nakaherbs.com/
Naked Juice Protein Juice Smoothie	http://www.nakedjuice.com/#
Nature's Plus Chocolate Berry Ultra Energy Acai Bar	http://www.naturesplus.com/products/productDetail.asp ?criteria=search&searchVar=99635&productnumbe r=99635&category=24
Nature's Plus Spiru-tein GOLD	http://www.naturesplus.com/
Neotame	http://www.neotame.com/
Nestle Boost Complete Nutritional Drink	http://www.boost.com/
Nestle Boost Kid Essentials	http://www.kidessentials.com/
NuGo Nutrition Bars	http://www.nugonutrition.com/
NuSalt	http://www.nusalt.com/
Nu-Tek	http://www.nu-teksalt.com/

(continued)

Table 21.1 (continued)

Product company and brand name	Website[a]
NuVim LactoMune LactoActin	http://nuvim.com/
Odwalla Super Protein & Protein Monster	http://www.odwalla.com/index.jsp
Old Home Foods SafflowerPower Yogurt	http://www.safflowerpower.com/
Olestra	http://www.pgfoodingredients.com/
Omega Pro eggs	http://www.burnbraefarms.com/consumer/our_products/omega_pro_liquid.htm
Optimum Nutrition Essential Amino Energy	http://www.optimumnutrition.com/products/essential-amino-energy-p-275.html
PaleoBars	http://www.designsforhealth.com/paleobar_choc.html
Pansalt	http://www.pansalt.ph/
Papetti Foods All Whites	http://www.allwhiteseggwhites.com/
Pharmavite SoyJoy fruit bar	http://www.otsuka.co.jp/en/product/
Pierre's Ice Cream Yovation	http://www.myyovation.com/
Pillsbury Sugar Free Cake Mix & Frosting	http://www.pillsburybaking.com/products/product_detail10.aspx?gclid=CLr8pubI_KcCFQnrKgodPGdLrg
POM Wonderful	http://www.pomwonderful.com/
PowerBar Performance Bar & Energy Bites	http://www.powerbar.com/
Pro Lab Lean Mass Creatine Bar.	http://www.prolab.com/products/protein
Probi ProViva	http://www.proviva.com/en/Magens-basta-van/History/Probi/
PROsnack Natural Foods Elevate Me	http://www.prosnack.com/
Protein Juice Smoothie,	http://www.nakedjuice.com/#
PureFit	http://www.purefit.com/
Red Bull	http://www.redbull.com/
Revival Crispy Bars	https://secure.revivalsoy.com/
Rubicon Food Products Papaya Drink with Lutein	http://www.rubiconexotic.com/
Saltrite	http://www.saltrite.com/
SkinCola	http://www.angelfire.com/ego2/skincola/
Skinny Water	http://skinnywater.com/
SmartSalt	http://www.smartsalt.com/
Snickers Marathon Protein Bar.	http://www.snickersmarathon.com/
Soylutions Earth Shake	http://www.soylutions.ca/eng/produits-originalEarth.html
Soylutions YU	http://www.soylutions.ca/eng/produits.html
SoZo Life & CoffeeBerry	http://www.sozolife.com/
Splenda	http://www.splenda.com/
Stoneyfield Farms	http://www.stonyfield.com/
Subway Omega-3 Bread	http://www.subwayfreshbuzz.com/
Sunsweet Prune Juice +	http://www.sunsweet.com/

(continued)

Table 21.1 (continued)

Product company and brand name	Website[a]
Supple Drink	http://supplebodies.com/?gclid=CPyjkY39g6gCFUa8K god9CkltQ
Sweet N'Low	http://www.sweetnlow.com/
SweetOne	http://www.sweetone.com/
Talin thaumatin	http://www.overseal.com/talin-natural-thaumatin.html
Think Products ThinkThin Bar	http://www.thinkproducts.com/2011/
Tip Top Up White Omega	http://www.tiptop.com.au/up/white-omega-3dha-thick
Tropicana Pure Premium Calcium + Vitamin D Orange Juice	http://www.tropicana.com/#/trop_home/home.swf
Truvia Rebiana	http://truvia.com/
Unilever Slim-Fast Meal Bar	http://www.slim-fast.com/products/mealbar/
Unilever Take Control	http://www.unileverfoodsolutions.com/
Unilever/Wall Moo frozen desert	http://www.unilever.com/innovation/buildingthefuture/ anunusualsourceofcalcium/
Vacation in a Bottle (ViB)	http://www.drinkvib.com/our-products/ingredients.html
Valio Gefilus and *Evolus*	http://www.valio.fi/portal/page/portal/valiocom/R_D/ Valion_tuotteet
*Valio Gefilus*cheese	http://ammattilaiset.valio.fi/portal/page/portal/ valiocom/Company_information/Products_ International_Sales/functional_prod- ucts18102006164724/ lactobacillus_gg19102006134654
Vanilla Taste of Beauty Embrace Beauty Drink	http://www.vanillabeauty.net/supreme.htm
Votre Vu Snapdragon Beauty Beverage	http://www.votrevu.com/corporate/public?page=/jsp/ home.jsp
Wellmune WGP	http://www.wellmune.com/
Wells Bluebunny Bomb Pop	http://www.bluebunny.com/ Products/d/Sugar_Free_Bomb_Pop
Westsoy SmartPlus	http://www.westsoy.biz/
WholeSoy Cultured Soy	http://www.wholesoyco.com/our-products/soy-yogurt/ item/soy-yogurt
WineTime	http://www.winetimebar.com/order.php
Xenadrine	http://www.xenadrine.com/products/drink/drinkmix. shtml
Yagua Beauty Juicer	http://www.yagua.nl/producten/beautyjuicer.php
Yagua Free Move	http://www.benseng.com/nieuws/ginseng/nieuws- ginseng-publicaties-2003.html#yagua
Yakult Provie & Pretio	http://www.yakultusa.com/
Yili Dairy Milk	http://www.yili.com/en/about_yili/history.shtml
Yoplait Yorik, Petits Filous, Nouriche & Yo-Plus	http://www.yoplait.com/

[a]Last accessed 20 June 2011

21.2 Functionality by Reduction and Replacement

Starting around the 1980s in the United States, issues relating health and the current food supply began to receive significant media attention. This raised public awareness and concern about the contents of common foods we eat. Certain ingredients in foods have come to be viewed as unhealthy and harmful when consumed in excess. Most often these are components found in processed foods, such as sodium, cholesterol, fat, *trans* fats, and sugar. Finding solutions to reduce and often replace these components has become a significant focus in the food industry, creating functional foods with both real and perceived health benefits. Some of the replacement ingredients have become controversial, leading to newer alternatives as this sector evolves. This impetus for change in the food industry has been brought forward by government regulation, improved packaging labels, food label claims, and improved consumer awareness and education.

21.2.1 Reduction of Sodium

High sodium content in manufactured foods, restaurant foods, children's foods, and in home cooking were noted as concerns in a 2007 World Health Organization study [3] and are also discussed in Chap. 11. These concerns result from data linking high sodium intake and hypertension, leading to government issued health claims such as the United States Food and Drug Administration (FDA) approved "diets low in sodium may reduce the risk of high blood pressure" [4]. Table salt has been recognized as a major point of individual control of sodium reduction in cooking and seasoning of food. With this market opportunity, many commercial tabletop salt substitutes have been created, a few examples being *Lite Salt, Nu-Tek, SmartSalt, Nu-Salt, AlsoSalt, Pansalt, SaltRite*, and *Lomasalt.* These various blends with reduced sodium have been formulated by using nonsodium mineral salts, such as potassium chloride and/or other flavor components with and without real salt. These are also being formulated into traditionally high-salt manufactured foods, such as cheese, processed meats, soups, and snacks. However, flavor changes generally become noticeable after a 25% reduction in the sodium content of the product. Alternatively, simple gradual reduction of salt in food products is also effective and over time individual taste perceptions adjust to accept reduced levels.

21.2.2 Reduction of Cholesterol

Cholesterol is only obtained in the diet from animal products. The clinical data behind the FDA claim that "diets low in saturated fat and cholesterol may reduce the risk of heart disease" [5] resulted in a number of animal-based whole foods that contain large amounts of cholesterol to come under attack: eggs, red meat, bacon, cheese, and other high-fat dairy products. Initial efforts to offer reduced-cholesterol products employed technology, such as supercritical fluid extraction [6] and steam stripping [7], to extract cholesterol. These processes proved effective but costly, so use has not been widespread. A current example utilizing this type of processing can be found in the *Benelact* line of dairy products. A cost-effective alternative was to remove the fat that contains the cholesterol. One innovation that has survived is extended shelf life (ESL) pasteurized liquid egg white sold in a carton, where the cholesterol-rich egg yolk has been removed. Examples of this include *ConAgra Foods Egg Beaters* and *Papetti Foods All Whites*. No need to crack eggs: the convenience, flavor, food safety, and health aspects have outweighed the added cost. Interestingly, this has created a safe option for consuming liquid (undenatured) egg protein for bodybuilders [8]. It is worthy to note that the industry has commissioned studies that indicate that for some products (e.g., eggs) consumer concerns about cholesterol content are not warranted, a topic also discussed in Chap. 9.

21.2.3 Reduction of Fat

High-fat foods have received attention as a health risk in part from the FDA claim stating "diets low in fat may reduce the risk of some cancers" [9]. Attempts in the 1990s were dramatic in removing fat completely from many foods. For higher fat foods, maintaining a desirable taste and texture as a "fat-free" product was a challenge. So a large commercial effort went into the development of fat-replacement ingredients. Many of these were digestible carbohydrates, so the caloric load of foods using these could still be deceptively high despite a "fat-free" label. These fat replacers included ingredients such as: (1) starches and maltodextrins from corn, tapioca, and potato; (2) fibers from oat, corn, rice, and pea; and (3) stabilizers such as pectin and microcrystalline cellulose. There was no "magic bullet" that could transform fat-laden foods into "fat-free" without obvious changes in flavor or texture. And the wisdom of removing fat from many foods was false as certain fats are essential and healthy in our diets. Yet certain categories of reduced fat products are commonplace in today's markets.

21.2.3.1 Reduced-Fat Dairy Foods

Manufacturers of high-fat dairy foods felt threatened by high fat concerns, so they countered with intense research efforts and a resulting host of new products. Skim milk was relabeled as "fat-free" and gained popularity over 3% fat whole milk. Many other high-fat dairy products had to be reformulated in more dramatic ways. Exemplifying the effort, *Kraft Foods* introduced lines of "reduced-fat" and "fat-free" dairy products that included cream cheese, processed cheese, and natural cheese. Consumer interest in many "fat-free" products declined as most products fell short in flavor and mouth feel. However, reduced-fat dairy products, especially fluid milk and yogurts, have done well.

21.2.3.2   Fat-Free Snack Foods

Looking at the huge snack food market in the United States, Proctor & Gamble developed *Olestra*: "the no-fat cooking oil with the full-fat flavor." A fat unlike any other, it is a nondigestible fat substitute that was approved for use in the USA in 1996. Still available, it has found a snack food niche, examples being *Frito-Lay's Light Potato Chips* and taco chip products and in *Pringles* fat-free chip products. But *Olestra* has had significant criticism since it was approved; most recently it has been found to interfere with the body's ability to regulate what it eats, resulting in overeating [10]. While it may have been a technological achievement to create a functional noncaloric fat, the product has limited appeal within the USA and has not been approved for use elsewhere.

21.2.3.3   Removal of *Trans* Fat in Foods

Trans fats are inadvertently formed when liquid oils undergo a synthetic process of partial hydrogenation to turn them into solid saturated fats that have a firm texture at room temperature. These fats have been used widely in processed foods for texture shelf stability, such as keeping peanut butter oil from separating and margarine spreadable. However, when health studies showed *trans* fats raised low-density lipoprotein (LDL or "bad" cholesterol) and increased risk of coronary heart disease, mandatory labeling for *trans* fats was initiated by the FDA in 2006 [11]. In the United States this initiated a massive effort to eliminate *trans* fats from foods such as margarine, bakery products, ice cream, and peanut butter, all traditionally using solid fats and shortenings. The search for replacements resulted in renewed use of palm, palm kernel, and coconut oils: a trio of tropical oils naturally high in saturated fats. These are the very oils that back in the 1980s businessman Phil Sokolof demonized after he survived a heart attack, resulting in a campaign that almost single-handedly influenced removal of saturated tropical oils from most food products in the US [12]. In response to removing tropical oils, hydrogenated fats were the replacement. So we have come full circle. The challenge now is to reduce saturated fats.

21.2.4   Reduction of Digestible Carbohydrates

Rising health alarms in the US of type 2 diabetes and obesity in adults and children [13, 14] have increased concern about the high use of caloric sweeteners in many beverages and foods. In 2002, dietary attention in the US became fixated on reduction of high levels of digestible carbohydrate in foods for weight loss, highlighted by the popularity of diets such as *Dr. Atkins Diet Revolution* [15], the *Zone Diet* [16], *Sugar Busters* [17], and the *South Beach Diet* [18]. While the trend of extreme reduction of all digestible carbohydrates in foods has not been sustained, the outcome has been a growth in awareness of the type and amount of carbohydrates in foods. Some countries have found a tool in the measurement of glycemic index (GI) as a means to differentiate food products for the rate at which they raise blood sugar after ingestion [19]. A measure of GI can be found on food labels in Australia, South Africa, and the United Kingdom.

Food carbohydrate caloric and GI reduction are achieved by a combination of means. Full or partial elimination of digestible sugars is paired with replacement by nonnutritive intense sweeteners and/or sugar alcohols to provide sweetness and reduction of carbohydrate calories. Replacement of a portion of any remaining carbohydrates with soluble and insoluble fibers provides further reduction of digestible carbohydrates.

21.2.4.1 Replacement with Intense Sweeteners

An "intense sweetener" is up to 600 times stronger in sweetness than table sugar and has little to no calories for the amount used. Common artificial intense sweeteners are *Sweet N'Low* (saccharin), *Equal* (aspartame), *Splenda* (sucralose), *Sweet One* (acesulfame K), *Neotame*, and cyclamate, where approved. Natural (not chemically synthesized) intense sweeteners include versions of stevia products: ranging from botanical stevia plant leaves (stevia or stevioside) to purified steviol glycoside rebaudioside A (Reb-A, *Rebiana*), and the blend *Truvia*, containing *Rebiana*, erythritol (sugar alcohol) and natural flavors. There are also the natural botanical extracts of licorice root (glycyrrhizin), the fruit based thaumatin (*Talin*), and Lo Han Kuo (*Fruit-Sweetness*, also known as Monk Fruit) that have intense sweetening properties.

21.2.4.2 Replacement with Sugar Alcohols

Sugar alcohols are available as both natural and synthetic derivatives of sugar, having fewer calories and lower GI than sugar due to slower and incomplete digestion [20, 21]. Less sweet than table sugar, they are synergistic with intense sweeteners to enhance sweetness and they provide bulk volume (it makes it easier for the common measure of "a spoonful" in blends such as *Truvia*). These include sorbitol, maltitol, isomaltitol, mannitol, lactilol, xylitol, erythritol, and glycerol. Sugar alcohols are not all similar in their blood sugar response or digestibility, including variable laxation effects [21]. The use of sugar alcohols has been critical to product functionality in development of numerous diabetic, low glycemic, no sugar added, and sugar-free ice creams, sweetened baked goods, chocolates, gums, and beverages. Examples include *HealthSmart Foods* sugarless *Chocolite* candy bars with erythritol, *Pillsbury Sugar Free* cake mixes and frostings with maltitol and sucralose, and *Blue Bunny* sugar free frozen novelty *Bomb Pop* with sorbitol.

21.2.4.3   Replacement with Dietary Fibers

Many diabetic, low glycemic, weight management, and/or prebiotic formulated foods use added dietary fiber. Soluble and insoluble fibers from many sources are available with variable fiber concentration. Commonly used, the soluble probiotic fibers known as inulin (natural) and fructo-ologosaccharides (natural and synthetic sources) are found in everything from beverages, soups, and yogurts to cereals, bars, and breads. Polydextrose is a sweet-flavored synthetic soluble fiber and bulking agent. A number of fibers, including inulin and polydextrose, are multifunctional as sweetness enhancers and carbohydrate-based fat replacers. Additional fibers include resistant (nondigestible) maltodextrins that are clear in beverages and resistant starches that are used in high-fiber cereals, breads, and pastas. Gums are also soluble fibers; particularly useful as fiber in foods are lower-viscosity products such as larch gum (aribinogalactan), low-viscosity guar gum, and gum arabic. Still important are the more traditional fibers found in various grains (wheat, soy), cereals (oat, barley, rice), fruits (apple, orange), vegetables (tomato, sugar beet, pea), seeds (cottonseed, flax, psyllium), and nuts (almond) that can be concentrated and prepared to very fine mesh powders.

21.3   Functionality by Addition

Dietary or nutritional supplements, also known as nutraceuticals, have no universal definition. They refer to vitamins, minerals, herbs, botanicals, concentrated protein, probiotics, fibers, and other specialty ingredients that are added to the diet for their purported health benefit. In the USA, food supplements are regulated by DSHEA (Dietary Supplement Health and Education Act of 1994) [22]. In Canada and the European Union, nutraceuticals and new ingredients, the latter by definition known as "Novel Foods" with no history of safe use prior to 1997 [23–25], are also regulated and must pass approval before they can be used in foods. What is available in one country may not be available in another, either due to marketing (the company making the product has not sought approval) or because the government did not approve its use in foods. The marketing of dietary supplements is discussed in greater detail in Chap. 20.

What follows is a sampling of foods and beverages, categorized by the form of supplement and/or health claims where they have been widely recognized. Due to the variation in regulations, many of the products described are considered as just "foods," while some are food supplements or nutraceuticals, depending on where it is available and what approvals the manufacturer has sought.

21.3.1 *Protein Addition*

Traditionally, concentrated protein in diets has come from meat, poultry, fish, egg, and dairy in the form of whole foods. Plant-based forms of whole food protein tend to be less concentrated and traditionally come from legumes, such as soybeans, chick peas, and lentils, that are amino acid complemented with cereal grains, such as corn, rice, and wheat. Isolating proteins from their sources is feasible with the technologies of ion exchange resin columns and/or membrane filtration (reverse osmosis, nanofiltration, ultrafiltration, and microfiltration) that allow commercial scale protein purification. These technologies allow commercial purification of many other ingredients and supplements too, often using only water as a solvent. Higher levels of protein supplementation are most common in beverage and bar type foods where they are often marketed as a healthy snack, a sports product, or a meal replacement. Increasingly, lower levels of protein supplementation are found in more traditional foods such as baked goods, frozen desserts, and puddings; these are reaching out to wider specialty markets, such as children, vegans, and persons needing gluten-free food.

21.3.1.1 Soy Protein

The FDA approved the following health claim for soy: "Diets low in saturated fat and cholesterol and that include soy protein may reduce the risk of heart disease" [26]. In addition, the natural presence of isoflavones, such as genistein and daidzein, in many soy protein foods are of interest to women's health as sources of phytoestrogens to relieve menopausal symptoms [27]. Soy protein isolates are used in nutritional beverages such as *Westsoy SmartPlus*, *Odwalla Super Protein*, *Soylution YU*, and *WholeSoy Cultured Soy*. Nutritional bars, such as *GeniSoy*, *PowerBar Performance Energy*, *SoyJoy*, *Revival Crispy Bars*, *PureFit*, *Clif Luna*, and *Builders*, may target for 6.25 g soy protein per serving to meet the heart healthy claim.

21.3.1.2 Other Plant Proteins

Alternative plant protein concentrates have become commercially available as vegetarian, gluten free, nonallergenic, and often organic options. Wheat protein, in the form of vital wheat gluten, is a classic meat alternative in the form of "Seitan," or it can be more concentrated as wheat protein isolate used in foods such as *Labradas* high protein *Cookie Roll*, *Dreamfields Low Carb Pasta*, and *La Tortilla Low Carb High Fiber Tortillas*. Rice protein concentrate is found in bars such as *Ultrameal Rice Bar*, *PaleoBars*, and *NuGo Nutrition Bars*, and beverage powders such as *Jarrow Formulas Brown Rice Protein*. Other protein concentrates that are commercially available include pea, hemp, flax, chia, and potato. These types of proteins are often found in various combinations, such as *Chia Greens Chia Energy Bar* and *Natures Plus Spiru-Tein GOLD Shake* with sprouted brown rice, chia, pea, and flax protein.

21.3.1.3 Dairy Protein

Concentrating fluid milk with ultrafiltration membranes can yield milk protein concentrates (MPC) and milk protein isolates (MPI), or if microfiltration is used, the major proteins are separated into micellar casein and native whey protein. Alternatively, fluid milk can be made into caseinates by a chemical process. Most whey proteins are obtained from whey (a by-product of cheese making) using ultrafiltration to make whey protein concentrate (WPC) and whey protein isolate (WPI). The high protein micellar caseins, caseinates, MPI, and WPI typically have lactose removed to the 1% or less range. Guidelines for categorizing dairy containing products as "lactose free" typically are less than 0.5 g lactose/serving.

Use of MPC and MPI provides a sustained protein digestion that may be beneficial to a variety of beverages and bars in areas of sports nutrition, meal replacement products (MRP), weight-loss products, or nutrition-focused snacks. Examples include sports beverage *MuscleMilk*, sports bar *Detour*, MRPs such as *MET-Rx MRP* and *EAS Myoplex*, and nutrition shakes such as the family of *Slim-Fast, Ensure*, and *Boost* products. Use of fast digesting, highly bioavailable WPI is popular in many sports nutrition supplements, yet WPI is also used in novel products such as a transparent beverage that looks like water called *Isopure*, a WPI fortified yogurt *Better Whey of Life* or a frozen dessert called *Arctic Zero*. Combinations of dairy and soy protein have become more common and found in a variety of products such as *Glucerna* nutritional products for diabetics, *Odwalla Protein Monster* protein shake, *Naked Juice Protein Juice Smoothie, Bolthouse Farms Protein Plus* drink, and bars such as *Zone Perfect, Balance Bar, Slim-Fast Meal Bar, ThinkThin* bar, and *Snickers Marathon Protein Bar.*

21.3.2 Ingredient Additions for Cardiovascular Health

Claims have been made for a variety of nutraceutical ingredients from different sources with regard to improving heart health. With heart disease still the leading cause of death in the United States [28], many look to these foods to improve their health. This is an immense topic that was also discussed in Chaps. 9–11.

21.3.2.1 Phytosterol Esters

Numerous clinical trials have shown that plant sterols can lower blood cholesterol [29]. The FDA allows a claim that "associates diets that include plant sterol/stanol esters with reduced risk of heart disease" [30]. Introduced in Finland, *Benecol* was the first margarine-type spread to contain stanol esters and is now available in 27 countries. It is found in beverages, yogurts, and candy chew type products. *Unilever's Take Control* is another margarine-type spread. Introduced in 2003, *Heart Wise* orange juice from *Minute Maid* has been supplemented with plant stanol esters and carries the claim "proven to help reduce cholesterol." Recently in Canada, *Danone* introduced *Danacol*, a drinkable yogurt with plant sterols.

21.3.2.2 Oat (1,4) Beta-Glucan

Another FDA-approved health claim addresses soluble fiber, including beta-glucan from whole oats: "Oat fibers may reduce the risk of heart disease" [31]. Aside from classic use in cereals, there are novel applications such as US based *First Foods* with *Oat Supreme* oat-based nondairy frozen desserts and an oat nondairy creamer. Popular oat-based products *Oatly Oat Drink* and *Oat Cream* are manufactured in Sweden by *Ceba Foods*, while *Biovelop AB* has their *PromOat* beta-glucan in *Super Juice* offered by *Marks & Spencer* in the UK.

21.3.2.3 Bioactive Milk Peptides

Bioactive peptides that have been shown to help control blood pressure can be made by hydrolysis of milk proteins using various lactic acid bacteria [32]. Functional beverages *Gefilus* and *Evolus* that promote reduced blood pressure are made by *Valio* dairy in Finland. These are based on patented fermented milk technology using probiotic strains of Lactobacillus GG. In Japan, dairy based *Yakult Pretio* is a fermented beverage to lower blood pressure.

21.3.2.4 *n*-3 Fatty Acids

Interest in *n*-3 fatty acids initially escalated with their correlation to reducing risk of coronary heart disease and a qualified (supportive but not conclusive research) FDA health claim that addressed this [33], a topic also discussed in Chap. 10. Potential health benefits have blossomed to include studies on stroke, arthritis, depression, and cancer. The major *n*-3 fatty acids are EPA (eicosapentanoic acid) and DHA (docosahexanoic acid), both primarily sourced from fish or krill oil, but also available from algae (*Martek*). Meanwhile, ALA (alpha-linolenic acid) is found in vegetable oils, seeds, and nuts, such as flax, hemp, chia, canola, and walnut [34]. Special chicken feed produces eggs enhanced with *n*-3 fatty acids. They are available across the US as *Eggland's Best Eggs*. Fortified *n*-3 bread is becoming widely available, from Australia with *Tip Top Up White Omega* fortified with encapsulated fish oil DHA to food chain *Subway Omega-3 Bread*. A patented margarine product, *Smartbalance* from *GFA Foods*, combines canola, soy, and olive oil with balanced *n*-6 and *n*-3 fatty acids for heart health.

21.3.3   Ingredient Additions for Bone and Joint Health

As populations age, such as the baby boomers in the United States, foods that assist in bone and joint health become more visible. However, calcium is critical for bone development at all ages, especially in growing children.

21.3.3.1   Calcium

Calcium is important in multiple health concerns. An FDA health claim relates the intake of calcium with a reduced risk of osteoporosis [35]. Emerging research is also finding that "dietary calcium plays a key role in the regulation of energy metabolism and obesity risk" and that dairy sources of calcium have an effect in decreasing body fat [36]. An ingredient innovation taking advantage of high-calcium by-products from dairy protein concentration has resulted in a natural, highly bioavailable calcium source called "milk minerals." It can be found in products such as *Yoplait's Petits Filous* yogurt for infants and toddlers. For those that are lactose intolerant, there is calcium fortified *Lactaid* fluid milk. While milk, yogurt, and cheese may have been traditional sources of dietary calcium, there are nondairy calcium-fortified products such as *Tropicana Pure Premium Calcium + Vitamin D* orange juice, *General Mills Total* cereal, and products directed toward children such as *Interstate Bakeries Wonder Bread* and *Unilever/Wall* Asian brand *Moo* frozen desserts.

21.3.3.2   Glucosamine and Chondroitin

Glucosamine and chondroitin are often used together to slow cartilage damage and alleviate arthritis [37]. Joint health beverages are becoming a popular vehicle to deliver both ingredients in a readily absorbed form. Product examples include *Joint Juice, Supple, Emergen-C Joint Health*, and *Logic's Juice4Joints*. A Dutch company, *Yagua*, manufactures a drink called *Free Move* with collagen hydrolysate and glucosamine to promote flexibility and relieve repetitive strain injury.

21.3.4   Ingredient Additions for Eye Health

Lutein and zeaxanthin, two carotenoid pigments, are antioxidants that have been associated with decreasing the risk of eye disease, such as age-related macular degeneration (AMD) [38]. Lutein can be found in beverages such as *Abbot Laboratories Glucerna* diabetic weight control meal replacement beverages *Sunsweet's Prune Juice +, Rubicon Food Products Papaya Drink with Lutein*, and *Glaceau Vitamin Water Focus*, and in bars such as *Flash-5 Energy Bars*, and *Ecco*

Bella' s Beauty Bliss chocolate bars. Lutein is also promoted in *Burnbrae Farms Break Free Omega Pro* liquid eggs.

21.3.5 Ingredient Additions for Boosting Gut Health and the Immune System

The terms prebiotic and probiotic are often found together as they are a synergistic combination for enhancing gut health and the immune system. Probiotic is defined as a "microbial culture beneficial to health by restoring balance to the intestinal flora," while prebiotics are "nondigestible oligosaccharides that support the growth of colonies of certain bacteria in the colon…so changing and possibly improving the colonic flora" [39]. But immune enhancing characteristics are not exclusive to probiotics and some other unique products are included in this category.

21.3.5.1 Prebiotics

Prebiotics are soluble fibers, meaning they are soluble in water, as opposed to insoluble fibers such as bran. There are many types of soluble fibers used in foods and all of these help to add bulk to the diet. However, only a limited number of these have been studied as prebiotics where they support the growth of bacteria in the gut. Of these, inulin and related fructo-oligosaccharides (including natural chicory root fiber) have become widespread in foods. The caution with these fibers is that if the dose is greater than about 5 g/serving, individuals that are not accustomed to high amounts of fiber in their diets may experience bloating and gas. The benefits outweigh the problems for most and these are found in combination with probiotics in many of the foods mentioned in the next section. They are also in foods without cultures, such as *Kellogg's FiberPlus Antioxidants* brand of bars, cereal, and waffles, *General Mills Fiber One* brand of bars, cereals, pancake mix, brownies, muffins, toaster pastries, cottage cheese, yogurt, and milkshake beverages, and natural cereal *Kashi U Black Currents & Walnuts* cereal.

21.3.5.2 Probiotics

Probiotic yogurts and beverages have been a success story in wellness foods. A Japanese fermented skimmed milk, *Yakult*, was introduced in Japan in 1935 and is now available in 30 countries. Originating in France, *Danone's* probiotic beverage *Actimel* has been introduced into 22 countries, while *Activia* yogurts and beverages are designed to help regulate the digestive system and are available in 30 countries. Another brand originating in France, *Yoplait*, is offered in eight countries with products such as *Yorik* fermented milk, *Nouriche* Smoothie, and *Yo-Plus* yogurt in the US with probiotic cultures and added calcium, vitamins A, D, and E, and fiber. In the US, *Stonyfield Farms'* line of yogurts and smoothie beverages contain six live cultures and include a soy yogurt, while *Lifeway Kefir* and *BioKefir* shots containing ten strains of bacterial cultures and two probiotics to support digestive health and immunity. *Lifeway* has expanded their line to include *Sweet Kiss* probiotic cheese. And in Finland, *Valio* was the first to gain worldwide license to probiotic culture Lactobacillus GG in 1987. Their most recent introduction in Belgium under the *Gefilus* label is a probiotic Emmantel-type cheese.

Not all probiotic products are dairy-based. *Attune Foods* offers a probiotic granola *Munch*. The Swedish firm *Probi* has licensed probiotic cultures to Skanemejerier that are found in fruit juice products under the name *ProViva*. And Kombucha Tea, made by culturing a sweetened black or green tea with what is called the "Kombucha mushroom" of yeast and bacteria, into a highly acidic brew that may be mixed with juice, such as *GT's Enlightened Organic Raw Kombucha*. In Norway, *BioSophia* makes *Graindrops* oat beverages. Featuring oat, rice, and spelt drinks, these cereal beverages are made by the Japanese Koji fermentation process creating organic "bio-dynamic" milk alternatives. In Finland, *Bioferme Ltd* makes probiotic oat products such as a yogurt like snack *Yosa*, a smoothie, and a food dip.

Thermal processing inactivates cultures, thereby losing probiotic functionality. However, there are two innovative solutions. First is *LifeTop* culture straws, developed between aseptic process developer *Tetra Pak* and *BioGaia AB*. The straw, internally coated with beneficial lactic acid bacterium, is wrapped and attached to the outside of beverage cartons. Upon consumption, the probiotic is released into the beverage. Products include *Orchard Maid* organic drinkable yogurt by *Farm Produce Marketing* in the UK, *Nestle's Boost Kid Essentials*, and *Yili Dairy* fluid milk in China. Second is a probiotic heat-resistant beneficial Bacillus spore, *GanedenBC*, patented by *Ganeden Labs*. This can be added directly to heat-treated beverages, baked goods, cereals, soups, candies, bars, etc., and the culture will survive processing. Among the growing number of products with this probiotic culture are *Agostoni Chocolate's* all natural probiotic dark chocolate, *Pierre's Ice Cream Yovation* frozen dessert, and *MojoMilk* probiotic chocolate milk drink mix, claiming ten times the active cultures found in yogurt.

21.3.5.3 Colostrum

Colostrum, a serum rich in immunoglobulin G (IgG) that comes from the first milk of mammals, is commercially isolated from cow's first milk [40]. Used as a boost to the immune system, it has been a popular supplement in Scandinavia, Saudi Arabia, and Asia. Beverages containing colostrums include a kefir beverage from *Lifeway* called *Basics Plus*, an antiaging juice drink from *Bomba Energy* called *SOL(e) UTION*, *NewLife Colostrum* from Australia, and New Zealand's *Bioactive Technologies Colostrum Sports Drink*.

21.3.5.4 Milk Micronutrients

NuVim is a unique milk-juice beverage supplemented with two patented micronutrients isolated from milk: *LactoMune* assists the human immune system and *LactoActin* promotes muscle flexibility and sturdy joints. These components are from the milk of select cows in New Zealand that are raised under a program that stimulates their immune system by using feeding and husbandry techniques. The overall impact of drinking *NuVim* is described as the maintenance of good health, more energy, and faster recovery times from physical activity.

21.3.5.5 Baker's Yeast (1,3) (1,6) Beta-Glucan (WGP 3–6)

Derived from baker's yeast, *Biothera* patented *Wellmune WGP* was 2011 front cover news in *Nature*, the international weekly journal of science, where research described how this biological response modifier activates immune receptors [41]. Not to be confused with oat (1,4) beta-glucan, the *Wellmune* product can be found in Canadian *Lassonde Oasis Immuniforce* juices, smoothies and soups, *Groupe Biscuits Leclerc Praeventia* cookies, *MonaVie (M)mun* fruit beverage, and *Fuze Healthy Infusions Defensify* beverage.

21.3.6 Ingredient Additions for Weight Maintenance, Muscle Mass, and Body Fitness

These products range from promoting satiety, to assisting in weight management, to promoting development of lean muscle mass. There is a crossover in many products between weight maintenance and body fitness.

21.3.6.1 Weight Management

The satiety effect of releasing hunger suppressing hormones such as cholecytstokinin (CCK) is suggested to be promoted by a trio of different ingredients. New ingredients include *Kemin's Slendesta*, a patented extract from potato protein and *Lipid Nutrition's PinnoThin*, an *n*-6 fatty acid, pinolenic acid, that is derived from pine nuts. A bioactive by-product from cheese making, glycomacropeptide (GMP), is a fragment of casein protein left in whey after cheese making that can be isolated. It is used as a component of weight-loss formulations, such as the beverage *Prycena* from *Immunotec Research*. Alternatively, GMP is one of the only known phenylalanine-free natural proteins and offers a protein option for individuals with phenylketonuria (PKU) [42].

21.3.6.2 Conjugated Linoleic Acid

Conjugated linoleic acid, a natural ingredient in dairy, beef, poultry, and eggs, yet commercially produced from safflower oil, is reported to decrease body fat and increase muscle mass [43]. It can be found in sports nutrition products, such as *Xenadrine* weight control bars, and beverages, such as *EAS Myoplex Deluxe*. It is also present in more mainstream products such as *Old Home Foods Safflower Power Yogurt* and *LiveGreat Foods Acclaim* chocolate milk.

21.3.6.3 Amino Acids

Amino acids, such as L-creatine, L-glutamine, L-carnitine, L-taurine, and branched chain amino acids (BCAA: L-leucene, L-valine, L-isoleucene), have become popular supplements in endurance, muscle building, and recovery sports products. Creatine is reported to increase lean muscle mass and have an "ergogenic effect" by storing energy in muscle. It is available in nutrition bars such as *ProLab's Lean Mass Creatine Bar*. Glutamine, reported to stimulate muscle growth and recovery, is added as glutamine peptides in products such as *BSN's* anabolic sustained release muscle building *Lean Dessert Protein*. *Optimum Nutrition* offers *Essential Amino Energy* with taurine, glutamine, arginine, leucine, isoleucine, valine, tyrosine, histidine, lysine, phenylalanine, threonine and methionine amino acids for a lean muscular physique. In *Power Bar Energy Bites*, BCAAs are added to the energy formulation and are found in a popular energy drink in Japan called *Amino Value*. An amino acid metabolite HMB (hydroxy-methylbutyrate) branded as *Revigor* is formulated with protein in *Abbott Laboratories Ensure Muscle Health*, a product to help rebuild muscle and strength naturally lost over time (as people age, decline, or are recovering from surgery).

21.3.7   Ingredient Additions for Improved Mood

The launch of the Austrian product *Red Bull* in 1987 initiated a category of energy beverages and shots that has grown in staggering proportions to hundreds of brands. With high consumption patterns in younger people, they are being criticized as harmful by the medical community [44, 45]. Ingredients, such as caffeine, taurine, glucoronolactone, inositol, guarana, yerba mate, L-carnitine, creatine, ginseng, and ginkgo biloba, along with vitamins and minerals, are combined in formulations to energize and stimulate.

A new category of beverages are the diametric opposite of energy drinks; these are known as relaxation beverages and are intended to help people de-stress and wind down, often with the aim of helping people to get to sleep. *Drank Extreme Relaxation Beverage* has melatonin and valerian root ingredients, similar to *iChill* relaxation shot made with melatonin, valerian root, B vitamins, and is stevia sweetened. In a slightly different vein, there are beverages that claim to help people unwind without drowsiness, such as *ViB* (vacation in a bottle), using ingredients from green tea extracts with B vitamins and amino acids L-threanine and L-threonine, or *MiniChill* with valerian root, GABA, L-threanine, and 5-HTP.

21.3.8 Ingredient Additions for Antiaging Benefits

Foods for healthy appearance initially were an extension of the cosmetics industry and were marketed like cosmetics. As more of these products adopted the use of antioxidants to promote health and beauty by slowing down cellular breakdown and inflammation, they have gained the attention of aging adults who want to look younger. This coincided with growing consumer awareness of phytochemicals in our diets and increased urging to consume more fruits and vegetables, such as the "My Plate" diet recommendations in the United States [46].

21.3.8.1 Beauty Foods

A concept popularized in Japan and Europe is the crossover of food and cosmetics, known as Cosmeceuticals or beauty foods, their function is to promote healthy appearance, especially of the skin, hair, and nails. Exemplified by the 2002 joint venture between French cosmetics firm *L'Oreal* and Swiss food giant *Nestle* to create the company *Laboratoires Inneov*, their objective is cosmetic nutritional supplements. The first food product from this venture has been *Inneov Firmness*, containing lacto-lycopene, vitamin C, and soy isoflavones, "targeted at women over 40 concerned with loss of cutaneous firmness." From Swiss cosmetics firm *Ecco Bella* is *Health by Chocolate*, "the delicious way to look beautiful" with cranberry seed oil, *n*-3 fatty acids, blueberry extract, lutein, lycopene, astaxanthin, and fiber. An American product introduced in 2003 to New York City is *SkinCola*, a beverage to hydrate the skin, being a flavorless, noncarbonated purified water with "activated oxygen," zinc, and vitamins, and now there is also *Borba Skin Balance Waters* in four versions that *Defy*, *Clarify*, *Firm*, or *Replenish* to provide "healthy skin from within." Dutch company *Yagua* offers a beverage *Beauty Juicer*, with grapefruit, ginger, white cocoa, collagen, and aloe vera to revitalize skin, hair, and nails. In France, *Lactel* created the product *Jour Apres Jour*, an antiaging milk with fiber, zinc, magnesium, selenium, and vitamins A, D, and E.

21.3.8.2 Polyphenols

Polyphenols, also called bioflavonoids, include a large number of bioactive plant substances that tend to be identified by color. Examples include resveratrol (as in wine), pterostilbene (as in blueberries), proanthocyanidins (as in cranberries), curcuminoid (as in turmeric), and catechins (as in green tea). These substances are ubiquitous in plants and are described as "an integral part of the human diet" [47].

These and other categories of polyphenols have gained attention for a myriad of potential health benefits including antioxidant activity, free radical scavenging properties, cardio-protective activity, reducing inflammation, and inhibiting cancer cells. However, the health implications of adding high levels of antioxidants to foods is being questioned as possibly increasing the risk of chronic disease by upsetting cell signaling pathways [48]. More research is needed. In the meantime, more foods are being marketed that contain high levels of polyphenols; examples include *WineTime* Bars, a dark chocolate with fiber and added resveratrol reported as equivalent to 50 glasses of wine, *Modjo for Life*, a rejuvenation beverage made from prickly pear cactus extract that claims five times more resveratrol than a liter of wine, *SoZo* functional fruit and vegetable extract beverage highlighting antioxidants from *CoffeeBerry* (the noncaffeinated fruit of coffee, not the bean) promoted as having 625 times more antioxidant value than blueberries, and *POM Wonderful* pomegranate juice that promotes more antioxidant potency than a glass of red wine. Originating from Japan, Matcha green tea powder has the equivalent of ten times the strength of standard green tea. Aside from the classic Matcha green tea beverage, it can be found in products such as *Earthling Organics Matcha Green Tea Bar* and *Elevate Me* Matcha green tea with cranberries, natural whey protein, and whole fruit bars. Lastly, combining a little bit of everything discussed in this chapter, *Good Cacao Superfood Chocolate* starts with 70% cacao and includes polyphenols from *AstaREAL* astaxanthin harvested from red microalgae, along with *Martek n-*3 DHA, *Wellmune WGP*, *FloraGlo* Lutein, *NutriHEART* plant sterols, and *Ganeden BC30* probiotics. Where will functional foods go from here?

21.4 Conclusion

The majority of these examples are unusual and nontraditional foods by historical standards; those in the first section (functionality by reduction and replacement) go back some 20 years, while those in the second section (functionality by addition) represent the more recent and evolving category most often referenced as functional food. In characterizing this broad range of functional foods, the American Dietetic Association (ADA) describes these as including "conventional foods" (e.g., probiotic yogurts), "modified foods" that have been fortified or enriched (e.g., calcium-fortified orange juice), "medical foods" that aid in the dietary management of a disease (e.g., diabetic beverage formulations), and "foods for special dietary use" sold at the retail level (e.g., gluten-free cookies) [49]. What has not been described in this chapter is the sector of conventional fresh whole foods, unprocessed, and unrefined. These are the simplest and most cost-effective forms of healthful functional foods that may be found in various grains, vegetables, fruits, seeds, nuts, milks, eggs, and fish.

Functional foods lack any standard definition despite their worldwide promotion and growth as a category of food, perhaps best fitting what the ADA describes as "a marketing rather than a regulatory idiom" [49]. By necessity, individual countries have had to create their working definitions to legally separate these health-focused foods from drugs. Functional foods, with their perceptions and claims of health benefits, are available at the retail level to consumers with no prescriptions. But lacking the regulatory controls of drugs, it is not clear where and if actual health benefits match claims. At the 2011 annual meeting of the Institute of Food Technologists, global consumer product and market research firm Mintel gave a report on functional foods, stating:

> .....along with the recession, growth in functional foods slowed, consumers were less likely to experiment with these products due to the cost...consumers don't really believe functional products are either effective or ineffective, and only one in five consumers see a benefit from the products. In addition, 68% of US consumers believe functional beverages should be tested by the FDA to make sure they actually do what they claim [50].

Food Business News highlighted their article on this by saying "functional foods [are] on life support." Other reports are more optimistic, with a UK Leatherhead Food Research group report pointing out that the "US functional food market [is] up 31% since 2006" and that the main health outcomes driving the market are antiaging, heart health, bone and joint health, weight management, gut health, energy/mood enhancing, and immune function [51].

Functional foods are a complex chapter in the history of food and what future it has in the marketplace will be vetted by further research and by reactions of consumers and government policies to what is learned.

References

1. FDA. Significant dates in U.S. Food and Drug law history. http://www.fda.gov/aboutfda/whatwedo/history/milestones/ucm128305.htm. Accessed 20 June 2011.
2. DeFelice S. Nutraceuticals: reflections. In: Pathak Y, editor. Handbook of nutraceuticals: ingredients, formulations, and applications, vol. 1. London: CRC Press; 2010. p. 356–7.
3. Elliot P, Brown I. Sodium intakes around the world. WHO. 2007. http://www.who.int/dietphysicalactivity/publications/sodium/en/index.html. Accessed 20 June 2011.
4. FDA. Code of Federal Regulations. 21CFR101.74. 2002;2:127–9.
5. FDA. Code of Federal Regulations. 21CFR101.75. 2002;2:129–31.
6. Hegenbart S. It's a gas. Food Prod Des. 1997(April):83–6.
7. Evaporator/Stripper. Artisan Industries Process Technologies. http://artisanind.com/ps/equipment/es_apps.html. Accessed 20 June 2011.
8. Egg Whites International. http://www.eggwhitesint.com/store.php?crn=66&action=show. Accessed 20 June 2011.
9. FDA. Code of Federal Regulations. 21CFR101.73. 2002;2:126–7.
10. Swithers SE, Ogden SB, Davidson TL. Fat substitutes promote weight gain in rats consuming high-fat diets. Behav Neurosci. 2011;125(4):512–8.
11. FDA. Trans fat now listed with saturated fat and cholesterol on the nutrition fats label. http://www.fda.gov/food/labelingnutrition/ConsumerInformation/ucm109832.htm. Accessed 20 June 2011.
12. Associated Press. 5 more companies to end use of tropical oils. Los Angeles Times, 17 Jan 1989.
13. FDA. The FDA's obesity working group report. 2006. http://www.fda.gov/Food/LabelingNutrition/ReportsResearch/ucm082094.htm. Accessed 20 June 2011.

14. CDC. Diabetes data and trends. 2011. http://apps.nccd.cdc.gov/DDTSTRS/default.aspx. Accessed 20 June 2011.

15. Atkins RC. Dr. Atkins diet revolution: the high calorie way to stay thin forever. New York: David McKay; 1972.

16. Sears B. The zone: a revolutionary life plan to put your body in total balance for permanent weight loss, higher energy, a happier state of mind, a healthier heart. New York: Harper Collins; 1995.

17. Stewart L. Sugar busters: cut sugar to trim fat. New York: Ballantine; 1998.

18. Agatston A. The South Beach Diet: the delicious, doctor designed, foolproof plan for fast and healthy weight loss. New York: Rodale Press; 2003.

19. University of Sydney. About glycemic index. http://www.glycemicindex.com/aboutGI.htm. Accessed 20 June 2011.

20. Calorie Control Council. Polyols. http://www.caloriecontrol.org/sweeteners-and-lite/polyols. Accessed 20 June 2011.

21. Part II. Reduced calorie sweeteners. In: O'Brien-Nabors L, editor. Alternative sweeteners. 3rd ed. New York: Marcel Dekker; 2001. p. 235–365.

22. FDA. Dietary supplements. http://www.fda.gov/food/dietarysupplements/default.htm. Accessed 20 June 2011.

23. Health Canada. Policy paper—nutraceuticals/functional foods and health claims on foods. http://www.hc-sc.gc.ca/fn-an/label-etiquet/claims-reclam/nutra-funct_foods-nutra-fonct_aliment-eng.php. Accessed 20 June 2011.

24. Health Canada. Guidelines for the safety assessment of novel foods. http://www.hc-sc.gc.ca/fn-an/legislation/guide-ld/nf-an/guidelines-lignesdirectrices-eng.php. Accessed 20 June 2011.

25. Europa. Novel foods—review of regulation (EC) 258/97. http://ec.europa.eu/food/food/biotechnology/novelfood/initiatives_en.htm. Accessed 20 June 2011.

26. FDA. Code of Federal Regulations. 21CFR101.82. 2002;2:144–6.

27. Han KK, Soares Jr JM, Haidar MA, et al. Benefits of soy isoflavone therapeutic regimen on menopausal symptoms. Obstet Gynecol. 2002;99:389–94.

28. CDC. Heart disease facts and statistics. http://www.cdc.gov/heartdisease/statistics.htm. Accessed 20 June 2011.

29. Institute of Food Science and Technology (UK). Information statement—phytosterol esters. http://www.ifst.org/science_technology_resources/for_food_professionals/information_statements/index/. Accessed 20 June 2011.

30. FDA. Code of Federal Regulations. 21CFR101.83. 2002;2:146–9.

31. FDA. Code of Federal Regulations. 21CFR101.81. 2002;2:141–4.

32. Gedes S, Harper WJ, Miller G. Bioactive components of whey and cardiovascular health. USDEC. 2001. www.wheyoflife.org/news/cardiohealth.pdf. Accessed 20 June 2011.

33. FDA. Summary of qualified health claims subject to enforcement discretion. http://www.fda.gov/Food/LabelingNutrition/LabelClaims/QualifiedHealthClaims/ucm073992.htm. Accessed 20 June 2011.

34. Rudra PK, Nair SSD, Leitch JW, Garg ML. Omega-3 polyunsaturated fatty acids and cardiac arrhythmias. In: Wildman EC, editor. Handbook of nutraceuticals and functional foods. Boca Raton, FL: CRC Press; 2001. p. 331–2.

35. FDA. Code of Federal Regulations. 21CFR101.72. 2002;2:124–6.

36. Zemel M, Miller S. Dietary calcium and dairy modulation of adiposity and obesity risk. Nutr Rev. 2004;62:125–31.

37. Arthritis Foundation. http://www.arthritistoday.org/treatments/supplement-guide/supplements/glucosamine.php. Accessed 20 June 2011.

38. Bartlett H, Eperjesi F. An ideal ocular nutritional supplement? Ophthalmic Physiol Opt. 2004;24:339–49.

39. Bender DA, Bender AE. In: Benders dictionary of nutrition and food technology. 7th ed. London: CRC Press; 1999. Prebiotic p. 325, Probiotic p. 326, Sugar alcohols p. 383.

40. Center for Nutritional Research. All about colostrum. http://www.icnr.org/all-about-colostrum/. Accessed 20 June 2011.

41. Wellmune. New research on biothera beta glucans is published in Nature. 2011. http://www.wellmune.com/en/2011/04/27/new-research-on-biothera-beta-glucans-is-published-in-nature/. Accessed 20 June 2011.
42. Van Calcar S, Gleason S, Hall A, Nelson K, Ney D. Using glycomacropeptide (GMP) for PKU diet treatment. National PKU News. Winter 2007. http://www.pkunews.org/research/UsingGMP.htm. Accessed 20 June 2011.
43. Gaullier J, Halse J, Hoye K, et al. Conjugated linoleic acid supplementation for 1 y reduces body fat mass in healthy overweight humans. Am J Clin Nutr. 2004;79:1118–25.
44. Sprinkle D. Energy drinks and the new food supplement continuum. Natural Products Insider, Dec 2010. http://www.naturalproductsinsider.com/articles/2010/12/energy-drinks-and-the-new-food-supplement-continuum.aspx. Accessed 20 June 2011.
45. Arria A, O'Brien M. The "high" risk of energy drinks. JAMA. 2011;305:600–1.
46. USDA. MyPlate. http://www.choosemyplate.gov/. Accessed 20 June 2011.
47. Ho C, Rafi M, Ghai G. 12.3 Health-promoting phytochemicals. In: Damodaran S, Parkin K, Fennema O, editors. Fennema's food chemistry. 4th ed. London: CRC Press; 2008. p. 753–64.
48. Finley J, Kong A, Hintze KJ, Jeffery EH, Ji LL, Lei XG. Antioxidants in foods: state of the science important to the food industry. J Agric Food Chem. 2011;59:6837–46.
49. ADA. Position of the American Dietetic Association: functional foods. J Am Diet Assoc. 2009;109:735–46.
50. Food Business News Consumer Trends. Mintel: functional foods 'on life support'. June 15, 2011. http://www.foodbusinessnews.net/News/News%20Home/Consumer%20Trends/2011/6/Mintel%20Functional%20foods%20on%20life%20support.aspx. Accessed 20 June 2011.
51. Nutraingredients-USA. US functional food market up 31% since 2006, says new report. June 13, 2011. http://www.nutraingredients-usa.com/Industry/US-Functional-food-market-up-31-since-2006-says-new-report. Accessed 20 June 2011.

Chapter 22
Food Industry and Political Influences on American Nutrition

Marion Nestle and Ted Wilson

Keywords Politics • Consumer rights • Public advocacy • Corporate self-regulation • Food and Drug Administration • US Department of Agriculture

Key Points

- The USDA is the government organization responsible for promoting healthy eating habits, but it is also under pressure to maintain consumption of foods purchased as commodities such as butter and sugar.
- The dollars spent on advertising by private industry far outweigh the dollars spent on public service nutrition information.
- Our food consumption patterns are influenced by commercial activities such as advertising. This starts in childhood where we develop the eating patterns we carry with us as adults.
- Legal measures (lawsuits) can be used by either public advocacy groups or private industry to influence the formation and implementation of nutritional policy.

M. Nestle, PhD, MPH(✉)
Department of Nutrition, Food Studies, and Public Health, New York University,
411 Lafayette Street, 5th Floor, New York, NY 10003-7035, USA

Department of Nutritional Sciences, Cornell University, Ithaca, New York
e-mail: marion.nestle@nyu.edu

T. Wilson, PhD
Department of Biology, Winona State University, Winona, MN 55987, USA
e-mail: twilson@winona.edu

N.J. Temple et al. (eds.), *Nutritional Health: Strategies for Disease Prevention*,
Nutrition and Health, DOI 10.1007/978-1-61779-894-8_22,
© Springer Science+Business Media, LLC 2012

22.1 "Influence: A Power Indirectly or Intangibly Affecting a Person or Course of Events"

We eat what we buy and decisions about what we buy are heavily influenced by the food industry, government policies, and our personal beliefs about food. A nutritionist's viewpoint with respect to eating is expected to be governed largely by objective, peer-reviewed science, but the beliefs of most people are insulated from that primary layer of information. In modern industrialized nations, most people's beliefs and behaviors about nutrition are influenced directly or indirectly by, amongst other factors, governmental nutritional policy. For all of us, appreciating the influence placed on our nutrition habits can be a tough pill to swallow.

Individual behavior and governmental policies are shaped by the pressures exerted by food companies. By exploring how this influence occurs, this chapter looks at the general nature of nutritional policy and its practical application by consumers in the United States. This chapter is based on the book *Food Politics: How the Food Industry Influences Nutrition and Health*, written by one of us (MN) [1]. The reader is directed to that book for a more detailed description of this history and examples of how US food policy has been manipulated by the food industry.

The starting point for any understanding of the food industry and its game plan is to appreciate that the primary goal for each and every food company is to expand market share and generate more profit. In that respect, the food industry is little different from most other industries. Take the car industry, for example. Ford regularly launches new models and then markets them to potential customers. Its goal is clearly to boost sales, increase its market share, and generate profits. In exactly the same way food companies launch new brands of breakfast cereals and then market them to the target groups of potential customers.

There is overwhelming evidence that food companies have very little interest in the effects of their products on the health of their customers. Sometimes the new product offers a health benefit to the consumer. For example, in recent years low-salt vegetable juices have been marketed. But more often new products are as unhealthy as the majority of other food products. The common denominator is that new foods are marketed if—and only if—higher sales might be generated.

The food industry is well aware that most of its sales and the lion's share of its profits come from the sale of unhealthy foods. For that reason, these are the ones that are most heavily advertised. But the food industry goes well beyond trying to increase sales by conventional marketing, it also actively lobbies governments so as to protect its commercial interests.

22.2 The Food Industry and Its Influence

Eating-related behavior, purchasing in particular, occurs in a very complicated social context in which marketers play a huge part. Advertising is the most obvious tool by which the food industry attempts to manipulate consumer habits. But other marketing strategies are also employed and these are usually much more subtle.

22.2.1   *Advertising and the Problem of Manufactured Food*

Tremendous amounts of money are spent on advertising to boost the consumption of particular products. In 1999 direct media spending (advertising that goes through agencies) by McDonald's, Burger King, Taco Bell, and Coke and Diet Coke was $627, $404, $207, and $174 million, respectively [2]. This total had increased to $5.6 billion in 2010 [3], with McDonald's contributing $1.2 billion of that sum. In addition, companies are investing heavily in social media. Coca-Cola's direct media expenditure of $758 million in 2010 [4] is likely to represent one-third to one-half of total marketing expenditures [5].

In light of the financial resources available for marketing, it is not surprising that the 2010 Dietary Guidelines and the new "MyPlate" program from the USDA have a lot of work to do if they wish to truly "change" the food habits of our nation. The federal government has done, and perhaps can do, little to redress this imbalance. As a result the quality of the American diet worsens and the prevalence of "diseases of the waistline" continues to grow, with diabetes now affecting approximately 10% of the population, a topic discussed in greater detail in Chap. 6 by Franz and Chap. 7 by Temple and Steyn.

Foods are consumed in their raw state that is commonly called a "whole food" or after processing. Processing in theory increases the palatability, appearance, marketability, and consumption of food products. The cost of the raw materials ("farm cost") is about 55% for eggs, 13% for frozen peas, 10% for corn flakes, 9% for canned tomatoes, and 4% for corn syrup [6]. The remainder of the consumer cost (45–90%) is for such things as transportation, labor, and packaging, and, most importantly, advertising to promote demand. On average, 80% of the cost of a food product went for these costs [7, 8], and this figure still holds [9].

Farm subsidies in the USA further exacerbate this problem by favoring corporate farmers producing corn and beans which have become "cheaper" in the last 40 years while largely unsubsidized fruits and vegetables have become more expensive [10].

For the manufacturer, it is enormously advantageous if the degree of food processing is increased because this allows huge price markups. The price per kilogram will be about three times higher for the manufactured foods than for the cost of the basic ingredients. Food corporations therefore have an obvious incentive to concentrate their advertising on manufactured food. If 10% of the product price is used for advertising and a raw apple costs 20 cents, then 2 cents would go to advertising the raw product. But if the apple is converted into an "instant microwaveable apple crisp" that sells for a dollar, then 10% means that 10 cents is now available for advertising. In other words, the funding available for advertising has jumped fivefold after the apple has been processed. And that, in brief, helps explain why so little money is spent on the advertising and promotion of raw food. Improving our nutrition may be as simple as teaching people to cook foods from raw ingredients [10], which explains the efforts of many advocates to restore cooking and home economics classes to public schools in the USA.

Food packaging includes FDA-approved labels that are "intended" to improve the consumers understanding of their nutritional "value," a problem highlighted in a recent commentary "Front-of-Package Food Labels: Public Health or Propaganda?" [11]. When people consume manufactured food, they generally have little idea of the nutritional value of their food and how well their diets conform to their nutrition guidelines. The content, convenience, and availability of these foods, and the money spent on their marketing, all hinder the implementation of healthful nutritional policy. The myriad of package claims is broad enough that Consumer Reports now has a special online tool for the consumer to understand what these claims are and determine if they carry any scientific merit [12]. The real nutritional value of food may be further obscured by the addition of supplemental vitamin and mineral "fortification." The FDA has recently sought to increase the documentation needed to support a claim and regularly sends warning letters [13], but it has also been met with resistance from industry arguing the First Amendment and their "right to free speech" on package labels.

22.2.2   Influential Association

One way the food industry influences people is through "image management." Consider, for example, a donation from a corporation that aggressively markets a product known to negatively affect the oral health of children. Should this act be judged as good for society or harmful? Illustrating this, in 2003 Coca-Cola donated a million dollars to the American Academy of Pediatric Dentistry (AAPD) for research [14]. The action runs the risk of undermining the science incriminating sugar-rich soda in poor dental health. It can potentially confuse the message of the AAPD which states that consumption of soft drinks may be harmful to dental health.

Following the successful drive to eliminate smoking, another movement was created that involved lawsuits blaming food companies for obesity. This lead to the introduction of campaigns and bills to prevent advertising to young people (e.g., HeLP or the Healthy Lifestyles and Prevention America Act of 2004), and schools banning soft drinks sales and pulling out of "pouring rights" and similar deals (in sum, these were exclusive arrangements to market soft drinks in schools in return for a payment). In 2005, the HeLP bill was introduced to congress, attempting to "improve the health of Americans and reduce health care costs by reorienting the Nation's health care system toward prevention, wellness, and self care," without success. Reintroduced in 2011, HeLP has been referred to the committee of finance but it not expected to pass this time either [15].

22.2.3   Schooling Young People to Become Adult Consumers

Pouring rights provide marketers with a valuable tool to influence brand name loyalty in children. This form of marketing undermines sound nutritional teaching and has been criticized as being exceptionally unethical. What many of these perceived healthy choices have in common is misleading. A fruit juice or fruit snack, for example, may contain little or no fruit, while its calories derive from added sugars or sweeteners. Acknowledging that very young people need help to make sound nutrition choices, Coca-Cola, in November 2003, vowed to improve its corporate image by, amongst other moves, refraining from promoting the sale of soft drinks to elementary students during the day [16]. While it has become less socially, and therefore financially, acceptable for beverage companies to enter into strict pouring rights agreements for soft drink distribution to junior and high school kids, universities are apparently another matter. Purdue University is one such school that recently entered into a pouring rights agreement with Coca-Cola [17].

School students are often targeted by food companies. Campbell's Soup, Pizza Hut, and McDonalds are just a few of the many manufacturers who create special "educational" materials that highlight the "selective nutritional value" of their products. These materials may focus on, for example, the variety of food types in a slice of pizza, but neglect to mention how the fat content and refined grains may contribute to obesity and diabetes in children who consume too much of them. These companies sometimes distribute alternative food guides that include pictures of their products and marketing symbols. Manufacturers know that children learn to recognize brand names, symbols, and cartoon/action figures as product advocates. The Keebler Elves' Cookies and Crackers, Tony Tiger's *GRRRRREAT* Frosted Corn Flakes, and Ronald McDonald the Clown are a few examples of cartoon personalities that carry a consumer from childhood products through to products marketed to adults.

In the USA, the Food and Trade Commission (FTC) is responsible for regulating advertising and, therefore, the marketing of food and beverages to children [18]. Companies are believed to spend at least $1.5 billion annually to market products to children. Food industry fears of the threat of direct government regulatory interference has led many to create FTC-approved "self-regulatory actions," such as the Children's Advertising Review Unit (CARU) by the advertising industry to prevent misleading children age 12 and under. The food industry response to concerns about federal interference in corporate policy has been robust with respect to preserving their interests [19]. Standards for improving food industry self-regulation have been suggested to include better transparency, meaningful objectives, accountability, and oversight [20].

22.3   Problems with the Food Supply

The USA is able to produce a surplus of food, something that is perhaps both a blessing and a curse for health. Historically, surplus agricultural commodities have been purchased and distributed to needy persons through a series of programs supervised by the USDA. These programs supported farm income and land values, and particularly distributed butter, cheese, meat, and other commodities.

Even though commodity distribution forms a smaller and smaller portion of federal food assistance, the USDA continues to be caught in a conflict of interest.

On the one hand, it is responsible for promoting "eating less" of foods such as meat, butter, and sugar. But, at the same time, it is under political pressure from various lobby groups to boost sales of the products it purchases as part of its price-support programs. Surplus commodities are distributed, in large part, through school lunch programs and to persons of low income. This may partly explain why the prevalence of obesity in the USA is inversely related to income [21]. School lunch programs are where many people learn some of their food preferences, and the poor nutritional habits observed when food commodities are dumped in this manner may well continue into adulthood. Similar problems occur in the behavior of adult consumers who want healthier food items on the menu, but typically choose less healthy options [22].

22.4 Influence of Special Interest Groups and Political Action Committees

There are a great number of special interest groups in the USA, and they exert a tremendous influence on nutritional health policy. Here is one example that implicitly makes this point. A press release from the Grocery Manufacturers of America (GMA) in 2002 reported that former Health and Human Services Secretary Tommy Thompson "encouraged GMA members to 'go on the offensive' against critics blaming the food industry for obesity" and "said the industry is 'doing wonderful things' to educate people about proper diet and exercise, but that GMA member companies should tell more people about them and implement wellness programs for their own employees" [23].

As explained earlier, the value added to food products during manufacturing and processing allows appreciably higher prices to be charged. The larger revenues generated thereby allow food corporations to influence the direction of nutritional health policy in both obvious and less obvious ways. Advertising was discussed earlier. Corporate funds are used to sponsor lobbyists who influence the writing and enforcement of laws. Special interest groups support lobbyists who work at the state and national levels of government. By giving contributions to campaign funds, corporations help ensure that politicians are sympathetic to the needs and wants of their donors. It has been estimated that in 1998 $2.7 million dollars were spent lobbying each US Senator and Representative [24]. In 2010, the $1.44 billion price tag from 1998 had more than doubled, reaching $3.5 billion [25].

22.4.1 The Sugar Industry Empire Strikes Back

The manipulation of policy is well-illustrated by looking at the activities of the sugar lobby. Federal nutritional policy guidelines for sugar consumption have changed in a clear direction of increasing obfuscation. In 1980 and 1985, the US *Dietary Guidelines* for sugar said, simply, and in just four words, "Avoid too much sugar." In 1990, it went to five words, "Use sugars only in moderation," and in 1995 to six: "Choose a diet moderate in sugars." In 2000, the scientific committee recommended that the guideline say, "Choose beverages and foods to limit your intake of sugars" (ten words), but even that was too strong. Under pressure from sugar lobbyists, the government agencies substituted the word "moderate" for "limit" so it read "Choose beverages and foods to moderate your intake of sugars." The committee working on the *Guidelines* for 2005 dropped the sugar recommendation entirely, and discussed the issue under the heading, "Choose carbohydrates wisely....," comments made by the USDA and DHHS, 2005 Report of the Dietary Guidelines Advisory Committee [26].

The 2010 Dietary Guidelines Advisory Committee report stated that "Americans of all ages … eat too much added sugars, solid fats, refined grains, and sodium." While this return to clarity will be helpful for many, the latter combination is regularly abbreviated as "SoFAS" within the report, potentially adding further confusion to their recommendations. As an example of double-speak, consider the recent Sugar Association reply: "The Sugar Association completed its review of the 112-page Guidelines policy document and has determined that the 2010 Dietary Guidelines do not include or establish an upper limit on sugar or added sugars intake. However, the Association does point out that throughout the text of the Guidelines, there is a call to reduce added sugars intake, which it believes is not science-based" [27].

By contrast, the American Heart Association recommends that added sugar should be limited to no more than 100 and 150 kcal/day for women and men, respectively [28]; this is equivalent to about 5–6% of energy. It is perhaps refreshing to see that the corporate interests for salt and sugar were unable to influence the design of the new MyPlate food guide which leaves no room for snacks and desserts.

The sugar industry's stated interpretation of the evidence has for many years stood in sharp contrast to the opinions of health professionals. A saga during 2003–2004 well illustrates this. WHO issued guidelines recommending a limit on sugar consumption to 10% of total caloric intake. This is an old recommendation, one used by many countries that issue dietary advice and the precise level recommended in the booklet that accompanies the USDA Food Guide Pyramid. Nevertheless, in a campaign characterized by the media and consumer groups as tantamount to blackmail, the year between the initial release and ratification of these guidelines saw vigorous attempts by the sugar industry to prevent their adoption [29]. The Sugar Association argued that the preponderance of evidence indicates that people can safely consume a quarter of their calories as sugar. The sugar lobby's efforts included demands that the report be removed from the internet and threats that the industry would be asking congressional appropriators to challenge the $406 million in US funding of WHO if the report was not withdrawn [30, 31].

22.4.2 The Influence of TV

When public service nutrition messages are played on TV, the station must be careful to avoid alienating their advertisers. Messages that promote moderation in calorie, fat, or sugar consumption can provide a valuable public health service, but airing these ads may negatively impact the station's relationships with advertisers from food and beverage companies. Profitability is required for the economic survival of private media outlets and the corporations they are part of. Predictably, a TV station in Iowa, "The Pork State," is likely to be reluctant to air a public service message that says: "…. reduce your fat intake ….eat less pork…."

By sponsoring the production and presentation of programs on private and public television, corporations also exert influence on the content of our programming. It would be naïve to believe that public TV programs, underwritten by food manufacturers, would unhesitatingly air a news item that was clearly damaging to the interests of their sponsors. As with private TV, there is no surer way to lose one's funding than to alienate one's sponsors.

22.4.3 Influence of Farm and Corporate Interests on Research

There are many commodity and farm/food product promotion boards and organizations in the USA. How do commodity groups generate the large amounts of money needed to influence nutritional health policy? Here is one example. The Beef Checkoff Program was established as part of the 1985 Farm Bill. The National Cattlemen's Beef Promotion and Research Board, which administers the national checkoff program, receives $1 per head on all cattle sold [32]. This program has been very successful in promoting the popular and familiar "Beef: It's What's for Dinner" program. While subject to USDA approval, the organization is a nongovernmental body and its revenues may be used for promotion, education, and research programs to improve the marketing climate for beef.

Marketing boards and corporate interests tend to fund research projects that have the potential to improve the marketability of their products. This seed funding has generated a literal explosion of research studies in areas of corporate/marketing interest. The following example illustrates this. A PubMed search for publications with the word "soy and human" found 78 hits for the 12 months of 1992, but a decade later this same period found 396 hits. Industrial giant Archer Daniels Midland (net sales of $30.7 billion in 2003, and $62 billion at the end of the fiscal 2010), and others like it, contribute millions of research dollars into selective projects related to soy proteins and isoflavones. By sponsoring advertisements in peer-reviewed journals, corporate soy advocates help ensure, directly or by indirect influence, that pro-soy publications will likely receive a more favorable passage through the peer-review process.

These selective projects generally have some features in common: if their findings are published, they will probably be positive, they will help ensure future grants, and the results will probably increase product marketability. The best evidence in support of this argument comes from the realm of medical research, especially research on drugs. A study of 332 randomized trials of drugs and other medical interventions published between 1999 and 2001 revealed that those that had received industry funding were 90% more likely to report statistically significant pro-industry findings [33]. This problem appears to be rampant with food research. An analysis was made of studies conducted between 1999 and 2003 on soft drinks, juice, and milk [34]. The findings of each study were classified as being favorable or unfavorable to the industry that sponsored the study. For interventional studies, none of the studies with industry financing reported a finding that was unfavorable whereas 37% of studies with no industry funding did so.

For all types of study (including observational studies and reviews), those with industry financing were 7.6 times more likely to report a finding favorable to industry than studies with no industry financing. Funding this type of research also places pressure on the USDA and other government sources to divert funds to validate and expand on what has already been done. This sequence of events means that the original grant, obtained from industry, has the potential to create an industry-friendly research agenda.

22.5   Using the Legal System to Influence Nutritional Policy

Corporate giants, such as ADM and Monsanto, have gone to great lengths to ensure that food labels do not provide information that could hurt product marketing [11]. If a corporation cannot influence nutrition, they can sometimes dictate it via the courts. Classic examples include attempts in recent years to improve consumer acceptance of genetically engineered grains with improved insect resistance and meats produced with bovine growth hormone. Another noteworthy episode concerns attempts to block the ability of the consumer to identify the source and methods used in the production of their food took the form of a lawsuit in Maine. Monsanto filed suit against Oakhurst Dairy, a small dairy company which placed the words: "Our Farmer's Pledge: No Artificial Growth Hormones Used" on its packaging. A settlement was reached where Oakhurst was permitted to keep their original claim as long as their packaging also included this statement: "FDA states: No significant difference in milk from cows treated with artificial growth hormones" [35]. Small companies often find the threat of legal action by a corporate giant is tantamount to blackmail, just by considering the costs of a legal defense.

Lawsuits are a two-way street. A variety of public interest organizations, such as the Center for Science in the Public Interest (CSPI), the Union of Concerned Scientists, and the Center for Food Safety, have turned to the legal system as a way to influence nutrition and food quality policy. By initiating lawsuits, food producers or government organizations can be influenced. The impact often comes more from the bad press that can be associated with a lawsuit, in addition to any fines that may result. More recently groups with little history of involvement in nutrition have weighed in, such as the Sierra Club and their dispute with the use of genetically modified grains in the USA and other countries.

Success in the USA with respect to controlling smoking and putting limits on advertising by the smoking industry has hinged in large part on the powers of State Attorney Generals. The national obesity epidemic presents a similar problem that is often managed at the level of a state. *Parens patriae* is the legal responsibility that State Attorney Generals have to promote the well-being of its residents and the interests of the states, although legal limits to this power differ widely between states [36]. They are particularly important for determining whether to enforce laws regarding Unfair and Deceptive Acts and Practices. State Attorney Generals can wield their power by directly filing lawsuits, by promoting/enforcing consumer protection, by drafting new legislation, and developing coordinated multistate actions to address a problem. Coordination between states and the federal government and the Federal Trade Commission (FTC) has also been an important aspect of control, especially with respect to food advertising directed toward children [18].

Self-regulatory bodies within areas of industry seek to limit the negative public image of industry and maintain independence from government interference. The Distilled Spirits Council represents one such self-regulatory body within the US alcohol industry which has had the most successful outcomes [20]. A key part of their "code" is the incorporation of responsible drinking statements in advertisements. However, their use of the phrase "drink responsibly" or "drink in moderation" at the end of a TV advertisement is of little consequence to the habits and actions of potential consumers.

22.6 Front Organizations to Influence Public Opinion

Front organizations provide another way for the food industry to influence lawmakers by creating and directing consumer hostility. Food taxes have been suggested as a way to improve the US diet and reduce sugar consumption [37, 38]. Americans Against Food Taxes is a classic "front organization". Their website [39] claims that they are:

> ..... a coalition of concerned citizens—responsible individuals, financially strapped families, small and large businesses in communities across the country—opposed to the government tax hikes on food and beverages, including soda, juice drinks, and flavored milks. The mission of the coalition is twofold: 1) To promote a healthy economy and healthy lifestyles by educating Americans about smart solutions that rely upon science, economic realities and common sense; and 2) To prevent the enactment of regressive and discriminatory taxes that will not teach our children how to live a healthy lifestyle, and will have no meaningful impact on public health, but will have a negative impact on American families struggling in this economy.

Perhaps we are cynical but with 499 corporate sponsors displayed on their website, there is little doubt as to their true motives.

22.7   British Food Policy Goes Bananas

Lest anyone think that the USA is the only place where the food industry has been allowed to hijack food policy, a look across the pond reveals some disturbing trends.

The UK has an agency—or at least it used to have—called the Food Standards Agency (FSA). This agency was set up by the government but had much autonomy. One of its great achievements was to invent the traffic lights food label which is described in Chap. 17 by Cuskelly, Woodside, and Temple. This food label system brought clarity and simplicity to the front of food packaging. In 2010, the UK elected a new government which implemented an economic policy similar to that advocated by the Republican Party in the USA. One of its first actions was to dismantle the FSA and incorporate its pieces into the Department of Health. By this means, the FSA was brought under more direct government control.

Is there a direct link between the traffic lights food label (and other excellent work of the FSA) and its demise? Quite simply, we don't know. But the following episode indicates what the food industry thinks about traffic lights food labels. In June 2010, the European Parliament was debating whether to implement this food labeling system within the European Union. The food industry spent an estimated $1.3 billion on lobbying to try and kill the plan [40].

Dismantling the FSA was merely the British government's first step in a reorientation of policies related to food and public health. In late 2010 the government announced their intention to set up several groups to provide advice on new policies, such as how best to tackle obesity. These groups would be dominated by fast food and processed food companies, including McDonald's, KFC, PepsiCo, and Mars [41]. For good measure, the lobby group representing the alcohol industry was made chair of the group in charge of formulating alcohol policy.

22.8   Conclusions

Today, Americans are locked into an ever-worsening spiral of obesity and diabetes, what are often called "diseases of the waistline." This tells us that American nutritional habits have become terribly misdirected. Vast amounts of money are now spent on advertising and other forms of marketing so as to generate consumer demand for food products much of which is unhealthy.

Lobbyists for the food industry have become entrenched in the political system; their primary function being to serve the narrow financial interests of particular farm and food industry interests. As a result, government food policy is often dictated by questions of economics rather than of health. Trade and political action groups also influence the availability of research funding and the direction of projects funded by otherwise "independent" universities and research centers. Finally, the legal system has become a tool where corporate interests can influence national health policy in such areas as food labeling. More recently, legal measures have also been implemented by consumer advocacy groups hoping to assert new policies for promoting consumer health.

A variety of voices will continue to influence the way our nutritional health is determined. What remains to be determined is how effective farm and industrial groups will be at influencing the formation of a policy that promotes economic success and profitability. What also remains to be determined is how effective consumer advocacy groups, such as the CSPI, can be at providing a counterbalance to corporate interests. What is most difficult to determine is the extent to which government organizations can be neutral with respect to setting nutritional health guidelines and legislation that ensures both the health of people and the economy.

It is also becoming apparent that transparency is critical with respect to the influence of funding to the nonprofit and advocacy agencies that make recommendations to government. The recent policy recommendations of the American Heart Association could go along ways toward creating a model for transparency in this respect [42].

Many people suppose that, knowing this, governments, like nutritionists, rely heavily on the best available health-related scientific evidence in formulating their nutrition policies. Alas, this is often far from the case. In practice, governmental nutritional policy translates nutrition science then tempers it within the current context of the relevant food market, which leaves it sometimes attempting to serve at once the disparate interests of individuals and economic interests of corporations. As such, the cart of nutritional policy oftentimes seems to be in the lead with the horse of scientific justification following some distance behind. This routinely seems to have been the case with how nutritional health policy was created, manipulated, and implemented in the USA.

What is the answer to our tempering the influence of the food and beverage industry on the formation of our government policy? Asking industry to "play nice" does not appear to be a realistic solution, neither does a strict Stalinist legal code. The solution is somewhere in between, and will be forged in part by the need of industry to maintain its "image" with respect to the risk of government regulatory interference, the need to maintain investor stock value, and the need to maintain enough public image to ensure that their products are not boycotted. The most important part of the solution is to recognize that influence can be a problem.

Acknowledgment   The authors thank Mr. David Barr for his contributions to this chapter.

References

1. Nestle M. Food politics: how the food industry influences nutrition and health. Berkeley: University of California Press; 2007.
2. Gallo AE. Food advertising in the United States. In: Frazao E, editor. America's eating habits: changes and consequences. Washington, DC: USDA; 1999.
3. Restaurant Ad Spending Tops $5.6 Billion in 2010. http://www.burgerbusiness.com/?p=6925. Accessed July 5, 2011.
4. Advertising Age. 100 Leading National Advertisers: 2010 Report. http://adage.com/datacenter/marketertrees2010. Accessed 1 June 2011.

5. Coca-Cola cut ad spend by 6.6% and invest more in social media. http://www.freshnetworks. com/blog/2011/03/coca-cola-cut-ad-spend-and-invest-more-in-socia-medi. Accessed 5 July 2011.
6. Dunham D. Farm costs from farm to retail in 1993. Washington, DC: USDA Economic Research Service; 1994.
7. Meade B, Rosen S. Income and diet differences greatly affect food spending around the globe. Food Review. 1996;19:39–44.
8. Haung KS. Prices and incomes affect nutrients consumed. Food Review. 1998;21:11–5.
9. Canning P. A revised and expanded food dollar series. USDA/ERS Report No. 114, Feb 2011. http://www.ers.usda.gov/Publications/ERR114. Accessed 5 July 2011.
10. Popkin B. Agricultural policies, food and public health. EMBO Rep. 2010;12:11–8.
11. Nestle M, Ludwig DS. Front-of-package food labels: public health or propaganda? JAMA. 2010;303:771–2.
12. Eco-labels center. http://www.greenerchoices.org/eco-labels. Accessed 1 July 2011.
13. Inspections, compliance, enforcement, and criminal investigations. http://www.fda.gov/ICECI/ EnforcementActions/WarningLetters/ucm261150.htm. Accessed 4 July 2011.
14. American Academy of Pediatric Dentistry Foundation. Campaign Quarterly: Healthy Smiles-Healthy Children. AAPDF, Chicago, IL, Fall 2003.
15. 174S: Healthy Lifestyles and Prevention America Act. http://www.govtrack.us/congress/bill. xpd?bill=s112-174. Accessed 3 July 2011.
16. Global School Beverage Guidelines. http://www.thecoca-colacompany.com/citizenship/ global_school_beverage_guidelines.html. Accessed 5 July 2011.
17. Coca-Cola Beverage Agreement: Frequently asked questions. http://www.eas.purdue.edu/ news/newsletter/Coke-FAQ_8-23-10.pdf. Accessed 5 July 2011.
18. Mello MM. Federal trade commission regulation of food advertising to children: possibilities for a reinvigorated role. J Health Polit Policy Law. 2010;35:227–76.
19. Brownell KD, Koplan JP. Front-of-package nutrition labeling—an abuse of trust by the food industry? N Engl J Med. 2011;364:2373–5.
20. Sharma LL, Teret SP, Brownell KD. The food industry and self-regulation: standards to promote success and to avoid public health failures. Am J Public Health. 2010;100:240–6.
21. Drewnowski A, Specter SE. Poverty and obesity: the role of energy density and energy costs. Am J Clin Nutr. 2004;79:6–16.
22. Nachay K. Would you like a salad with that? Food Technol. 2011;65:29–41.
23. Grocery Manufacturers of America, Inc. Top administration officials brief GMA board. Thompson, Hubbard, McClellan Give Views. GMA News Release. November 12, 2002. http:// www.gmabrands.com/news/docs/NewsRelease.cfm?DocID=1028. Accessed 29 July 2004.
24. Stout D. Tab of Washington Lobbying: $1.42 billion. New York Times, July 29, 1999: A14.
25. Lobbying database. http://www.opensecrets.org/lobby/index.php. Accessed 5 July 2011.
26. USDA and DHHS, 2005. Report of the Dietary Guidelines Advisory Committee. http://www. health.cov/dietaryguidelines. Accessed 7 Jan 2005.
27. Sugar Association Responds to 2010 Dietary Guidelines. http://www.iscnewsroom. com/2011/02/10/sugar-association-responds-to-2010-dietary-guidelines. Accessed 5 July 2011.
28. Johnson RK, Appel LJ, Brands M, et al; American Heart Association Nutrition Committee of the Council on Nutrition, Physical Activity, and Metabolism and the Council on Epidemiology and Prevention. Dietary sugars intake and cardiovascular health: a scientific statement from the American Heart Association. Circulation 2009;120:1011–1020.
29. The Sugar Association, letter addressed to Gro Harlem Brundtland, WHO, dated April 14, 2003, available July 27, 2004. http://www.commercialalert.org/sugarthreat.pdf. Accessed 5 July 2011.
30. Boseley S. Sugar industry threatens to scupper WHO. Guardian Unlimited: The Guardian (UK), April 21, 2003.
31. Sibbald B. Sugar industry sour on WHO report. CMAJ. 2003;168:1585.
32. Beef check-off. http://www.beefboard.org/, http://www.beefboard.org/producer/ CBBFinalUnderstandingBrochure.pdf. Accessed 5 July 2011.

33. Bhandari M, Busse JW, Jackowski D. Association between industry funding and statistically significant pro-industry findings in medical and surgical randomized trials. CMAJ. 2004;170: 477–80.
34. Lesser LI, Ebbeling CB, Goozner M, Wypij D, Ludwig DS. Relationship between funding source and conclusion among nutrition-related scientific articles. PLoS Med. 2007;4:e5.
35. Oakhurst slogan settlement was no win for Monsanto. Portland Press Herald. Dec 29, 2003; 10A.
36. Pomeranz JL, Brownell KD. Advancing public health obesity policy through state attorneys general. Am J Public Health. 2011;101:425–31.
37. Andreyeva T, Chaloupka FJ, Brownell KD. Estimating the potential of taxes on sugar-sweetened beverages to reduce consumption and generate revenue. Prev Med. 2011;52:413–6.
38. Brownell KD, Farley T, Willett WC, et al. The public health and economic benefits of taxing sugar-sweetened beverages. N Engl J Med. 2009;361:1599–605.
39. Americans Against Food Taxes. http://www.nofoodtaxes.com. Accessed 8 July 2011.
40. Food Standards Agency to be abolished by health secretary. http://www.guardian.co.uk/politics/2010/jul/11/food-standards-agency-abolished-health-secretary. Accessed 5 July 2011.
41. McDonald's and PepsiCo to help write UK health policy. http://www.guardian.co.uk/politics/2010/nov/12/mcdonalds-pepsico-help-health-policy. Accessed 5 July 2011.
42. Goldstein LB, Whitsel LP, Meltzer N, et al. American Heart Association and nonprofit advocacy: past, present, and future. A policy recommendation from the American Heart Association. Circulation. 2011;123:816–32.

Chapter 23
Nutrition Policy by Governments for the Prevention of Disease: Issues of Cost-effectiveness

Norman J. Temple

Keywords Cost-effectiveness • Nutrition policy • Government policy • Public health • Salt • Vitamin D • *Trans* fatty acids • Health promotion • Food labels • Food advertising • Food prices • Taxes on food price • Subsidies on food

Key Points

- Governments should implement nutrition policies that will improve population health.
- These policies include reducing the salt content of processed foods, use of dietary supplements of proven value, and eliminating hydrogenated oils that contain *trans* fatty acids from food.
- Implementation of the proposed policies would cost relatively little and should achieve significant health benefits within a few years.
- These policies therefore have a very attractive cost-effectiveness (i.e., they generate quality-adjusted life years (QALYs) at a fraction of the cost of many types of conventional medical treatment).

23.1 Introduction

It has been well established since the 1970s that dietary factors play a major role in the causation and prevention of a spectrum of diseases. These diseases have been referred to by different names, including Western diseases, noncommunicable diseases, and chronic diseases of lifestyle (CDL). The major CDL include most forms

N.J. Temple, PhD (✉)
Centre for Science, Athabasca University, Athabasca, AB, T9S 3A3, Canada
e-mail: normant@athabascau.ca

N.J. Temple et al. (eds.), *Nutritional Health: Strategies for Disease Prevention*,
Nutrition and Health, DOI 10.1007/978-1-61779-894-8_23,
© Springer Science+Business Media, LLC 2012

of cardiovascular disease (CVD) (including coronary heart disease (CHD), stroke, and hypertension), obesity, type 2 diabetes, and several major types of cancer.

The high prevalence of these diseases across the Western world has created immense pressures on health-care systems. This crisis is most severe in the United States where the cost of health care now exceeds $2 trillion and accounts for more than 16% of GDP. This level of spending has created great strain on both public and private finances. Unless drastic measures are taken, this spending is set to increase sharply over the next two decades. One factor driving this is that millions of baby boomers are now moving into their 60s. Another major factor is that the epidemic of obesity that has swept the world since the 1980s is now precipitating other health conditions such as type 2 diabetes. Compounding these problems, the relentless cost inflation of medical treatment has shown little sign of being brought under control.

The economic crisis that swept the world in late 2008, the worst since the 1930s, created enormous budget deficits for governments in many countries. Thus, while the cost of health care is on an ever-rising trajectory, governments have a reduced capacity to pay these costs. This crisis of over-spending is equally severe where medical costs are paid by individuals or by insurance companies: the individuals and companies who must pay the bills, whether directly or via insurance premiums, are also under much financial pressure resulting from the economic crisis.

There was one positive outcome from the economic crisis: a large section of the population woke up to the realization that the capitalist system requires careful government oversight. For many years, banks and investment companies operated with complete freedom to conduct business as they pleased with minimal government regulation or oversight.

But what has this to do with nutrition policy? Actually, there are strong parallels between the problems of the financial system and that of the national diet. The economic interests of the food industry have been the dominant driver of nutrition policy in most countries, including the USA. This has frequently occurred in disregard of the health impact of these policies [1]. This has directly led to many of the health problems that are so prevalent in today's society, and their huge economic consequences. In brief, a lack of government oversight of the banks and investment companies played a major role in the economic meltdown, while the willingness of governments to place the economic interests of the food industry above the health interests of the general population has caused a deterioration in public health and the consequent expanded cost of medical treatment.

This chapter argues the case for a strategic shift on nutrition policy in relation to population health. It is essential to see the consequences of nutrition policy, not merely in terms of the economic interests of the food industry, but also in terms of the huge cost of medical treatment resulting from diseases that can be prevented by a healthy diet. In many cases, it is possible to make a direct comparison between the cost-effectiveness of these policies with those of conventional medical treatments that target the same conditions. As we shall see, a nutrition policy approach often achieves far more benefit per dollar of expenditure than does medical spending.

23.2   The Cost-Effectiveness of Medicine

Many analyses have been made regarding the cost-effectiveness of medicine [2]. Benefits achieved as a result of medical interventions are often quantified based on how many quality-adjusted life years (QALY) are generated. The use of QALY allows all types of interventions to be directly compared, both those that prevent deaths and those that improve the quality of life. The cost-effectiveness of medical interventions can then be estimated based on cost per QALY. In the USA interventions that cost in the range of $50,000–100,000 per QALY are often regarded as being acceptable, though there is a lack of consensus on this. Other countries, such as the UK, often use lower cost thresholds.

The cost-effectiveness of medical interventions covers an extremely wide range. Here, we shall briefly examine some clinical preventative services, using estimates based on costs in the USA. Some such interventions are cost-saving. Examples are aspirin therapy for the prevention of CVD in persons at elevated risk, immunization of children, and screening for tobacco use followed by a brief intervention [3]. Nicotine replacement therapy costs less than $5000 per QALY [4], clearly making it highly cost-effective. Drug treatment of hypertension for nondiabetics is estimated to cost approximately $53,000 per QALY [5]. Statins are a family of drugs widely prescribed for the prevention of CHD. Their cost-effectiveness depends greatly on the level of risk of the patients being treated: the higher the risk, the lower the cost per QALY. Thus, for patients at high risk of CHD, their cost-effectiveness translates to a rather modest $20,000, or so, per QALY [6]. But this figure becomes tenfold higher for patients at intermediate risk of CHD [6]. For that reason, statins are justified only for people at high risk of CHD [7]. Nevertheless, these drugs are heavily marketed for patients at intermediate risk of CHD as this allows pharmaceutical sellers to hugely increase total sales. These costs for drug treatment of hypertension and for use of statins are based on the cost of brand-name medications. However, the use of generic drugs would reduce the cost per QALY by a factor of about five to eight (to about $7,800 for hypertension and $9,600–47,000, per QALY, for CHD) [8]. Many medical interventions are far more costly: around 1 in 11 preventative interventions cost more than a quarter million dollars per QALY [9].

While the focus above is clinical preventative services, the same problem of widespread use of medical procedures that exceed reasonable cost limits is seen in many other areas of American medicine. While this issue is much discussed in academic journals, there is seldom any serious attempt to bring this discussion into the public arena. A major reason for this is widespread opposition by politicians and others in the USA to the use of cost as a criterion for decision making [10].

Other countries take a very different approach on this issue. In the UK, a cost-effectiveness analysis (CEA) is routinely performed as part of the process of determining whether particular interventions can be employed by the National Health Service and therefore paid for using tax money. Such evaluations are done by the National Institute for Health and Clinical Excellence (NICE) [11].

23.3   Nutrition Policy and Public Health

The lesson from the previous discussion is that unless serious efforts are made to constrain costs, medical interventions can often be extremely expensive; many cost in the range $10,000–200,000 per QALY. A very different picture emerges when we examine what happens when governments implement policies designed to improve public health. These can often achieve widespread benefit at remarkably low cost. For example, a CEA has estimated that the mandatory use of daytime running lights and of motorcycle helmets is cost-saving [12], while a smoke-free workplace can prevent disease and do so at an estimated cost of a mere $500 per QALY [4]. Much the same is seen with regard to nutrition policies: a strong body of evidence reveals that they can also deliver major health benefits at relatively low cost. Unfortunately, few CEAs have been carried out in this area.

Here, we consider the nutrition policies where the evidence is strongest. The proposed interventions have been arranged in approximate order of cost-effectiveness, starting with the lowest cost. The policies have been broken into two groups:

Group A. There is a high probability that implementation of these policies will achieve significant health benefits within a few years and are cost-effective.

Group B. Here, the cost-effectiveness is much less clear and/or the health benefits may be delayed by many years.

23.3.1   Group A Nutrition Policies

23.3.1.1   Cutting the Salt Content of Food

A substantial body of evidence reveals that the great majority of people across the Western world consume a grossly excessive quantity of salt in their diets and that this plays a major role in the causation of hypertension [13, 14] and CVD [14, 15]. Salt is also believed to be an important causative factor in stomach cancer [16].

In order to substantially reduce the salt intake of the general population, it is necessary to cut the salt content of processed foods by at least half as this is where 75–80% of the salt in the diet comes from [17]. Feeding studies demonstrate that consumers have little problem accepting food with a much reduced salt content [18, 19]. As it would cost extremely little to implement this policy and the potential benefits are so large, it clearly follows that the cost-effectiveness would be highly favorable. Based on an Australian estimate, this policy would cost approximately US $1,180 per QALY [20]. According to an American estimate, cutting the sodium content of the diet to 2,300 mg/day would save around $18 billion/year in health-care costs [21]. Another study estimated that a population-wide reduction in dietary salt of 3 g/day (1,200 mg of sodium) would prevent between 44,000 and 92,000 deaths per year [22].

Should the governments of the USA and other countries implement this policy? To do otherwise would be heartless and a no-brainer! Alas, despite the strong case for action [23] there has been only inertia in the USA. As a result there was no change in the salt content of food between 2005 and 2008 [24]. This contrasts with the UK where policy implementation has lead to modest progress. In that country, the Food Standards Agency (FSA) embarked on a campaign to lower salt intake from 9.5 g/day in 2000/2001 to <6 g/day (roughly 2,300 mg sodium) by 2010. Actual intake by adults had fallen to 8.6 g/day in 2008 [25].

23.3.1.2 Dietary Supplements

There are several dietary supplements that have proven to be a highly effective, safe, and low-cost means to prevent disease. Iodine and fluoride are well-known examples. The prevention of spina bifida by the fortification of grain products with folic acid, a policy introduced in 1996, provides another illustration of the potential value of this strategy.

A strong argument can be made that vitamin D has the potential to repeat this success. The case for this is presented in Chap. 17. In brief, there is convincing evidence that the vitamin is of significant value in the prevention of osteoporosis. Strong evidence has also emerged in recent years that the risk of cancer is significantly reduced when circulating vitamin D levels are optimal. This benefit may also extend to CHD. Based on this evidence, a large section of the population would likely benefit from supplemental intake. This applies especially to people at risk of poor vitamin D status, notably people with darker skin color and inhabitants of northern latitudes, such as northern Europe, the northern states of the USA, and the whole of Canada.

According to a 2003 estimate, vitamin D supplements decrease fracture rates at a cost of $800 or less per vertebral fracture avoided [26]. This indicates that vitamin D supplements are highly cost-effective, especially for people aged over 40 or 50 who may be vulnerable to poor vitamin D status.

23.3.1.3 Trans Fatty Acids

Partially hydrogenated oils contain *trans* fatty acids. Major food sources of these fats include hard margarine, cakes, donuts, cookies, pastry, and deep-fried foods. *Trans* fatty acids are now recognized as adversely affecting multiple cardiovascular risk factors and contribute significantly to increased risk of CHD [27]. By one estimate if these fats were removed from the American diet, CHD rates would be reduced by 3–6%, possibly by as much as 12–22% [28].

There is nothing to stop governments implementing a policy requiring that hydrogenated oils containing *trans* fatty acids be removed from food. But as is often the case with important issues of public health, governments in several countries have chosen the path of relying on voluntary efforts by industry to reduce population intake of these fats [29]. Serious attempts have, however, been made in several jurisdictions to reduce consumption. Denmark stands as a fine example: that country passed laws that resulted, over the past decade, in the virtual elimination of these fats from food [30]. New York City and San Francisco have enacted regulations banning these fats from food sold in restaurants while British Columbia (Canada) has gone further and imposed a ban that affects not only restaurants but also bakeries and other food outlets.

Implementation of a policy that leads to the removal of *trans* fatty acids should prevent at least 13,000 CHD deaths per year in the USA. According to an estimate made in 2003 there would be a one-time industry cost of $139–275 million for testing, re-labeling, and reformulation [30]. The above information leaves little doubt that such a policy, calculated as dollars per QALY, would deliver improved health at bargain-basement prices.

23.3.1.4 Health Promotion

Since the 1970s, many health promotion interventions have been carried out in the USA and around the world. This subject is discussed in Chap. 18. Various types of health promotion interventions have been done in varied settings, including schools, worksites, physician offices, and in the community. Some have focused on just one or two lifestyle changes while others have been more wide ranging. The most common goals have been to reduce excess weight, lower the blood cholesterol and blood pressure, and encourage people to quit smoking and exercise more. Overall, the results have been mixed; typically, progress has amounted to no more than a few percentage points. This might be expected to reduce the risk of CHD by about 5–15%. Despite this limited impact, well-designed health promotion campaigns can be a cost-effective way to improve lifestyles and thereby positively affect the health of large numbers of people [31–33].

Much health promotion is carried out independently of governments. However, governments have the resources and authority required to implement health promotion programs across diverse settings. For that reason, the expansion of health promotion programs is best done as government policy.

23.3.1.5 Improved Food Labels

Food labels used in the USA and many other countries leave much to be desired; many consumers find them confusing. This issue is explored in Chap. 17.

A special issue of food labels concerns restaurants. These are where a substantial part of the population consumes much of their diet. Dinners in restaurants often supply huge amounts of food energy, as much as 1,000–1,500 kcal for a main course, but most people have little realization of this. Menus in restaurants are therefore another area where nutrition information is needed, especially the energy content of meals [34, 35]. As the cost of implementing this policy is quite low and it may help many people to control their energy intake, the policy is likely to be cost-effective.

Support for this policy approach has gathered speed in recent years. Legislation has been passed to implement this policy in several cities (including New York, San Francisco, and Philadelphia) and several states (including California, Oregon, and Massachusetts). The health-care law passed by Congress in 2010 turned this policy approach into federal law. It applies to all restaurant chains with 20 or more outlets.

23.3.2   Group B Nutrition Policies

We now examine nutrition policies where the cost-effectiveness is much less clear. This is often because the health benefits may be delayed by many years.

23.3.2.1   Nutrition Policy, Children, and Adolescents

Several important nutritional policies concern children and adolescents.

Food advertising on TV that targets children is overwhelmingly (80–90%) for unhealthy food choices or for fast-food restaurants [36, 37]. Not surprisingly, such advertising is often successful in inducing children to consume the advertised foods [38, 39] and is strongly associated with the risk of obesity in children and adolescents [40]. As the advertising of unhealthy foods is clearly detrimental to health, the obvious remedy is an outright ban. This has been done in Quebec (Canada), Sweden, and Norway [41]. A much more common approach, despite its limited effectiveness, is based on voluntary agreements and self-regulation by the food industry [41].

A study was recently carried out on 395 American public schools [42]. Vending machines were present in 82 and 97% of middle and high schools, respectively. Among schools where food was sold, approximately five in six of them permitted the sale of foods or beverages that are nutrient poor but high in energy (i.e., "junk food"). When schools permit the sale of unhealthy food, they are—implicitly—conveying an educational message that is the diametric opposite of the one stated in food guidelines. Schools should therefore be compelled to restrict the sale of unhealthy food. Likewise, where meals are served in schools, these should be of high nutritional quality. This can be justified as a means to directly improve the nutrition of children. One jurisdiction where such a policy has been implemented is Abu Dhabi (part of the United Arab Emirates). In 2010, it announced that it would implement a policy banning unhealthy food from schools [43]. Spain followed suit in 2011 [44].

The cost of these policies—on food advertising, on food sold in schools, and on school meals—is difficult to estimate. For example, many school authorities may oppose restrictions on the sale of unhealthy food as those sales can be a valuable source of extra income. However, from a societal perspective, the true cost of the proposed policies is likely to be reasonably low as spending lost in one area will be directed to other areas. In particular, as the sale of less healthy foods declines, sales of healthier foods will rise, and so will advertising for it.

The proposed policies should bring about a reduced prevalence of obesity. Another important benefit is that improved dietary habits of children and adolescents will, at least to some extent, carry over to middle age. Clearly, most of the health benefits of the proposed policies will accrue many years (~50 years) after the initial expenditures. As much is unknown with regard to both the cost and health benefits of these policies, their cost-effectiveness cannot be estimated with any accuracy. Nevertheless, much like environmental protection, expenditures today are justified based on their long-term benefits.

23.3.2.2 Food Prices: Using Taxes and Subsidies

Changing food prices by means of taxes and subsidies is an attractive means to encourage healthier eating patterns. The price structure of food in the USA and other Western countries means that less healthy foods with a high energy density (energy per gram) are generally cheaper than healthier food choices. Refined cereals and foods with added sugar and fat are amongst the cheapest sources of energy, whereas the more nutrient-dense foods, such as fish, lean meat, vegetables, and fruit, are generally much more expensive when the price is expressed as the cost of food energy (dollars per 1,000 kcal) [45–47]. For that reason, people with a low income are pressured to select relatively less healthy diets with a low content of several micronutrients (such as vitamin C and β-carotene) and a high energy density [48]. This is probably an important reason why the poorest people are often the least healthy.

An important factor responsible for the current price structure of food is that government subsidies are paid to agricultural enterprises with little regard for their nutritional consequences. For example, the subsides paid to American corn producers have made high-fructose corn syrup a cheap energy source. Between 1983 and 2009, the consumer price index in the USA rose by 210%. During those years, the price of carbonated beverages increased by only 152% while the price of fruit and vegetables jumped by 325% [49]. These price trends help explain the huge increase in consumption of soft drinks in recent decades [50]. Likewise, the Common Agricultural Policy of the European Union gives much higher subsidies to farmers for production of full-fat milk than for skimmed milk, while large quantities of fruit and vegetables are withdrawn from the market and destroyed so as to maintain high prices [51].

Studies on both smoking and alcohol have revealed "price elasticity" (i.e., consumption falls in response to a rise in price) [52–55]. The effect is stronger among the lower socioeconomic groups. What applies to tobacco and alcohol also applies to food [56]. A detailed analysis across the USA concluded that as the price of fast food rises, children and adolescents eat more fruit and vegetables and a generally healthier diet [57]. We can reasonably assume that the judicious use of taxes and subsidies can shift eating patterns in a healthier direction. This was explored by Jeffery, French, and colleagues in a series of studies conducted at worksites and in high schools in the USA [58–60]. They observed that halving the prices of healthier food choices (low-fat snacks sold in vending machines and of fruit and salad ingredients sold in cafeterias) lead to a doubling or trebling of sales. Consistent with these findings when shoppers in New Zealand were given price discounts of 12.5% on healthier foods, sales of those foods rose by 10 or 11% [61].

The above evidence points to the potential of government policies concerning prices of food to be an effective means to bring about desirable changes in eating patterns: the healthier choice must also be the affordable one. Taxes and subsidies could be used as tools to persuade people, for example, to consume whole grain bread rather than white, low-fat milk rather than full-fat milk, and chicken rather than beef. Researchers in the UK have made computer simulations of the effects of how this could be done and the likely effect [62, 63]. American advocates of this approach have proposed a tax on sugar-sweetened beverages (SSBs) [64]. Denmark has recently implemented such a policy with a 25% tax on ice cream, chocolate, and candy [65]. Both Denmark and France are planning a tax on SSBs, while Denmark also plans to impose a tax on foods containing >2.3% saturated fat.

The cost of the proposed changes to food prices may be quite high in the short term due to disruptions in patterns of agriculture and food production. However, in the longer term the cost should be minimal; depending on how the program is planned, extra costs in one area are cancelled out by savings in other areas. As is the case with several of the other policy proposals discussed here, the long-term health benefits are very difficult to estimate. For these reasons, more research is required before a realistic CEA can be made. This strategy is potentially one of the most powerful nutrition policy approaches but is also one the most challenging. There is likely to be much opposition from industrial lobbies.

A variation of a strategy based on direct changes to food prices is the provision of healthy foods to people by more direct means. The US government gives food assistance to the more needy members of its population by way of several different programs. The largest of these is the Supplemental Nutrition Assistance Program (SNAP; formerly the Food Stamp Program). These programs are a potential vehicle by which people could be encouraged to consume healthier foods [66]. As an example of this, studies in both the UK [67] and USA [68] reported an increased intake of fruit and vegetables when low-income women were given vouchers that could be exchanged for these foods.

23.4 Conclusion

The policy proposals discussed here fall into two distinct groups:

Group A. There is a high probability that implementation of these policies will achieve significant health benefits within a few years and are highly cost-effective. These policies include reducing the salt content of processed foods, use of dietary supplements of proven value (perhaps starting with vitamin D for selected population groups, such as middle-aged and elderly people who have a lack of sunshine exposure), eliminating *trans* fatty acids from food, carefully selected health promotion interventions, and improving food labels so that they present clear information on the health value of food (e.g., traffic lights labels and adding nutrition information to menus in restaurants, especially the energy content of meals). These policies are likely to cost well under $1,000 per QALY.

Group B. Here, the cost-effectiveness is much less clear and/or the health benefits may be delayed by many years. These policies include nutritional policies that target children and adolescents and changes to food prices by means of taxes and subsidies.

The proposals made here, especially those in Group A, should be seen as the basis of a new overall nutrition strategy. They should help counter the burden of CDL. This proposed strategy is best seen in the wider context: the objective of all policies and actions that impact on human health, whether carried out by government or the health-care industry, can be viewed as simply a means to generate QALYs. This encompasses policies as varied as improved road safety, removing hazardous chemicals from the environment, protecting the population from epidemics of infectious disease, giving dietary advice to the general population, and treating people with drugs to treat hypertension or cancer. Seen in this context, the proposed nutrition policies are a logical development of the above policies and actions.

Measured as dollars per QALY—or bang for the buck—the proposed nutrition policies generate QALYs far more cheaply than is achieved by many types of conventional medical treatment. By making comparisons as dollars per QALY it becomes clear that conventional medical interventions, such as the use of drugs for lowering blood cholesterol and controlling hypertension, typically cost many times more than nutrition policies to achieve comparable benefit: whereas medical interventions mostly cost in the range $10,000–200,000 per QALY, the proposed nutrition policies are likely to cost well under $1,000 per QALY.

There is one important barrier that stands in the way of the implementation of the policies discussed here, namely that health care is typically viewed as an essential service whereas improved nutrition policies are seen as having a much lower priority. We see this in the high priority politicians give to maintaining first class health care, at least for the majority of the population. Costs are seen as something that should be reduced where possible. Nutrition policies, by contrast, are discussed narrowly in terms of improving population health and have a far lower priority for most governments. One important factor responsible for this is pressure on governments by commercial interests: on the one hand, the pharmaceutical industry reaps vast profits from treating disease while the food industry has a vested interest in ignoring the health impact of its products. I argue here that it makes far more sense to evaluate the costs and potential value of health care and nutrition policies with the goalposts in a fixed position. Clearly, we need a paradigm shift.

Based on current trends, it is likely that in coming years the total cost of health care will steadily become higher and suck in an ever-greater proportion of the national economic pie. A logical response to this will be policy initiatives to reject the use of medical procedures that exceed preset limits, measured as dollars per QALY. As mentioned earlier, this policy has already been implemented in the UK. Proposals along these lines have been made for the USA [69]. The bottom line is that what makes obvious sense is to direct finite resources to where they can be most usefully deployed; to do otherwise should be seen as irrational. The proposed strategic shift on nutrition policy and health should be viewed from that perspective.

References

1. Nestle M. Food Politics. Berkeley, CA: University of California Press; 2002.
2. Neumann PJ. Using cost-effectiveness analysis to improve health care. Opportunities and barriers. Oxford: Oxford University Press; 2005.
3. Maciosek MV, Coffield AB, Edwards NM, Flottemesch TJ, Goodman MJ, Solberg LI. Priorities among effective clinical preventive services: results of a systematic review and analysis. Am J Prev Med. 2006;31:52–61.
4. Ong MK, Glantz SA. Free nicotine replacement therapy programs vs implementing smoke-free workplaces: a cost-effectiveness comparison. Am J Public Health. 2005;95:969–75.
5. Kahn R, Robertson RM, Smith R, Eddy D. The impact of prevention on reducing the burden of cardiovascular disease. Circulation. 2008;118:576–85.
6. Franco OH, Peeters A, Looman CW, Bonneux L. Cost effectiveness of statins in coronary heart disease. J Epidemiol Community Health. 2005;59:927–33.
7. Thompson A, Temple NJ. The case for statins: Has it really been made? J R Soc Med. 2004;97:461–4.
8. Shrank WH, Choudhry NK, Liberman JN, Brennan TA. The use of generic drugs in prevention of chronic disease is far more cost-effective than thought, and may save money. Health Aff. 2011;30:1351–7.
9. Cohen JT, Neumann PJ, Weinstein MC. Does preventive care save money? Health economics and the presidential candidates. N Engl J Med. 2008;358:661–3.
10. Grosse SD, Teutsch SM, Haddix AC. Lessons from cost-effectiveness research for United States public health policy. Annu Rev Public Health. 2007;28:365–91.
11. Pearson SD, Rawlins MD. Quality, innovation, and value for money: NICE and the British National Health Service. JAMA. 2005;294:2618–22.
12. Graham JD, Corso PS, Morris JM, Segui-Gomez M, Weinstein MC. Evaluating the cost-effectiveness of clinical and public health measures. Annu Rev Public Health. 1998;19:125–52.
13. Sacks FM, Svetkey LP, Vollmer WM, et al. for the DASH-Sodium Collaborative Research Group. Effects on blood pressure of reduced dietary sodium and the Dietary Approaches to Stop Hypertension (DASH) diet. DASH-Sodium Collaborative Research Group. N Engl J Med. 2001;344:3–10.
14. He FJ, MacGregor GA. A comprehensive review on salt and health and current experience of worldwide salt reduction programmes. J Hum Hypertens. 2009;23:363–84.
15. Cook NR, Cutler JA, Obarzanek E, et al. Long term effects of dietary sodium reduction on cardiovascular disease outcomes: observational follow-up of the trials of hypertension prevention (TOHP). BMJ. 2007;334:885–8.
16. World Cancer Research Fund/American Institute for Cancer Research. Food, nutrition, physical activity, and the prevention of cancer: a global perspective. Washington, DC: AICR; 2007.
17. Hooper L, Bartlett C, Davey SG, Ebrahim S. Advice to reduce dietary salt for prevention of cardiovascular disease. Cochrane Database Syst Rev. 2004;1:CD003656.
18. Beauchamp GK, Bertino M, Engelman K. Failure to compensate decreased dietary sodium with increased table salt usage. JAMA. 1987;258:3275–8.
19. Girgis S, Neal B, Prescott J, et al. A one-quarter reduction in the salt content of bread can be made without detection. Eur J Clin Nutr. 2003;57:616–20.
20. Neal B. The effectiveness and costs of population interventions to reduce salt consumption. WHO, Geneva, 2007. http://www.who.int/dietphysicalactivity/Neal_saltpaper_2006.pdf. Accessed 22 Oct 2009.
21. Palar K, Sturm R. Potential societal savings from reduced sodium consumption in the U.S. adult population. Am J Health Promot. 2009;24:49–57.
22. Bibbins-Domingo K, Chertow GM, Coxson PG, et al. Projected effect of dietary salt reductions on future cardiovascular disease. N Engl J Med. 2010;362:590–9.
23. Temple NJ. Population strategies to reduce sodium intake: the right way and the wrong way. Nutrition. 2011;27:387.

24. Center for Science in the Public Interest. Industry not lowering sodium in processed foods, despite public health concerns, 2008. http://www.cspinet.org/new/200812041.html. Accessed 6 Jan 2009.

25. Food Standards Agency. Dietary sodium levels surveys. 22 July 2008. http://www.food.gov.uk/science/dietarysurveys/urinary. Accessed 9 Nov 2008.

26. Buckley LM, Hillner BE. A cost effectiveness analysis of calcium and vitamin D supplementation, etidronate, and alendronate in the prevention of vertebral fractures in women treated with glucocorticoids. J Rheumatol. 2003;30:132–8.

27. Mozaffarian D, Aro A, Willett WC. Health effects of trans-fatty acids: experimental and observational evidence. Eur J Clin Nutr. 2009;63 Suppl 2:S5–21.

28. Mozaffarian D, Katan MB, Ascherio A, Stampfer MJ, Willett WC. Trans fatty acids and cardiovascular disease. N Engl J Med. 2006;354:1601–13.

29. L'Abbé MR, Stender S, Skeaff CM. Ghafoorunissa, Tavella M. Approaches to removing *trans* fats from the food supply in industrialized and developing countries. Eur J Clin Nutr. 2009;63:S50–67.

30. U.S. Department of Health and Human Services, 2003. Food labeling: trans fatty acids in nutrition labeling, nutrient content claims, and health claims, Final Rule. Fed Regist. 2003;68(133):41434–506.

31. Aldana SG. Financial impact of health promotion programs: a comprehensive review of the literature. Am J Health Promot. 2001;15:296–320.

32. Golaszewski T. Shining lights: studies that have most influenced the understanding of health promotion's financial impact. Am J Health Promot. 2001;15:332–40.

33. Carnethon M. Whitsel LP, Franklin BA, et al; American Heart Association Advocacy Coordinating Committee; Council on Epidemiology and Prevention; Council on the Kidney in Cardiovascular Disease; Council on Nutrition. Physical Activity and Metabolism Worksite wellness programs for cardiovascular disease prevention: a policy statement from the American Heart Association. Circulation. 2009;120:1725–41.

34. Berman M, Lavizzo-Mourey R. Obesity prevention in the information age: caloric information at the point of purchase. JAMA. 2008;300:433–5.

35. Roberto CA, Schwartz MB, Brownell KD. Rationale and evidence for menu-labeling legislation. Am J Prev Med. 2009;37:546–51.

36. Harrison K, Marske AL. Nutritional content of foods advertised during the television programs children watch most. Am J Public Health. 2005;95:1568–74.

37. Batada A, Seitz MD, Wootan MG, Story M. Nine out of 10 food advertisements shown during Saturday morning children's television programming are for foods high in fat, sodium, or added sugars, or low in nutrients. J Am Diet Assoc. 2008;108:673–8.

38. Coon KA, Tucker KL. Television and children's consumption patterns. A review of the literature. Minerva Pediatr. 2002;54:423–36.

39. Wiecha JL, Peterson KE, Ludwig DS, Kim J, Sobol A, Gortmaker SL. When children eat what they watch: impact of television viewing on dietary intake in youth. Arch Pediatr Adolesc Med. 2006;160:436–42.

40. Chou S, Rashad I, Grossman M. Fast-food restaurant advertising on television and its influence on childhood obesity. J Law Econ. 2008;51:599–618.

41. Caraher M, Landon J, Dalmeny K. Television advertising and children: lessons from policy development. Public Health Nutr. 2006;9:596–605.

42. Finkelstein DM, Hill EL, Whitaker RC. School food environments and policies in US public schools. Pediatrics. 2008;122:e251–9.

43. El Shammaa D. Authority pushes for healthier school meals. Gulf News, 14 November 2010.

44. de Lago M. Spain bans sale of unhealthy food in schools in bid to tackle obesity. BMJ. 2011;342:4073.

45. Drewnowski A, Darmon N. Food choices and diet costs: an economic analysis. J Nutr. 2005;135:900–4.

46. Jetter KM, Cassady DL. The availability and cost of healthier food alternatives. Am J Prev Med. 2006;30:38–44.

47. Drewnowski A, Monsivais P, Maillot M, Darmon N. Low-energy-density diets are associated with higher diet quality and higher diet costs in French adults. J Am Diet Assoc. 2007;107:1028–32.
48. Darmon N, Drewnowski A. Does social class predict diet quality? Am J Clin Nutr. 2008;87: 1107–17.
49. Brownell KD, Frieden TR. Ounces of prevention—the public policy case for taxes on sugared beverages. N Engl J Med. 2009;360:1805–8.
50. Popkin BM, Kiyah Duffey K. Sugar and artificial sweeteners: seeking the sweet truth. In: Wilson T, Temple NJ, Bray GA, Boyle Struble M, editors. Nutrition guide for physicians. New York: Humana; 2010. p. 25–38.
51. Elinder LS, The EU. Common agricultural policy from a health perspective. Eurohealth. 2004;10:13–6.
52. Meier KJ, Licari MJ. The effect of cigarette taxes on cigarette consumption, 1955 through 1994. Am J Public Health. 1997;87:1126–30.
53. Anderson P, Lehto G. Prevention policies. Br Med Bull. 1994;50:171–85.
54. Herttua K, Mäkelä P, Martikainen P. Changes in alcohol-related mortality and its socioeconomic differences after a large reduction in alcohol prices: a natural experiment based on register data. Am J Epidemiol. 2008;168:1110–8.
55. Anderson P, Chisholm D, Fuhr DC. Effectiveness and cost-effectiveness of policies and programmes to reduce the harm caused by alcohol. Lancet. 2009;373:2234–46.
56. Andreyeva T, Long MW, Brownell KD. The impact of food prices on consumption: a systematic review of research on the price elasticity of demand for food. Am J Public Health. 2010;100:216–22.
57. Beydoun MA, Powell LM, Chen X, Wang Y. Food prices are associated with dietary quality, fast food consumption, and body mass index among U.S. children and adolescents. J Nutr. 2011;141:304–11.
58. French SA, Story M, Jeffery RW, et al. Pricing strategy to promote fruit and vegetable purchase in high school cafeterias. J Am Diet Assoc. 1997;97:1008–10.
59. French SA, Jeffery RW, Story M, et al. Pricing and promotion effects on low-fat vending snack purchases: the CHIPS Study. Am J Public Health. 2001;91:112–7.
60. Jeffery RW, French SA, Raether C, Baxter JE. An environmental intervention to increase fruit and salad purchases in a cafeteria. Prev Med. 1994;23:788–92.
61. Ni Mhurchu C, Blakely T, Jiang Y, Eyles HC, Rodgers A. Effects of price discounts and tailored nutrition education on supermarket purchases: a randomized controlled trial. Am J Clin Nutr. 2010;91:736–47.
62. Nnoaham KE, Sacks G, Rayner M, Mytton O, Gray A. Modelling income group differences in the health and economic impacts of targeted food taxes and subsidies. Int J Epidemiol. 2009;38:1324–33.
63. Tiffin R, Arnoult M. The public health impacts of a fat tax. Eur J Clin Nutr. 2011;65:427–33.
64. Brownell KD, Farley T, Willett WC, et al. The public health and economic benefits of taxing sugar-sweetened beverages. N Engl J Med. 2009;361:1599–605.
65. Wilkins R. Danes impose 25% tax increases on ice cream, chocolate, and sweets to curb disease. BMJ. 2010;341:c3592.
66. Finkelstein DM, French S, Variyam JN, Haines PS. Pros and cons of proposed interventions to promote healthy eating. Am J Prev Med. 2004;27(3 Suppl):163–71.
67. Burr ML, Trembeth J, Jones KB, Geen J, Lynch LA, Roberts ZE. The effects of dietary advice and vouchers on the intake of fruit and fruit juice by pregnant women in a deprived area: a controlled trial. Public Health Nutr. 2007;10:559–65.
68. Herman DR, Harrison GG, Jenks E. Choices made by low-income women provided with an economic supplement for fresh fruit and vegetable purchase. J Am Diet Assoc. 2006;106:740–4.
69. Temple NJ, Thompson A, editors. Excessive medical spending: facing the challenge. Oxford: Radcliffe; 2007.

Chapter 24
Use of Biotechnology to Increase Food Production and Nutritional Value

Scott P. Segal, Travis J. Knight, and Donald C. Beitz

Keywords Biotechnology • Genetically modified organisms • Food safety • Genetically modified plants • Genetically modified animals • Food production

Key Points

- Today's technology allows for detailed genetic manipulation of plants and animals.
- Biotechnology includes the use of other methods, such as the production and in vitro purification of recombinant peptides and antisense technology.
- Adding genes or knocking down their expression in plants or animals can result in significantly increased food production as well as foods with improved nutritional or processing characteristics.
- There is no clear scientific evidence to support the idea that transgenic proteins could harm consumers or cause an increased chance of allergic reaction when compared with native proteins in common foods.
- Many concerns exist with regards to the environmental safety of using genetically modified organisms (GMOs) in large-scale crop and livestock production.
- Although metabolomics is attempting to solve this problem, there is no clear scientific data to support the idea that GMOs have the same effects on human health as do the unmodified organisms.

S.P. Segal, PhD (✉)
Department of Biology, Winona State University, 240 Pasteur Hall, Winona, MN 55987, USA
e-mail: ssegal@winona.edu

T.J. Knight, PhD
Iowa Department of Agriculture and Land Stewardship, 2230 South Ankeny Boulevard, Ankeny, IA 51023, USA

D.C. Beitz, PhD
Iowa State University, 313 Kildee Hall, Ames, IA 50011, USA

N.J. Temple et al. (eds.), *Nutritional Health: Strategies for Disease Prevention*,
Nutrition and Health, DOI 10.1007/978-1-61779-894-8_24,
© Springer Science+Business Media, LLC 2012

24.1   Introduction

The term biotechnology is often used synonymously with genetically modified organisms (GMOs) in agriculture to produce plants and animals that will either increase the efficiency of food production or increase the nutritional content. Biotechnology uses a number of different genetic tools, beyond modification, including the use of antisense technology, and the exogenous synthesis of important enzymes. This chapter describes the use of biotechnology by the food industry with its main focus towards genetic modification, although other forms of biotechnology will also be described. Some benefits of biotechnology can be summarized as follows [1]:

1. Plant production

    (a) Higher yielding varieties because of herbicide resistance or better nutrient uptake
    (b) Improved resistance to diseases, pests, and adverse conditions
    (c) Improved nutritional content
    (d) Decreased need, in many cases, for fertilizers and other chemical treatments

2. Animal production

    (a) Improved efficiency of converting feeds into useful animal products
    (b) Greater control of animal diseases
    (c) Modified composition of foods derived from animals

Although the use of biotechnology presents some simple and exciting benefits for food production, it comes with some significant safety concerns which could either easily offset increases in production and nutritional content or cause unintended damage to the environment. Some significant safety concerns are listed below, but are not limited to:

1. Environmental damage due to increased fertilizer/pesticide usage
2. Unintended creation of organisms resistant to herbicides/pesticides
3. Unintended spread of GMOs into neighboring fields or into the wild
4. Vertical and horizontal gene transfer (HGT)
5. Increased allergenicity of food products

24.2   Using Biotechnology to Improve Food Production by Plants

Modification of plants, animals, and microbes to produce more desirable traits began nearly 10,000 years ago through the selective breeding of plants and animals containing desirable traits. Such traits include increased yield, resistance to pests, resistance to disease, and increased ability to cope with adverse conditions [2]. During the past 50 years, tremendous success has been achieved in food production by creating many varieties of crops through the diversification of strains using traditional breeding practices [3]. Within the last 30 years, biotechnological producers catalyzed another revolution in crop production by creating genetically modified (GM) plants and animals; these have desired traits through the insertion of foreign genes into their genomes. The first generation of GM crops focused on creating varieties that would increase yield. By 2000, a second generation of crops became commercially available which are modified to increase nutritional content.

24.2.1   Extent of Use of Genetically Modified Crops

Agricultural biotechnology companies such as Monsanto, Dupont, Novartis, and Dow Chemical have invested billions of dollars in the development of GM crops. Monsanto introduced "Roundup Ready" soybeans in 1996 as the first GM crop available for commercial production. Globally, 167 million acres of GM crops were planted in 2003 according to the International Service of the Acquisition of Agri-Biotech Applications (ISAAA; www.isaaa.org). This encompasses 15.4 million farmers in 29 countries. In 2011, the National Agricultural Statistics Service (NASS) of the US Department of Agriculture (USDA) estimated that 94% of US soybean acreage is planted to herbicide-resistant and insect-resistant GM varieties. GM corn (herbicide-resistant and insect-resistant) constitutes 72% and 65%, respectively, of corn production in the United States. Nearly three-fourths of all cotton grown in the USA is GM that is insect or herbicide resistant.

The acceptance of GM crops continues to grow because of continued safety testing [4] and economic advantages [5], but scientific, moral, ethical, and emotional objections to and support for biotechnology are being expressed openly and vigorously as regulatory agencies, governments, and consumer groups debate the health, environmental, and commercial risks and benefits of this technology. Research on the acceptance of foods resulting from GM crops has indicated that consumers are less likely to purchase those foods because of information supplied by environmental groups, but third-party, verifiable information usually completely dissipates the negative influences of the environmental groups [6].

24.2.2   Generation of Genetically Modified Crops

There are two commonly used methods for the genetic modification of plants, namely transformation by particle bombardment (also known as the gene gun technique) and transformation mediated by the plant pathogen, *Agrobacterium tumefaciens* [7–9]. The *Agrobacterium* transformation is the preferred method due to its ease of use, as well as the fact that the number of insertion events is markedly reduced, allowing for a reduced copy number [10]. Moreover, strains of *A. tumerfaciens* have been engineered to allow for transformation of monocotyledons by this method [11–13].

A. tumefaciens is a bacterium found in soil that normally infects wound sites in the roots of dicotyledonous plants causing crown gall tumors [14]. *A. tumefaciens* transforms the plant by transferring a small DNA segment, known as T-DNA into the nuclei of infected plant cells. Upon transfer, the T-DNA becomes stably integrated into the host cell genome, allowing for tumor formation. The T-DNA is found in the Ti plasmid, which is carried by the *A. tumefaciens* [9]. Versions of the Ti plasmid have been engineered in which the tumor formation genes have been removed from the T-DNA, and sequences have been added to allow for insertion of both a gene of agronomic interest as well as a marker gene for transformation [14].

24.2.3 Modifying Plants to Increase Production

Although enough food is being produced to feed the world's population of approximately seven billion people, still one in every six in developing countries continues to suffer from hunger [15]. Exacerbating this situation, the world's population is expected to grow to nine billion, with most of that growth occurring in the developing world. With the global population becoming more urban and developing countries becoming increasingly wealthy, food demand is only expected to increase. Improved food production using traditionally selected crops was the focus of Norman Borlaug's now famous "Green Revolution" of the 1960s and 1970s. To this end, initial varieties of GM crops were those that allowed increased yield through resistance to herbicides or pests [15].

The first GMO to become commercially available was the "Roundup Ready" soybean in 1996, which is resistant to glyphosate, the active component in the Roundup herbicide [16, 17]. Roundup was developed as a ubiquitous postgermination herbicide to allow for a quicker and simpler procedure for prevention of weed growth, thereby leading to increased crop production [16]. Roundup ready crop lines are modified to contain the CP4 gene from *Agrobacterium sp.* which encodes the glyphosate-resistant form of the 5-enolpyruvyl-shikimate-3-phosphate synthase (EPSPS) [17]. Development of Roundup ready crops allows farmers to use Roundup herbicide without killing their crops in the process. In addition to soybeans, several other commercially available crops have now been modified to produce the glyphosate-resistant EPSPS; these include alfalfa, corn, cotton, spring canola, winter canola, and sugar beets [18].

Lepidopteran pests (butterflies and moths), including the European corn borer, can significantly reduce crop yields. A number of different crops, also known as Bt crops, have been modified to express the *cry1Ab* gene from *Bacillus thuringiensis* (Bt) [19]. The *cry1Ab* gene expresses a Bt delta endotoxin that kills lepidoteran larvae [19]. Crops modified with the *B. thuringiensis cry1Ab* gene include corn, rice tomato, potato, and soybean [20].

Interestingly, viral resistance has been engineered into cassava, an important crop in Africa [21]. Cassava containing AC3 transgene from African cassava mosaic virus has been produced. This transgene provides cassava with resistance to the African cassava mosaic virus [21]. Beyond the Bt crops and virally resistant cassava, improved resistance of potatoes, as well as a number of other crops, to several viral, fungal, and bacterial diseases will be continually researched.

24.2.4 Modifying Plants to Change Nutrition Value

The health and well-being of humans is entirely dependent on foods derived directly from plants or indirectly when plants are consumed by food animals. Plants supply macronutrients (carbohydrates, lipids, and proteins) and many essential dietary micronutrients (vitamins and minerals). They also supply nonessential compounds including a wide variety of organic phytochemicals that are linked to the promotion of good health (see Chap. 13).

Modification of the content of macronutrients, micronutrients, and phytochemicals in plant foods is an urgent need worldwide, but especially in developing countries where people exist on a few staple foods that are typically deficient in specific essential nutrients [22]. For example, deficiencies of vitamin A and iron are especially common. Even in developed countries where food is abundant and caloric intake is often excessive, deficiencies of micronutrients, such as iron, are prevalent because of poor eating habits. Hence, genetic modification of nutrient content of plants can have a significant impact on nutritional status and human health [22]. The tools of biotechnology allow manipulation of plant food composition. Genes for the synthesis of flavonoids, isoflavonoids, carotenoids, biotin, thiamin, and vitamin E and for increased iron uptake are cloned and are either being used or could potentially be used for plant transformation [23–29]. Genes for other nutrients and phytochemicals are being cloned and studied as tools to improve human nutrition.

Billions of people in developing countries depend on rice as a food staple, with about 400 million people suffering from a vitamin A deficiency and 3.7 billion people, especially women, having iron deficiency anemia [30]. Towards this end, research was undertaken to introduce into rice the genetic capability of synthesizing β-carotene and of increasing iron content [24, 30]. Rice does not have the ability to produce β-carotene, a substance that can be converted by humans into vitamin A. However, rice does have the ability to synthesize the early intermediate geranylgeranyl diphosphate [24, 31], which can be used as a precursor. Introduction of four genes encoding enzymes within the β-carotene synthetic pathway allows for production of β-carotene [24, 30]. This genetic modification gives the rice a golden color, hence the name "golden rice." About 300 g of this modified rice would meet the daily needs for vitamin A for an adult [30]. The use of golden rice has steadily increased over time due to the fact that developers, Peter Beyer and Ingo Potrykus, insisted that the licensing of the crop be free for humanitarian use and subsistence farming [25, 32]. As of 2008, 13.3 million farmers in 25 countries, 90% of them smallholders in less developed countries, were growing the GM crop on 763 million acres (www.goldenrice.org).

Iron in rice is normally unavailable for absorption because of the presence of phytate. However, introduction of a heat-stable phytase, which can break down the phytate, improves iron absorption. In addition, introduction of the ferritin gene allows for doubling the amount of iron sequestered in the rice, and introducing a gene encoding a metallothionein-like protein increases its bioavailability [26, 30]. Consumption of modified rice with these three genes has great potential to decrease iron malnutrition throughout the developing world [30].

In contrast to the situation with the essential nutrients described above, primary evidence for health-promoting roles of nonessential nutrients, such as phytochemicals and inulins, are more difficult to prove because of their complex interaction with food constituents. In fact, the exact identity of many active phytochemicals still needs to be determined. Below are selected examples of nonessential nutrients that are candidates to be produced in plants through genetic modification [22, 33]:

1. Carotenoids such as lycopene in tomatoes and lutein in kale and spinach
2. Glucosinolates such as glucoraphanin in broccoli and broccoli sprouts
3. Phytoestrogens such as genistein and diadzein in soybeans, tofu, and other soy products
4. Phenolics such as resveratrol in red wine and red grapes
5. Inulins in potatoes

24.2.5   Other Engineered Traits of Economic Importance

Along with diseases, drought, and pests, plant growth is retarded by low nitrogen content as well as excesses of specific metals in soils. Nitrogen deficiency in rice production systems has a strong negative impact on yield. Nitrogen fixation by rice is one possible approach to help increase rice production by 60% over the next 30 years, which is the expected need to maintain global food security [34]. As yet, no transgenic plants have been converted to nitrogen fixers. Incorporation of metal-resistant genes into plants and other organisms can allow plant cells to sequester metal complexes, such as phytochelatins, thereby making food-producing plants more tolerant to excess aluminum, mercury, copper, or cadmium in soil [35]. This technology could improve the acreage of land available for food production.

Drought resistance in rice also has been addressed by Capell et al. [36] when they studied polyamine biosynthetic pathways in transgenic rice. For example, transgenic rice expressing arginine decarboxylase from *Datura stramonium* (jimson weed) has greater concentrations of putrescine when stressed by drought, which promoted spermidine and spermine synthesis and ultimate protection from drought [36].

Enteric bacterial pathogens are the most common cause of infectious diarrhea in the developing world [37]. As a method of prevention of bacteria-caused diarrheal disease, GM plants have been created that express human lactoferrin and lysozyme (an enzyme usually found in tears, saliva, and breast milk) (www.checkbiotech.org). These antimicrobial proteins have been expressed in barley.

24.2.6 Modified Lipid Composition of Selected Plants

Vegetable oils are of much importance in human nutrition. Biotechnology has made it possible to tailor the composition of plant-derived lipids with respect to food functionality and human dietary needs [3]. Plant breeders have taken advantage of the natural diversity in fatty acids that exists among plant varieties in order to develop plants that produce unique oils. For example, soy oils that are relatively resistant to the development of rancidity are now commercially available, and were produced through traditional breeding practices [38]. These modified oils have decreased concentrations of linolenic and palmitic acids. More importantly, they have increased levels of linoleic acid, which gives the oil its resistance to rancidity [38].

Directed genetic modifications leading to altered plant lipid content depends on whether genes encoding enzymes necessary for their production are cloned. For instance, the following genes have been cloned: fatty acyl desaturase genes which encode enzymes for synthesis of both α- and γ-linolenate; a Δ^5-desaturase for the synthesis of eicosapentaenate [39]; and genes for fatty acid elongases for synthesis of eicosamonoenoate (20:1) and docosamonoenoate (22:1) from oleate. Several genes to control chain length and degree of unsaturation have been cloned to allow plants to produce unique and novel oils. As plant oils may form oxidized products that are potentially toxic, increasing monounsaturated fatty acids at the expense of polyunsaturated fatty acids may be desirable for improved heat stability and longer shelf life.

Another possible application for use of biotechnology in the production of oils is to create plant strains with increased synthesis of the very-long-chain fatty acids with 20 and 22 carbons [39]. These fatty acids are important constituents of animal cell membranes. In addition, the 20-carbon acids serve as precursors for eicosanoids, such as prostaglandins, leukotrienes, and thromboxanes; these substances play vital roles in the body. Intake of these very-long-chain fatty acids is vital for infants during their first months of life, and is beneficial for renal function, retinal development, brain development, as well as the prevention of cardiovascular disease [39].

Beyond direct genetic modification, other biotechnologies may also be used to modify plant lipid composition. For instance, antisense technology, which is used to knock down expression of specific genes, has been used to produce soy oil with greater than 80% oleate and about 11% saturated fatty acids and to decrease linolenate from 8 to 2% [39]. In addition, using antisense technology (oleate desaturase), oleic acid-rich (>80%) canola oil has been produced [40]. Investigators continue to study ways to change fatty acid composition of oils (e.g., increased stearate content) to decrease formation of *trans* fatty acids during hydrogenation. This research is of clinical significance for decreasing the intake of the atherogenic *trans* fatty acids by humans.

24.3 Use of Biotechnology to Improve Food Production by Animals

For thousands of years, people have attempted to improve animal genetics through selective breeding of animals with desired or superior phenotypes. Success depends on identifying traits and transmissibility of traits to offspring. Improvements are limited by naturally occurring variations and mutations within the species of interest. With the advent of recombinant DNA technology, a variety of new technologies became available that allowed the acceleration and refinement of genetic manipulation of animals [41].

Insertion of modified gene constructs into livestock can be used to create "designer production animals" that possess improved disease resistance and, therefore, improved productivity [41]. Producing useful proteins, tissues, and organs for pharmaceutical and biomedical use is likely to be another future use of GM animals. In general, scientists hope to use this technology to produce animals that are larger, leaner, grow faster, and more efficiently produce more healthful foods, and are more resistant to diseases [41]. Additionally, the cloning of mammals through somatic nuclear transfer has opened new opportunities [42–45]. Thus, generation of a large number of identical animals from a single donor with a desired genotype is possible.

An excellent review on commercializing animal biotechnology has been written by Faber et al. [46]. It outlines the merits and pitfalls of utilizing animals to produce biomedical products and the modification of food-producing animals. The review points out three factors—economics, societal values, and regulatory agencies—that will drive the intensity to which animals will be modified genetically. The following examples illustrate the breadth of applications of animal transgenesis for improved food and even pharmaceutical production [1, 47, 48]:

1. Meat animals, including fish, with increased copy numbers of genes encoding somatotropin or insulin-like growth factor, allowing for improved growth efficiency.
2. Food-producing animals with greater resistance to viral and bacterial diseases for improved productivity.
3. Lactating cattle with specific genes that increase efficiency of milk production.
4. Lactating animals with the capability of secreting novel proteins into milk, such as human albumin, α_1-antitrypsin, α-glucosidase, antibodies, antithrombin III, collagen, factor IX, fibrinogen, hemoglobin, lactoferrin, protein C, tissue plasminogen activator, and cystic fibrosis transmembrane conductance regulator (CTFR).
5. Genetic modification of animals for use in human organ replacement.

24.3.1 Modification of Digestive Tract Microorganisms to Improve Efficiency of Food Production by Animals

An important application of biotechnology to animal agriculture is the genetic modification of microbes that inhabit the rumen and cecum of food animals. For example, development of a modified microbe population that could be maintained in the rumen and/or cecum and that increases lignin degradation in fibrous feeds would improve digestibility of dietary fibers [49]. Degradation of lignin would, in most cases, improve the digestion of cellulose and thus increase the quantity of useful nutrients and nutraceuticals absorbed from a given quantity of consumed feed [49]. Genetic modifications for improvement of cellulose digestion and protein utilization by microbes merit greater research emphasis because, first, feed represents a major cost of food production by animals and, second, because there is sufficient margin for improvement in digestibility of dietary fibers and protein.

There are animal health and food safety problems that could be addressed by administering recombinant antimicrobial peptides, produced exogenously, such as alamethicin, cecropin, melittin, and magainin [50, 51]. Lazarev et al. [50] prevented mycoplasma infection in chickens by treating animals with a melittin-expressing plasmid. Other infectious pathogens, including common food pathogens such as *Escherichia coli* O157:H7 or Salmonella, could also be prevented with antimicrobial peptide technology. Nandiwada et al. [51] used colicin Hu194 to decrease and, in some cases, eliminate *E. coli* O157:H7 from alfalfa seeds for use in growing alfalfa sprouts.

24.3.2 Production of Low-Lactose Dairy Products

Currently, the accepted method for treatment of lactose intolerance is to supplement the diet with in vitro purified lactase [52]. One possible method to assist lactose-intolerant humans with consumption of dairy foods is to create transgenic animals that express lactase (encoded by the *lacZ* gene). These animals would then produce milk with a lower lactose content [52, 53]. The capability of expressing a lactose-hydrolyzing enzyme in the mammary gland of mice has been developed [53]. These transgenic female mice secreted lactase into milk that contained 50–85% less lactose, but with no changes in fat or protein concentrations. Thus, creating transgenic cows expressing lactase could potentially be a way to produce milk with lower lactose composition for lactose-intolerant humans, allowing them to be free of following strict treatment regimen or avoiding most dairy products.

24.3.3 BioPharming with Cloned Animals

One potential technology is to use somatic nuclear transfer, using a nucleus from an already transgenic animal, to propagate a herd of animals containing a desired transgene, in a process known as biopharming [54]. This method ensures that all offspring contain and likely express the transgene. Sheep, cows, mice, and goats, among others, have been cloned by somatic cell nuclear transfer [42–44]. Interestingly, nuclei from quiescent fetal cells of transgenic goats were transferred into enucleated oocytes and the manipulated embryos were implanted into surrogate mothers [54]. The fetal cells were derived from a goat that had been mated to a transgenic male containing a human antithrombin transgene. One of the cloned offspring that began lactating produced milk with 3.7–5.8 g/L of antithrombin in her milk, which was a similar amount to that of her female ancestors on her father's side [54]. Cloned animals, therefore, still synthesized a foreign protein at the expected level. If the efficiency of somatic nuclear transfer ever increases, this technology indicates a potential method to produce a herd of lactating animals from one transgenic animal for production of a useful nutritional or pharmaceutical compound.

24.3.4   Use of Recombinant Somatotropin for Production of Dairy Foods

In 1937, scientists documented that injections of crude extracts of pituitary glands stimulated milk production by dairy cows. This initial discovery led to a classic study at Cornell University in which dairy cows produced about 40% more milk during a 6-month period when injected daily with purified bovine somatotropin [55]. Numerous follow-up studies at many research institutions demonstrated efficacy and safety of the technology to increase milk production with natural or recombinantly derived bovine somatotropin. The following is a summary of the findings [56–58]:

1. Milk production increases
2. Feed intake increases to compensate for increased milk production
3. Amount of milk produced per unit of feed consumed increases
4. Milk composition does not change
5. Milk quality does not change

Milk produced by cows treated with recombinant bovine somatotropin are considered safe because [59]:

1. The hormone is a protein and is degraded during normal digestive processes.
2. The hormone is species specific and thus is inactive in humans.
3. Milk from treated cows contains normal concentrations of the hormone.
4. The hormone is inactivated by the commonly used pasteurization process.
5. All animal-derived foods contain small amounts of natural somatotropin.

Monsanto received Food and Drug Administration (FDA) approval in 1995 for commercialization of their product, called Posilac®, which is an injectable and slow release preparation of recombinant bovine somatotropin. According to the company website, in 2004, approximately 35% of the dairy cows in the United States were treated with Posilac® to improve profitability (www.monsantodairy.com). In a study of the use of recombinant somatotropin in the northeastern United States, scientists concluded that this technology improves lactation yield and persistency over the 4-year postapproval period with no effects on cow stayability and herd life [56].

24.3.5   Use of Recombinant Somatotropin for Meat Production

In 1934, it was noted that injections of pituitary gland extracts into rats stimulated growth and produced carcasses with more protein and less fat. Hence, administration of exogenous somatotropin to meat-producing animals seemed to merit additional study. Subsequently, it was demonstrated that injections of bovine somatotropin into growing beef cattle improved growth rates and slightly increased the lean-to-fat ratio of carcasses. Such improvements, however, have not yet proved economical, and thus commercial adoption of this technology has not occurred [60]. Similarly, injections of porcine somatotropin into young growing pigs caused marked increases in the lean-to-fat ratio of pig carcasses [60]. Efficiency of growth improved as well. The color of lean pork was slightly paler because of the treatment; firmness, juiciness, and flavor were not changed; but tenderness tended to be slightly decreased. Especially important for fat-conscious consumers, marbling or fat content of the pork was decreased markedly. To date, however, economics and injection protocol have not justified commercialization of administration of recombinant porcine somatotropin to growing pigs, even though the pork from treated pigs seems more desirable for human health. Administration of recombinant somatotropin does not show promise in poultry meat production because no significant improvements in growth efficiency or carcass composition were observed.

24.4   Safety of Genetically Modified Foods

The use of biotechnology in food production provides some exciting possibilities for feeding a growing world population through increasing production as well as improved nutritional content of many staple crops. However, it is possible that some of the gains created by biotechnology in the short term could be offset by long-term safety concerns [61]. Most of the concerns with the use of GMOs come from the potential of environmental damage; this may result in reduced crop production in the long term. In addition, GMOs in the United States are not subjected to rigorous testing to ensure that they are safe for human consumption. Given the fact that GM crops are expressing foreign proteins, it is currently unclear as to how these proteins may impact human health. These foreign proteins could have the potential to cause harmful, if not fatal, allergic reactions. Moreover, unexpected gene interactions could lead to the expression of previously unexpressed toxins. In brief, "the law of unintended consequences" might come into play.

24.4.1 Environmental Safety Concerns

Assessing the environmental impact of the use of GMOs is necessary to ensure that they do not pose significant environmental risk. Such risks can be looked upon as direct or indirect. A direct risk is posed when a product produced by a GM crop is toxic to nontarget organisms, whereas an indirect risk is posed when agricultural practices are changed due to the use of GM crops [62]. The most common type of direct risk is when a nontarget organism ingests the toxin by feeding on the intended target organism [63, 64]. For example, research has shown that development of common green lacewing larvae may be delayed due to the use of Bt maize. The green lacewing larvae do not directly feed on the maize but instead prey on corn borer larvae. Therefore, the cause of the delayed development could be due to the ingestion of the endotoxin produced by the Bt maize or from feeding on sick, poisoned larvae [63]. Similarly, 2-spot ladybirds suffered from reduced egg viability when they were fed aphids, which ate transgenic potatoes expressing the anti-aphid snowdrop lectin [64].

An indirect risk is posed when changes in agricultural practices due to the use of a GM crop affect the soil or a large number of organisms [61]. It may be that the indirect risks of using GMOs may be the most significant of all risks. Many of the GM crops currently on the market have the feature of enhanced production over their non-GM counterparts. Enhanced production can come in the form of having resistance to glyphosate or being pest resistant. However, the enhanced production could come at a cost. The increased use of glyphosate introduces a significant toxin into the environment which can spread beyond the field intended to be treated; this may then reduce plant diversity in nearby ecosystems. In addition, as production increases, a significant amount of nutrients, including fixed nitrogen and phosphorus, are removed from the soil. The loss of these nutrients could potentially affect the biodiversity of soil-dwelling microbes. To combat the loss of nutrients, increased amounts of fertilizers are used. Runoff from increased fertilizer use has been documented to cause significant damage to aquatic ecosystems [65, 66].

Another issue of considerable concern is the pollen-mediated transfer of transgenes from fields where GM crops are grown to other fields, or from GM crops to related wild species resulting in genetic pollution of the environment. In 2002, Rieger et al. [67] followed the pollen-mediated movement of herbicide-resistant genes in canola to nearby fields where the GM canola was not being grown. They found a small but significant amount of pollen-mediated movement of the herbicide-resistant genes to nearby fields, but only those fields within a 3-km radius [67]. Even with a small spread radius, the GM canola could become widespread over time, if left unchecked. To reduce the spread, it is recommended that appropriate separation distances should be maintained between fields containing GM crops and those with non-GM crops.

24.4.2 Possible Health Risks

If GM food can be characterized as substantially equivalent to its "natural" anteced-ent, it can be assumed to pose no new health risks to consumers and hence to be acceptable for commercial use. This concept is known as "substantial equivalence" and has been used for approval of GM food by the FDA [68, 69]. The concept was introduced in 1993 by the Organization for Economic Cooperation and Development and was endorsed by the Food and Agricultural Organization and the World Health Organization in 1996 [68]. The concept was applied to the approval of glyphosate-tolerant soybeans. These GM soybeans, although clearly different because of the newly acquired biochemical trait, are not different from their unmodified counter-parts in terms of amounts of protein, carbohydrates, vitamins, minerals, amino acids, fatty acids, fiber, isoflavones, and lecithins [68]. The GM soybeans were therefore deemed substantially equivalent and acceptable. Moreover, the recombi-nant proteins expressed by the GMO should pose no more risk of being allergenic than would naturally occurring proteins [4].

Some have questioned the use and definition of this concept and encouraged the use of more biochemical and toxicological testing before commercial use is allowed [4, 70–72]. This may be for good reason. For instance, expressing foreign proteins can result in a number of unintended effects. Expression of transgenic proteins could result in the unintended expression of endogenous genes which may produce potentially toxic or allergenic proteins [73].

The idea of using full-scale metabolite profiling (metabolomics) to study the safety of plant-derived GM food is one possible method to ensure that there are no unin-tended consequences [69–71]. One such consequence that would easily be detected is the buildup of a different, unintended compound that could be potentially toxic. Use of metabolomics would ensure that only the intended compound would increase in concentration in the GM food, as compared to the conventional food [69–71].

The National Academy of Science has published a book entitled *Safety of Genetically Engineered Foods—Approaches to Assessing Unintended Health Effects*. It discusses potential hazards of using GMOs, such as allergenicity, and provides evidence for their safety [74]. Although not commonly thought about, tra-ditional breeding practices can also produce varieties with negative consequences to human health. One such example is the selection of celery plants that were bred to have elevated concentrations of psoralens. Psoralen production, at reasonable lev-els, provides resistance to insects and disease. Yet, in some cases, farm workers harvesting celery containing elevated psoralen concentrations may develop skin rashes due to a psoralen allergy [74]. Another example of unintended change is kiwi fruit, which were initially unpalatable. By using traditional breeding programs, kiwi developed into the more palatable fruits that are currently available, with the unintended negative consequence of it being hyperallergenic [74]. Similar unintended changes can be predicted for recombinant organisms and thus testing and safety assurance programs need to be in place to evaluate the potential risk of both newly established "natural" foods and recombinant foods.

There is a paramount need for testing for the spread of recombinant DNA from commercially used plants and animals by both vertical and horizontal transmission. Simple and complex polymerase chain reaction analysis methods have been developed to screen for the presence of the commonly used genes in GM crops [75]. A thorough review on detection and traceability of recombinant DNA in foods has been prepared by Miraglia et al. [76] as a summary of the Working Group IV discussions that were part of the ENTRANSFOOD Thematic Network on the Safety Assessment of Genetically Modified Food Crops. Two points of major concern are the transfer of GMOs to fields where GMOs are not being grown (vertical transfer), and the transfer of antibiotic resistance markers to soil-dwelling bacteria (horizontal transfer) [77–80]. van den Eede et al. [80] reviewed the concern of HGT from GM organisms (plant or animal) to humans, which makes it a significant concern for human health. Generally, it is thought that most DNA will be significantly degraded during its passage through the digestive tract before it arrives at what is thought to be critical sites for HGT in the lower part of the small intestine, cecum, and colon [80].

24.5 Summary

Scientific discoveries before modern biotechnology became available led to improved agricultural productivity and improved food composition. The application of biotechnology, however, has accelerated the rate of those improvements and will continue to do so in the future. Biotechnology has permitted the development of transgenic bacteria, plants, and animals that have a wide variety of new capacities. For example, new plant varieties have been developed that are highly productive as they have enhanced resistance to infectious diseases, insects, herbicides, and adverse environmental conditions. Other new plant varieties allow crops to be grown with more desirable composition for consumption by humans and animals. Additionally, bacteria and yeast can be altered genetically to produce dietary constituents, biologically active compounds such as insulin and somatotropin, and a variety of other protein products that are used directly by humans or are used in food production by animals. Moreover, transgenic animals show promise of increased efficiency of production of foods and perhaps even of pharmaceutical compounds by the mammary gland of lactating transgenic animals. However, with the use of biotechnology comes a number of safety concerns. These include the transfer of integrated transgenes from GMOs to non-GMOs through vertical transfer, horizontal transfer of transgenes to soil-dwelling microbes, and increased allergenicity or toxicity of the GMO as compared to conventional varieties.

References

1. Madden D. Food biotechnology: an introduction. Washington, DC: ILSI; 1995.
2. National Research Council and Institute of Medicine of the National Academies. Safety of genetically engineered foods—approaches to assessing unintended health effects. Washington, DC: National Academy of Science; 2004.
3. Foods D. Animal product options in the marketplace. Washington, DC: National Academy Press; 1988.
4. Helm RM. Food biotechnology: is this good or bad? Implication to allergic disease. Ann Allergy Asthma Immunol. 2003;90:90–8.
5. Chong M. Acceptance of golden rice in the Philippine 'rice bowl'. Nat Biotechnol. 2003;21:971–2.
6. Huffman WE, Rousu M, Shogren JF, Tegene A. Controversies over the adoption of genetically modified organisms. J Agric Food Ind Organ. 2004;2(8).
7. Klein TM, Fromm M, Weissinger A, Tomes D, Schaaf S, Sletten M, Sanford JC. Transfer of foreign genes into intact maize cells with high-velocity microprojectiles. PROC NATL ACAD SCI U S A. 1988;85:4305–9.
8. Klein TM, Harper EC, Svab Z, Sanford JC, Fromm ME, Maliga P. Stable genetic transformation of intact *Nicotiana* cells by the particle bombardment process. PROC NATL ACAD SCI U S A. 1988;85:8502–205.
9. Bevan M. *Agrobacterium* vectors for plant transformation. Nucleic Acids Res. 1984;12:8711–21.
10. Birch RG. Plant transformation: problems and strategies for practical application. Annu Rev Plant Physiol Plant Mol Bio. 1997;48:297–326.
11. Hiei Y, Ohta S, Komari T, Kumashiro T. Efficient transformation of rice (*Oryza sativa*) mediated by *Agrobacterium* and sequence analysis of the boundaries of T-DNA. The Plant J. 1994;6:271–82.
12. Cheng XY, Sardana R, Kaplan H, Altosaar I. *Agrobacterium*-transformed rice expressing synthetic cry1ab and cry1ac genes are highly toxic to striped stem borer and yellow stem borer. PROC NATL ACAD SCI U S A. 1998;95:2767–72.
13. Ishida Y, Saito H, Ohta S, Hiei Y, Komari T, Kumashiro T. High efficiency transformation of maize (*Zea mayz* L.) mediated by *Agrobacterium tumefaciens*. Nature Biotechnol. 1996;4: 745–50.
14. de la Riva GA, Gonzalez-Cabrera J, Vazquez-Padron R, Ayra-Parda C. *Agrobacterium tumefaciens*: a natural tool for plant transformation. Electronic J Biochem. 1998;1:118–33.
15. Ruane J, Sonnino A. Agricultural biotechnologies in developing countries and their possible contribution to food security. J Biotechnol. 2011;156(4):356–63.
16. Malik J, Barry G, Kishore G. The herbicide glyphosate. Biofactors. 1998;2:17–25.
17. Padgette SR, Kolacz KH, Delannay X, et al. Development, identification, and characterization of a glyphosphate-tolerant soybean line. Crop Sci. 1995;35:1461–7.
18. Brooks G, Barfoot P. Global impact of biotech crops: socio-economic and environmental effect in the first ten years of commercial use. AgBioForum. 2006;9:139–51.
19. Lynch RE. European corn borer: yield losses in relation to hybrid and stage of corn development. J Econ Entomol. 1980;73:159–64.
20. Qiam M, Zilberman D. Yield effects of genetically modified crops in developing countries. Science. 2003;299:900–2.
21. Taylor N, Chavarriaga P, Raemakers K, Siritunga D, Zhang P. Development and application of transgenic technologies in cassava. Plant Mol Biol. 2004;56:671–88.
22. DellaPenna D. Nutritional genomics: Manipulating plant micronutrients to improve human health. Science. 1999;285:375–9.
23. Forkmann G, Martens S. Metabolic engineering and applications of flavonoids. Curr Opin Biotechnol. 2001;12:155–60.
24. Ye X, Al-Babili S, Kloti A, Zhang J, Lucca P, Beyer P, Potrykus I. Engineering of provitamin A (β-carotene) biosynthetic pathway into (carotenoid-free) rice endosperm. Science. 2000;287:303–5.

25. Dixon R, Steele C. Flavonoids and isoflavonoids—a gold mine for metabolic engineering. Trends Plant Sci. 1999;4:394–400.

26. Lucca P, Hurrell R, Potrykus I. Genetic engineering to improve the bioavailability and the level of iron in rice grains. Theor Appl Genet. 2001;102:392–7.

27. Norris SR, Shen X, DellaPenna D. Complementation of the *Arabidopsis pds1* mutation with the gene encoding p-hydroxyphenylpyruvate dioxygenase. Plant Physiol. 1998;117:1317–23.

28. Baldet P, Alban C, Douce R. Biotin synthesis in higher plants: purification and characterization of bioB gene product equivalent from *Arabidopsis thaliana* overexpressed in *Escherichia coli* and its subcellular localization in pea leaf cells. FEBS Lett. 1997;419:206–10.

29. Belanger FC, Leustek T, Chu B, Kriz AL. Evidence for the thiamine biosynthetic pathway in higher-plant plastids and its developmental regulation. Plant Mol Biol. 1995;29:809–21.

30. Gura T. New genes boost rice nutrients Science. 1999;285:994–5.

31. Burkhardt P, Beyer P, Wunn J, et al. Transgenic rice (*Oryza sativa*) endosperm expressing daffodil (*Narcissus pseudonarcissus*) phytoene synthase accumulates phytoene, a key intermediate of provitamin a biosynthesis. The Plant J. 1997;11:1071–8.

32. Dobson R. Royalty-free licenses for genetically modified rice made available to developing countries. Bull World Health Organ. 2000;78:1281.

33. Hellewege E, Czapla S, Jahnke A, Willmitzer L, Heyer A. Transgenic potato (*Solanum tuberosum*) tubers synthesize the full spectrum of inulin molecules naturally occurring in globe artichoke (*Cynara scolymus*) roots. Proc Natl Acad Sci U S A. 2000;97:8699–704.

34. Britto DT, Kronzucker HJ. Bioengineering nitrogen acquisition in rice: can novel initiatives in rice genomics and physiology contribute to global food security? BioEssays. 2004;26:683–92.

35. Moffatt AS. Engineering plants to cope with metals. Science. 1999;285:369–70.

36. Capell T, Bassie L, Christou P. Modulation of the polyamine biosynthetic pathway in transgenic rice confers tolerance to drought stress. PROC NATL ACAD SCI U S A. 2004;101:9909–14.

37. Hodges K, Gill R. Infectious diarrhea cellular and molecular mechanisms. Gut Microbes. 2010;1:4–21.

38. Hidetoshi S, Pokorny J. The development and application of novel vegetable oils tailor-made for specific human dietary needs. Eur J Lipid Sci Technol. 2003;105:769–78.

39. Broun P, Gettner S, Somerville C. Genetic engineering of plant lipids. Annu Rev Nutr. 1999;19:197–216.

40. Hitz WD, Yadav NS, Reiter RS, Mauvais CJ, Kinney AJ. Reducing polyunsaturation in oils of transgenic canola and soybean. In: Kader J, Mazliak P, editors. Plant lipid metabolism. Dordrecht: Kluwer; 1995. p. 506–8.

41. Wheeler M. Production of transgenic livestock: promise fulfilled. J Animal Sci. 2003;81 Suppl 3:32–7.

42. Hammer RE, Pursel VG, et al. Production of transgenic rabbits, sheep, and pigs by microinjection. Nature. 1985;315:680–3.

43. Wilmut I. Viable offspring derived from fetal and adult mammalian cells. Nature. 1997;385:810–3.

44. Wilmut I, Schnieke E, McWhir J, Kind AJ, Colman A, Campbell KHS. Nuclear transfer in the production of transgenic farm animals. New York: CABI; 1999.

45. Wilmut I, Schnieke A, McWhir J, Kind A, Campbell K. Viable offspring derived from fetal and adult mammalian cells. Nature. 1997;385:810–3.

46. Faber DC, Molina JA, Ohlrichs CL, Vander Zwaag DF, Ferre LB. Commercialization of animal biotechnology. Theriogenology. 2003;59:125–38.

47. National Research Council (U.S.). Committee on a National Strategy for Biotechnology in Agriculture. Agricultural biotechnology: strategies for national competitiveness. Washington, DC: National Academy Press; 1987.

48. Schnieke AE. Human factor IX transgenic sheep produced by transfer of nuclei form transfected fetal fibroblasts. Science. 1997;278:2130–3.

49. Russell JB, Wilson DB. Potential opportunities and problems for genetically altered rumen microorganisms. J Nutr. 1988;118:274–9.

50. Lazarev VN, Stipkovits L, Biro J, et al. Induced expression of the antimicrobial peptide melittin inhibits experimental infection by *Mycoplasma gallisepticum* in chickens. Microbes Infect. 2004;6:536–41.

51. Nandiwada LS, Schamberger GP, Schafer HW, Diez-Gonzalez F. Characterisation of an E2-type colicin and its application to treat alfalfa seeds to reduce *Escherichia coli* 0157:H7. Int J Food Microbiol. 2004;93:267–79.

52. Swaggerty D, Walling A, Klein R. Lactose intolerance. Am Fam Physician. 2002;65:1845–60.

53. Jost B, Vilotte JL, Duluc I, Rodeau JL, Freund JN. Production of low-lactose milk by ectopic expression of intestinal lactase in the mouse mammary gland. Nat Biotechnol. 1999;17:180–4.

54. Baguisi A. Production of goats by somatic cell nuclear transfer. Nat Biotechnol. 1999;17:456–61.

55. Bauman DE, Eppard PJ, DeGeeter MJ, Lanza GM. Response of high producing dairy cows to long-term treatment with pituitary somatotropin and recombinant somatotropin. J Dairy Sci 1985; 1352–62.

56. Bauman DE, Everett RW, Weiland WH, Collier RJ. Production responses to bovine somatotropin in Northeast dairy herds. J Dairy Sci. 1999;82:2564–73.

57. McGuffy RK, Green HB, Easson RP, Ferguson TH. Lactation response of dairy cows receiving bovine somatotropin via daily injections or in a sustained-release vehicle. J Dairy Sci. 1990;73:763–71.

58. Soderholm CG, Otterby DE, Ehle FR, Linn JG, Hansen WR, Annerstad RJ. Effects of recombinant bovine somatotropin on milk production, body composition, and physiological parameters. J Dairy Sci. 1988;71:355–65.

59. Juskevich JC, Guyer CG. Bovine growth hormone: human food safety evaluation. Science. 1990;249:875–84.

60. Etherton TD, Wiggins JP, Chung CS, Evock CM, Rebhun JF, Walton PE. Stimulation of pig growth performance by porcine somatotropin and growth hormone releasing factor. J Anim Sci. 1986;63:1389–99.

61. Petty J. The rapid emergence of genetic modification in world agriculture: contested risks and benefits. Environ Conserv. 2001;28:248–62.

62. Poppy G. GM crops: environmental risks and non-target effects. Trends Plant Sci. 2000;5:4–6.

63. Hilbeck A, Baumgartner M, Fried PM, Bigler F. Effects of transgenic *Bacillus thuringiensis* corn-fed prey on mortality and development time of immature *Chrysolperla carnea*. Environmen Entomology. 1998;27:480–7.

64. Birch ANE, Geoghegan IE, Majerus MEN, McNicol JW, Hackett CA, Gatehouse AMR, Gatehouse JA. Tri-trophic interactions involving pest aphids, predatory 2-spot ladybirds and transgenic potatoes expressing snowdrop lectin for aphid resistance. Molecular Breeding. 1999;5:75–83.

65. Eghball B, Gilley JE, Baltensperger DD, Blumenthal JM. Long-term manure and fertilizer application effects on phosphorus and nitrogen in runoff. Transactions ASAE. 2002;45:687–94.

66. Diaz RJ, Rosenberg R. Spreading dead zones and consequences for marine ecosystems. Science. 2008;321:926–9.

67. Rieger MA, Lamond M, Preston C, Powles SB, Roush RT. Pollen-mediated movement of herbicide resistance between commercial canola fields. Science. 2002;296:2386–8.

68. Millstone E, Brunner E. Mayer S. Beyond substantial equivalence Nature. 1999;401:525–6.

69. van der Voet H, Perry J, Paoletti C. A statistical assessment of differences and equivalences between genetically modified and reference plant varieties. BMC Biotech. 2011;11:1–20.

70. Rischer H, Oksman-Caldentey K. Unintended effects in genetically modified crops: revealed by metabolomics? Trends Biotech. 2006;3:102–4.

71. Jenkins H, Hardy N, Beckmann M, et al. A proposed framework for the description of plant metabolomics experiments and their results. Nature Biotech. 2004;22:1601–6.

72. Davies H, Shepherd LV, Steward D, Frank T, Rohlig R, Engel K. Metabolome variability in crop plant species-when, where, how much and so what? Regul Toxicol Pharmacol. 2010;58 (3 Suppl):S54–61.

73. Hug K. Genetically modified organisms: do the benefits outweigh the risks? Medicina (Kaunas). 2008;44:87–99.

74. Committee on Identifying and Assessing Unintended Effects of Genetically Engineered Foods on Human Health, Board of Life Sciences, Food and Nutrition Board, Board on Agriculture and Natural Resources, Institute of Medicine, National Research Council of the National Academies. Safety of genetically engineered foods; approaches to assessing unintended health effects. Washington DC: National Academies Press; 2004.

75. James D, Schmidt AM, Wall E, Green M, Masri S. Reliable detection and identification of genetically modified maize, soybean, and canola by multiplex PCR analysis. J Agric Food Chem. 2003;51:5829–34.

76. Miraglia M, Berdal KG, Brera C, et al. Detection and traceability of genetically modified organisms in the food production chain. Food Chem Toxicol. 2004;42:1157–80.

77. Mikkelsen TR, Andersen B, Jorgensen RB. The risk of crop transgene spread. Nature. 1996;380:31.

78. Chevre AM, Eber F, Baranger A, Renard M. Gene flow from transgenic crops. Nature. 1997;389:924.

79. De Vries J, Wackernagel W. Detection of *nptII* (kanamycin resistance) in genomes of transgenic plants by marker-rescue transformation. Mol Gen Genetics. 1998;257:606–13.

80. van den Eede G, Aarts H, Buhk H, et al. The relevance of gene transfer to the safety of food and feed derived from genetically modified (GM) plants. Food Chem Toxicol. 2004;42:1127–56.

Chapter 25
Core Concepts in Nutritional Anthropology

Sera L. Young and Gretel H. Pelto

Keywords Anthropology • Biocultural • Ecological model • Pica • Adaptation • Emic • Etic

Key Points

- Nutritional anthropology uses a holistic, biocultural approach to studying nutrition. It is fundamentally concerned with understanding the interrelationships of biological and social forces that shape human food use and the resulting nutritional status of individuals and populations.
- The ecological model provides a framework for understanding and modeling the human–food environment interactions that influence nutritional status.
- In this chapter we use the example of pica, the craving and consumption of nonfood substances, to illustrate the utility of the ecological framework.
- Attention to the adaptive process—how humans cope and adjust either genetically, physiologically, or socioculturally to meet material needs—is fundamental to research in nutritional anthropology.
- When describing nutrition-related practices, anthropologists try to present both the cultural insiders' ("emic") views of what people do and why they do it, as well as a more scientific, external ("etic") view. This is done because people's beliefs about food play a central role in their nutrition-related behavior.

S.L. Young, PhD (✉) • G.H. Pelto, PhD
Division of Nutritional Sciences, Cornell University, Savage Hall, Ithaca, NY 14853, USA
e-mail: sera.young@cornell.edu

N.J. Temple et al. (eds.), *Nutritional Health: Strategies for Disease Prevention*,
Nutrition and Health, DOI 10.1007/978-1-61779-894-8_25,
© Springer Science+Business Media, LLC 2012

- Methodologically, nutritional anthropological studies draw on research tools and techniques from multiple disciplines, ranging from nutritional biochemistry, physiology, genetics, and epidemiology to the social and policy sciences.

25.1 Introduction

As a discipline whose aim is to understand the human animal and its place in the natural order of things, a hallmark of anthropology is that its practitioners often engage in research that has the effect of *making the familiar strange, and the strange familiar*. For example, nutritional anthropologists examine practices in contemporary Euro-American societies that are taken for granted as simply "normal" or "natural" and reveal how culture-bound they actually are. The structure of meals, in which foods are served sequentially with soup first and dessert last, strikes people in other parts of the world as quite peculiar. They also seek to understand and "make sense" out of culinary practices that at first encounter appear to be irrational, such as the prohibition of beef consumption in food-scarce, poor Hindu villages. On more careful study, this prohibition turns out to be ecologically sound because of the complex energetic relationships of animals, humans, fuel, and agricultural production in South Asia [1].

Another objective of anthropology is to elucidate the *variability across cultures* in relation to human universals. Thus, nutritional anthropologists study food patterns, cultural practices related to food, and food production systems in various societies in order to understand how they meet or fail to meet nutritional requirements, and the health and social consequences of these nutritional decisions.

This chapter will outline the core concepts in nutritional anthropology that are used to accomplish these objectives.

25.2 Nutritional Anthropology: A Biocultural Approach to Understanding Nutrition

For human beings, food has both sociocultural and biological dimensions. The symbolic meanings of food vary by society, as does the very definition of what is acceptable to consume. The analysis of food choice behavior, including its determinants and social consequences, involves the application of social science theories and methods. After it is consumed, the characteristics of food become the province of biological sciences, which are used to reveal how it is used for growth and maintenance of the body [2]. The field of nutritional anthropology attempts to integrate studies of human behavior and social organization (i.e., the sociocultural or "predental" aspects of food), with those of nutritional status, nutrient requirements, and growth (i.e., the biological or "postdental" aspects of food). Nutritional anthropology is *fundamentally concerned with understanding the interrelationships of biological and social forces in shaping human food use and the nutritional status of individuals and populations* [3]. Since it is focused particularly on the interactions of social and biological factors, nutritional anthropology is fundamentally biocultural in its approach.

The aim of much of the research conducted by nutritional anthropologists is to understand how the physical well-being of humans is affected by their food systems. This aim contrasts with the approach of cultural anthropologists, sociologists, and historians, who use food as a vehicle for understanding how social and cultural systems work. Both orientations, however, yield knowledge about food and society, through time and across space.

25.3 An Ecological Model for Nutrition

The ecological model was first introduced into nutritional anthropology by Jerome et al. [4]. It attempts to identify the multiple social and environmental factors that affect the nutrition of a population in a simple schematic, but holistic, manner. While aspects of society are not as easily compartmentalized as Fig. 25.1 might imply, it is a heuristic tool that is useful for drawing attention to and organizing the complexities of the context of human nutrition. It aims to encompass biological and cultural aspects of nutrition through their linkage with diet.

The component labeled "physical environment" refers to the climate, soil characteristics, water resources, flora and fauna, land availability, pathogens, and other features that establish the conditions for food procurement and production. "Technology" includes the range of tools and techniques used for production,

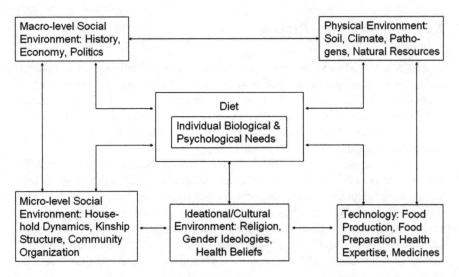

Fig. 25.1 An ecological model of nutrition, adapted from [4]

distribution, acquisition, storage, and preparation of all that is nutritionally valuable, including food and medicines. The "ideational environment" refers to cultural features, including beliefs about the role of food in well-being, cultural expectations related to health, definitions of food, gender ideologies, food taboos, and religious influences on diet. The "macro-level social setting" refers to politics, economics, and history, while the "micro-level social setting" refers to household dynamics, community organization, and kinship structure. All of these affect the diet of an individual, which, in turn, influences and is influenced by biological characteristics.

The ecological model could be considered as a roadmap of a food system. "Food systems" is a concept in nutritional anthropology that refers to the totality of activities, social institutions, material inputs and outputs, and cultural beliefs within a social group that are involved in the production, distribution, and consumption of food [3]. It is often a useful exercise to map out the food system that is being studied to identify the patterns of interaction among its components.

Methodologically, nutritional anthropological studies draw on research tools and techniques from multiple disciplines, ranging from nutritional biochemistry, physiology, genetics, and epidemiology to the social and policy sciences. Occasionally, they also turn to humanistic scholarship as a source of insights into the cultural and historical aspects of food. Thus, multi-disciplinary collaboration with scientists in other social and biological fields commonly occurs in anthropological research in nutrition.

Typically, studies utilize open-ended interviews and participant observation to generate descriptive, qualitative data. These are coupled with highly structured data collection techniques such as surveys and physical measurements that yield data amenable to statistical analysis. Taken together, these complementary methodologies help to integrate multiple perspectives on human behavior and experience to explain nutritional conditions in a population.

25.4 The Adaptive Process

Attention to the adaptive process—how humans cope and adjust to meet material needs—is fundamental to research in nutritional anthropology [3]. Three levels of adaptation are distinguished: (1) genetic adaptation, (2) physiological adaptation, and (3) sociocultural adaptation. Nutritional anthropologists tend to focus on the latter two because they are relatively more common and more rapid: humans are constantly responding to environmental change.

Classic examples of physiological adaptations are decreased basal metabolic rates during periods of starvation and expanded lung capacity when living at high altitudes. Sociocultural adaptations involve behavioral and technological innovations rather than biological adjustments that improve people's ability to successfully exploit food resources. Examples of sociocultural adaptations include food prohibitions that serve to protect or regulate animal food sources and the development of methods for preparing foods that remove toxic substances and make them safe for human consumption.

When one finds long-standing food patterns, it is wise to investigate them to determine whether they represent positive cultural adaptations that optimize nutritional well-being. The traditional method of preparing maize in Mexico and Central America by soaking it in lime and water is a case in point. Examination of this practice revealed that the soak in an alkaline solution chemically frees up the niacin that would otherwise be metabolically unavailable, and therefore permits maize to be used as the primary food stable without putting the population at risk of pellagra [5]. In other parts of the world where this preparation was not adopted with the adoption of maize as a staple food, pellagra became a major public health problem.

25.5 Differentiating Emic and Etic Perspectives

"Emic" and "etic" are key concepts in the entire field of anthropology, not only in nutritional anthropology. "Emic" refers to the perspective of the "insiders" in a culture, the definitions and interpretations of reality as seen through local eyes and ideas. The term "etic" is used to refer to the external, analytic perspective that is used by scientists when they are studying cultural and social phenomena. These concepts are important both methodologically and theoretically for nutritional anthropologists. With respect to description of nutrition-related practices, anthropologists try to obtain the emic perspectives of the people they are studying, to present the insiders' views of what they do and why. From a theoretical standpoint, etic analysis in nutritional anthropology depends heavily on concepts drawn from nutritional biochemistry, as well as on theories of biological and ecological adaptation.

From an etic perspective, some cultural features related to nutrition have negative consequences for health. For example, the cultural practice of restricting food intake during pregnancy, which is common in South Asian communities, contributes to the widespread prevalence of dangerously low birth-weight. However, larger infant size at birth, for which greater food intake is a significant determinant, increases the risk of serious problems in delivery that threaten the life of both mother and baby. Thus, there appears to be some logic to such a practice, although it clearly has a negative component.

For the purpose of programmatic decisions, Jelliffe and Bennett [6] suggested that food practices could be classified into four categories from the perspective of their impact on health: practices that are harmful, those that are beneficial, neutral practices and practices whose health impact is unknown. While an emic explanation might be "wrong" from the current scientific point of view, in many situations there is not enough knowledge about a particular practice to determine whether it is beneficial, harmful, or neutral in that particular setting, a fact that was clearly recognized by Jelliffe and Bennett [6]. The etic perspective is also incomplete and changes over time with the development of new theories and empirical findings.

The two perspectives, emic and etic, are not necessarily mutually exclusive; there is room for both explanations in the ecological model outlined above. Taken together, the concepts of emic and etic help to better characterize food practices.

Because people's beliefs about food play a central role in their nutrition-related behavior, understanding the emic perspective is usually an essential step in the design and development of interventions to reduce the burden of human disease through improved nutrition.

25.6 Applying an Ecological Perspective to Understand Ingestion Practices: Pica on Pemba Island, Zanzibar

To illustrate the utility of the concepts outlined above, we have modeled how the eco-logical perspective can be applied to understand a seemingly aberrant nutritional behavior that was encountered in the course of the fieldwork by one author (SLY) [7]:

SLY: *What else do you eat when you're pregnant?*

Pemban woman: *Umm...every day, twice a day, I eat some earth scraped from the house walls.*

25.6.1 What Is Pica?

Pica is defined as "the craving and purposive consumption of substances that the consumer does not define as food for more than 1 month" [8]. Pica takes its name from the magpie, *Pica pica*, a bird known for its indiscriminate appetite [9]. Yet those who ingest pica substances do not consume them indiscriminately; only cer-tain nonfood substances are craved and subsequently consumed, particularly earth (geophagy), raw starches (amylophagy, e.g., cornstarch, laundry starch, uncooked rice), and large quantities of ice (pagophagy). Pica has been observed in hundreds of cultures and on every inhabited continent [10]. It is not a habit that has disap-peared with "modernity," as some researchers had predicted [11]. The practice remains common in rural and urban settings around the world and is ubiquitous among pregnant women [8]. Prevalence of pica in obstetric populations has been documented to be as high as 77% in the United States [12] and even higher in devel-oping countries (cf. ref. [8], Supplementary Materials). In some parts of the world, the consumption of earth is so tightly coupled with pregnancy that it is even consid-ered to be a "symptom" of pregnancy (e.g., [13, 14]). In short, the consumption of substances few would consider as food is not an uncommon behavior.

25.6.2 Pica in Pemba

Pemba is the second biggest island in the Zanzibar archipelago, which lies in the Indian Ocean, just off the coast of Tanzania. It is a lush tropical island; clove and seaweed farming are the main economic activities. Swahili is the predominant language, Islam is the most commonly practiced religion, and the culture is similar to the rest of coastal East Africa. The social system is patriarchal, polygynous, and patrilineal.

In Pemba, nonfood substances are predominantly eaten by pregnant women. The three most commonly ingested nonfood items are *mchele* (husked but uncooked rice), three types of earth (*udongo*, *ufue*, and *vitango pepeta*), and ice [15]. Most types of uncooked rice are acceptable. In contrast, earth consumed is chosen very carefully. It is sought out from specific locations, e.g., a pit located outside the village, the foundations of houses, and termite mounds. It is frequently heated in a cooking fire or dried in the sun before being consumed [16]. All of the earths were notably smooth and were obtained from places not likely to be contaminated by human or animal excretions. Ice consumption is not always pica; a few chunks of ice are not a nutritional aberration. However, Pemban women ate as many as 12 glasses per day. Less commonly mentioned substances were charcoal, chalk, baby powder, ashes, dust, and powdered shells [15].

25.6.3 The Emic Perspective on Pica in Pemba

There is an element of embarrassment or shyness about pica; most informants switched to the third person when beginning to discuss it. "*They* like it because they are pregnant." "*They* just like it too much." "*They* don't do it again after giving birth." But after a bit of nonjudgmental conversation, women warmed to the subject and began to enjoy discussing the motivations for their cravings [17].

Pembans explained pregnant women's pica cravings in similar ways. "Everyone eats this when they are pregnant." "It is just a sign of pregnancy." "Eating that stuff is just a habit of pregnant women." When further probed about why pica was a habit, they could only explain that it is a *kileo*, a craving or addiction. Pica outside of pregnancy was seen as more problematic, less healthy, and even indicative of mental illness.

25.6.4 Etic Perspective: Pica as an Adaptation

Pica has long been characterized as a harmful, "depraved appetite" (e.g., [18]) and held responsible for a long list of detrimental side effects including dental damage, intestinal blockage, constipation, peritonitis, caloric displacement, iron deficiency, fatigue, lead poisoning, geohelminth infection, lead poisoning, toxemia, hypertension, tetanus, and heightened susceptibility to infection [8].

The way in which pica is practiced in Pemba makes it difficult to associate with many of these negative effects. Pregnant women consume about 40 g/day; an amount unlikely to cause intestinal blockage. The earths that are eaten are chosen because of their soft texture and easy dissolvability; therefore they are not likely to harm the mouth. Furthermore, earth is usually collected far from contaminated sites and is an unlikely vector for helminth infection [19]. There is, however, a strong association with anemia, although causality has not been demonstrated [15].

Using the ecological model outlined above, it becomes possible to hypothesize that pica is a positive sociocultural adaptation to the condition of pregnancy. Several useful effects of pica have been suggested [8]. Many, but not all, ingested earths have biologically important quantities of minerals, especially iron and calcium. The bioavailability of these and other micronutrients is not presently known and needs to be examined on a sample-by-sample basis. To that end, chemical analysis of pica substances on Pemba is currently being undertaken. Mineral supplementation of the diet may occur not only as a consequence of the mineral content of the earths that are consumed, but also because their consumption slows the motility of the gastrointestinal tract, thereby allowing more time for nutrients to be absorbed before being excreted.

Pica in Pemba may also be useful for relieving gastrointestinal distress (i.e., nausea, vomiting, and diarrhea). Our analysis of 12 representative samples of geophagic materials indicated they all contained kaolin (from which the antidiarrheal medicine, Kaopectate®, takes its name), which has a soothing effect on the intestinal tract [16]. Quelling nausea enables women to consume (more) food. Furthermore, preventing diarrhea and vomiting permits the nutrients that are consumed to remain in the digestive tract long enough to be absorbed.

Several researchers have proposed other adaptive benefits of geophagy, not related to the mineral content of earth, but to the earth's capacity as a detoxifier [20, 21]. Based on their biochemical analyses of geophageous samples, Johns and Duquette argue that although geophagy can be a source of nutrients, "detoxification broadly defined" is the most satisfactory explanation for this practice [22]. This protection may occur via two mechanisms (Fig. 25.2). Earth may form a protective coating on the mucous membrane of the digestive tract that may protect the mucosal cells from damage by toxins. It may also bind to the harmful substances before they can pass through the intestinal wall and enter the bloodstream.

We are currently working to test if nonearth pica substances have the capacity to bind environmental toxins. We believe this idea holds promise because it helps to explain why expectant mothers consume a number of other substances that have little, if any, nutritional value, including uncooked rice, charcoal, ashes, and dust.

Seen in the holistic context that the ecological model helps to delineate, our understanding of picas as a nutritional adaptation is beginning to emerge. For Pemban women, who have limited access to medicines and adequate health care, who are frequently undernourished and repeatedly exposed to environmental toxins that endanger the developing fetus, and who commonly experience pregnancy-related nausea and vomiting, pica begins to "make sense."

Finally, it should be noted that the emic and etic explanations of pica in Pemba are complementary. While a Pemban husband explained his wife's geophagy as "a sign of her pregnancy," an American dietitian attributed the expectant Pemban woman's consumption of earth to a mineral deficiency. There is a place for both types of explanations; the Pemban woman may be nutritionally deficient *because of* the added demands of pregnancy.

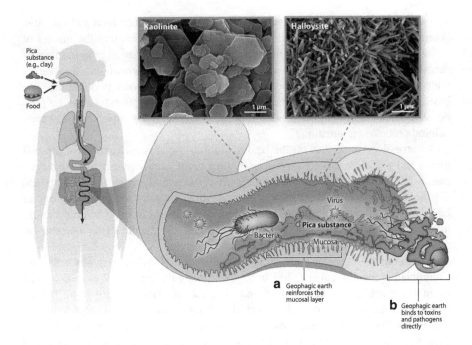

Young, Sera L. 2010.
Annu. Rev. Nutr. 30: In press.

Fig. 25.2 Geophagic earth may protect against toxins and pathogens by (**a**) strengthening the mucosal layer by binding with mucin and/or stimulating mucin production, thereby reducing the permeability of the gut wall, and (**b**) binding to toxins and pathogens directly, thereby rendering them unabsorbable by the gut. (Scanning electron microscope photo insets kindly provided by Evelyne Delbos of the Macaulay Institute.) Reprinted with permission from ref. [8]

25.7 Types of Research in Nutritional Anthropology

The types of research undertaken by nutritional anthropologists can be classified into the following main categories: (1) sociocultural processes and nutrition; (2) social epidemiology of nutrition; (3) cultural/ideational systems and nutrition; (4) physiological adaptation, population genetics, and nutrition; and (5) applied research for nutrition programs [23].

25.7.1 Sociocultural Processes and Nutrition

Investigations that can be categorized as studies of "sociocultural processes and nutrition" often focus on large-scale processes of change, such as globalization, modernization, urbanization, women's changing roles, and technological change.

The studies are aimed at understanding how these processes affect food and nutrition. While many investigators conduct studies in which they examine the effects of broad social processes in a specific location (e.g., rural to urban migration in a particular developing country), others are concerned with understanding how large-scale changes have affected nutritional conditions across many populations. The basic structure of the questions that nutritional anthropologists ask about the relationship of sociocultural processes to nutrition is: "What is the impact of X (a sociocultural process) on nutrition?"

Anthropological studies on the health and nutritional consequences of the historical and global shift from a foraging-hunting mode of subsistence to settled agriculture food systems are examples of research whose underlying aim is to understand the impact of a sociocultural process on nutrition. Investigators use a variety of biological techniques to analyze skeletal and plant remains from archeological sites in order to evaluate changes in nutritional status and other health status indicators in relation to information on changes in diet and lifestyle.

For example, Goodman et al. made detailed studies of skeletal remains from Pre-Columbian archeological sites in Illinois, comparing the types of lesions produced by malnutrition and disease before and after the adoption of agriculture, which spread into the region from Meso-America many centuries before the arrival of Europeans [24]. They found substantial evidence to suggest that the people who lived by foraging and hunting had fewer signs of disease and poor nutrition and greater longevity. Similar evidence has been obtained from many other parts of the world [25]. This is part of the process of the nutrition transition, a topic discussed in more detail in Chap. 5. The differences in health cannot, of course, be attributed only to dietary differences because other factors, including increasing population density and disease exposure, also played important roles.

25.7.2   Social Epidemiology of Nutrition

Nutritional anthropological research that falls into the category of "social epidemiology of nutrition" resembles other types of epidemiological studies in nutrition. Investigators examine relationships between sociodemographic factors, which are conceptualized as determinants of food intake and nutritional outcomes. When such studies are conducted by anthropologists, the emphasis is on cultural factors as well as on other social and biological characteristics. The research question takes the general form: "What are the determinants or factors associated with Y (dietary practice or nutritional condition) in a particular population group or between different populations?" Among the topics that have attracted attention are social and ecological determinants of micronutrient deficiencies, interactions of socioeconomic and cultural factors in infant and young child growth faltering, and the role of cultural perceptions about health and beauty as predictors of obesity.

Anthropologist Daniel Sellen's work among the Datoga, a pastoral group in Tanzania, is illustrative of social epidemiological investigations in nutrition [26]. He was interested in the determinants of nutritional status in children in a small-scale society whose economic mainstay is livestock. One of the factors he considered was the role of polygyny. In a culture in which having several wives is a social goal for men and indicates their relative economic superiority, one would expect that children of polygynous marriages would be better off nutritionally than children whose mother was the only wife. Contrary to this expectation, Sellen found that the anthropometric status of children in families with multiple wives was poorer than for single-wife families, even when the overall wealth of the family was taken into account. This finding, an example of "social epidemiology of nutrition," led him to a further ethnographic examination of resource use in Datoga households.

Incorporating sociocultural factors in baseline epidemiological studies that are undertaken prior to the design of public health interventions is another way in which this type of research is practiced by nutritional anthropologists. Research by Gittelsohn et al. is illustrative of this approach [27]. In poor populations throughout the world, fatness is culturally valued because it indicates that the individual has enough to eat and is well provided for by his or her family. On the other hand, the information that overweight and obesity present serious threats to health is also becoming more and more widely known. Therefore, at the beginning of a project to prevent obesity and diabetes in a native American population in the southwestern part of the United States, the senior anthropologist in the research team, Joel Gittelsohn, felt it was important to assess beliefs about healthy body size before the intervention was designed.

Gittelsohn first conducted ethnographic research to develop an understanding of local people's perceptions about healthy body size and eating, including their views about how these applied to young children. This information was used to design questions for a large-scale survey that included data on food intake, body mass index (BMI), health status, sociodemographic characteristics, and cultural/ideational variables. Using epidemiological statistical methods, the investigators examined the relationship of caregivers' perceptions about the importance of controlling food intake of preschoolers and concerns about weight with the children's BMI, an indicator of child size and fatness. They found that caregivers who expressed greater worry about their children's weight and restricted their eating also had heavier children. This relationship remained statistically significant, even when parental BMI, sociodemographic variables, and the frequency of consumption of specific foods were taken into account. Thus, the epidemiological investigation showed that in spite of the poverty and history of hunger the community has experienced, many families are well aware of the dangers of being overweight and are trying to take steps to deal with it in their young children. The researchers were able to conclude that what the community needed was better knowledge and behavioral strategies for managing the problem.

25.7.3   Ideational Systems and Nutrition

Studies in the area of cultural/ideational systems and nutrition are often aimed at understanding how particular beliefs relate to food selection, including food prescriptions and proscriptions as well as responses to food and nutrient-related illness. The ways in which culturally-structured food avoidances in pregnancy or in childhood illness affect health outcomes are among the topics investigated by nutritional anthropologists who link their work to public health issues. The general format of the basic research question that characterizes studies on "ideational systems and nutrition" is: "What is the relationship of X (beliefs) to Y (food practice, nutritional or health condition)?"

An investigation of vitamin A deficiency in two areas of Niger in West Africa by anthropologist Lauren Blum illustrates this type of research [28]. The first phase of her study involved in-depth interviews with mothers and traditional healers. In an arid northern region, where there was reason to expect people to be familiar with the signs of vitamin A deficiency, many people mentioned traditional cures for night blindness (a sign of vitamin A deficiency) that involved consuming liver (an ideal practice because it is rich in vitamin A). In-depth interviews led Blum to hypothesize that these home remedies were not the result of earlier nutrition education by outside public health activities, but were indigenous practices that had been passed down through the generations. In another part of the country, where higher rainfall supported more fruit and vegetable varieties, the people whom she interviewed did not describe these traditional remedies.

The second phase of her study was designed to assess how people respond to signs and symptoms of vitamin A deficiency and where they would turn for help in managing these problems. To obtain the data she used a "scenario method" in which a sample of mothers was presented with a several different scenarios illustrating different manifestations of vitamin A deficiency in children and pregnant women. As each "story" was presented, the respondent was asked a series of questions concerning the causes of the signs and symptoms and recommended treatment strategies. In the arid northern region, the majority of the sample said that both the child and the pregnant woman with night blindness were suffering from a food-related malady and felt that home remedies that involved liver should be used to treat it. In the tropical region, some respondents felt that night blindness was due to a disease, while others thought it was the result of witchcraft. Depending on their views of the etiology, they recommended that the afflicted person should be taken either to a medical doctor or to a traditional specialist who deals with witchcraft. From this example, we see that the interpretation of what is clearly a nutritional problem from an etic point of view is highly influenced by local cultural knowledge and has several emic explanations.

25.7.4 Population Genetics and Nutrition

Another area of research in nutritional anthropology is the relationship between genetic variability in populations and food consumption patterns. Nutritional anthropologists who are interested in the relationship of nutrition to biological experiences of populations often conduct research to elucidate how the nutritional history of a population has shaped or influenced its physiological or genetic characteristics.

Studies on population distributions of lactose tolerance in adults are illustrative of this type of research interest. Anthropologists have sought to understand the genetic distribution of the trait that makes it possible for adults in some populations to consume milk when the common pattern is for humans to lose their capacity to digest lactose after childhood. The role of this genetic trait has been explored in relation to the development of dairy-based food economies in northern Europe and some regions in Africa [29, 30]. It is likely that there will be an expansion of this type of research with the development of new techniques and knowledge in nutritional genomics.

25.7.5   Applied Research for Nutrition Programs

In addition to conducting basic research, nutritional anthropologists also engage in applied research, which is undertaken in direct support of public health activities. Research in this category often involves community-level investigations, although applied nutritional anthropology may also be carried out for the purpose of informing national-level or international nutrition policy and planning. Community-level studies may focus on identifying the sociocultural factors that need to be taken into account in instituting intervention activities (i.e., "formative research") or in process evaluations to see how these factors are affecting the utilization of programs.

Another type of applied research activity is the development of tools that can be used by nutrition and health programs to quickly and efficiently collect and analyze data on local conditions in order to plan or reorganize intervention activities. The Rapid Assessment Procedures (RAP) Manual [31, 32] and the guidelines for focused ethnographic studies of vitamin A [33] are examples of such tools; they have been utilized across a wide range of countries and situations.

25.8   Conclusions: Implications of the Nutritional Anthropological Perspective

Nutritional anthropology is a discipline that approaches the analysis of nutritional conditions within and across populations in a holistic fashion. The ecological model exemplifies this holism and provides a general framework for understanding and modeling the human–food–environment interactions that influence nutritional status. The holistic, biocultural approach requires competency in several types of methodologies, as well as familiarity with both biological and cultural aspects of human nutrition. As a consequence, the studies that nutritional anthropologists conduct are usually multi-disciplinary and collaborative, bringing together investigators from several different fields. The inclusion of cultural/ideational elements in their explanatory frameworks often requires researchers to step into the worldview of the people whom they are studying in order to obtain emic interpretations as well as testing etic theories.

"Strategies for disease prevention" is the subject of this volume. In this chapter, we have focused on nutritional anthropology as a research strategy that is useful for generating a holistic understanding of nutritional behaviors. We used the example of pica on Pemba to show how this perspective can lead to new ways of conceptualizing the analysis of cultural food practices. Once regarded with derision, pica is beginning to make sense as an adaptive strategy when contextualized in terms of the social and physiological burdens pregnant Pemban women face.

Finally, in addition to illustrating the range and types of research questions that nutritional anthropologists address, we have also attempted to indicate the value of incorporating anthropological approaches into research that is undertaken to help solve public health problems and facilitate the translation of basic nutrition knowledge into actions to improve people's health.

References

1. Harris M. India's sacred cow. Hum Nat. 1978;28–36.
2. Quandt SA. Nutrition in medical anthropology. In: Sargent CF, Johnson TM, editors. Medical anthropology. Westport: Praeger; 1996. p. 272–89.
3. Pelto GH, Goodman AH, Dufour DL. The biocultural perspective in nutritional anthropology. In: Goodman AH, Dufour DL, Pelto GH, editors. Nutritional anthropology. Mountain View, CA: Mayfield Publishing Company; 2000. p. 1–9.
4. Jerome NW, Kandel RF, Pelto GH. An ecological approach to nutritional anthropology. In: Jerome NW, Kandel RF, Pelto GH, editors. Nutritional anthropology: contemporary approaches to diet and culture. New York: Redgrave Publishing Company; 1980. p. 13–45.
5. Katz SH, Hediger ML, Valleroy LA. Traditional maize processing techniques in the new world. Science. 1975;184:765–73.
6. Jelliffe DB, Bennett FJ. Cultural and anthropological factors in infant and maternal nutrition. Fed Proc. 1961;20:185–8.
7. Young SL. Craving earth: understanding pica. New York: Columbia University Press; 2010.
8. Young SL. Pica in pregnancy: new ideas about an old condition. Annu Rev Nutr. 2010;30: 403–22.
9. Aetius of Amida, Ricci JV. Aetios of Amida, the gynaecology and obstetrics of the sixth century. Philadelphia: Blakiston Company; 1542/1950.
10. Laufer B. Geophagy. Field museum of natural history. Anthropol Ser. 1930;18(2):97–198.
11. Horner R, Lackey C, Kolasa K, Warren K. Pica practices of pregnant women. J Am Diet Assoc. 1991;91:34–8.
12. Rainville AJ. Pica practices of pregnant women are associated with lower maternal hemoglobin level at delivery. J Am Diet Assoc. 1998;98:293–6.
13. Vermeer DE. Geophagy among the Tiv of Nigeria. Ann Assoc Am Geogr. 1966;56:197–204.
14. Hunter JM. Macroterme geophagy and pregnancy clays in southern Africa. J Cult Geogr. 1993;14:69–92.
15. Young SL, Khalfan S, Farag T, et al. Pica is associated with anemia and gastrointestinal distress among pregnant Zanzibari women. Am J Trop Med Hyg. 2010;83:144–51.
16. Young SL, Wilson MJ, Hillier S, Delbos E, Ali SM, Stoltzfus R. Differences and commonalities in physical, chemical, and mineralogical properties of Zanzibari geophagic soils. J Chem Ecol. 2010;36:129–40.
17. Young S, Wilson M, Miller D, Hillier S. Toward a comprehensive approach to the collection and analysis of pica substances, with emphasis on geophagic materials. PLoS One. 2008;3:e3147.

18. Riverius L, Culpeper N, Cole A. The practice of physick wherein is plainly set forth the nature, cause, differences, and several sorts of signs: together with the cure of all diseases in the body of man. With many additions in several places never printed before. printed by Peter Cole printer and book-seller at the sign of the Printing-press in Cornhil near the Royal Exchange. London; 1663.
19. Young SL, Goodman D, Farag T, et al. Geophagia is not associated with Trichuris or hookworm transmission in Zanzibar, Tanzania. Trans R Soc Trop Med Hyg. 2007;101:766–72.
20. Profet M. Pregnancy sickness as adaptation: a deterrent to maternal ingestion of teratogens. In: Barkow JH, Cosmides L, Tooby J, editors. The adapted mind: evolutionary psychology and the generation of culture. Oxford: Oxford University Press; 1992. p. 327–66.
21. Johns T. The chemical ecology of human ingestive behaviors. Annu Rev Anthropol. 1999;28:27–50.
22. Johns T, Duquette M. Detoxification and mineral supplementation as functions of geophagy. Am J Clin Nutr. 1991;53:448–56.
23. Pelto GH. Nutritional anthropology. In: Solomon HK and William WW, editors. Encyclopedia of food and culture. New York: Scribners; 2002. p. 595–6.
24. Goodman AH, Lallo J, Armelagos GJ, Rose JC, Cohen MN, Armelagos GJ. Health changes at Dickson Mounds, Illinois (AD 950–1300). In: Cohen MN, Armelagos GJ, editors. Paleopathology at the origins of agriculture. New York: Academic Press; 1984. p. 271–306.
25. Cohen MN. Health and the rise of civilizations. New Haven: Yale University Press; 1989.
26. Sellen D. Polygyny and child growth in a traditional pastoral society: the case of the Datoga of Tanzania. Hum Nat. 1999;10:329–71.
27. Gittelsohn J, Davis SM, Steckler A, et al. Pathways: lessons learned and future directions for school-based interventions among American Indians. Prev Med. 2003;37:S107–12.
28. Blum L, Pelto G, Pelto P. Coping with a nutrient deficiency: cultural models of vitamin A deficiency in northern Niger. Med Anthropol. 2004;23:195–227.
29. McCracken RD. Lactase deficiency: an example of dietary evolution. Curr Anthropol. 1971;12:479–517.
30. Bloom G, Sherman P. Dairying barriers affect the distribution of lactose malabsorption. Evol Hum Behav. 2005;26:301–12.
31. Scrimshaw N, Gleason G, editors. Rapid assessment procedures: qualitative methodologies for planning and evaluation of health related programmes. Boston: International Nutrition Foundation for Developing Countries; 1992.
32. Scrimshaw S, Hurtado E. Rapid Assessment Procedures for nutrition and primary health care: anthropological approaches to improving programme effectiveness. Los Angeles: Latin American Studies Center; 1987.
33. Blum L, Pelto P, Pelto G, Kuhnlein H. Community assessment of natural food sources of vitamin A: guidelines for an ethnographic protocol. Boston: International Nutrition Foundation for Developing Countries and International Development Research Center; 1997.

Chapter 26
Postscript: Strategic Nutrition—A Vision for the Twenty-First Century

Norman J. Temple, Ted Wilson, and David R. Jacobs, Jr.

Key Points

- Nutrition science in the twentieth century was mainly concerned with the individual substances found in food. Beyond correcting deficiency diseases, this strategy has not met expectations. Most obviously, the obesity and diabetes epidemic has occurred despite population-wide nutrition effort.
- It is established that dietary pattern relates to chronic disease. Study of dietary patterns provides a window for nutrition progress, from which information about individual foods and nutrients may be derived.
- Government, industry, scientists, and food consumers all play key roles in nutrition. There is no obvious alternative to delivering food by industry, guided by a profit motive, yet the food industry has consistently spent enormous sums of money for a market niche, no matter what they are selling. Just as society supplies clean water, society should make good food available, rather than requiring every member of the public to be a nutrition expert.
- The missing link seems to be information. Contrary to the balanced and effective research program for development and delivery of medicines, no such infrastructure exists for food. Yet food is ubiquitous and far more complex than any drug. Each food is a nonrandom mixture of thousands of constituents, determined through evolution.

N.J. Temple, PhD (✉)
Centre for Science, Athabasca University, Athabasca, AB, T9S 3A3, Canada
e-mail: normant@athabascau.ca

T. Wilson, PhD
Department of Biology, Winona State University, Winona, MN 54603, USA

D.R. Jacobs, Jr., PhD
Division of Epidemiology and Community Health, School of Public Health,
University of Minnesota, 1300 South Second Street, Suite 300, Minneapolis, MN 55454, USA

N.J. Temple et al. (eds.), *Nutritional Health: Strategies for Disease Prevention*,
Nutrition and Health, DOI 10.1007/978-1-61779-894-8_26,
© Springer Science+Business Media, LLC 2012

- We need a nutrition research program for the twenty-first century focused initially on dietary patterns and foods and many times greater than what we currently have. It have would provide the information needed. We think government, industry, and consumers would be responsive.

Alberta is a province in Canada where one of us lives (NT). Its rivers flow into three different seas. One river flows north into the Arctic Ocean, several rivers flow east into Hudson Bay and thence into the Atlantic, while one minor river in the south flows into the Mississippi and then into the Gulf of Mexico. The world of nutrition is like these rivers: many streams can merge into major rivers of thought and action, but these rivers can flow in opposite directions and interact in complex ways with the total landscape of nutritional health.

Nutrition in the twentieth century was mainly concerned with the individual substances found in food. Up until the 1960s the major focus was on vitamins, minerals, protein, carbohydrates, and fats. In the 1970s dietary fiber received much attention, followed later by *trans* fatty acids, *n*-3 fatty acids, and phytochemicals. At the same time it became firmly established that the disease pattern seen across the Western world is a direct result of lifestyle, and that diet plays a major role in this. This information has been of tremendous value in showing how health may be enhanced and diet-related diseases prevented. Indeed, it is no exaggeration to say that these discoveries in nutrition science represent one of the most important advances in the field of public health during the past half century.

Many efforts have been made to turn this information into nutrition education for the general public. This has taken various forms, such as food guides and health promotion. The results of these efforts have been mixed. Some sections of the population have paid attention to the nutrition messages, have followed the advice, and gained benefit. But much more often the information has been ignored or ineffective. The worsening obesity epidemic is the clearest demonstration of this problem. Hundreds of books on diverse topics related to nutrition have been published in the recent decades, but a large proportion of them are worse than useless.

Over the last 2 decades several new streams have merged with the river. These are seen in the chapters of this book:

- Chapters 6–12 provide a detailed survey of our present knowledge on the relationship between diet and the prevention and therapy of diabetes, obesity, heart disease, hypertension, and cancer.
- Chapter 14 by Jacobs and Temple explains how the concept of food synergy is the surest route to a better understanding of the relationship between diet and disease. The key aspect of this is a focus on foods and dietary patterns rather than single substances.
- Many new teaching tools have been developed to facilitate educating the public on how to eat more healthily. Chapter 17 by Cuskelly, Woodside, and Temple discusses recent innovations in food guides, food-based dietary guidelines, and improved systems of food labeling. This is followed by Chap. 18 by Temple and Nestle that critically evaluates the field of health promotion.

- Chapter 21 by Rippe provides an overview of thousands of new foods that are now available and which are claimed to be healthier, as a result of either addition or removal of various bioactive substances.
- Chapter 23 by Temple explores the huge potential value of nutrition policy approaches for the prevention of disease. This covers such topics as lowering the salt content of food, improved food labels, and making healthy foods more affordable. The advantages of this strategy become clear when viewed in terms of cost-effectiveness.
- Chapter 24 by Segal, Knight, and Beitz examines the highly controversial subject of how genetic engineering is changing the DNA of the foods we eat.

Where are we now? When we take a broad overview of where nutrition science and nutrition practice stands in the early years of the twenty-first century, we see overwhelming evidence that the application of our present knowledge can deliver major advances in public health. What this means is that nutrition in this century can now move to a new higher level. We call this strategic nutrition.

So what is blocking progress? Alas, some major rivers are flowing in the wrong direction, as detailed in several chapters. The food industry spends enormous sums of money advertising unhealthy foods to get that market niche, no matter what they are selling. At the same time that industry regularly opposes any regulatory change by governments whenever they suspect that this might threaten their profits (Chap. 22 by Nestle and Wilson), even if this means that opportunities to improve population health are lost. For example, sections of the food industry have been energetic in their opposition to such proposals as lowering the salt content of food, modifying food guides so that they are based solely on the best scientific evidence, and stating calorie values on meals in restaurants. Adding to these problems, the dietary supplement industry also shows little hesitation in displaying disreputable behavior when this serves its interest, as described by Temple in Chap. 20.

Can we change the course of these rivers? The simple answer is yes.

When we look at the constellation of actors and forces involved in the provision of food, one can point the finger at industry, at government (inadequate guidance and regulation), and at the consumer (no brains, only wallet and taste buds). Perhaps it does not make sense to put the food supply of the populace in the hands of industry when food is so critical for keeping people healthy. Actually, however, in a highly developed society there is no way around delivering food except by way of the profit motive. This is just the way it works. Moreover, governments have tried regulating, but the policy initiatives have not always been substantively correct. Working with individual consumers is hugely frustrating; a well-organized society would not present bad food alternatives in the first place and we personally would prefer to ensure healthy food centrally, while still supplying people with the information to help make sensible choices in food selection. We do not demand of consumers that they be responsible for the healthy aspects of their water supply; we should follow this model with food, although the situation with food is, of course, much more complex. The missing link seems to be information. Our society does a reasonably good job with drugs, with a superlative research program that allows a balance between probable benefit and course correction when more is learned about the drug's good or bad effects. Food is ubiquitous and far, far more complex than any drug. Each food is a mixture of many constituents, at least thousands considering all variations in the molecules contained. These mixtures are not random: they evolved for the lives of the organisms eaten, and perhaps in co-evolution for the lives of the organisms doing the eating. We think that nutrition science is still in its early stages; there is much, much more to learn. We recommend a strong focus on dietary patterns that have been repeatedly shown to be associated with health benefits in observational epidemiologic studies, followed by decomposition of these patterns into their food and nutrient parts. Our society is responsive to information and also would be to information about diet. The central problem is that the funds allocated to nutritional research are tiny when compared to the need. A nutrition research program that is many times greater than what we currently have would provide the information needed. We think government, industry, and consumers would be responsive. This is a tall order, but not impossible. Therefore, we need increased and much improved educational efforts, including health promotion campaigns. Closely related to this we need better designs for food guides and food labels. And, of especial importance, we need governments to implement policies geared more towards population health rather than the commercial demands of the food industry.

This third edition of Nutritional Health provides the reader with a synopsis of where the different directions of nutritional thought are going and how they interact. The authors have indicated the impediments that restrict the downward flow of this information to the consumer. This book demonstrates some solutions to improving flow. We have laid down the groundwork that is needed for implementing nutritional policies that are strategic and all encompassing. Changes in nutrition are coming and this book will help the reader understand these changes and the need for implementing them.

Index

N.J. Temple et al. (eds.), *Nutritional Health: Strategies for Disease Prevention*,
Nutrition and Health, DOI 10.1007/978-1-61779-894-8,
© Springer Science+Business Media, LLC 2012

About the Series Editor

Dr. Adrianne Bendich has recently retired as Director of Medical Affairs at GlaxoSmithKline (GSK) Consumer Healthcare where she was responsible for leading the innovation and medical programs in support of many well-known brands including TUMS and Os-Cal. Dr. Bendich had primary responsibility for GSK's support for the Women's Health Initiative (WHI) intervention study. Prior to joining GSK, Dr. Bendich was at Roche Vitamins Inc. and was involved with the groundbreaking clinical studies showing that folic acid-containing multivitamins significantly reduced major classes of birth defects. Dr. Bendich has coauthored over 100 major clinical research studies in the area of preventive nutrition. Dr. Bendich is recognized as a leading authority on antioxidants, nutrition and immunity and pregnancy outcomes, vitamin safety, and the cost-effectiveness of vitamin/mineral supplementation.

Dr. Bendich, who is now President of Consultants in Consumer Healthcare LLC, is the editor of ten books including *Preventive Nutrition*: *The Comprehensive Guide*

N.J. Temple et al. (eds.), *Nutritional Health: Strategies for Disease Prevention*,
Nutrition and Health, DOI 10.1007/978-1-61779-894-8,
© Springer Science+Business Media, LLC 2012

for Health Professionals, fourth edition coedited with Dr. Richard Deckelbaum, and is Series Editor of *Nutrition and Health* for Springer/Humana Press (www.springer. com/series/7659). The Series contains 40 published volumes—major new editions in 2010–2011 include *Vitamin D*, second edition edited by Dr. Michael Holick; *Dietary Components and Immune Function* edited by Dr. Ronald Ross Watson, Dr. Sherma Zibadi, and Dr. Victor R. Preedy; *Bioactive Compounds and Cancer* edited by Dr. John A. Milner and Dr. Donato F. Romagnolo; *Modern Dietary Fat Intakes in Disease Promotion* edited by Dr. Fabien DeMeester, Dr. Sherma Zibadi, and Dr. Ronald Ross Watson; *Iron Deficiency and Overload* edited by Dr. Shlomo Yehuda and Dr. David Mostofsky; *Nutrition Guide for Physicians* edited by Dr. Edward Wilson, Dr. George A. Bray, Dr. Norman Temple, and Dr. Mary Struble; *Nutrition and Metabolism* edited by Dr. Christos Mantzoros; and *Fluid and Electrolytes in Pediatrics* edited by Leonard Feld and Dr. Frederick Kaskel. Recent volumes include: *Handbook of Drug-Nutrient Interactions* edited by Dr. Joseph Boullata and Dr. Vincent Armenti; *Probiotics in Pediatric Medicine* edited by Dr. Sonia Michail and Dr. Philip Sherman; *Handbook of Nutrition and Pregnancy* edited by Dr. Carol Lammi-Keefe, Dr. Sarah Couch, and Dr. Elliot Philipson; *Nutrition and Rheumatic Disease* edited by Dr. Laura Coleman; *Nutrition and Kidney Disease* edited by Dr. Laura Byham-Grey, Dr. Jerrilynn Burrowes, and Dr. Glenn Chertow; *Nutrition and Health in Developing Countries* edited by Dr. Richard Semba and Dr. Martin Bloem; *Calcium in Human Health* edited by Dr. Robert Heaney and Dr. Connie Weaver; and *Nutrition and Bone Health* edited by Dr. Michael Holick and Dr. Bess Dawson-Hughes.

Dr. Bendich served as Associate Editor for *Nutrition* the International Journal, served on the Editorial Board of the *Journal of Women's Health and Gender-based Medicine*, and was a member of the Board of Directors of the American College of Nutrition.

Dr. Bendich was the recipient of the Roche Research Award, was a *Tribute to Women and Industry* Awardee, and was a recipient of the Burroughs Wellcome Visiting Professorship in Basic Medical Sciences, 2000–2001. In 2008, Dr. Bendich was given the Council for Responsible Nutrition (CRN) Apple Award in recognition of her many contributions to the scientific understanding of dietary supplements. Dr. Bendich holds academic appointments as Adjunct Professor in the Department of Preventive Medicine and Community Health at UMDNJ and has an adjunct appointment at the Institute of Nutrition, Columbia University P&S, and is an Adjunct Research Professor, Rutgers University, Newark Campus. She is listed in *Who's Who in American Women*.

About the Editors

David R. Jacobs, Jr., PhD, holds the degree of PhD in Mathematical Statistics (1971) from The Johns Hopkins University. He has been on the faculty of the School of Public Health, University of Minnesota since 1974, and has held the rank of Professor of Epidemiology since 1989. He concurrently holds a guest professorship at the Department of Nutrition at the University of Oslo, Norway (1999–present). He is a fellow of the American Heart Association and the American College of Nutrition. He was Deputy Editor of the *British Journal of Nutrition* (2006–2011) and is on the editorial boards of *Clinical Chemistry and Preventive Medicine*.

He has written over 700 articles on various topics concerning the epidemiology of chronic diseases and their risk factors, including the epidemiology of specific molecules, and particularly those relating to cardiovascular diseases and diabetes. Topics of interest include monitoring of cardiovascular disease and its risk factors, the relation of high cholesterol with atherosclerotic diseases and the relation of low cholesterol with nonatherosclerotic diseases; measures related to oxidative stress and damage, serum antioxidant vitamins, gamma glutamyl transferase, plasma F2-isoprostanes, C-reactive protein, adiponectin, persistent organic pollutants, and arterial elasticity. Since 1994, he has focused extensively on whole grain intake and health. His work was influential in the 2000 decision of the USDA Dietary Guidelines Advisory Committee to add a specific guideline to "eat a variety of grains, especially whole grains," and in the strengthening of this message in the 2005 USDA Dietary Guidelines. Recent work has included study of periodontal disease, as a model of infectious disease, and its implications for cardiovascular disease. He has published intriguing cross-sectional findings relating background exposure of persistent organic pollutants to diabetes. He has written several articles on the health implications of synergies of different plant foods and dietary patterns. He is also principal investigator of a study of arterial elasticity and is expert in diverse subclinical markers of cardiovascular disease. He is an unpaid consultant to the California Walnut Commission.

Norman J. Temple, PhD, is the professor of nutrition at Athabasca University in Alberta, Canada. He has published more than 60 papers, mainly in the area of nutrition in relation to health. He has also published 11 books previously. Together with Denis Burkitt he coedited *Western Diseases: Their Dietary Prevention and Reversibility* (1994). This continued and extended Burkitt's pioneering work on the role of dietary fiber in chronic diseases of lifestyle. With Ted Wilson he coedited *Beverages in Nutrition and Health* (2004) and *Nutrition Guide for Physicians* (2010) as well as the two previous editions of *Nutritional Health*. He also coedited *Excessive Medical Spending: Facing the Challenge* (2007). He conducts collaborative research in Cape Town on the role of the changing diet in South Africa and on the pattern of diseases in that country, such as obesity, diabetes, and heart disease.

Dr. Ted Wilson received his PhD from Iowa State University and currently works in the Department of Biology at Winona State University in Winona, Minnesota. He has taught courses in Nutrition, Physiology, Anatomy, and Cell Biology. His research looks at human nutrition, high-altitude physiology, cancer, cardiovascular disease, and cardiac rehabilitation. Validation of food-health claims and the verification of measurable physiological effects caused by changes in the diet is an important part of this work. He has examined dietary affects on platelet aggregation, lipoprotein oxidation, arterial vasodilation, mechanisms for urinary tract infection, and nitric oxide formation. He has studied many foods and dietary supplements including cranberry juice, pomegranate juice, apple juice, grape juice, wine, resveratrol, creatine phosphate, walnuts, pistachios, and energy drinks. In addition to editing this third edition of *Nutritional Health*, he has also edited *Nutrition Guide for Physicians* (2010) and *Beverages in Health and Nutrition* (2005).

CPSIA information can be obtained
at www.ICGtesting.com
Printed in the USA
LVHW06*1932120918
589926LV00001B/7/P